IET HEALTHCARE TECHNOLOGIES SERIES 7

Semiconductor Lasers and Diode-based Light Sources for Biophotonics

Other volumes in this series:

Volume 1	**Nanobiosensors for Personalized and Onsite Biomedical Diagnosis** P. Chandra (Editor)
Volume 2	**Machine Learning for Healthcare Technologies** D.A. Clifton (Editor)
Volume 3	**Portable Biosensors and Point-of-Care Systems** S.E. Kintzios (Editor)
Volume 4	**Biomedical Nanomaterials: From design to implementation** T.J. Webster and H. Yazici (Editors)
Volume 6	**Active and Assisted Living: Technologies and applications** F. Florez-Revuelta and A.A. Chaaraoui (Editors)
Volume 9	**Human Monitoring, Smart Health and Assisted Living: Techniques and technologies** S. Longhi, A. Monteriù and A. Freddi (Editors)
Volume 13	**Handbook of Speckle Filtering and Tracking in Cardiovascular Ultrasound Imaging and Video** C.P. Loizou, C.S. Pattichis and J. D'hooge (Editors)
Volume 14	**Soft Robots for Healthcare Applications: Design, modelling, and control** S. Xie, M. Zhang and W. Meng

Semiconductor Lasers and Diode-based Light Sources for Biophotonics

Edited by
Peter E. Andersen and Paul Michael Petersen

The Institution of Engineering and Technology

Published by The Institution of Engineering and Technology, London, United Kingdom

The Institution of Engineering and Technology is registered as a Charity in England & Wales (no. 211014) and Scotland (no. SC038698).

First published 2018

The Institution of Engineering and Technology
Michael Faraday House
Six Hills Way, Stevenage
Herts, SG1 2AY, United Kingdom

www.theiet.org

British Library Cataloguing in Publication Data
A catalogue record for this product is available from the British Library

ISBN 978-1-78561-272-5 (hardback)
ISBN 978-1-78561-273-2 (PDF)

Typeset in India by MPS Limited

Contents

Preface

In terms of performance and modest production costs, diode lasers have become increasingly attractive for a plethora of applications. Due to direct electrical pumping, diode lasers and light emitting diodes are by far the most efficient light sources currently available. Being based on semiconductor chip technology, they can be manufactured in high numbers and at low cost. Their sub-millimetre scale of dimensions enables compact laser systems. Accordingly, these features increase their application potential, particularly within biophotonics, where their applications range from imaging and diagnostics to treatment of disease.

The present book covers semiconductor lasers, diode-based light sources and their application in biophotonics. We have aimed at selecting topics such that the entire book covers the basics, the technology and prominent applications, respectively. We have decided to include a brief introduction to semiconductor light sources and the basic physical properties of semiconductors, serving as a general introduction to succeeding chapters addressing the various aspects of the source technology. In turn, these chapters lead into discussing several established or emerging biomedical applications. Each chapter is written in a tutorial style, i.e. describing the subject matter in a self-consistent manner, and comprises adequate state-of-the art reviews, which will appeal to a broad readership ranging from skilled persons in the field to students who look for an introduction to the various topics.

During the course of preparing this handbook, we realized that the range of applications facilitated by these light sources is continuously expanding, and therefore, this exciting area continues to grow. It is expected that this specific application area will grow significantly in the future and become an increasingly important part of the semiconductor industry.

We would like to extend our sincere gratitude to all contributing authors and acknowledge your dedicated and careful efforts in preparing each of your chapters. We fully realise that without your devotion, this book would not have been published.

We hope that readers find this book an indispensable entry into the field of diode lasers and light emitting diodes in biophotonics, a field in rapid development and continuous expansion.

Enjoy reading!

Sincerely,
Peter E. Andersen
Paul Michael Petersen
October 2018
Roskilde, Denmark

Chapter 1

Introduction to semiconductor light sources and their biomedical applications

Peter E. Andersen[1] *and Paul Michael Petersen*[1]

The present chapter serves as a brief introduction to semiconductor light sources and the basic physical properties of semiconductors. The chapter also discusses the main reasons for semiconductor light sources being attractive for biomedical applications. It is expected that this specific application area will grow significantly in the future, and that it will be an increasingly important part of the semiconductor industry.

1.1 Brief history and semiconductor physics background

The initial research in semiconductors started in 1874 when the German physicist Karl Ferdinand Braun [1] discovered that a metal-semiconductor point contact rectifies alternating current. Another important contribution to semiconductors was made by the English engineer Henry J. Round [2] in 1907 with the discovery of electroluminescence in the semiconductor material SiC. He observed that light was emitted from the crystal when a voltage was applied to the SiC semiconductor, and this was in fact the first observation of light emission from an LED. In 1947, Bardeen and Brattain [3] made important steps forward towards the semiconductor technology that we use today when they invented the point contact transistor. Another milestone in the history of semiconductors came two years later, in 1949, when Shockley [4] presented the ideas for the p-n junction and the bipolar transistors.

In the following, we introduce the basic concepts of semiconductors physics. For a more detailed analysis of the semiconductor physics, we refer to Smith [5], Moss [6], Wang [7] and Sze *et al.* [8] In semiconductors, the energy levels are collected in energy bands with a large number of electrons in each band. Energy bands in semiconductors are formed due to the collection of interacting atoms. This band structure may be described theoretically by the Schrödinger Equation. The solution to the Schrödinger Equation gives energy bands where electrons are

[1]DTU Fotonik, Technical University of Denmark, Denmark

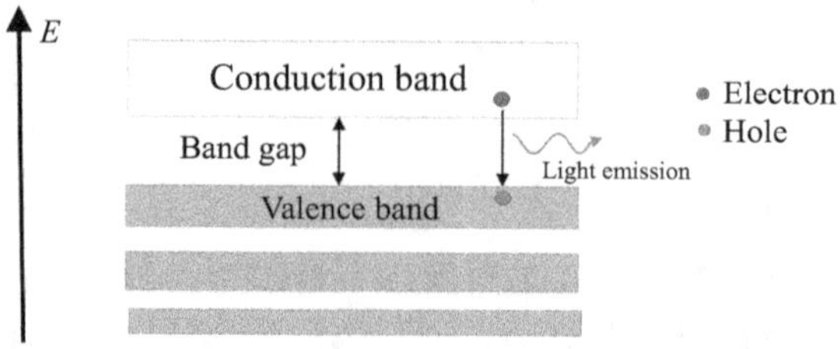

Figure 1.1 The energy level in the semiconductor where a band gap separates the conduction band and the valence band. Light is emitted when an electron in the conduction band that recombines with a hole in the valence band

confined in specific energy states and forbidden energy bands where electron states are not allowed since the solutions decay exponentially.

Figure 1.1 shows the band structure of the semiconductor. The conduction band is the lowest energy band that is fully unoccupied by electrons and the valence band is the highest lying fully occupied energy band. These two energy bands are separated by an empty band gap with energy E_g. An electron in the valence band may be excited into the conduction band by optical, electronic or thermal excitation. It can be seen from Figure 1.1 that once excited into the conduction band the electrons may recombine into an empty state (hole) in the valence band. This process is called electron-hole recombination, and it is the origin of light emission in semiconductor light sources. The conditions for light emission are that the emission process must fulfill the conditions of conservation of both energy and momentum. The band gap energy E_g is an important parameter for the semiconductor light sources, as it determines the wavelength of the emitted light from the semiconductor.

1.2 Direct and indirect bandgap semiconductors

Figure 1.2 shows the energy–momentum relation for a direct gap semiconductor. Near the bottom of the conduction band the energy band may be approximated by a simple parabolic relation between the energy E and the momentum $p = \hbar k$:

$$E = E_c + \frac{\hbar^2 k^2}{2m_c} \tag{1.1}$$

where E_c is the energy at the bottom of the conduction band, m_c is the effective mass of electrons in the conduction band, $\hbar$ is the Planck's constant divided by 2π, and k is the magnitude of the wave vector.

Near the top of the valence band, the E versus k relation is given by:

$$E = E_v - \frac{\hbar^2 k^2}{2m_v} \tag{1.2}$$

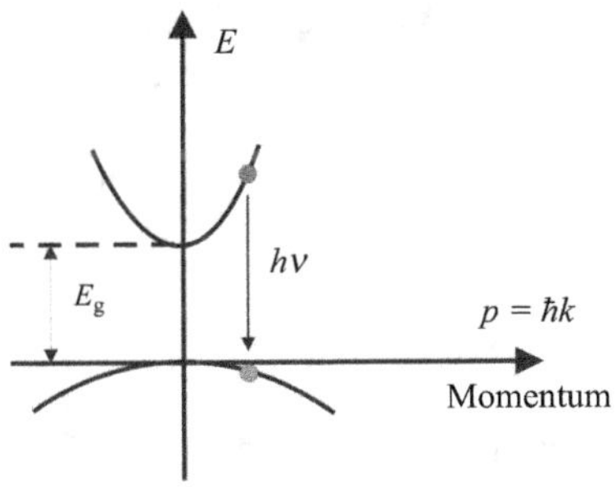

Figure 1.2 The energy (E) versus momentum (p) relation for a direct bandgap semiconductor

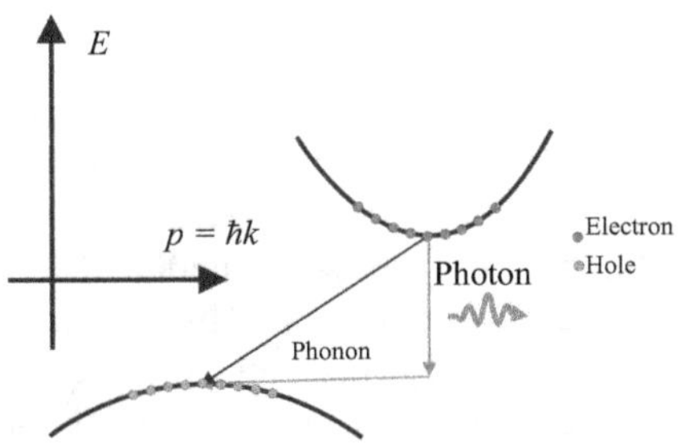

Figure 1.3 Energy momentum diagram for an indirect bandgap semiconductor. The momentum conservation is only fulfilled if interaction with another particle occurs simultaneously

where E_v is the energy at the top of the valence band and m_v is the effective mass of holes in the valence band. From the energy bands the effective masses m_c and m_v can be determined by:

$$\frac{1}{m_i} = \frac{1}{\hbar^2}\frac{d^2E}{dk^2}, \text{where } i = c, v \tag{1.3}$$

The semiconductor is called a direct bandgap semiconductor when the maximum energy of the valence band and the minimum energy of the conduction band correspond to the same momentum. If this is not the case, the semiconductor is called an indirect bandgap semiconductor.

Figure 1.3 shows the energy–momentum relation for an indirect bandgap semiconductor. The energy conservation condition can be fulfilled by emission of light. However, since the momentum of the photon is small, conservation is only fulfilled if interaction with other particles than photons occurs simultaneously. Therefore, light emission from indirect bandgap semiconductors is much less likely than light emissions from direct bandgap semiconductors. One example of such interaction with a third particle is simultaneous interaction with phonons as shown in Figure 1.3.

The probability for occupation of energy states by electrons in the semiconductor is given by Fermi–Dirac statistics. For electrons in the conduction band, the probability of an electron occupying a particular energy level E at an absolute temperature T is given by the Fermi–Dirac distribution function:

$$f(E) = \frac{1}{e^{(E-E_{\mathrm{F}})/k_{\mathrm{B}}T} + 1} \tag{1.4}$$

where k_{B} is the Boltzmann constant, T is the absolute temperature in Kelvin, and E_{F} is Fermi energy that describe the energy level where the probability of finding an electron is 50%.

1.3 Doping of semiconductors and p-n junctions

Doping of semiconductors by adding small concentrations of donors and acceptors impurities can change the optical and electrical properties of semiconductors. These impurities can change the concentration of mobile electrons or holes significantly and the process is important for developing efficient light emitting semiconductor sources. Figure 1.4(a) shows an undoped semiconductor, where the Fermi energy is in the middle of the band gap. This semiconductor is referred to as an intrinsic semiconductor. Figure 1.4(b) shows the p-type semiconductor where the semiconductor has been doped with acceptors with deficiency of electrons. The

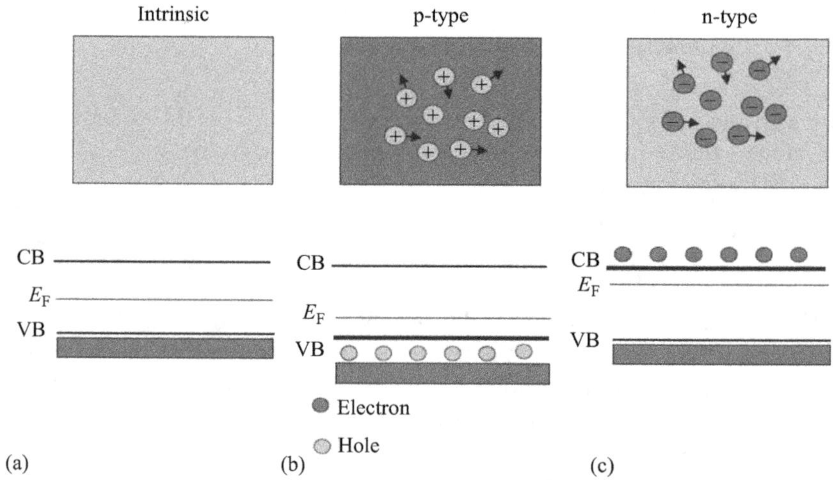

Figure 1.4 The configuration for the (a) intrinsic semiconductor; (b) p-type semiconductor; and (c) n-type semiconductor. In the intrinsic semiconductor, the Fermi energy E_F is in the middle of the energy band gap at $T = 0$ K. The Fermi energy is shifted towards the valence band in the p-type semiconductor and towards the conduction band in the n-type semiconductor

n-type semiconductor is formed, as shown in Figure 1.4(c), by doping with donors with excess valence electrons. The excess of electrons in the n-type material and the deficiency of electrons in the p-type material are the key elements in the p-n junction used in most light emitting semiconductors.

The p-n junction shown in Figure 1.5(a) consists of a p-type semiconductor and an n-type semiconductor placed next to each other. The charge carrier concentration near the junction interface causes electrons from the n-type to diffuse into the p-type semiconductor. This is shown in Figure 1.5, when a forward bias voltage is applied to the p-n junction. The emission of light takes place near the p-n junction and the thickness of the light emitting layer is relatively large due to diffusion of the carriers involved in the emission process. One way to increase the efficiency of light emission is to improve the confinement of electrons and holes in the light emitting active region. This can be achieved as shown in Figure 1.6 by placing the active semiconductor layer between two cladding layers with higher band gap energy than the bandgap of the active layer. This configuration is called a double heterojunction structure, see Figure 1.6(a), and the configuration has significantly improved the lifetime and the efficiency of both LEDs and semiconductor laser diodes. The advantages of the hetero structure are that both carriers and light emission are confined to the active region.

Another configuration that offers additionally improved performance of semiconductor light sources is the quantum well (QW) configuration [9,10]. The QW diode laser is somewhat similar to the double heterojunction structure, yet the thickness of the active layer is typically much smaller. The thickness d of the QW layer is comparable to the de Broglie wavelength $\lambda = h/p$, where h is the Planck's constant. Figure 1.6(b) shows the energy level of the QW. Laser diodes based on QW structures have advantages in terms of low threshold for the diode laser current and more narrow linewidth of the emitted laser light. Both laser diodes and LEDs based on QW structures benefit from high efficiency and long lifetime.

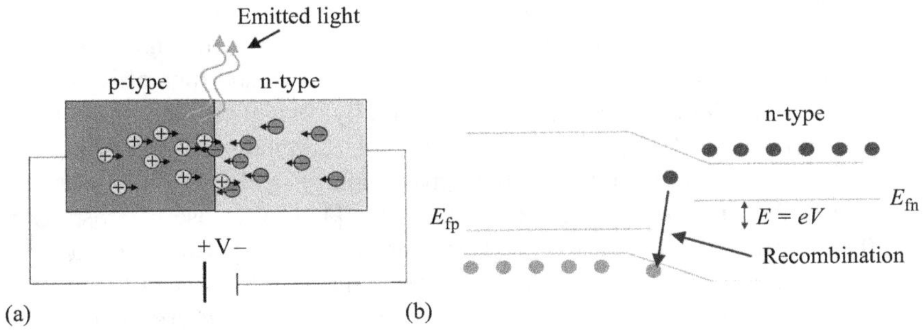

Figure 1.5 (a) The configuration of the p-n junction under forward bias; and (b) the energy band in the p-n junction. When a forward bias voltage is applied, light is emitted due to electron-hole recombination. E_{fn} and E_{fp} are the quasi-Fermi levels of the n-type and p-type semiconductor, respectively

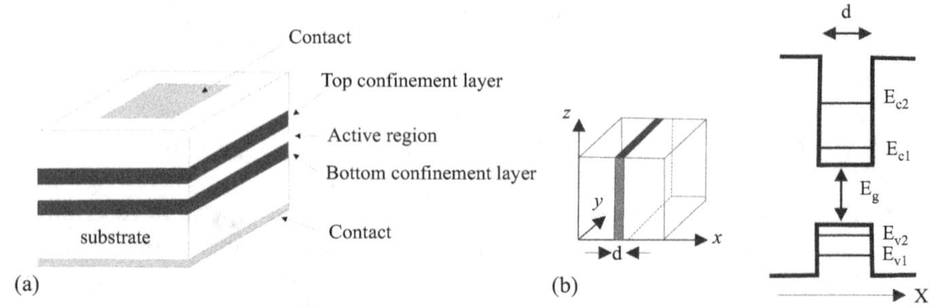

Figure 1.6 *(a) The double hetero semiconductor structure; and (b) the QW structure with quantum well energies E_{c1} and E_{c2} in the conduction band and E_{v1} and E_{v2} in the valence band. These semiconductor hetero structures have been extremely important for the development of semiconductor light sources*

1.4 Laser diodes and light-emitting diodes

Important semiconductors for laser diodes and LEDs are binary III-V compounds semiconductors formed by group III elements such as aluminum (Al), gallium (Ga), and indium (In) combined with phosphorus (P), arsenic (As) or nitrogen (N) from group V elements. Examples of these III–V compounds are GaAs and GaN. Ternary III–V compound semiconductors such as $Ga_{1-x}Al_xAs$ and $GaAs_{1-x}P_x$ are important because they can change the bandgap by varying the alloy composition x. Silicon (Si) and germanium (Ge) are some of the most important materials in semiconductor electronics and photonics. However, due to their indirect bandgaps they are not suited for photonic light sources.

Semiconductor laser diodes provide wavelength from ultraviolet to the far infrared region and they have several advantages compared to conventional lasers, such as solid-state lasers and gas lasers. The advantages of the semiconductor laser are high power, small size, simple power supply, long lifetime, and high efficiency. The first semiconductor lasers were developed in 1962 by several research institutions at General Electric [11], IBM [12], and MIT's Lincoln Laboratories [13]. These lasers were based on $Ga_{1-x}Al_xAs$ grown by liquid phase epitaxy (LPE). In 1963, these diode laser were further improved when it was suggested [14–16] to use heterostructures, where the laser material was sandwiched between two cladding semiconductor layers with higher band gab energy. These structures significantly reduced the threshold current for lasing. The laser structure is shown in Figure 1.7, where the thin active GaAs is placed between two $Ga_{1-x}Al_xAs$ layers. In 1969, the first room temperature operation of a double-hetero-structure GaAs laser was obtained [16,17].

The semiconductor diode laser structure uses heavily doped p-n junction in direct-gap semiconductors. When a forward bias voltage is applied to the diode

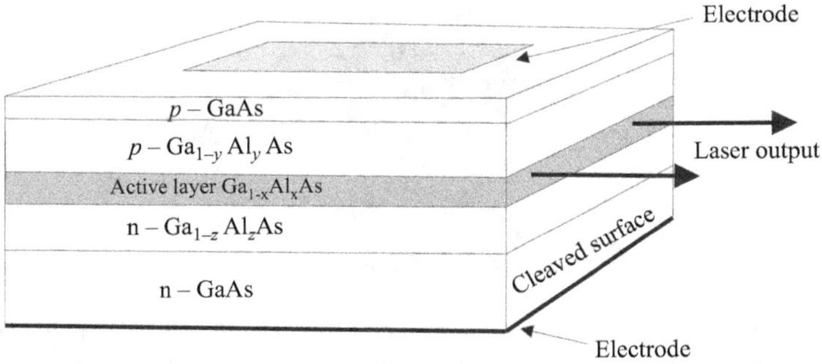

Figure 1.7 The double heterojunction semiconductor laser structure

laser population inversion takes place in a narrow region of the in the p-n junction. In addition to the gain also a feedback mechanism is needed in order to obtain laser oscillation. The two cleaved end surfaces serve as cavity (feedback) mirrors to obtain stimulated emission, i.e., lasing. One important parameter is the threshold current density J_{th}, i.e., the value to overcome in order to obtain lasing. The advantages of the heterojunction in Figure 1.7 are the low threshold current and high efficiency. These properties can be further improved using semiconductor quantum dot technology, where the active region is reduced into one dimension. The density of states in the quantum dot laser is enhanced with the consequence that the peak gain is increased and threshold current density is reduced even further [18].

The LED is a highly efficient light source that converts electrical power very efficiently into optical power. Within solid-state lighting LED light sources have efficiencies in research laboratories of more than 300 Lumen/W [19]. In contrast to conventional light sources such as incandescent light sources, halogen lamp, and fluorescent tubes the LED is a compact light source and the spectrum can be tailored from almost any wavelength in the UV to the infrared. In addition, the LED has a much longer lifetime up to 50,000–100,000 h and offers new possibilities within medical applications.

The first LED was invented by Round in 1907 and at that time, the origin of the light emission was not well understood. The light that was generated was due to a Schottky contact that was formed when a metal electrode was applied to the SiC crystal. During 1924–28, the Russian scientist Oleg Losev made the first comprehensive investigations [20] of light generated by electroluminescence from the SiC metal-semiconductor interface. He concluded that the emitted light was very different from incandescent light and he discovered that the light could be turned on and off very fast. The invention of the transistor [3] in 1947 at the Bell Laboratories was a major step forward for the development of the LED technology. John Barden, William Shockley and Walter Brattain received the Nobel Prize (physics) in 1956 for the development of the transistor. Although the development of the transistor did not focus on LED technology, it provided an improved understanding of

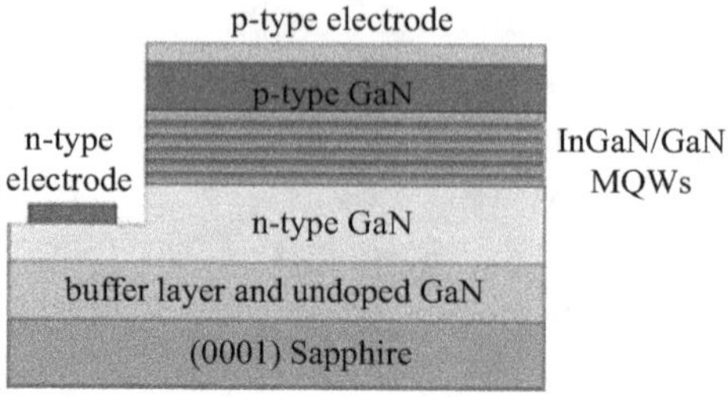

Figure 1.8 The blue LED that has been very important for the development of the LED technology. The geometry consists of InGaN/GaN multiple QWs structures that are sandwiched between two p-type and n-type layers. From Ref. [23]

semiconductor materials that lead to the development of the LED technology. In 1962, Nick Holonyak Jr. [21] produced the first commercial red diodes in GaAsP and an important step forward towards the white light LED technology was made in 1994 when Shuji Nakamura [22] developed the blue LED based on GaN. Together with two Japanese scientists, Isamu Akasaki and Hiroshi Amano, Nakamura received the Nobel Prize (physics) in 2014 for the research breakthrough in the early 1990s. This scientific breakthrough has been extremely important for the development of the LED technology that we use today. The basic structure of modern GaN LEDs is shown in Figure 1.8. It is based on p-n junctions with multiple QWs InGaN/GaN structures. The blue LED has been important for the development of light sources based on LEDs. The main reason is that the blue LED in combination with phosphor coatings, such as $Y_3Al_5O_{12}$:Ce, enable emission of white light.

1.5 Motivation for semiconductor sources in biomedical optics

In terms of performance and potential modest production costs, diode lasers also have become increasingly attractive for a plethora of applications. Due to direct electrical pumping, diode lasers and LEDs are by far the most efficient light sources currently available [24,25]. Being based on semiconductor chip technology, they can be manufactured in high numbers and at low cost. Their dimensions of only a few mm^3 enable very compact laser systems. Accordingly, these features increase their application potential, particularly within biomedical applications. Applications range from imaging and diagnostics, e.g., optical coherence tomography (Chapter 13), fluorescence lifetime imaging [26], diffuse optical imaging [27], or Raman spectroscopy (Chapter 11), to direct treatment such as photocoagulation [28], photodynamic therapy [29] or biomodulation and bioactivation [30]. Several of these topics are among the topics covered in this book.

Compared to lasers limited to specific atomic transitions, diode lasers cover a much wider spectral range. As detailed in the previous section, the emission wavelengths of typical III-V compound semiconductors range from blue to near infrared (400 nm–2 μm) [31], depending on the used compound semiconductors and their composition. In addition, some wavelengths, especially in the visible range, may not be achievable due to lack of available laser structures. One option also to achieve these wavelengths, or to increase the output power at a certain spectral region, is nonlinear frequency conversion [32], as discussed in Chapter 7.

Despite the number of wavelengths that can be accessed with diode lasers, the output power might not always meet the requirements, i.e., it is too low. Other options are optically pumped semiconductor lasers [33] or solid-state lasers [34], although the latter topic is not within our scope. Due to the optical excitation, optically pumped lasers show reduced optical efficiencies compared to electrically pumped diode lasers [35].

The output power and the beam propagation parameters of the diode lasers strongly depend on the design of the semiconductor structures. Nearly diffraction-limited beams are obtained with ridge-waveguide and tapered diode lasers. While the output power of the former lasers is limited to 1–2 W, more than 10 W is achieved with tapered lasers [36]. High-power emission is also obtained with broad area diode lasers [37] or diode laser bars and stacks [38]. However, these devices typically show reduced beam qualities, but for applications where shear power is needed such lasers are useful.

All of the above-mentioned devices are the so-called edge emitting diode lasers, i.e., the propagation of the generated laser emission is in plane with the substrate surface [39]. In surface emitting diode lasers, also known as vertical cavity surface-emitting lasers (VCSEL), the propagation direction is normal to the substrate surface. Their optical cavities are short and the facet reflectivities are high, resulting in low threshold currents [39]. The output power is typically in the milliwatt range but can be increased significantly by optical pumping in the so-called vertical external cavity surface-emitting lasers [40]. In comparison to edge-emitting diode lasers, the challenge towards high-power, near diffraction-limited emission from VCSELs is the proper heat removal from the active region [41].

In addition to continuous wave emission, diode lasers may also be operated in pulsed mode. Pulsed emission is achieved by mode-locking [42], Q-switching [43] or in a more direct manner by gain-switching [44]. These techniques enable the generation of pico- to femtosecond pulses with repetition rates in the GHz range. Compared to other mode-locked or Q-switched lasers, the lower upper-state lifetime of nanoseconds [45] reduces the obtained pulse energies of diode lasers [42].

The above-mentioned characteristics, i.e., output power, beam properties, wavelength spectral coverage and tunability, compactness and low cost, make diode laser technology versatile and increasingly applicable particularly in the biomedical field.

1.6 Organization of this book

Semiconductor-based light sources holds great promise for translation of innovative methods and technology into a wide range of applications. In particular, the fields of biology and clinical applications would benefit from these versatile sources. This handbook covers semiconductor lasers and diode-based light sources and their application in biophotonics and biomedical optics. The book is naturally divided into three parts covering fundamentals, technology and applications, respectively. Each chapter is written in a tutorial style, i.e., describing the subject matter in a self-consistent manner, and comprises adequate state-of-the art reviews, which will appeal to the skilled person in the field and to students introduced to the topics.

Leading on directly from this brief introduction to semiconductor physics, the fundamentals on modeling edge-emitting devices is thoroughly described in Chapter 2. It provides basic understanding of device operation and it contains a recipe for setting up simple numerical models of devices. Chapter 3 covers the basic principles of Q-switching and model-locked devices, i.e., detailed explanations of the various modes of operation, choice of material (QW vs. quantum dot devices), and configurations. The chapter finishes with a brief outlook of applications within biophotonics, although covered in detail in subsequent chapters. In some applications, light sources such as LEDs are the optimum choice due to their compactness, versatile wavelength, and low cost. Chapter 4 reviews the basic principles of LED operation and the potential improvement in efficiency. In lighting applications as well as biophotonic applications, efficiency is paramount. Chapter 5 describes the fundamentals of GaN-devices, emitting in the blue spectral region, of high interest of several biophotonic applications. Quantum cascade lasers have the potential to become compact, yet powerful devices for the mid-infrared region. Chapter 6 reviews the fundamentals of these devices, including fabrication, characterization, and performance. Sources in the mid-infrared is important for THz applications, and spectroscopy tapping into the spectral fingerprint region addressing relevant bio-molecules. Chapter 7 demonstrates that the addressable spectral region of lasers can be extended by introducing nonlinear optics. By nonlinear frequency conversion, it is possible to extend lasing into new wavelengths maintaining beam quality and linewidth, i.e., single frequency, and the basic principles are covered in this chapter. In terms of power scaling, the so-called disk lasers, or optically pumped lasers, show huge potential. Chapter 8 reviews the basics of said lasers, addressing power scaling and thermal management, and linking to the previous chapter introduces intracavity nonlinear frequency conversion. The chapter also briefly reviews some biomedical applications, taking advantage of the available power at specific visible wavelengths, in, for example, dermatologic therapeutic applications.

The remaining chapters targets biophotonic applications of light, where, for example, translation into said application is facilitated by semiconductor lasers. Chapter 9 addresses use of up converting nanoparticles as a means of preclinical diagnostic applications. Given that the particles may have excitation cross sections at specific wavelengths, not coinciding with atomic transitions offered by, e.g., solid-state lasers, their application would require lasers with tailor-made wavelengths, i.e., diode lasers. In near-infra spectroscopy, the need for a wide wavelength coverage is apparent. In addition, compactness and low cost are two important factors that render diode-based source well suited for this application. Chapter 10 reviews diffuse optical tomography and spectroscopy, and demonstrates its clinical potential. Moving from imaging of relatively large tissues or volumes, Chapter 11 demonstrates the important impact of nonlinear microscopy within biology. The chapter reviews a plethora of light sources and of various complexity, including semiconductor disk lasers offering ultra-high performance, relatively low degree of complexity and compactness. Raman spectroscopy is a powerful spectroscopic tool, which currently seems to expand into the field of biomedical diagnostics and fundamental biological studies. Excitation requires narrow linewidth lasers (and efficient detector units), and, in addition, compactness. Chapter 12 reviews this important biomedical application underlining the importance of semiconductor lasers for its huge potential across a wide range of applications. Chapter 13 reviews the basics of optical coherence tomography, a cross-sectional imaging modality, which successfully translated into several clinical applications. It is often highlighted as one of the most successful optical imaging modalities. The chapter demonstrates the important characteristics of semiconductors (wavelength agility, tuning speed, and low noise) and their impact on achieving the imaging performance level needed, i.e., spatial resolution, image quality and image acquisition speed.

It is our intention that the present handbook will allow the readership to get a first understanding of the underlying physics of semiconductor-based light sources, their performance, and their advantages as well as disadvantages. The feasibility of these lasers is demonstrated through selected, prominent biophotonic or biomedical applications. In the course of preparing this handbook, we realized that the range the range of applications facilitated by these lasers is continuously expanding, and therefore this exciting area continues to grow.

References

[1] F. Braun, "Über die Stromleitung durch Schwefelmetalle", Ann. Phys. Chem., 153, 556 (1874).

[2] H.J. Round, "A note on Carborundum", Electronic World, 19, 309 (1907).

[3] J. Bardeen and W.H. Brattain, "The transistor, a Semiconductor Triode", Phys. Rev., 1, 230 (1948).

[4] W. Shockley, The Theory of p-n Junctions in Semiconductors and p-n Junctions Transistors", Bell Syst. Tech J., 28, 435 (1949).

[5] R.A. Smith, Semiconductors, 2nd Ed. Cambridge University Press, London, 1979.

[6] T.S. Moss, Ed. "Handbook on Semiconductors", Vol. 1–4, North-Holland, Amsterdam, 1980.

[7] S. Wang, "Fundamentals of Semiconductor Theory and Device Physics", Prentice-Hall, Englewood Cliffs, New Jersey, 1989.

[8] S. Sze and M. LEE, Semiconductor Devices: Physics and Technology, 3rd edition, Wiley, 2012.

[9] N. Holonyak, Jr., R.M. Kolbas, R.D. Dupuis, and P.D. Dapkus, "Quantum Well Heterostructure Laser", IEEE J. Quant. Electron, QE-16, 170 (1980).

[10] K. Kasukawa, Y. Imajo, and T. Makino, "1.3μm GaInAsP/InP Buried Heterostructure Graded Index Separate Confinement Multiple Quantum Well Lasers Entirely Grown by MOCVD", Electron. Lett., 25, 104 (1989).

[11] R. Hall, G. Fenner, J. Kingsley, T. Soltys, and R. Carlson "Coherent Light Emission from GaAs Junction," Phys. Rev. Letters, 9, 366 (1962).

[12] M.I. Nathan, W.P. Dumke, G. Burns, F.H. Dill, and G. Lasher, "Stimulated Emission of Radiation from GaAs p-n Junctions," Appl. Phys. Lett., 1, 62 (1962).

[13] T.M. Quist, R.H. Rediker, R.J. Keyes, *et al.*, "Semiconductor Maser of GaAs," Appl. Phys. Lett., 1, 91 (1962).

[14] H. Kroemer, "A Proposed Class of Heterojunction Injection Lasers," Proc. IEEE, 51, 1782 (1962).

[15] Zh.I. Alferov, and R. F. Kazarinov, "Semiconductor with Electric Pumping," Soviet Union Patent, Authors certifica, 181737 (U.S.S.R), 1963.

[16] Zh.I. Alferov, V.M. Andreev, E.L. Portnoi, and M.K. Trukan, "AlAs–GaAs Hetero Junction Injection Lasers with a Low Room Temperature Threshold," Sov. Phys. Semiconductor, 3, 1107 (1970).

[17] N.N. Ledentsov, M. Grundmann, F. Heinrichsdorff, *et al.*, "Quantum-Dot Heterostructure Lasers," IEEE J. Selected Topics Auan. Electr., 6, 439 (2000).

[18] M. Asada, Y. Miyamoto, and Y. Suematsu, "Gain and the Threshold of Three-Dimensional Quantum-Box Laser," IEEE J. Quantum Electron, QE-22, 1915 (1986).

[19] http://www.cree.com/News-and-Events/Cree-News/Press-Releases/2014/March/300LPW-LED-barrier

[20] O.V. Lossev, "Luminous Carborundum Detector and Detection Effect and Oscillations with Crystals," Philosophical Magazine, 6, 1024 (1928).

[21] N. Holonyak Jr. and S. Bevacqua, "Coherent (visible) Light Emission from Ga(As1-xPx) Junctions," Appl. Phys., 1, 82 (1962).

[22] S. Nakamura, T. Mukai, and M. Senoh, "Candela Class High Brightness InGaN/AlGaN Double Hetero Structure Blue Light Emitting Diodes," Appl. Phys. Lett., 64, 1687 (1994). View online: http://dx.doi.org/10.1063/1.111832

[23] X. Sun, D. Li, H. Song, *et al.*, "Short-Wavelength Light Beam In Situ Monitoring Growth of InGaN/GaN Green LEDs by MOCVD," Nanoscale Res. Lett., 7, 1 (2012).

[24] A. Knigge, G. Erbert, J. Jonsson, *et al.*, "Passively Cooled 940 nm Laser Bars with 73% Wall-Plug Efficiency at 70 W and 25°C," Electron. Lett., 41, 250–251 (2005).

[25] M. Kanskar, T. Earles, T. Goodnough, E. Stiers, D. Botez, and L. Mawst, "73% CW Power Conversion Efficiency at 50 W from 970 nm Diode Laser Bars," Electron. Lett., 41, 245–247 (2005).

[26] K.W. Berndt, I. Gryczynski, and J.R. Lakowicz, "Phase-Modulation Fluorometry using a Frequency-Doubled Pulsed Laser Diode Light Source," Rev. Sci. Instrum., 61, 1816–1820 (1990).

[27] L.V. Wang and H. Wu, Biomedical Imaging: Principles and Imaging (Wiley, New Jersey, 2007), chap. 11, pp. 249–281.

[28] A. Puliafito, T.F. Deutsch, J. Boll, and K. To, "Semiconductor Laser Endophotocoagulation of the Retina," Arch. Ophthalmol., 105, 424–427 (1987).

[29] T.J. Dougherty, "Photodynamic Therapy – New Approaches [review]," Sem. Surg., Oncol., 5, 6–16 (1989).

[30] L. Longo, S. Evangelista, G. Tinacci, and A.G. Sesti, "Effect of Diodes-Laser Silver Arsenide-Aluminium (Ga-Al-As) 904 nm on Healing of Experimental Wounds," Las. Surg. Med., 7, 444–447 (1987).

[31] G. Erbert and R. März, in: F. Bachmann, P. Loosen, and R. Poprawe (eds.), High Power Diode Lasers: Technology and Applications (Springer, Berlin, 2007), chap. 2.1, pp. 5–7.

[32] R.W. Boyd, Nonlinear Optics, 2nd Edition (Academic Press, San Diego, USA, 2003), chap. 1.2, pp. 4–16.

[33] E.A. Saleh and M.C. Teich, in: J.W. Goodman (ed.), Fundamentals of Photonics (Wiley-Interscience Publication, Hoboken, USA, 1991), chap. 16.2B, p. 615.

[34] A. Tünnermann, H. Zellmer, W. Schöne, A. Giesen, and K. Contag, in: R. Diehl (ed.), High Power Diode Lasers – Fundamentals, Technology, Applications, with contributions by numerous experts (Springer, Berlin, 2000), chap. 10.1, pp. 369–373.

[35] J.A. Rossi, S.R. Chinn, J.J. Hsieh, and M.C. Finn, "Comparison of Optical to Injection Excitation in GaAs Heterostructure Lasers," J. Appl. Phys., 45, 5383–5388 (1974).

[36] B. Sumpf, K.-H. Hasler, P. Adamiec, *et al.*, "High-Brightness Quantum Well Tapered Lasers," IEEE J. Sel. Top. Quantum Electron., 15, 1009–1020 (2009).

[37] K. Paschke, S. Einfeldt, A. Ginolas, *et al.*, "15-W Reliable Operation of 96-mm Aperture Broad-Area Diode Lasers Emitting at 980 nm," in CLEO/QELS 2008 Conference Digest p. CMN4 (2008).

[38] W. Pittroff, G. Erbert, B. Eppich, C. Fiebig, K. Vogel, and G. Tränkle, "Conductively Cooled 1 kW-QCW Diode Laser Stacks Enabling Simple Fiber Couplin," IEEE Trans. Compon. Packag. Technol., 33, 206–214 (2010).

[39] P. Unger, in: R. Diehl (ed.), High Power Diode Lasers – Fundamentals, Technology, Applications, with contributions by numerous experts (Springer, Berlin, 2000), chap. 1.4, pp. 13–15.

[40] M. Kuznetsov, F. Hakimi, R. Sprague, and A. Mooradian, "High-Power (>0.5-W Cw) Diode-Pumped Vertical-External-Cavity Surface-Emitting Semiconductor Lasers with Circular Tem/Sub 00/ Beams," IEEE Photon. Technol. Lett., 9, 1063–1065 (1997).

[41] W.J. Alford, T.D. Raymond, and A.A. Allerman, "High Power and Good Beam Quality at 980 nm from a Vertical External-Cavity Surface-Emitting Laser," J. Opt. Soc. Am. B, 19, 663–666 (2002).

[42] D.J. Derickson, R.J. Helkey, A. Mar, J.R. Karin, J.G. Wasserbauer, and J.E. Bowers, "Short Pulse Generation using Multisegment Mode-Locked Semiconductor Lasers," IEEE J. Quantum Electron., 28, 2186–2202 (1992).

[43] M.B. Flynn, L. O'Faolain, and T.F. Krauss, "Ultracompact And Low-Power Optical Switch Based On Silicon Photonic Crystals", J. Opt. Soc. Am. B, 22, 792–795 (2005).

[44] C. Lin, P.L. Liu, T.C. Darnell, D.J. Eilenberger, and R.L. Hartman, "Simple Picosecond Pulse Generation Scheme for Injection Lasers," Electron. Lett., 16, 600–602 (1980).

[45] J. Ohtsubo, Semiconductor Lasers – Stability, Instability and Chaos, 2nd Enlarged Edition (Springer, Berlin, 2007), chap. 5.1.4, pp. 93–95.

Chapter 2

Fundamentals on modeling of edge-emitting semiconductor lasers and superluminescent light-emitting diodes

Mariangela Gioannini[1], Paolo Bardella[1] and Ivo Montrosset[1]

2.1 Introduction

This chapter is intended as a tutorial on the modeling for the simulation of semiconductor light sources as lasers [1], superluminescent diodes (SLDs) and optical amplifiers (SOAs) [2].

We address this chapter to students or researchers not specialized in the field of device modeling, but interested in setting up their own simplified models for an understanding of the device performance. As in a recipe, we introduce the ingredients that should be used and we describe how to merge them together to build up a model that will be not rigorous but useful to get indications on the main device characteristics. The fundamental ingredients that one should include in the model are:

- the electro-optical characteristics of the semiconductor active material
- the structure of the optical waveguide
- the carrier-photon interaction
- the device layout (multiple sections, tapered waveguide, etc.).

The various sections of the chapter are dedicated to the discussion of each of the ingredients, presenting them in the way they enter in the model setup. In the final section we will consider as an example the case of the SLD simulation. The references, provided to books and journal papers, should be considered only as examples and not as a comprehensive list.

2.2 Electro-optical characteristics of the semiconductor active material

The electro-optical characteristics of the semiconductor material forming the active waveguide are determined by the structure used to realize the quantum confinement

[1]Department of Electronics and Telecommunications, Politecnico di Torino, Italy

of the carriers: bulk, quantum well (QW), quantum wire/dash or quantum dot (QD). As predicted in [3], the increase of the quantum confinement of carriers leads to an increase of the material gain as well as a reduction of the current required to get a given gain value at a selected wavelength. For this reason, there has been continuous attempt in increasing the carrier confinement to obtain higher gain. Despite this, the material most used in commercial applications is still nowadays the QW material for its very good electro-optic characteristics and reliability due to the simpler control of the parameters of the material growth (size, composition, accumulated strain...). Quantum dashes (QDashes) and QDs (grown without any patterning effect [4]) can be obtained only by a self-assembled process that causes the nanostructures to be spread in size and composition [5]. The inhomogeneity leads to a significant reduction of the gain respect to what is ideally predicted, because stimulated emission occurs from several slightly different energy states giving an inhomogeneously broadened gain. We also have to remember that the actual *gain* of interest in a waveguide is not the *local gain* or *material gain* but the *modal gain* (see Section 2.3.2); therefore, a reduction of the dimensions of the region, where stimulated emission takes place, gives as a drawback the reduction of the effective interaction of the photons, confined by the waveguide, with the active material confining the electrical carriers. This requires a partial compensation by increasing the density of the nanostructures or the number of nano-structured layers. For these reasons, QDashes and QDs have actually found application in just a limited range of guided-wave devices where they have demonstrated to be actually superior to the QW counterpart.

2.2.1 Optical gain and spontaneous emission

A simple expression typically used to compute the material gain and spontaneous emission spectrum in the case of the QW materials is derived from the detailed balance between absorption and stimulated emission processes and writes as [6]:

$$g(\hbar\omega) = \frac{1}{\hbar\omega}\frac{\pi e^2 \hbar n_g}{\varepsilon_0 c m_0^2 n^2}|M_T|^2\left\{\frac{\rho_{red}(\hbar\omega)}{d_{QW}}(f_e + f_h - 1) * L(\hbar\omega)\right\} \tag{2.1}$$

where $\hbar\omega$ is the photon energy, m_0 is the free electron mass, c is the light velocity, e is the electron charge, n_g is the group refractive index, n is the refraction index of the active material, M_T is the momentum matrix element including the overlap integral between the electron and hole envelope wavefunctions, ρ_{red} is the QW reduced density of states per unit of area, d_{QW} is the well thickness, E_g is the energy gap and $f_{e,h}$ are the Fermi functions evaluated at the energy $E_{e,h}$ of the electron (e) and hole (h) states involved in the transition such that $E_e - E_h = \hbar\omega$. The convolution with a Lorentzian function (or other more appropriate broadening functions) accounts for the homogeneous broadening of emission line due to polarization dephasing time. Consistently with expression (2.1), the rate of spontaneous emission is written as:

$$R_{sp}(\hbar\omega) = \rho_{opt}(\hbar\omega)\frac{1}{\hbar\omega}\frac{\pi e^2 \hbar}{\varepsilon_0 m_0^2 n^2}\frac{1}{3}\sum|M_T|^2\frac{\rho_{red}(\hbar\omega)}{d_{QW}}f_e f_h * L(\hbar\omega) \tag{2.2}$$

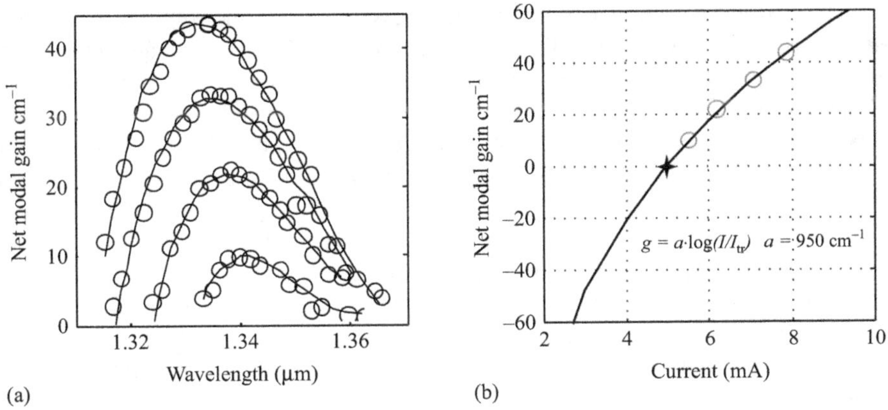

Figure 2.1 *(a) Measured gain spectrum (circles) and fitting with cubic polynomial expression [14] and (b) measured gain peak versus current (circles) and fitting with logarithm expression. The star symbol indicates the measured transparency current*

where ρ_{opt} is the density of the optical modes and the summation is over the three polarizations. The quantitative matching of the measured gain with expression (2.1) is often not sufficiently accurate because not all the microscopic effects (as, for example, many-body effects) are included. More sophisticate and cumbersome microscopic models have been developed [7] and have been successfully applied to predict the electro-optical properties of several QW materials [8]. The development of such tools is however not a trivial task; the huge effort is often not justified when simple device models are needed. In this case a simplified solution consists in using the gain (or spontaneous emission spectrum) as an input parameter of the device simulator in the form of simplified polynomial expressions taken from the fitting of the measured data. Several methods may be adopted for measuring the gain such as the Hakki–Paoli method [9], its improved version called the Fourier Transform method [10] and the segmented contact method [11]. Whereas the first two techniques limit the gain measurement up to the laser threshold, the last one is preferred because it provides gain over a larger current range; in addition, the internal modal loss, the absorption and the spontaneous emission spectrum can also be measured [12]. The fitting of the measured absorption spectrum is very useful to extract the dipole moment [13] that enters in the modeling expression (2.1). Once the gain has been measured cubic polynomials were used in [14] to fit with accuracy the measured results; we show in Figure 2.1(a) an example of application of this method. The peak gain versus current (often used in modeling single mode lasers) can be fitted with a logarithmic function (Figure 2.1(b)) for a QW material and with a linear function for bulk materials provided that a measurement of the transparency current (I_{tr}) is available. These fitting functions are defined in the literature as dependent of the active volume carrier density N_v. This requires relating N_v to the injection current [15] by the measurements of the recombination parameters defined in Section 2.2.2.

The simplified approach presented above can provide input optical gain parameters in a device modeling approach based on the rate equations (REs) (Section 2.4), because the carriers (electrons or holes) are treated as a unique ensemble in quasi-equilibrium distribution. This approximation is valid when the inter-band relaxation processes, much faster than radiative and non-radiative recombination and stimulated emission rate, can be neglected. To include the finite inter-band relaxation, a microscopic model [16] should replace the REs and the gain should be calculated from first principles. For this reason when we deal with modeling of QD-based devices, the simple polynomial fitting is no more suitable to describe the gain because of the longer (with respect to the bulk or QW case) inter-sub band relaxation dynamics. Furthermore, the inhomogeneity of the emission from different QDs should also be included in the model setup [17,18]. In this case, an expression similar to (2.1) reads as

$$g(\hbar\omega) = \frac{1}{\hbar\omega}\frac{\pi e^2 \hbar n_g}{\varepsilon_0 c m_0^2 n^2}\frac{N_D}{d_{QD}}\sum_k \mu_k |M_{T,k}|^2 \sum_m G_m \left(\rho^e_{k,m} + \rho^h_{k,m} - 1\right) L\left(\hbar\omega - E_{k,m}\right) \quad (2.3)$$

where N_D is the areal density of the QD layer with thickness d_{QD}. The index k is for the QD confined state involved in the transition (i.e., ground state, GS, and excited states, ES) with degeneracy μ_k, whereas the index m account for the m-th QD subgroup of the inhomogeneous ensemble [17]. We assume, for the recombination energies ($E_{k\,=\,GS,m}$ and $E_{k\,=\,ES,m}$) of the ensemble, a Gaussian distribution whose width can be, for example, estimated from low temperature photoluminescent spectra; G_m is the integral of the Gaussian distribution over the spectral range of the $m-th$ QD subgroup [17]. The state occupation of electrons/holes is $\rho^{e,h}_{k,m}$ and the homogeneous broadening of the $E_{k,m}$ is approximated with a Lorentzian function whose width depends on the dephasing time. An example of application of expression (2.3) is shown in Figure 2.2(a); as evidenced in Figure 2.2(b) this method reproduces quite well the measured gain at low current injection, but it significantly overestimates the gain at high currents. For this reason additional effects such as free carrier absorption [19], dependence of homogeneous broadening linewidth with carrier density [20] and shift of the recombination energies due to Coulomb interaction [21] should be included at least using phenomenological expressions with parameters tuned to fit the measurements (as done in Figure 1.2(b)) or from first-principle calculations [22].

2.2.2 *Carrier recombination and carrier life-time*

Other electro-optical parameters of the active section to be included in the phenomenological RE approach are the recombination parameters of Shockley–Read–Hall, radiative and Auger interband recombination rates. As for the optical gain, these parameters should be calculated from microscopic modeling in the frame of the semiconductor Bloch equations [23], but they are often simply quantified through the semi-empirical ABC power law:

$$R(N_v) = AN_v + BN_v^2 + CN_v^3 \quad (2.4)$$

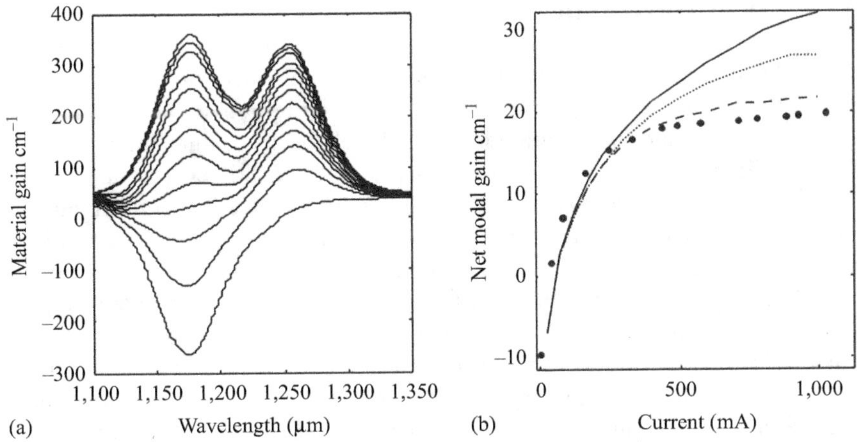

Figure 2.2 *(a) QD material gain calculated applying expression (2.3) and (b) GS peak gain versus current from measurements (circles), from expression (2.3) in solid line, with phenomenological inclusion of free carrier absorption (dotted line) and homogeneous broadening linewidth dependent on carrier density (dashed line)*

where the terms on the right hand side is associated to the Shockley–Read–Hall, the spontaneous emission and the Auger recombination rates, respectively; N_v is the volume carrier density in the active region and the input ABC coefficients can be extracted from measurements [24]. If the model adopts a gain expression similar to (2.1), the spontaneous emission rate should be written, self-consistently with the gain, as:

$$BN_v^2 = \int R_{sp}(\hbar\omega)\mathrm{d}(\hbar\omega) \tag{2.5}$$

where the integral extends all over the energy range of $R_{sp}(\omega)$.

The reader must be aware that this simple ABC model (2.4) may lead to significant errors at high carrier densities [25] and is not very suitable for modeling recombination in low dimensionality materials such as QDs or QDashes [26].

In even more simplified approaches the total carrier loss due to both radiative and non-radiative recombination are frequently represented with a single loss rate:

$$R_{\mathrm{loss}}(N_{\mathrm{v}}) = \frac{N_{\mathrm{v}}}{\tau_{\mathrm{N}}} \tag{2.6}$$

with the time constant τ_N defined as a carrier lifetime; the main advantage is the use of just one single parameter, τ_N, that can be extracted from various measurements such as impedance measurements [27], optical modulation techniques [28] or from the long-lived dynamics of the gain recovery trace obtained in pump-probe experiments [29].

2.3 Optical waveguides

Semiconductor optical waveguides are usually complicated structures due to the variety of materials used for the different layers (the active region, the photon confinement layer and the cladding layer) and the various designs implemented for the lateral field confinement. The devices presented in this chapter are based on optical waveguides of the types shown in Figure 2.3: Figure 2.3(a) schematizes the structure of a buried heterostructure (BH) waveguide where carriers are confined in the active layer (gray area) by mean of the double heterostructure, whereas photons are confined by the index contrast between the active and surrounding medium. Figure 2.3(b) presents a typical ridge waveguide: carriers and photons are confined by the heterojunction in the vertical direction, whereas in the lateral direction the photon confinement is provided by etched narrow ridge. The carrier confinement in the lateral direction is only guaranteed by the reduced current spreading and by the limited carrier diffusion in the active layer lateral to the ridge.

2.3.1 Transverse optical mode calculation

When, as shown in the examples of Figure 2.3, the optical field confinement is determined by the refractive index variation between the cladding and confinement layers in the vertical and horizontal direction (index guided structures), the optical waveguide can be analyzed by calculating the modes assuming zero gain and loss. The stimulated emission and the free carrier absorption, being a slight perturbation of the real and imaginary part of the dielectric constant, do not usually determine, in index guided structures, significant changes in the field distribution and their effects can be evaluated separately. Other waveguide losses (for example, due to photon scattering and waveguide imperfections) are usually included through an average intrinsic loss parameter indicated in the following as α_i.

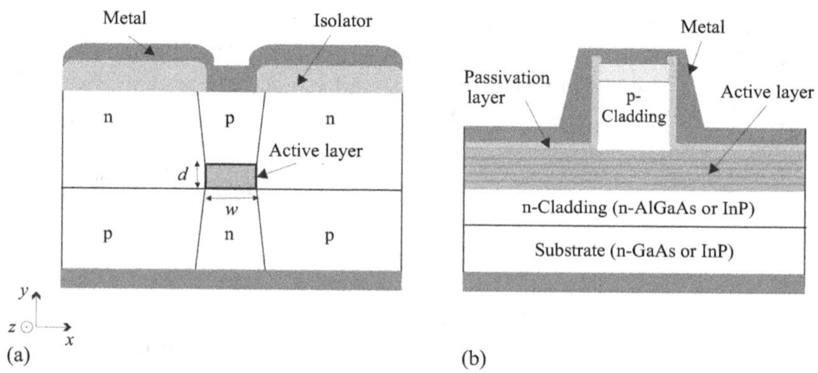

Figure 2.3 Schematic of (a) buried heterostructure (BH) waveguide and (b) ridge (RW) waveguide

Since we are looking for the Maxwell's equation solutions with the field propagating along the z-direction and confined in the $x-y$ plane, the electric and magnetic fields are assumed to have the form:

$$\mathbf{E},\mathbf{H}(x,y,z) = \mathbf{E},\mathbf{H}(x,y)\mathrm{e}^{j(\omega t-\beta z)} \tag{2.7}$$

where ω is the angular frequency and β is the mode propagation constant along z.

The finite element method (FEM) is typically used to evaluate the modal fields $\mathbf{E}$, $\mathbf{H}(x,y)$ of such structures. A rigorous numerical solution can be obtained using vectorial formulations as in the pioneering papers [30,31]. In the case of weakly guiding devices considered in this chapter, a very good approximation to evaluate the parameters, necessary for the device modeling, is the semi-vectorial approximation [32]. This is based on the assumption that the modes can be approximated as quasi-TE or quasi-TM characterized only by the two main components of the E and H fields in the plane perpendicular with respect to the propagation direction z. With the approximation $H_x = 0$ for the q-TE and $H_y = 0$ for the q-TM fields, the following scalar wave equation is obtained for the main components of the H field:

$$\frac{\partial^2\Psi}{\partial x^2} + \frac{\partial^2\Psi}{\partial y^2} + \left[k^2 n(x,y)^2 - \beta^2\right]\Psi^2 = 0 \tag{2.8}$$

where $\Psi = H_y$ in the q-TE case and $\Psi = H_x$ in the q-TM case, $k = \omega/c$ and $n(x,y)$ the refractive index profile.

From the solution of this equation, one can then recover all the other components of the electromagnetic field. As a consequence of this approximation, the most relevant transversal components of the electric field are E_x for the q-TE and E_y for q-TM cases. From this result (and also from the relation between E_x and H_y and between E_y and H_x) we can see that the solutions are TE in depth and TM in the transversal direction for the q-TE fields and vice versa for the q-TM. The scalar equation (2.8) can be easily solved with a FEM solver.

A more simplified approach is the effective index method (EIM) [33] which is frequently used for analyzing simple 2D structures such as the one in Figure 2.3 and in Figure 2.4(a). This method is less rigorous than the previous one, but it is

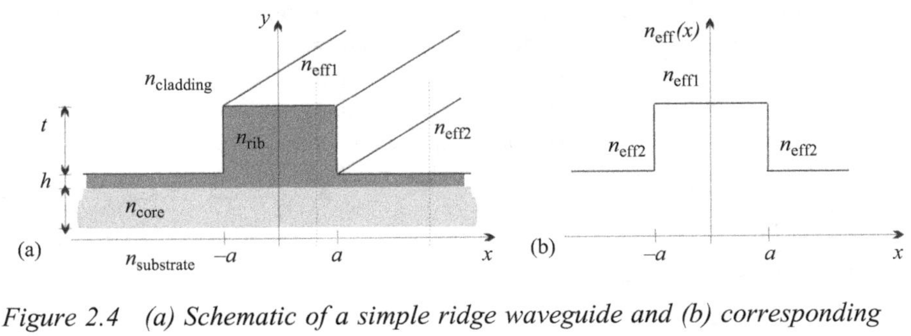

Figure 2.4 (a) Schematic of a simple ridge waveguide and (b) corresponding solution of (2.12)

sufficient for a quick analysis or design of the waveguides. The method is based on the assumption that the solution of the previous equation can be factorized as:

$$\Psi(x,y) = X(x)Y(y;x) \tag{2.9}$$

Substituting this expression in the wave equation, dividing it by $X \cdot Y$ and neglecting the x derivative of $Y(y; x)$, we obtain:

$$\frac{1}{X}\frac{\mathrm{d}X^2}{\mathrm{d}x^2} + \frac{1}{Y}\frac{\mathrm{d}Y^2}{\mathrm{d}y^2} + \left[k^2n^2(x) - \beta^2\right] = 0 \tag{2.10}$$

Adding and subtracting a y-independent term $k^2n_{\mathrm{eff}}^2(x)$ in the square brackets, we can approximately separate equation (2.10) into two independent equations:

$$\frac{1}{Y}\frac{\mathrm{d}Y^2}{\mathrm{d}y^2} + \left[k^2n^2(x,y) - k^2n_{\mathrm{eff}}^2(x)\right] = 0 \tag{2.11}$$

$$\frac{1}{X}\frac{\mathrm{d}X^2}{\mathrm{d}x^2} + \left[k^2n_{\mathrm{eff}}^2(x,y) - \beta^2\right] = 0 \tag{2.12}$$

The two-dimensional problem is therefore reduced to the sequential solution of two one-dimensional problems. Solving first (2.12) we get $Y(y; x)$ and $n_{\mathrm{eff}}(x)$; the latter is defined as *effective refractive index*. Mathematically $n_{\mathrm{eff}}^2(x)$ assumes the role of the variable separation constant in this approximate method. Solving (2.12) we get the propagation constant of the mode of the ridge waveguide β and the wavefunction $X(x)$. In order to keep the q-TE/TM field polarization characteristics, the equation in y should be solved assuming TE/TM boundary conditions and vice versa for the equation in x. The $n_{\mathrm{eff}}(x)$ solution of the waveguide in Figure 2.4(a) is depicted in Figure 2.4(b). The solution obtained with this method is a good approximation of the full 2D structure (except when the modes are close to the cutoff) and allows computing with an acceptable accuracy the parameters for the simulation and design of the devices considered in this chapter.

2.3.2 The relevant parameters of the modal field

The most relevant parameters of the modal field to be used as inputs in the device modeling are the effective refractive index ($\overline{n}_{\mathrm{eff}}$), the transverse optical confinement factor (Γ_{xy}) and the near and far field distribution.

The **effective refractive index** of a mode is defined as $\overline{n}_{\mathrm{eff}} = \beta/k$, where β is the propagation constant obtained from the waveguide analysis previously discussed.

The **optical confinement factor** in the active region (i.e., the part of the waveguide where stimulated emission occurs) is defined as:

$$\Gamma_{xy} = \frac{\iint_{A_{\mathrm{active}}} S(x,y)\mathrm{d}x\mathrm{d}y}{\iint_{-\infty}^{+\infty} S(x,y)\mathrm{d}x\mathrm{d}y} \tag{2.13}$$

where $S(x,y)$ is the photon density that can be approximated as proportional to the square electric field modulus ($|E(x,y)|^2$). The optical confinement factor is therefore used to quantify the fraction of the photon density distribution in the transversal ($x-y$ plane) active area (A_{active}) respect to the whole distribution of the photon density. For this reason it enters in the expression of the modal gain g_{mod} as:

$$g_{\text{mod}} = \Gamma_{xy} g \tag{2.14}$$

where g is the material gain as defined in (2.1). If the EIM is used, Γ_{xy} can be approximated with $\Gamma_{xy} \simeq \Gamma_x \Gamma_y$, where Γ_x and Γ_y are the optical confinement factor applying the definition in (2.13) and using the factorized solution (i.e., $X(x) \cdot Y(y;x)$). This conceptual separation between the *material* and *modal gain* is widely used in papers and textbooks as a retention from the modeling of bulk active materials [1]. In the case of QW or QD structures the definition of *material gain* is actually not appropriate since we can only refer to a *modal gain*, being the optical gain the consequence of the interaction of a given electric field profile (the modal field profile in this case) with the QW or QD layer considered [6]. For this reason a more appropriate and general expression of the modal gain has been proposed in [6]. In the models presented in the following we will however continue to follow the formulation in (2.14), since it is the most widely adopted and it remains valid when the field profile is practically constant respect to the wavefunction profile of electrons and holes.

The definition of the confinement factor in (2.13) is correct only when the active region is well defined and is possible to consider a uniform carrier distribution in it. Making reference to the examples of waveguides structure shown in Figure 2.3, this definition is correct for the BH where A_{active} is, for example, the gray area of Figure 2.3(a), but it does not apply for the ridge waveguide in Figure 2.3(b). In this more general case, the modal gain is calculated as:

$$g_{\text{mod}} = \frac{\iint_{-\infty}^{+\infty} g(x,y) S(x,y) \mathrm{d}x\mathrm{d}y}{\iint_{-\infty}^{+\infty} S(x,y) \mathrm{d}x\mathrm{d}y} \tag{2.15}$$

An expression similar to (2.15) can be used to evaluate the modal losses (for example, due to doping which is usually not uniform in the waveguide) substituting the gain with the loss distribution ($\alpha(x,y)$).

The general expression (2.15) for the modal gain can be further simplified considering that the active layer in the y-direction is usually very thin. The carrier distribution (and so the gain) in the active area can therefore be considered constant in y and the integration over y can be eliminated introducing the vertical confinement factor Γ_y. Attention should be given to the choice of the active layer thickness when dealing with active waveguide with nano-structures (i.e., QW, QD, Dots in a Well...): the choice of the thickness of the active region must be consistent with the thickness chosen to calculate the density of states per unit of volume entering in the expression of the material gain in (2.1) and (2.3) [6]. With self-assembled QDs, for example, we define d_{QD} as the equivalent thickness of the QD layer (typically the average thickness estimated from TEM images) and then we calculate Γ_y accordingly. It therefore turns out the important result that, even if the material gain in (2.1) and (2.3) scales as one over the layer thickness (apparently thinner QDs or QWs should give higher material gain!), the modal gain (i.e., the actual amplification of the field propagating along the waveguide) is independent on the equivalent thickness the QD layer (because the Γ_y scales proportionally with the thickness) [6]. In the QD case the modal gain remains therefore only proportional to the density of QDs per layer and on the number of QD layers; in the QW case it depends on the QW thickness only through the overlap integral of the electron and hole wavefunctions (being these ones also dependent on the well thickness).

Other important parameters from the practical point of view are the near field and the far field. The near field is usually defined as the field distribution profile on the aperture, and the far field as the angular intensity distribution far from the laser facet. The far field is obtained as the intensity of the Fourier transform of the aperture field multiplied by the obliquity factor $\cos(\theta)$ (where θ is the angle respect to the z-direction). Typically the aperture field is approximated with the incident modal field at the air interface and the far field is computed simply using it [33].

2.4 Carrier-photon interaction

2.4.1 A general overview

A very simple way to account for the carrier-photon interaction in the active medium is the phenomenological RE approach which can be employed for modeling semiconductor lasers (edge-emitting and vertical cavity-emitting lasers [34,1]), optical amplifiers [2] and superluminescent light-emitting diodes. It is very easy to set up an RE model, simulations are usually not time consuming and they can capture the most relevant characteristics of the device operation. Furthermore, with this approach, several measured data (as discussed in Section 2.2) can be used as input parameters or, reversely, from the analytical solution of RE and the comparison with the experimental results, we can find methods for extracting various parameters of the device under test. The traditional RE system [1] is a set of equations (non-linear first-order differential equations in the time variable) for the average carrier density (with no distinction between electrons and holes) and for the average photon density in the active region. The equation of the photon density is substituted with the equation of the complex electro-optic field when interference effects become important, as in the study of external optical feedback effects [35] or in the traveling wave (TW) model (see Section 2.4.6) when longitudinal distribution of carrier and photons must be considered.

Based on the general principle of the RE approach, several variants were developed to account for various effects peculiar to the active materials with nanostructures. For example, in the case of multi quantum well (MQW) active region, the RE of the carrier density in each QW were coupled with the drift-diffusion equations to include the modeling of electron and hole transport in the separate confinement layer ([36–38]). In the case of quantum dash or quantum wire materials, the complicated carrier and photon dynamics has been studied through RE in [39]. In the case of QD material, various versions of the RE approach have been developed to include the competitive emission from both ground state and excited state [40], the interaction of the inhomogeneously broadened gain medium with many lasing longitudinal modes [17] and the different time scales of the electron and hole dynamics [41–44]. The approach presented in [17] includes also gain dispersion due to inhomogeneous gain broadening and the spectral hole burning due to finite inter-band relaxation time of carriers from upper states to the lasing state. Following the same approach of [17], the dependence of the refractive index change on wavelength and carriers was included in RE model for the QD lasers in [45] eliminating the use of the linewidth enhancement factor (LEF) parameter (or α-parameter [46,47]) required to relate the refractive index change to the optical gain variation.

Opposite to simple RE modeling, we have the rigorous approach based on the Semiconductor Maxwell-Bloch Equations [48,49] consisting in a closed set of integro-differential equations for the quantum dynamics of the polarization and of the electron and hole distribution coupled with the Maxwell equations driven by the polarization. The active material parameters such as the carrier lifetime, the polarization decay time and the inter-band relaxation time are calculated self-consistently from the microscopic modeling of carrier-carrier and carrier-phonon collisions [8,16] or inserted in a parameterized way [50]. This approach, applied to the QD case, has been reviewed in [51].

The rigorous analysis of the carriers in the active waveguide requires a self-consistent solution of the continuity and drift-diffusion equations in the x–y plane, coupled with the Maxwell equation for the modal field; moreover, temperature effects should also be included. This is a complicated numerical problem that is treated by advanced numerical models as [52] or available commercial software as done in [53]. In this chapter we describe the simpler RE model, starting with its basic formulation and introducing step by step more elaborate models. In Sections 2.4.2 and 2.4.3, we discuss how the RE equation can be applied to light-emitting devices with a cavity as, for example, the edge emitting lasers in Figure 2.5(a) and with different transversal waveguide structures. In Section 2.4.4, we discuss how the RE system can be coupled with the classical diode continuity

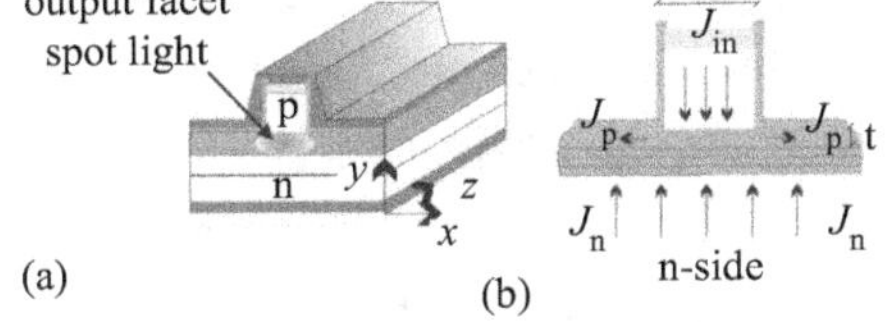

Figure 2.5 (a) Structure of a laser cavity realized with a ridge waveguide and cleaved facets; (b) later carrier diffusion under the ridge

equations; whereas in Section 2.4.5 we present how the RE approach can be extended to analyze SOAs and SLDs. In Section 2.4.6, we discuss how RE can be modified to simulate more complicated laser dynamic regimes as, for example, the generation of light pulses.

2.4.2 The case of laser cavity with BH waveguide

When dealing with BH waveguides, the carriers are very well confined in the active region and the carrier photon interaction can be well represented with two differential equations for the averaged values of the carrier and photon density in the active region. These equations are often reported in the literature with respect to the average quantities in the active region volume (V) (as, for example, in [1]) or with respect to the average quantity in the junction active area (A) (as, for example, [54]). These quantities will be indicated here with N_v and $N_{p,V}$ for carrier and photon density per unit of volume, and with N and N_p for carrier and photon density per unit of area. The formulations of the RE differ only for the presence or the absence of some constants related to the different normalization used. Here we will use the normalization with respect to the active area and the quantities will be expressed per unit of area; this choice returns more advantageous when we need to couple together several rate-equations, to represent the carriers in different volumes of the waveguide or, in the case of nano-structured layers, the carriers in 0D, 2D and 3D states. Typical examples are when we need to couple the bulk carriers with QW carriers (see, for example, Section 2.4.4) or when we write RE for a QD active waveguide. In this last case we have carriers confined in the 0D states of the QDs, in the 2D wetting layer state and also in the 3D barrier state [45].

The RE system for bulk active material and for the MQW case (considering only the carriers in the QWs) writes as:

$$\frac{dN}{dt} = \frac{\eta_i I}{eA} - R(N) - v_g g_{mod}(N, N_p) N_p \tag{2.16}$$

$$\frac{dN_p}{dt} = \beta_{sp} R_{sp}(N) + \left(v_g g_{mod}(N, N_p) - \frac{1}{\tau_p} \right) N_p \tag{2.17}$$

These equations quantify phenomenologically how fast carrier and photon density changes with time due to the rate of: carrier injected from the emitter $\left(\frac{\eta_i I}{eA}\right)$, carrier loss due to radiative and non-radiative recombination ($R(N)$), photon generation (or carrier loss) due to stimulated emission ($v_g g_{mod}(N, S) N_p$) and photon loss at the cavity facets and due to internal modal loss $\left(\frac{N_p}{\tau_p}\right)$.

In particular,

- I is the current injection and η_i is the internal quantum efficiency accounting for the carrier leakage. Routinely η_i and the waveguide internal loss (α_i) can be extracted from the measurements of the external quantum efficiency (i.e., the slope of the light vs. current curve) of FP lasers with different cavity lengths [1,34].

- A is the area of the active region in the y–z plane given by the waveguide width w times the active section cavity length.
- $R(N)$ is the carrier loss discussed in Section 2.2 and written here as rate per unit of area.
- v_g is the group velocity; the value of the group refractive index can be, for example, extracted from the separation of the peaks of the optical spectrum of an FP laser below threshold or applying on this spectrum the Fourier Transform method [10]
- τ_p is the photon lifetime.
- $g_{mod}(N, N_p)$ is the modal gain as defined in (2.14); the gain dependence as function of the volume carrier density can enter in the model through the empirical expressions discussed in Section 2.2.1. At this modeling level, the non-linearities due to the dependence of the gain on photon density are included through a phenomenological gain compression factor coefficient (ε) as:

$$g_{mod}(N, N_p) = \frac{\tilde{g}_{mod}(N)}{(1 + \varepsilon N_p)} \tag{2.18}$$

  where $\tilde{g}_{mod}(N)$ is the optical gain as function of the sole carrier density at low photon density. This phenomenological expression can account for several microscopic effects such as the spectral-hole burning due to finite inter-sub-band relaxation time [55,56], the carrier heating [56] and the carrier transport [57] and also for the standing wave of the lasing field and the spatial diffusion of carriers [58].
- β_{sp} is a coefficient accounting for the fraction of photons, emitted by spontaneous emission, which couples with the considered longitudinal cavity mode with photon density N_p.
- $R_{sp}(N)$ is the spontaneous emission rate discussed in Section 2.2.1.

This simple RE approach has been used in many textbooks (e.g., [1,34,35]) to analyze the basic continuous wave (CW) characteristics as well as small and large signal modulation response of semiconductor lasers. It is worth mentioning that these REs have been frequently applied to laser diodes with other waveguide structures (not just buried, for example, ridge or gain-guided) because, from the analytical solution and the comparison with experimental results, it is simple to extract (or estimate) useful device parameters (internal modal loss, internal quantum efficiency, gain compression factor, etc.). When these REs are applied to other waveguide structures, in which the area occupied by the carriers is not simply defined by the active waveguide volume, an effective injection active area (A_{eff}), greater than the previously defined geometrical area, is introduced accounting for the current spreading and the lateral diffusion of carriers.

In this section, as an application example of the RE, we discuss the concept of laser threshold current, usually estimated from the cavity round trip gain condition or experimentally estimated from the extrapolation to zero emission power of the linear part of the light–current (L–I) characteristic. This concept has been reviewed in [59–62] showing that the threshold current, measured in this way, cannot be considered a distinctive feature of the onset of the laser emission in the case of micro- and nano-cavity lasers. In Figure 2.6, we plot the normalized photon density

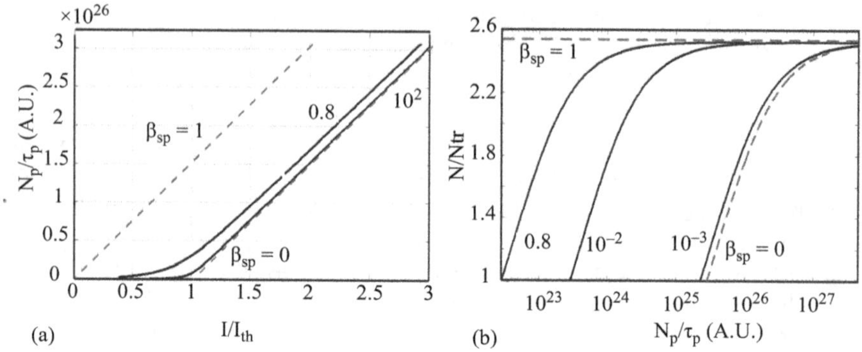

Figure 2.6 *(a) Normalized photon density versus normalized threshold current I_{th}; I_{th} is defined as the threshold current with β_{sp} = 0 and (b) normalized carrier density versus photon density. The dashed lines in (a) and (b) draw the characteristics for β_{sp} = 0 and β_{sp} = 1, respectively*

vs. normalized injected current (Figure 2.6(a)) and the normalized carrier density versus normalized photon density (Figure 2.6(b)) obtained by the analytical solution of the RE (2.16) and (2.17) for various values of β_{sp} (i.e., from very small values, typical of edge emitting lasers, to high values typical of micro-cavity lasers). From the analysis of these characteristics, it is obvious that the precise measure of a threshold current is only possible for $\beta_{sp} \rightarrow 0$, whereas the determination of the threshold gets more and more unprecise when β_{sp} gets larger. Indeed for $\beta_{sp} \rightarrow 1$, there is apparently no threshold current, but the true onset of coherent laser emission is for N_p very large, when the carrier density (and therefore the round trip gain) equalizes the cavity loss [60]. This analysis, carried on in [60], lead to the conclusion that the lasing threshold should not be defined with respect to the L–I characteristics but with respect to the field coherence properties [61]. This concept is of particular importance in the case of micro- or nano-cavity lasers when the photon density is very low and the β_{sp} becomes closer to 1.

2.4.3 *The case of laser cavity with ridge waveguide*

As shown in Figure 2.5(b), even if the ridge structure can well confine the injected current (J_{in}) in the region w_r, there is some lateral carrier diffusion under the ridge in the layer above the active region (thickness t in the Figure) and in the barrier layers separating the QWs. The rigorous treatment of the lateral diffusion of electrons and holes requires to solve the two-dimensional steady-state continuity equation in the $x-y$ plane; this is a complicated numerical problem solved by several commercial software as shown in [53]. Here we address a simplified approach based on the fact that the ridge structure can well confine the injected current in the region below the ridge because of the very small thickness t. Assuming current injected only in the strip w_r and carrier distribution constant along y (due to the very small thickness of the active region), the lateral carrier distribution can be evaluated considering only the lateral diffusion process. By using an effective diffusion rate approximation [63], we can determine the carrier profile in the active region using the equation [64]:

$$D_{\text{eff}} \frac{dN^2}{dx^2} = \frac{\eta_i J_{\text{in}}(x)}{e} - R(N(x)) - v_g \Gamma_y g(N, N_p) N_p(x) \tag{2.19}$$

where $J_{in}(x)$ is the injected current density assumed constant under the ridge and zero elsewhere, D_{eff} is the effective diffusion coefficient [63]. With this model, we can quantify the carrier diffusion in the ridge waveguide [64]; we can account for the transversal spatial hole burning effect caused by the lasing mode profile and, given the carrier profile $N(x)$, we can quantify the modal gain of higher-order optical modes to estimate how far they are from threshold. Rigorously, (2.19) should be solved self consistently with the field (2.12) including the perturbation due to the complex effective refractive index variation due to carrier distribution N(x). However, for laterally index guided structures, it is typically assumed that the modal field profile and the photon distribution remain unchanged respect to that of the unpumped waveguide.

It is clear that the previous approaches based on RE neglect the hetero-junction characteristic in the y-direction and the distribution of carriers and photons in the longitudinal direction z. In the following section we will discuss how these effects can be included starting from the RE approach.

2.4.4 Inclusion of the carrier transport

Most of the active layer structures of edge emitting lasers are based on MQW active materials; several new devices based on QDs have also appeared in the last ten years. As conceptually shown in Figure 2.7 the carriers feed the active layers through a capture process from the barrier, wherein carriers are injected from the p- and n-side of the junction [37]. To account for the transport of the bulk carriers along the barrier, the simplest approach makes use of an additional RE for the carrier population in the barrier state and all the transport properties are represented through an empirical capture time constant [65]. This model is limited by the impossibility of distinguishing the transport characteristics of electrons and holes (due to their different mobility and diffusion coefficients) and also does not consider modifications of the potential profile caused by different electron and hole concentration in the barriers. The more accurate approach consists in coupling the classical diode transport equations for electrons and holes in the bulk material, with the RE system of the carriers in the MQW active layer and photon density [37]. In this model each nano-structure active layer (QW or QD layer) is located in position y_i; carriers are captured in the layer via a phonon-assisted or carrier-carrier scattering process or they can escape out of the active layer in the barrier states. The transport of 3-D carriers is modeled by drift-diffusion equations relating the electron/hole volume densities in the barrier state ($n_B(y)$ and $p_B(y)$) to the sheet electron/hole density in the active layer ($N,P(y_i)$). We discuss here the approach presented in [37], whereas more sophisticate models based on the coupling the

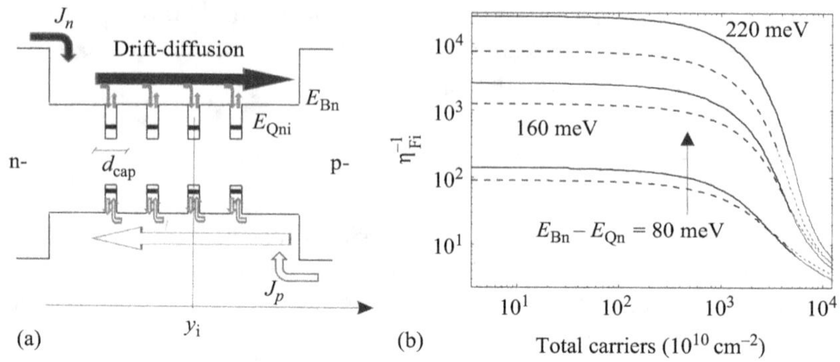

Figure 2.7 (a) Conceptual energy diagram of a nano-structure (QW or QD) active layer with schematic of the carrier injection processes: carrier capture in the active layers from the bulk state and escape out are indicated with the arrows in and out the QW; (b) calculated η_{Fi}^{-1} parameter for different energy separation ($E_{Bn} - E_{Qni}$) between QW and barrier state and at lattice temperature T_L = 300 K (solid line) and T_L = 350 K

transport equations with a microscopic treatment of carrier-carrier and carrier-phonon interaction have been treated in [57,66].

The set of equations [37] (reported here for electrons only) reads as:

$$\frac{\partial n_B(y)}{\partial t} = \frac{1}{q}\frac{\partial J_n}{\partial y} - R_n^B - \sum_i \left(R_{n,\mathrm{CAP}_i} - R_{n,\mathrm{ESC}_i}\right)\delta(y - y_i) \tag{2.20}$$

$$\frac{\partial^2 \phi}{\partial y^2} = -\frac{q}{\varepsilon}\left[p_B(y) - n_B(y) + N_{\mathrm{dop}} + \sum_i \left(P(y_i) - N(y_i)\right)\delta(y - y_i)\right] \tag{2.21}$$

where $n_B/p_B(y)$ is the electron/hole volume density in the bulk state of the heterostructure, $R_n^B(y)$ is the carrier loss rate due to band-to-band radiative and non-radiative recombination, $R_{n,\mathrm{CAP/ESC}}$ is the capture/escape rate (per unit of area) treated as local event taking place at the i-th QW layer in position y_i; ϕ is the electrostatic potential and N_{dop} is the barrier net doping (present in the case of modulation doped or δ-doped) structures. The electron current density J_n is given as:

$$J_n = -q\mu_n n_B \frac{\partial \phi}{\partial y} + qD_n \frac{\partial n_B}{\partial y} \tag{2.22}$$

where μ_n and D_n are the electron mobility and diffusivity, which are related through the Einstein relation.

The classical drift diffusion equations reported above are coupled with the RE of the sheet electron/hole carrier density (N_i, P_i) in each QW layer through the capture and escape rates as:

$$\frac{\mathrm{d}N_i}{\mathrm{d}t} = \left(R_{n,\mathrm{CAP}_i} - R_{n,\mathrm{ESC}_i}\right) - R(N_i) - v_g g_{\mathrm{mod}_i}\left(N_i, N_p\right)N_p \tag{2.23}$$

where g_{mod_i} is the modal gain of the i-th layer and $R(N_i)$ is the carrier loss term.

The capture rate is included phenomenologically through a capture time constant $\tau_{n,\text{cap}}$:

$$R_{n,\text{CAP}_i} = \frac{N_{B,\text{eff}_i}}{\tau_{n,\text{cap}}} \tag{2.24}$$

where N_{B,eff_i} is the sheet carrier density of bulk carriers in the capture range of thickness d_{cap} around the well (see Figure 2.7(a)): $N_{B,\text{eff}_i} = \int_{d_{\text{cap}}} n_B(y)\text{d}y$. It has to be noted that the capture rate is only apparently dependent on the parameter d_{cap} because a different choice of the value of d_{cap} corresponds to the re-scaling of $\tau_{n,\text{cap}}$. In [57], d_{cap} is defined as the distance over which the bound wave-functions are normalized and the capture probability in the QW is calculated self consistently with the adopted wave-function normalization. In [67], d_{cap} is assumed equal to three times the QW width. The escape rate is:

$$R_{n,\text{ESC}_i} = \eta_{Fi} \frac{N_{Q,i}}{\tau_{n,\text{cap}}} \tag{2.25}$$

where η_{Fi} is the ratio between the escape time from the well and the capture time; η_{Fi} is calculated under the assumption of a single quasi-Fermi level in the capture region at the steady state and at the lattice temperature T_L [37]. In the numerical implementation, one has therefore to calculate first the total carriers in the capture range:

$$\begin{aligned} &N_{\text{total},i} = N_{B,\text{eff}_i} + N_{Q,i} = \\ &d_{\text{cap}} \int_{E_{B,n}} \rho_{B,n}(E) f(E, E_{f,n}, T_L)\text{d}E + \int_{E_{Qi,n}} \rho_{Qi,n}(E) f(E, E_{f,n}, T_L)\text{d}E \end{aligned} \tag{2.26}$$

and get from this the electron (hole) quasi-Fermi level $E_{f,n}$; the η_{Fi} is then given by:

$$\frac{1}{\eta_{Fi}} = \frac{\int_{E_{Q,n}} \rho_{Q,n}(E) f(E, E_{f,n}, T_L)\text{d}E}{d_{\text{cap}} \int_{E_{B,n}} \rho_{B,n}(E) f(E, E_{f,n}, T_L)\text{d}E} \tag{2.27}$$

As an example, we show in Figure 2.7(b) the parameter η_{Fi} calculated applying, for the electrons, equation (2.27); it is evident that the escape rate gets smaller as the well depth increases and/or temperature decreases.

This approach can account for the different carrier filling in the various QW or QD layers of the hetero-structure and can quantify how much the transport effects impact on the laser response to current modulation [36,65].

2.4.5 Inclusion of the longitudinal and spectral distribution of photons

In the case of an SOA or an SLD source, longitudinal spatial hole burning effects and the emitted amplified spontaneous emission (ASE) spectrum are of interest;

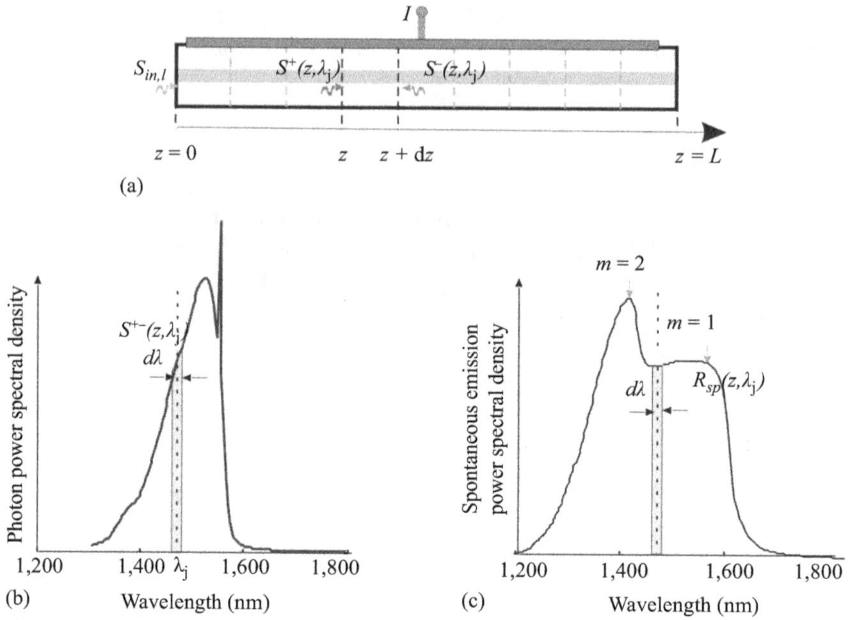

Figure 2.8 (a) Longitudinal description of the device and (b) conceptual definition of $S^{\pm}(z, \lambda_j)$ including ASE and signal. (b) spectrum of the spontaneous emission rate $R_{sp}(N,\lambda_j)$; m =1 and m = 2 indicate emission from the first and second confined states of the QW, respectively

these device characteristics can be simulated modifying the previous RE model as we will show in this section.

We can account for the spatial distribution of carriers and photons in the longitudinal direction z, by modifying the RE (2.17) as follows:

$$\frac{\mathrm{d}S^{\pm}(z,\lambda_j)}{\mathrm{d}z} = \left(\Gamma_{xy}\frac{g(N,\lambda_j)}{1+\varepsilon(S^+ + S^-)} - \alpha_i\right)S^{\pm}(z,\lambda_j) + \frac{\beta_{\mathrm{sp}}}{2}R_{\mathrm{sp}}(N,\lambda_j) \qquad (2.28)$$

Referring to Figure 2.8, $S^{\pm}(z, \lambda_j)\mathrm{d}z$ is the steady-state total photon density in the slice dz and in the wavelength range $[\lambda_j - \mathrm{d}\lambda/2, \lambda_j + \mathrm{d}\lambda/2]$, associated to the forward (+ sign) and backward (– sign) propagation of the optical field; $d\lambda$ is the spectral width we divide the active material emission spectrum (Figure 2.8(b) and (c)). As shown in Figure 2.8, the spontaneous emission rate appearing in (2.28) corresponds to the spectral integration of the spontaneous emission spectral density over the same wavelength interval. The boundary conditions are $S^+(z = 0, \lambda_j) = R_0(\lambda_j)S^-(z = 0, \lambda_j)$ and $S^+(z = L, \lambda_j) = R_L(\lambda_j)S^-(z = L, \lambda_j)$ where $R_{0,L}$ are the residual facet reflectivity due to anti-reflection (AR) facets of SOA or SLD structures; the dependence of this reflectivity with the wavelength can account for the actual reflectivity profile of the AR. In the SOA case and CW optical input signal at wavelength λ_{in} and photon density $S_{in}(\lambda_{in})$, we must add the following boundary condition:

$$S^+(z = 0, \lambda_{in}) = S_{in}(\lambda_{in}) \cdot (1 - R_0(\lambda_{in})) \tag{2.29}$$

Equation (2.28) is coupled with the steady-state solution of the carrier RE that assumes the following form:

$$\frac{\eta_i I}{eA} - R(N,z) - v_g \Gamma_{xy} \int_{\infty}^{-\infty} (S^+(z,\lambda) + S^-(z,\lambda)) \frac{g(\lambda, N)}{1 + \varepsilon(S^+ S^-)} \mathrm{d}\lambda = 0 \tag{2.30}$$

This model requires the inclusion of the wavelength dependence of the gain and spontaneous emission; in the QW case (and neglecting finite intra-band relaxation time) this can be done applying expressions 2.1 and 2.2 where the quasi-Fermi levels entering in the occupation probability are extracted from the total carrier density and assuming the neutrality condition:

$$N(z) = P(z) = \sum_m \int_{E_{Qm,n/p}} \rho_{2D}^{n/p}(E, E_{Qm,n/p}) f(E, E_{f,n/p}, T_L) \mathrm{d}E \tag{2.31}$$

The sum in (2.31) is over the m electron/hole confined states in the QW with energy $E_{Qm,n/p}$; $\rho_{2D}^{n/p}(E)$ is the two-dimensional density of states (including number of well layers) and $f(E,E_{f,n/p},T_L)$ is the electron/hole Fermi function at lattice temperature T_L. The total carrier loss due to spontaneous emission entering in (2.30) is:

$$R_{sp_{tot}} = \sum_m \int_{\infty}^{-\infty} R_{sp,m}(\lambda) \mathrm{d}\lambda \tag{2.32}$$

The reader must be aware that the present approach does not allow the inclusion of non-linear effects such as four-wave mixing (FWM) and cross gain modulation resulting from the microscopic (finite inter sub-band relaxation, carrier heating...) and macroscopic (carrier pulsation) carrier dynamics. The rigorous treatment of these effects (of particular importance when the non-linear properties of the SOA are exploited for all-optical signal processing) is reported, for example, in [56].

The advantage of this simplified approach is the fast computation time required to simulate both SLDs and SOAs CW characteristics, including, in the SOA case, the noise due to ASE and the corresponding saturation of the signal gain due to the signal and the ASE [68]. Figure 2.9 depicts an example of analysis of a 1 mm long SLD with InGaAs/InP MQW active region: we show the calculated ASE output spectrum (Figure 2.9(a)) and the corresponding longitudinal photon and gain distributions at the two wavelength corresponding to the emission peak from the first ($m = 1$) and second ($m = 2$) confined state in the QWs.

Figure 2.10 provides an example for the SOA case: the same MQW device of Figure 2.9 is now simulated as an SOA adding in $z = 0$ a CW monochromatic signal at $\lambda_{in} = 1{,}550$ nm. We compare in Figure 2.10(a) the chip gain $G(\lambda_{in}) = S_{in}(z = L)/S_{in}(z = 0)$ calculated including or not the effect of ASE, whereas Figure 2.10(b) shows the optical power spectrum out of the SOA. The role of ASE in reducing the chip gain is evident at

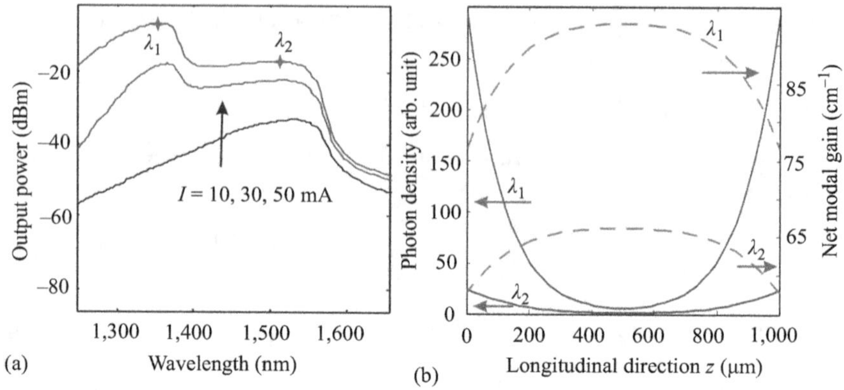

Figure 2.9 Example of simulation of MQW SLD device: (a) output power spectra at different currents and (b) photon (continuous lines) and net modal gain (dashed lines) longitudinal distribution at current of 50 mA and at the wavelengths indicated by the stars in (a)

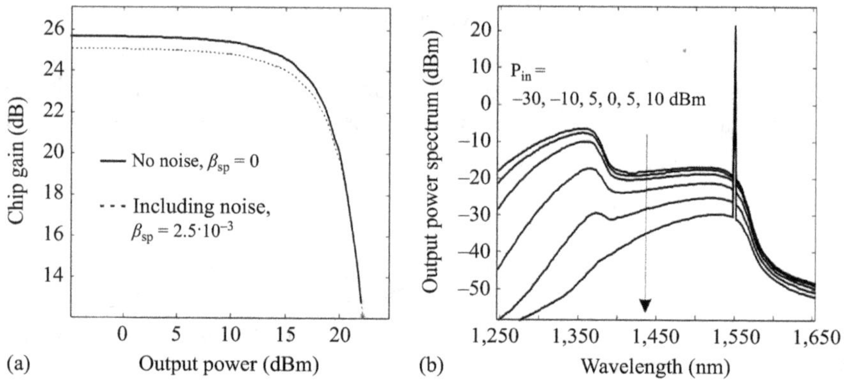

Figure 2.10 Example of simulation of MQW SOA device with monochromatic input signal at in, $\lambda_{in,l}$ = 1,550 nm: (a) simulated chip gain with injection current of 50 mA and (b) simulated SOA output power spectrum including the ASE contribution

small input power, because the carriers, not yet saturated by the signal, produce a significant spontaneous emission that is then further amplified by the high gain during the propagation. For low input power the signal gain is therefore more saturated by the ASE than by the signal itself.

This general approach can also be applied to a QD active material provided that the RE of (2.30) is substituted with the carrier RE of the QD material. Examples are in [69], where the model was used for the analysis of the CW characteristics of QD-SOAs or [70] and [71] where a similar approach was applied to analyze and design various SLDs structures.

2.4.6 The time domain TW model

The formulations previously presented are not exhaustive when we need to study complex dynamics such as the generation of optical pulses in Q-switching or Mode-Locking regime or the propagation of short pulses (few picoseconds) in SOAs. In this section, we introduce the time domain traveling wave (TDTW) model that, under the condition of waveguide monomodality, allows representing propagation effects in the cavity longitudinal direction including the possibility of simulating multi-section devices and dynamic responses [72]. In the TDTW model the longitudinal saturation effects are automatically included; we can also account for spontaneous emission noise, distributed feedback (DFB), etc. The previous REs for the photons are substituted by the TW equations for the forward ($F(z,t)$) and backward ($R(z,t)$) propagating component of the slowly (in t and z) varying electric field. The optical electric field is written as:

$$E(x,y,z,t) = \Phi(x,y)\left[F(z,t)e^{-j\beta_0 z} + R(z,t)e^{+j\beta_0 z}\right]e^{j\omega_0 t} + c.c. \tag{2.33}$$

where $\Phi(x,y)$ is the guided mode transversal profile, ω_0 is a reference pulsation ($\omega_0 = 2\pi c/\lambda_0$) and β_0 is the extracted propagation constant. For example β_0 can be set equal $\frac{2\pi}{\lambda_0} n_{\mathrm{eff0}_{\mathrm{ref}}}$, where $n_{\mathrm{eff0}_{\mathrm{ref}}}$ can be chosen equal to the effective refractive index at the wavelength λ_0 and at a reference carrier density N_{ref}. The TW equation system reads as follows [73]:

$$\begin{cases} \left[\dfrac{1}{v_g}\dfrac{\partial}{\partial t} + \dfrac{\partial}{\partial z}\right]F(z,t) & = [g_{\mathrm{net}}(z,t) - j\delta(z,t)]F(z,t) + jk_{\mathrm{RF}}(z)R(z,t) \\ & \quad + G_{\mathrm{sp}}(z,t) \\ \left[\dfrac{1}{v_g}\dfrac{\partial}{\partial t} - \dfrac{\partial}{\partial z}\right]R(z,t) & = [g_{\mathrm{net}}(z,t) - j\delta(z,t)]R(z,t) + jk_{\mathrm{FR}}(z)R(z,t) \\ & \quad + G_{\mathrm{sp}}(z,t) \end{cases} \tag{2.34}$$

where v_{g} is the group velocity (assumed constant in the wavelength range of interest), g_{net} is the modal gain in the active section, δ is the detuning factor, k_{RF} and k_{FR} are the coupling coefficients between forward and backward field in the case of structures with integrated gratings. G_{sp} is the spontaneous emission source that operates as the driving source for the set up of the lasing. In [73], G_{sp} is modeled as a zero-mean noise source, Gaussian distributed and with autocorrelation function proportional to the rate of spontaneous emission coupled with the guided mode [73]. The TDTW equations (2.34) are coupled with the RE (2.16) of carriers to be solved in each point z of the longitudinal grid; the photon density entering in the stimulated emission rate of the RE is normalized such that $N_p(z,t) = |F(z,t)|^2 + |R(z,t)|^2$. The field net modal gain $g_{\mathrm{net}}(z,t)$ is given by:

$$g_{\mathrm{net}}(z,t) = \frac{1}{2}\left(\frac{g_{\mathrm{mod}}(N,z)}{1+\varepsilon N_p(z,t)} - \alpha_i(N,z)\right)$$

The internal modal loss α_i may also depend on the carrier density to take into account for the free carrier absorption. The detuning factor δ is given by:

$$\delta = \frac{2\pi}{\lambda_0} n_{\text{eff}}(N) - \beta_0 \tag{2.35}$$

where

$$n_{\text{eff}}(N) = n_{\text{eff0}}(N_{\text{ref}}) + \Delta n_{\text{eff}}(N) \tag{2.36}$$

In (2.35), $n_{\text{eff}}(N)$ is the effective refractive index considering the effect of the carrier density N and it is given by the reference effective refractive index ($n_{\text{eff}\,\text{f0}_{\text{ref}}}$) plus the variation caused by the additional carrier injection and/or carrier saturation effects. In the most simple formulation $\Delta n_{\text{eff}_i}(N)$ can be approximated making use of the LEF parameter (often called also α-parameter) [46,47]:

$$\Delta n_{\text{eff}}(N) = -\frac{ac}{2\omega_0}(N - N_{\text{ref}})\alpha_{\text{LEF}}$$

where a and α_{LEF} are the differential modal gain and the LEF at the operation point N_{ref}.

The TW equation models the optical electric field over a broad spectral range because the field is seeded by the spontaneous emission which is spectrally broad. The inclusion of boundary conditions due to reflectivity at the cavity facets automatically permits the resolution of the longitudinal spectral modes when we deal with cavities. A critical point of the TDTW approach is the correct inclusion of the gain dispersion over the simulation spectral bandwidth. For solving this issue various solutions have been proposed such as a numerical filtering [73], inclusion of the polarization dynamics [74] or spatial digital filtering [75]. The simpler approach is based on the numerical filtering technique; as shown in Figure 2.11(a) the simplest filter is an infinite impulse response filter whose block diagram is schematized in the inset of Figure 2.11(a); the transfer function between the unfiltered ($\tilde{F}(z,t)$) and filtered ($F(z,t)$) field is:

$$H(\omega) = \frac{1 - f_c}{1 - f_c e^{-j\Delta t\omega}} \tag{2.37}$$

where $f_c < 1$ is the parameter tuned to fit the gain bandwidth around the gain peak as shown in 2.11a. Δt is the time step used in the numerical finite difference (FD) solution satisfying the constrain $\Delta z = v_g \Delta t$. From Nyquist theorem it also follows that the numerical simulation bandwidth is limited to $\pm\frac{\pi}{\Delta t}$ around the central pulsation ω_0. It therefore turns out that simulations of an emission bandwidth of about 100 nm requires very tiny Δt, of the order of 10 fs, [76] and development of fast algorithms [77] for increasing the computation speed. Figure 2.11(a) evidences however the significant limitation of this filtering technique in reproducing the realistic dispersion of the gain curve. This is due to the fact that this simplified TDTW model does not take into account the Fermi energy distribution of the

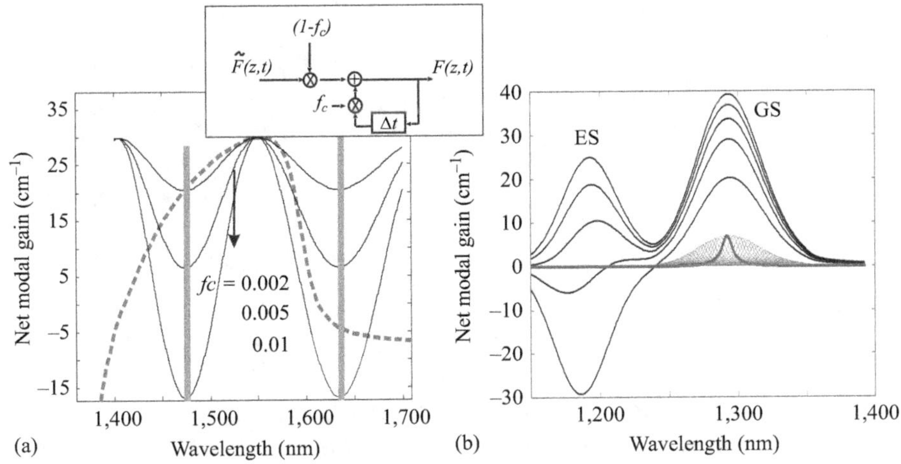

Figure 2.11 *(a) Frequency response (solid line) of an infinite impulse response filter (block diagram in the inset) used to model the gain spectral distribution (dashed line) of a broad band MQW material; the vertical gray lines delimit the simulation bandwidth determined by Δt = 50 fs (b) gain curves obtained as superposition of 51 Lorentzian filters (thin lines are for GS homogeneous broadening, one of them evidenced in bold) modeling the GS and ES polarization dynamics of a QD material; the simulation bandwidth, with Δt = 10 fs, is the wavelength range shown in the figure [86]*

carriers. The system (2.34) can be solved with a split-step algorithm [78]: the coupled-wave equations are split into two sets of equations; one set contains only the propagation factors and time derivatives, and the other one contains only the coupling terms. These separate set of equations are then solved exactly in their split form successively [78]. This method, respect to other numerical schemes based on FD approach ([73]), reduces the numerical errors even for high detuning factor (δ) and large sampling step (dz and dt). For this reason the split-step algorithm is particularly suitable for the static and dynamic simulation of long multi-section lasers and of widely tunable lasers, where the detuning factor in the passive gratings can be quite high due to the large refractive index variation necessary for the tuning.

The TW approach, presented here, has been applied to the analysis of bulk or MQW active waveguides where the TW equations of the slowly varying electric field are coupled with the RE discussed in this chapter. A great advantage of the TW approach is indeed the possibility of simulating with just one tool both CW and dynamic characteristics of many different devices such as SLDs and SOAs, DFB lasers [73], various structures of multisection lasers designed for wavelength tuning [79], microwave signal generation through self-pulsations [80], pulse generation in passive modelocking [81], ring laser configuration [82], etc.

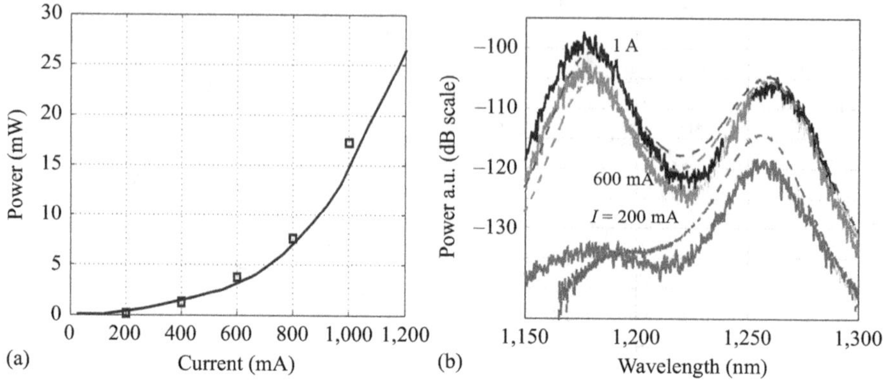

Figure 2.12 ASE out of a QD-SOA studied with a TDTW model: (a) calculated (squares) and measured (solid line) L-I curve and (b) calculated (solid line) and measured (dashed line) ASE spectra [87]

A comprehensive review of the TW approach capabilities is published in [72], whereas free simulation software are also available [83,84]. A new algorithm to improve the simulation speed, though relaxing the constrain $dz = v_g dt$, has been proposed in [77].

To simulate with a TDTW approach laser with QD active material, the inclusion of the polarization dynamics solved in the adiabatic approximation has been proposed [85,86,76]. As an example, we present in Figure 2.11(b) the gain spectrum obtained with this method, applied to the case of a wide band QD material. These gain curves, numerically calculated as superposition of 51 Lorentzian filters centered at the GS and ES emission wavelengths, can well reproduce measured gain spectra and their variation with current injection [86].

This more complicated TDTW mode has been applied in [85] to study passive mode-locking in QD lasers, in [86] to simulate QD-DFB lasers and in [76] to study QD-SOAs.

We apply here the QD-TDTW model to the study of the ASE of SOAs and SLDs structures. This approach is alternative (but more cumbersome) to the one discussed in Section 2.4.5. As an example, we show in Figure 2.12 the calculated L-I characteristic and the ASE spectra out of a QD-SOA without any injected input signal (SLD behavior) [87].

2.5 Device layout

Many devices, such as high power lasers, SOAs in master optical power amplifier configurations or high power SLDs, have a longitudinal waveguide layout properly designed to maximize the output power and, in the case of SOAs and SLDs, minimize the back reflections at the device facets. An example of various layout geometries is

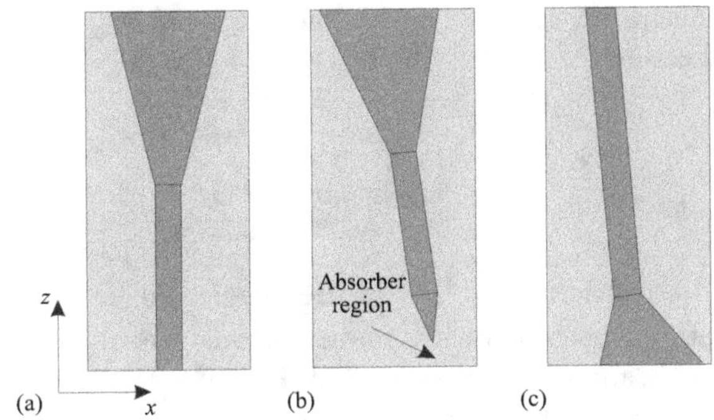

Figure 2.13 Schematic view of various waveguide layout: (a) flared geometry [89]; (b) tapered waveguide [90] optimized for absorption of backward propagating field; and (c) tilted-tapered layout [91]

shown in Figure 2.13: Figure 2.13(a) is the typical flared waveguide layout commonly used in the design of high power lasers [88] and SOAs [89], whereas Figures 2.13(b) and (c) are examples of layouts specifically designed for high-power SLDs [90,91].

The models discussed in the previous Sections 2.4.5 and 2.4.6 can be applied to analyze the structures presented above if spatially varying parameters (waveguide width, confinement factor, etc.) are included in the simulations. This simplified analysis is however only valid when there is no mode conversion from the fundamental guided mode to higher-order modes, because there is an adiabatic transformation of the fundamental modal field of the narrow waveguide in the longitudinal direction. This condition in practice imposes a slowly longitudinal variation of the waveguide layout.

A more accurate analysis and design of these structures imply the use of numerical tools that should be able to account for the waveguide variation and of the carrier-photon interaction also in the transversal section. The general formulation should embody the variation of the refractive index and of the gain in all the three spatial directions, but in the case of the structures presented in Section 2.3, the problem can be reduced to the two dimensions: the x and z directions in our notation. This transformation can be done easily using the EIM to define an equivalent effective index distribution $n_{\text{eff}}(x,z)$ for the passive waveguide structure and adding, as perturbation, all the interaction effects as done in 2.3 for the active waveguide analysis. The two-dimensional equation has the form:

$$\frac{\partial^2 E}{\partial x^2} + \frac{\partial^2 E}{\partial z^2} + k_0^2 \tilde{n}_{\text{eff}}^2(x,z,N)E + 2jk_0 n_0 \Gamma_y g\left(N(x,z), |E|^2\right)E(x,z) = 0 \tag{2.38}$$

where $\tilde{n}_{\text{eff}}(x,z,N)$ is the effective index distribution written as the sum of the effective index of corresponding passive structure (n_{eff}) plus the variation $\Delta n_{\text{eff}}(N)$; $g(x,z)$ and $\Delta n_{\text{eff}}(N)$ are the material gain and refractive index variation due to carrier distribution in the active layer, Γ_y the confinement factor in the junction direction. By extracting the fast varying phase term in z, the electric field is written as $E(x,z) = \Phi(x,z)e^{-jk_0 n_0 z}$ with $k_0 = \omega_0/c$; (2.38) becomes:

$$\frac{\partial \Phi}{\partial z} = -j\frac{1}{2k_0 n_0}\frac{\partial^2 \Phi}{\partial x^2} + \Gamma_y g\left(N(x,z), |\Phi|^2\right)\Phi(x,z) - j\frac{k_0}{2n_0}\left[\tilde{n}_{\text{eff}}^2(x,z) - n_0^2\right]\Phi \quad (2.39)$$

where the second-order derivative $\frac{\partial^2 \Phi}{\partial z^2}$ has been neglected.

There are various methods for the numerical analysis of this equation and the most popular is the beam propagation method (BPM); for a detailed presentation of the BPM techniques, see, for example [33]. The numerical method implemented in BPM solution is generally the FD approach using the Crank–Nicolson scheme that is accurate to the second order, unconditionally stable and provides accurate solutions in case of the weakly guiding structures we consider in this chapter. The full structure is uniformly segmented in the longitudinal direction z; in the propagation between two longitudinal sections the FD approach is used. Using the FD numerical scheme, one has to define a computational window and to impose consequently boundary conditions. Since in this kind of structures the main interest is generally on the guided field propagation, the transversal dimension is assumed infinite. Therefore, at the lateral boundaries of the computational window, only fields radiating outside (to the infinite) are physical possible solutions. Forcing transparent boundary conditions (i.e., radiation condition to the computed field in the extreme nodes of the computational window), we can avoid the artifacts of not physical reflected field from the boundary of the computational window in the numerical result [33,92].

The full numerical formulation is computationally very efficient because the solution requires at each propagation step the solution of a tridiagonal linear system of N (number of discretization points in the transversal direction) equations. An example of the effectiveness of this approach is reported in the Figure 2.14, where a bimodal active waveguide has been excited with a couple of anti-phase point sources (set in $y = \pm 1.5$ μm, $z = 0$) that lead to the formation of the second-order guided mode. It is clear also in this extreme case, respect to radiation effects, the absence of a reflected field component at the boundaries; the interference effects (visible as light/dark fringes in the figure) is due to the interference of the field radiated from the two sources.

In Figure 2.15(a), we present, as a second example of BPM analysis, the simulation of the effects of the coupling losses when the input field excitation (dotted line in Figure 2.15(b)) differs from that of the fundamental guided mode of a symmetrical waveguide. Figure 2.15(a) evidences a strong radiated field in the first part of the structure, whereas in the second part ($z \geq 250$ μm) the fundamental mode is completely excited and no further radiation effects are present. In Figure 2.15(b) cuts of the input field and of the computed fields in the central and output sections are reported; these two fields perfectly overlap with the modal field.

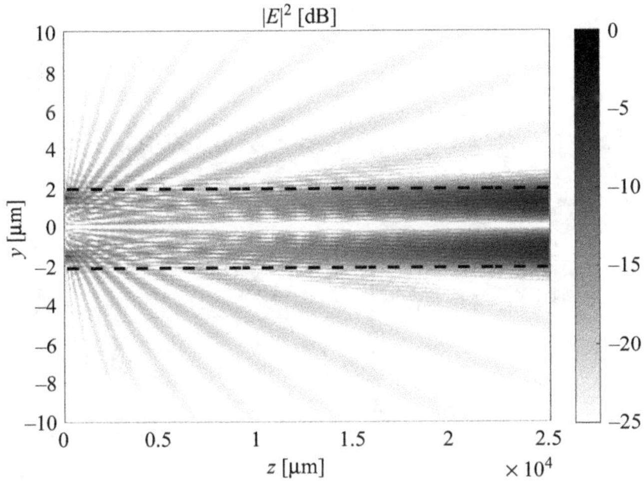

Figure 2.14 BPM solution of passive waveguide excitation with a couple of anti-phase optical sources in $y = \pm 1.5$ μm, $z = 0$. The formation of the second-order guided mode, the interference pattern due to the two sources, and the absence of reflections at the boundaries are clearly visible

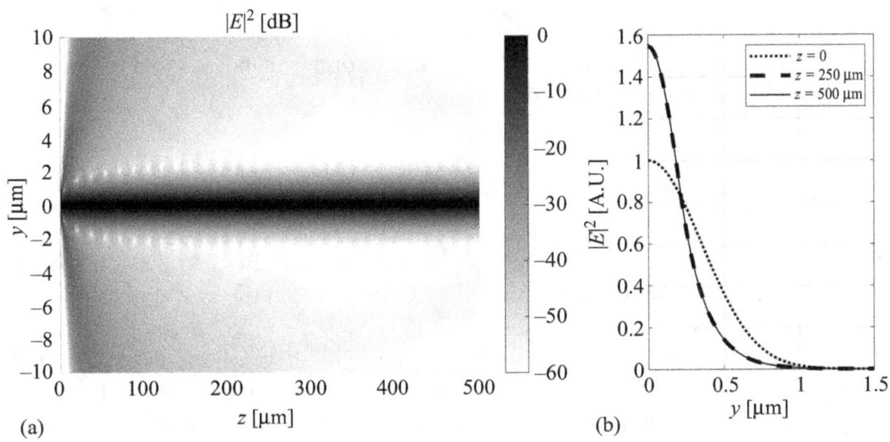

Figure 2.15 BPM solution of the field propagation in a passive waveguide, excited in $z = 0$ with a field intensity profile shown in (b) with red line. Lines in (b) are cuts of the field intensity profile in the center and at the termination

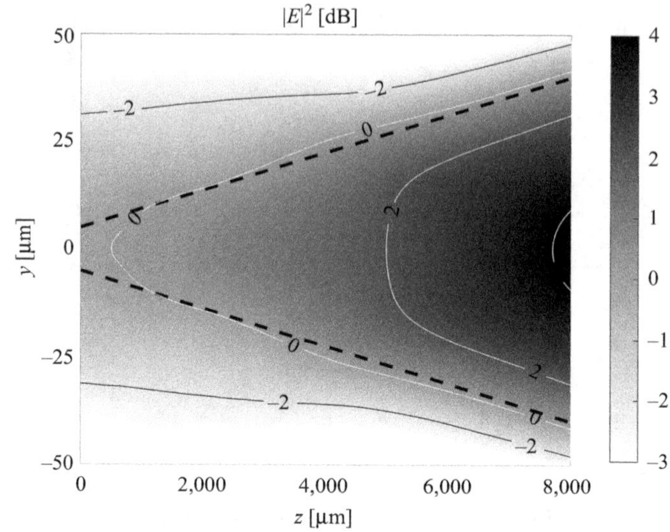

Figure 2.16 BPM solution of the field propagation in a flared gain-guided SOA, with input CW signal with intensity profile matching the modal field distribution. Dashed lines: border of the waveguide active region

In the case of a semiconductor material [93] the refractive index variation and the gain/absorption due to the carrier concentration can be computed from the carrier density solving the equation:

$$D_{\mathrm{eff}}\nabla^2 N(x,z) = \frac{\eta_i J_{\mathrm{in}}(x,z)}{\mathrm{e}} - R\big(N(x,z)\big) - v_{\mathrm{g}}\Gamma_y g\big(N,N_{\mathrm{p}}\big)N_{\mathrm{p}}(x,z) \tag{2.40}$$

where $N_{\mathrm{p}} = |\Phi|^2$ with proper normalization of the electric field. If the longitudinal diffusion of carriers can be neglected (for example, when the waveguide layout is slowly changing in z), (2.40) reduces in (2.19).

As an example shown in Figure 2.16, the color map of the computed field intensity for a flared gain-guided SOA with layout qualitatively similar to the one in Figure 2.13(a) and with an input Gaussian beam excitation matching the modal field. The used computational procedure is the following:

- the field in section $(n-1)$ is propagated in section (n) taking into account the gain and refractive index variation due to the carrier distribution obtained from the solution of (2.39) using the field distribution in section $(n-1)$;
- with the estimated field, the new complex refractive distribution to be used for the propagation in section $(n+1)$ is obtained;
- these two steps are repeated all along the structure.

Modification in the order of the previous operations does not change the results if the longitudinal step is properly chosen. Similar analysis has been extended also to QD SOA including the inhomogeneous broadening of the gain spectrum typical of the QD material [94].

2.6 An example of device modeling: SLDs

As an example of application of the simulation techniques previously discussed, we consider here the case of the SLD. SLDs have applications in various fields such as fiber optic gyroscope, sensors, source for broad band optical testing, telecom, etc. In the context of the biophotonic applications, they are the light source of choice for spectral domain optical coherence tomography (SD-OCT) imaging. Indeed, thanks to the broad optical emission spectrum, the SLD emitted light has a short coherence length that allows small axial resolution in interferometric experiments. Assuming a Gaussian distribution of the output spectrum, the resolution can be calculated as $0.44\lambda_c^2/\Delta\lambda$, with λ_c the optical spectrum central wavelength and $\Delta\lambda$ the FWHM spectral width. It is therefore clear the need of large emission spectrum SLDs, accompanied also by high output power required to detect a sufficiently intense backscattered light.

In the following, through the use of the modeling techniques discussed in this chapter, we will analyze the SLD characteristics (output power and optical spectrum) addressing the problem of the wide emission bandwidth (2.6.1), of the design of the anti-reflection coatings necessary at the output facets (2.6.2) and of the optimized design of the device layout (2.6.3). Further theoretical and experimental results addressed to the optimization of realized structures can be found in a following chapter of this book and in [95]; a full rigorous model for SLDs is presented in [96].

2.6.1 Wide optical bandwidth

The problem of widening the optical emission bandwidth is almost addressed only at the material level by a proper choice and design of the semiconductor material. Semiconductor materials, respect to other active materials, present inherently high spectral bandwidth for both spontaneous emission and optical gain and they are good candidate for realizing sources with a very large optical spectrum. The advent of III-V nano-structure materials such as QW or QDs has also offered various degrees of freedom for engineering the spectral emission.

From this point of view, interesting results have been obtained by:

- increasing the emission spectrum by growing in the active region QW or QD layers with different energy gap;
- combining the output of two or more SLDs with properly shifted central emission wavelength.

Since the second technique concerns more the system optimization rather than the device design, we discuss here briefly the first one which has been implemented in both QW and QD technology. The MQW material, thanks to the reduced dimensionality, presents a quite large optical bandwidth obtainable at more

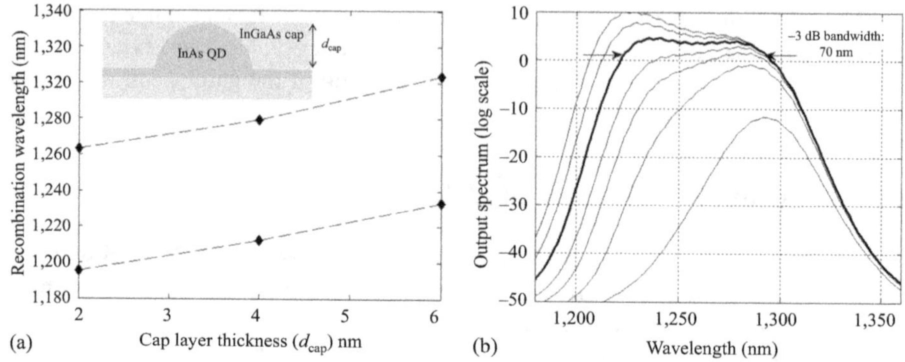

Figure 2.17 (a) Calculated GS and ES emission wavelength for the InAs/GaAs QD structure shown in the inset versus InGaAs cap thickness and (b) calculated ASE spectra of the QD-SLD with 3×3 chirped layers with d_{cap} *= 2, 4, 6 nm [108]*

moderate current density respect to the bulk counterpart; it can be also engineered to further improve the width of the optical emission. A way to increase the optical band is to vary (chirp) the energy gap of the QW inside the active region by chirping the well width and/or composition [97,98]. The QD material has also been considered for realization of broad band SLDs. The QDs are inherently narrow band due their ideally δ-like density of states, but the inhomogeneous broadening, associated to their growth process, makes them also interesting for wide band emission applications. As for the QW case, the spectrum can be widened by growing layers of QD with different emission wavelengths [99,70,100,101], by using selective area growth technique [102,103] or by exploiting the simultaneous emission from the GS and ES in QD emitting at 1,300 nm [104,105]. A further increase of the optical spectrum has been achieved with the combinations of QW and QD layers within the same intrinsic region [106].

The models presented in Section 2.4.5, modified to account also for the QD case, have been used to calculate the ASE output spectrum and L-I characteristics of this chirped SLD structures in [70] and in [107]. We show, for example, in Figure 2.17(a) how the variation of the thickness of the cap layer above the QD can be used to tune the GS and ES emission wavelength. The calculated ASE spectra of the corresponding SLD with chirped QD active layer (3×3 layers, with cap thickness of 2 nm, 4 nm, and 6 nm, respectively) are reported in Figure 2.17 for increasing current injection, the most broad and flat spectrum is evidenced by the thick line [108].

2.6.2 Ultralow reflectivity terminations

The light out of a SLD is determined by the amplification of the SE (ASE) generated in the structure. Without accounting for spectral and spatial gain saturation, the output intensity (I_{out}) is proportional to:

$$I_{out} \propto \frac{G_0}{g_{net} L} \tag{2.41}$$

where $G_0 = e^{g_{net}L}$ is the unsaturated single pass gain. Therefore, in order to obtain high output power, G_0 must be very high and furthermore the output spectrum should be very flat with very low residual ripple. The modulation causes indeed unwanted lateral peaks in the first-order coherence response affecting the quality of the OCT measurement. The modulation in the output spectrum is caused by the round trip gain ripple due to the non-zero facet reflectivity; it is equal to $\Delta G = \frac{(1+G_0 R)^2}{(1-G_0 R)^2}$, where R is the product of the two reflectivity at the terminations. Typical request, in the design of SLDs for OCT, is to limit ΔG to less than few dB. This implies an ultra-low end facet reflectivity in particular when the SLD is just a uniform tilted waveguide with output cleaved facets. For reducing further the effect of one cleaved termination (i.e., reduce the effective value of R), the unused SLD output termination can be properly designed to radiate the ASE in a region without pumping current acting as an absorber (for example, as in the structure of Figure 2.13(b)). However, even in this case, particular care should be taken to avoid the reflectivity coming from the discontinuity between pumped and not pumped sections; a solution consists in tapering or flaring the waveguide termination as in Figure 2.13(b) [90] and applying also a reverse bias voltage to avoid the saturation of the absorber at high power [91,109].

The usual way to realize a very low reflection termination is by tilting the waveguide respect to the cleaved facet and by depositing an antireflection coating (ARC) on the facet. The tilt reflects the incoming field away from the waveguide region, but part of it is coupled back into the waveguide; the presence of the ARC reduces further the reflected part of the field and simultaneously increases the transmitted part. The reflected field, coupling backward with the guided mode, contributes to the so called modal reflectivity, that entering in R, still contributes to the gain ripple ΔG. To compute the modal reflectivity with a reasonable accuracy we must take into account simultaneously the effect of the tilt, of the deposited ARC and the fact the waveguide is a two-dimensional structure with a well-confined modal field. Various models have been proposed to estimate the modal reflectivity for one-dimensional waveguides starting from [110] for cleaved facets to [111] to estimate only the effect of the waveguide tilt in the junction plane. The two results can be combined together and used to give an indication of the modal reflectivity also for two-dimensional waveguides. Various approaches have been used for the full problem of actual interest. Here we present a method that is based on a spectral domain approach [112] and on a semi-vectorial approximation of the modal field. From our experience, this technique produces results with good accuracy, comparable to more rigorous methods and also in good agreement with experimental results. The approach is very simple: (1) the modal field is expanded in plane wave; (2) for each plane wave and its corresponding reflection coefficient, the reflected component is computed; and (3) finally the modal reflection coefficient is obtained by projecting the obtained plane wave expansion of the reflected field on that of the field propagating back into the waveguide; this field is the field of the incoming waveguide spectrally shifted in the tilt plane. The obtained final expression is obtained as:

$$r_{\mathrm{qTE}} = \frac{\iint_{k_x,k_y} E(k_x + k_{\mathrm{tilt}}, k_y) E^{+}(k_x - k_{\mathrm{tilt}}, k_y) r_{x,y}(k_x, k_y) \mathrm{d}x\mathrm{d}y}{\iint_{k_x,k_y} E(k_x, k_y) E^{+}(k_x, k_y) r_{x,y}(k_x, k_y) \mathrm{d}x\mathrm{d}y} \tag{2.42}$$

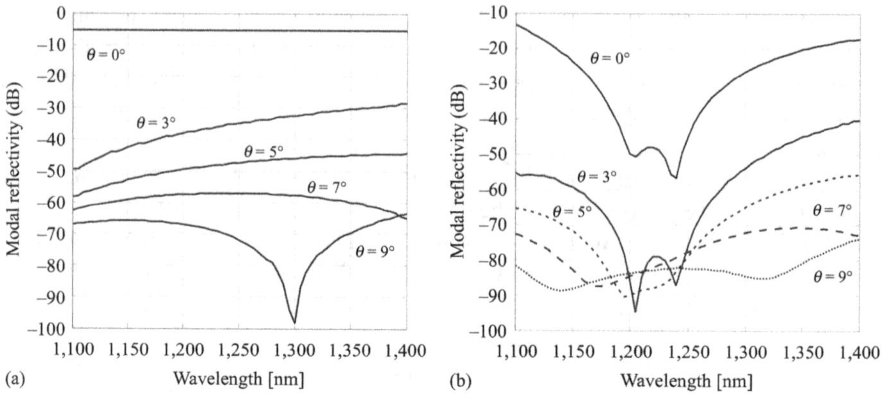

Figure 2.18 Typical spectral behavior of the modal reflectivity in case of: (a) tilted waveguide and cleaved facets and (b) same as (a), with the addition of an optimized two-layer ARC

where $E(k_x \pm k_{\text{tilt}}, ky)$ are the plane wave representation of the modal incident and reflected field; the plane wave reflection coefficient $r_{x,y}$ for each plane wave is given by

$$r_{x,y}(k_x, k_y) = \frac{k_y^2}{\left(k_x^2 + k_y^2\right)} r_{TM} + \frac{k_x^2}{\left(k_x^2 + k_y^2\right)} r_{TE} \tag{2.43}$$

where $r_{\text{TE,TM}}$ are the Fresnel reflection coefficients for the TE and TM polarized fields from the ARC or the cleaved facet [113]. Results from this analysis are reported in Figure 2.18 where we plot the calculated modal reflectivity as a function of the wavelength and of the tilt angle, θ, for the case of a tilted waveguide without ARC (Figure 2.18(a)) and with an optimized two layers ARC (Figure 2.18 (a)). The typical value of the tilt angle used in practice is in the range of 5 to 8 degrees because wider tilt angles complicate the output fiber coupling. The results in the figures show the weak wavelength sensitivity of the waveguide tilt (smooth dependence with wavelength for $\theta = 0°, 3°, 5°, 7°$ in Figure 2.18(a)), the further reduction of the modal reflectivity due to the ARC with the drawback of an increased wavelength selectivity (Figure 2.18(b)).

2.6.3 SLD layout

The longitudinal layout of the SLD structure can also be properly designed in order to achieve high output power together with large spectral bandwidth. This is obtained by a careful control of the longitudinal saturation effects. The proposed and realized experimental structures are typically multiple sections with a tapered termination [114] and the possibility of properly adjusting the separate currents of the electrodes [91]. The simulation approach in [70] has been used to design two-section QD-SLD structures with the aim of optimizing the flatness and bandwidth

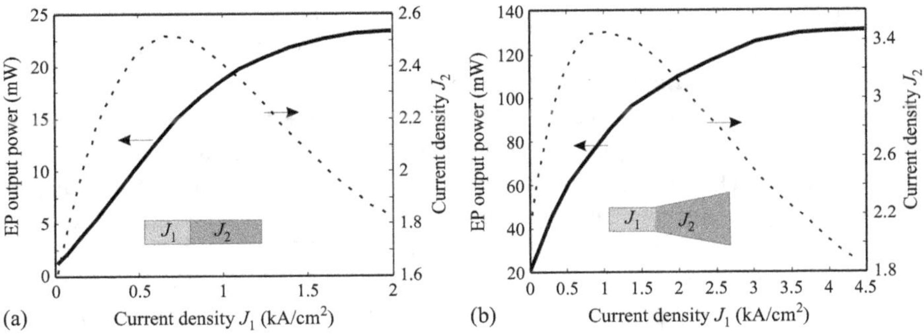

Figure 2.19 Output power at GS and ES EP condition of a two-section QD-SLD designed in [115]: (a) straight geometry with ideal AR coating and section 1 equal to 0.4 the total length of 1 cm and (b) flared geometry with ideal AR coating

of the emission spectrum through the simultaneous equal power (EP) emission from both GS and ES [115]. An example of design results is shown in Figure 2.19: Figure 2.19(a) depicts the calculated EP emission of a two-section SLD with optimized section ratio of 0.4, in this example the two-section structure assures power tunability (through the currents J_1 and J_2) whilst keeping the maximum emission bandwidth through the EP condition [115]. Figure 2.19(b) shows the design of a two-section flared structure optimized to maximize also the value of the power in the EP point [115].

References

[1] Coldren LA, Mashanovitch ML, Corzine SW. Diode Lasers and Photonic Integrated Circuits, Second edition. New York: John Wiley and Sons; 2012.

[2] Dutta NK, Wang Q. Semiconductor Optical Amplifiers, Second Edition. Singapore: World Scientific Publishing; 2013.

[3] Asada M, Miyamoto Y, Suematsu Y. Gain and the threshold of three-dimensional quantum-box lasers. IEEE Journal of Quantum Electronics. 1986;22(9):1915–1921.

[4] Coleman JJ, Young JD, Garg A. Semiconductor quantum dot lasers: A tutorial. Journal of Lightwave Technology. 2011;29(4):499–510.

[5] Dieter Bimberg MG, Ledentsov NN. Quantum Dot Heterostructures. New York: Wiley; 1998.

[6] Blood P. On the dimensionality of optical absorption, gain, and recombination in quantum-confined structures. IEEE Journal of Quantum Electronics. 2000;36(3):354–362.

[7] Haug H, Koch SW. Quantum theory of the optical and electronic properties of semiconductors. Singapore: World Scientific; 2003.

[8] Hader J, Moloney JV, Koch SW, Chow WW. Microscopic modeling of gain and luminescence in semiconductors. Selected Topics in IEEE Journal of Quantum Electronics. 2003;9(3):688–697.

[9] Hakki BW, Paoli TL. Gain spectra in GaAs double-heterostructure injection lasers. Journal of Applied Physics. 1975;46(3):1299–1306.

[10] Hofstetter D, Faist J. Measurement of semiconductor laser gain and dispersion curves utilizing Fourier transforms of the emission spectra. Photonics Technology Letters, IEEE. 1999;11(11):1372–1374.

[11] Thomson JD, Summers HD, Hulyer PJ, Smowton PM, Blood P. Determination of single-pass optical gain and internal loss using a multisection device. Applied Physics Letters. 1999;75(17):2527–2529.

[12] Blood P, Lewis GM, Smowton PM, Summers H, Thomson J, Lutti J. Characterization of semiconductor laser gain media by the segmented contact method. Selected Topics in IEEE Journal of Quantum Electronics. 2003;9 (5):1275–1282.

[13] Silverman KL, Mirin RP, Cundiff ST, Norman AG. Direct measurement of polarization resolved transition dipole moment in InGaAs/GaAs quantum dots. Applied Physics Letters. 2003;82(25):4552–4554.

[14] Leuthold J, Mayer M, Eckner J, Guekos G, Melchior H, Zellweger C. Material gain of bulk 1.55 μm InGaAsP/InP semiconductor optical amplifiers approximated by a polynomial model. Journal of Applied Physics. 2000;87 (1):618–620. Available from: http://scitation.aip.org/content/aip/journal/jap/87/1/10.1063/1.371909.

[15] Rideout W, Yu B, LaCourse J, York PK, Beernink KJ, Coleman JJ. Measurement of the carrier dependence of differential gain, refractive index, and linewidth enhancement factor in strained layer quantum well lasers. Applied Physics Letters. 1990;56(8):706–708.

[16] Chow WW, Schneider HC, Koch SW, Chang CH, Chrostowski L, Chang-Hasnain CJ. Nonequilibrium model for semiconductor laser modulation response. IEEE Journal of Quantum Electronics. 2002;38(4):402–409.

[17] Sugawara M, Mukai K, Nakata Y, Ishikawa H, Sakamoto A. Effect of homogeneous broadening of optical gain on lasing spectra in self-assembled $In_xGa_{1-x}As$/GaAs quantum dot lasers. Physical Review B. 2001;61(11): 7595–7603.

[18] Kim J, Laemmlin M, Meuer C, Bimberg D, Eisenstein G. Theoretical and experimental study of high-speed small-signal cross-gain modulation of quantum-dot semiconductor optical amplifiers. IEEE Journal of Quantum Electronics. 2009;45(3):240–248.

[19] Uskov AV, O'Reilly EP, McPeake D, Ledentsov NN, Bimberg D, Huyet G. Carrier-induced refractive index in quantum dot structures due to transitions from discrete quantum dot levels to continuum states. Applied Physics Letters. 2004;84(2):272–274. Available from: http://scitation.aip.org/content/aip/journal/apl/84/2/10.1063/1.1639933.

[20] Nilsson HH, Zhang JZ, Galbraith I. Homogeneous broadening in quantum dots due to Auger scattering with wetting layer carriers. Physical Review B.

2005;72:205331. Available from: http://link.aps.org/doi/10.1103/PhysRevB.72.205331.

[21] O'Driscoll I, Hutchings M, Smowton PM, Blood P. Many-body effects in InAs/GaAs quantum dot laser structures. Applied Physics Letters. 2010;97 (14):–. Available from: http://scitation.aip.org/content/aip/journal/apl/97/14/10.1063/1.3496011.

[22] Chow WW, Lorke M, Jahnke F. Will quantum dots replace quantum wells as the active medium of choice in future semiconductor lasers? Selected Topics in IEEE Journal of Quantum Electronics. 2011;17(5):1349–1355.

[23] Hader J, Moloney JV, Koch SW. Microscopic evaluation of spontaneous emission- and Auger-processes in semiconductor lasers. IEEE Journal of Quantum Electronics. 2005;41(10):1217–1226.

[24] Fehse R, Tomic S, Adams AR, *et al.* A quantitative study of radiative, Auger, and defect related recombination processes in 1.3-μm GaInNAs-based quantum-well lasers. Selected Topics in IEEE Journal of Quantum Electronics. 2002;8(4):801–810.

[25] Hader J, Moloney JV, Koch SW, Fan L, Fallahi M. Carrier recombination in semiconductor lasers: Beyond the ABC. In: Numerical Simulation of Semiconductor Optoelectronic Devices, 2006. NUSOD '06. International Conference on; Singapore, 2006. p. 39–40.

[26] Blood P, Pask H, Summers HD, Sandall I. Localized auger recombination in quantum-dot lasers. IEEE Journal of Quantum Electronics. 2007;43(12): 1140–1146.

[27] Weisser S, Esquivias I, Tasker PJ, Ralston JD, Romero B, Rosenzweig J. Impedance characteristics of quantum-well lasers. Photonics Technology Letters, IEEE. 1994;6(12):1421–1423.

[28] Paoletti R, Meliga M, Montrosset I. Optical modulation technique for carrier lifetime measurement in semiconductor lasers. Photonics Technology Letters, IEEE. 1996;8(11):1447–1449.

[29] Zilkie AJ, Meier J, Mojahedi M, *et al.* Carrier dynamics of quantum-dot, quantum-dash, and quantum-well semiconductor optical amplifiers operating at 1.55 μm. IEEE Journal of Quantum Electronics. 2007;43(11):982–991.

[30] Koshiba M, Maruyama S, Hirayama E. A vector finite element method with the high-order mixed-interpolation-type triangular elements for optical wave-guiding problems. Journal of Lightwave Technology. 1994;12(3):495–502.

[31] Rahman BMA, Davies JB. Vectorial finite element solution of GaAs/GaAlAs rib waveguides. Optoelectronics, IEE Proceedings Journal. 1985;132(6): 349–353.

[32] Stern MS. Semivectorial polarised H field solutions for dielectric wave-guides with arbitrary index profiles. Optoelectronics, IEE Proceedings Journal. 1988;135(5):333–338.

[33] Okamoto F. Fundamental of Optical Waveguides. Cambridge, MA: Academic Press; 2000.

[34] Agrawal GP, Dutta NK. Semiconductor Lasers, 2nd Edition. New York: Van Nostrand Reinhold.

[35] Petermann K. Laser Diode Modulation and Noise. Alphen aan den Rijn: Kluwer Academic Publisher; 1991.
[36] Nagarajan R, Ishikawa M, Fukushima T, Geels RS, Bowers JE. High speed quantum-well lasers and carrier transport effects. IEEE Journal of Quantum Electronics. 1992;28(10):1990–2008.
[37] Tessler N, Eisenstein G. On carrier injection and gain dynamics in quantum well lasers. IEEE Journal of Quantum Electronics. 1993;29:1586–1594.
[38] Gready D, Eisenstein G. Carrier dynamics in tunneling injection quantum dot lasers. IEEE Journal of Quantum Electronics. 2010;46(11):1611–1618.
[39] Dery H, Eisenstein G. Self-consistent rate equations of self-assembly quantum wire lasers. IEEE Journal of Quantum Electronics. 2004;40(10):1398–1409.
[40] Markus A, Chen JX, Provost OGLJG, Paranthoen C, Fiore A. Impact of intraband relaxation on the performance of a quantum-dot laser. IEEE Journal of Selected Topics in Quantum Electronics. 2003;9(5):1308–1314.
[41] Viktorov EA, Mandel P, Tanguy Y, Houlihan J, Huyet G. Electron-hole asymmetry and two-state lasing in quantum dot lasers. Applied Physics Letters. 2005;87(5):–. Available from: http://scitation.aip.org/content/aip/journal/apl/87/5/10.1063/1.1995947.
[42] Meuer C, Kim J, Laemmlin M, *et al.* High-speed small-signal cross-gain modulation in quantum-dot semiconductor optical amplifiers at 1.3 μm. Selected Topics in IEEE Journal of Quantum Electronics. 2009;15(3):749–756.
[43] Gioannini M. Ground-state power quenching in two-state lasing quantum dot lasers. Journal of Applied Physics. 2012;111(4): 043108–043108–9.
[44] Korenev VV, Savelyev AV, Zhukov AE, Omelchenko AV, Maximov MV. Effect of carrier dynamics and temperature on two-state lasing in semiconductor quantum dot lasers. Semiconductors. 2013;47(10):1397–1404.
[45] Gioannini M, Sevega A, Montrosset I. Simulations of differential gain and linewidth enhancement factor of quantum dot semiconductor lasers. Optical and Quantum Electronics. 2006;38(4–6):381–394. Available from: http://dx.doi.org/10.1007/s11082-006-0038-1.
[46] Henry CH. Theory of the linewidth of semiconductor lasers. IEEE Journal of Quantum Electronics. 1982;18(2):259–264.
[47] Osinski M, Buus J. Linewidth broadening factor in semiconductor lasers–An overview. IEEE Journal of Quantum Electronics. 1987;23(1):9–29.
[48] Chow WW, Koch S. Semiconductor Laser Fundamentals. Berlin: Springer; 1999.
[49] Chow WW, Koch S, Sargent M. Semiconductor Laser Physics. Berlin: Springer-Verlag; 1993.
[50] Ning CZ, Indik RA, Moloney JV. Effective Bloch equations for semiconductor lasers and amplifiers. IEEE Journal of Quantum Electronics. 1997;33(9):1543–1550.
[51] Chow WW, Jahnke F. On the physics of semiconductor quantum dots for applications in lasers and quantum optics. Progress in Quantum Electronics. 2013;37(3):109–184. Available from: http://www.sciencedirect.com/science/article/pii/S0079672713000098.

[52] Witzigmann B, Hybertsen MS, Reynolds J C L, Belenky GL, Shterengas L, Shtengel GE. Microscopic simulation of the temperature dependence of static and dynamic 1.3-μm multi-quantum-well laser performance. IEEE Journal of Quantum Electronics. 2003;39(1):120–129.

[53] Piprek J, Abraham P, Bowers JE. Self-consistent analysis of high-temperature effects on strained-layer multiquantum-well InGaAsP-InP lasers. IEEE Journal of Quantum Electronics. 2000;36(3):366–374.

[54] Blood P. Principles of semiconductor lasers, in Semiconductor Lasers: Fundamentals and Applications. Baranov A, Tournie E, editors. Cambridge: Woodhead Publishing; 2013.

[55] Agrawal GP. Gain nonlinearities in semiconductor lasers: Theory and application to distributed feedback lasers. IEEE Journal of Quantum Electronics. 1987;23(6):860–868.

[56] Uskov A, Mork J, Mark J. Wave mixing in semiconductor laser amplifiers due to carrier heating and spectral-hole burning. IEEE Journal of Quantum Electronics. 1994;30(8):1769–1781.

[57] Grupen M, Hess K. Simulation of carrier transport and nonlinearities in quantum-well laser diodes. IEEE Journal of Quantum Electronics. 1998;34(1): 120–140.

[58] Yamada M, Suematsu Y. Analysis of gain suppression in undoped injection lasers. Journal of Applied Physics. 1981;52(4).

[59] Blood P. The laser diode: 50 Years On. Selected Topics in IEEE Journal of Quantum Electronics. 2013;19(4):1503201–1503201.

[60] Ning CZ. What is laser threshold? Selected Topics in IEEE Journal of Quantum Electronics. 2013;19(4):1503604–1503604.

[61] Chow WW. Are nonclassical light sources lasers? Selected Topics in IEEE Journal of Quantum Electronics. 2013;19(4):1503402–1503402.

[62] Coldren LA. What is a diode laser oscillator? Selected Topics in IEEE Journal of Quantum Electronics. 2013;19(4):1503503–1503503.

[63] Joyce WB. Carrier transport in double-heterostructure active layers. Journal of Applied Physics. 1982;53(11):7235–7239. Available from: http://scitation.aip.org/content/aip/journal/jap/53/11/10.1063/1.331621.

[64] Letal GJ, Simmons JG, Evans JD, Li GP. Determination of active-region leakage currents in ridge-waveguide strained-layer quantum-well lasers by varying the ridge width. IEEE Journal of Quantum Electronics. 1998;34 (3):512–518.

[65] Ishikawa M, Nagarajan R, Fukushima T, Wasserbauer JG, Bowers JE. Long wavelength high-speed semiconductor lasers with carrier transport effects. IEEE Journal of Quantum Electronics. 1992;28(10):2230–2241.

[66] Koprucki T, Wilms A, Knorr A, Bandelow U. Modeling of quantum dot lasers with microscopic treatment of Coulomb effects. IEEE Journal of Optical and Quantum Electronics. 2011;42(11–13):777–783. Available from: http://dx.doi.org/10.1007/s11082-011-9479-2.

[67] Tessler N, Eistenstein G. On carrier injection and gain dynamics in quantum well lasers. IEEE Journal of Quantum Electronics. 1993;29(6):1586–1595.

[68] Connelly MJ. Wideband semiconductor optical amplifier steady-state numerical model. IEEE Journal of Quantum Electronics. 2001;31(3): 439–447.
[69] Berg TW, Mork J. Saturation and noise properties of quantum-dot optical amplifiers. IEEE Journal of Quantum Electronics. 2004;40(11):1527–1539.
[70] Bardella P, Rossetti M, Montrosset I. Modeling of broadband chirped quantum-dot super-luminescent diodes. Selected Topics in IEEE Journal of Quantum Electronics. 2009;15(3):785–791.
[71] Rossetti M, Li L, Markus A, *et al.* Characterization and modeling of broad spectrum InAs/GaAs quantum-dot superluminescent diodes emitting at 1.2 μm; 1.3 μm. IEEE Journal of Quantum Electronics. 2007;43(8):676–686.
[72] Wunsche HJ, Radziunas M. Traveling-wave modeling of semiconductor lasers and amplifiers. In: Numerical Simulation of Optoelectronic Devices Conference, Berlin: 2005. NUSOD '05; 2005. p. 25–25.
[73] Zhang LM, Carroll JE. Large-signal dynamic model of the DFB laser. IEEE Journal of Quantum Electronics. 1992;28(3):604–611.
[74] Bendelow U, Radziunas M, Sieber J, Wolfrum M. Impact of gain dispersion on the spatio-temporal dynamics of multisection lasers. IEEE Journal of Quantum Electronics. 2001;37(2):183–188.
[75] Kolesik M, Moloney JV. A spatial digital filter method for broad-band simulation of semiconductor lasers. IEEE Journal of Quantum Electronics. 2001;37(7):936–944.
[76] Puris D, Schmidt-Langhorst C, Lüdge K, Majer N, Schöll E, Petermann K. Time-domain model of quantum-dot semiconductor optical amplifiers for wideband optical signals. Optics Express. 2012;20(24):27265–27282. Available from: http://www.opticsexpress.org/abstract.cfm?URI=oe-20-24-27265.
[77] Javaloyes J, Balle S. Multimode dynamics in bidirectional laser cavities by folding space into time delay. Optics Express. 2012;20(8):8496–8502. Available from: http://www.opticsexpress.org/abstract.cfm?URI=oe-20-8-8496.
[78] Kim BS, Chung Y, Lee JS. An efficient split-step time-domain dynamic modeling of DFB/DBR laser diodes. IEEE Journal of Quantum Electronics. 2000;36(7):787–794.
[79] Dong L, Zhang R, Wang D, *et al.* Modeling widely tunable sampled-grating DBR lasers using traveling-wave model with digital filter approach. Journal of Lightwave Technology. 2009;27(15):3181–3188.
[80] Vizzino A, Gioannini M, Montrosset I. Dynamic simulation of clock recovery with self-pulsating three-section distributed-feedback lasers. IEEE Journal of Quantum Electronics. 2002;38(12):1580–1586.
[81] Bandelow U, Radziunas M, Vladimirov A, Hüttl B, Kaiser R. 40 GHz mode-locked semiconductor lasers: Theory, simulations and experiment. Optical and Quantum Electronics. 2006;48(4–6):495–512.
[82] Javaloyes J, Balle S. Emission directionality of semiconductor ring lasers: A traveling-wave description. IEEE Journal of Quantum Electronics. 2009;45(5):431–438.

[83] Javaloyes J, Balle S. Freetwm: A simulation tool for semiconductor lasers. "http://onl.uib.es/Softwares/Freetwm/".

[84] Radziunas M. Longitudinal dynamics in multisection semiconductor lasers. "http://www.wias-berlin.de/software/ldsl/".

[85] Rossetti M, Bardella P, Montrosset I. Time-domain travelling-wave model for quantum dot passively mode-locked lasers. IEEE Journal of Quantum Electronics. 2011;47(2):139–150.

[86] Gioannini M, Rossetti M. Time-domain traveling wave model of quantum dot DFB lasers. Selected Topics in IEEE Journal of Quantum Electronics. 2011;17(5):1318–1326.

[87] Gioannini M, Rossetti M, Montrosset I, *et al.* Modeling of InAsGaAs QD-SOAs for amplification of ultra-short high power pulses. In: Numerical Simulation of Optoelectronic Devices (NUSOD), 2011 11th International Conference on; Rome, 2011, p. 171–172.

[88] Sujecki S, Borruel L, Wykes J, *et al.* Nonlinear properties of tapered laser cavities. Selected Topics in IEEE Journal of Quantum Electronics. 2003;9(3): 823–834.

[89] O'Brien S, Welch DF, Parke RA, *et al.* Operating characteristics of a high-power monolithically integrated flared amplifier master oscillator power amplifier. IEEE Journal of Quantum Electronics. 1993;29(6):2052–2057.

[90] Zhang ZY, Hogg RA, Xu B, Jin P, Wang ZG. Realization of extremely broadband quantum-dot superluminescent light-emitting diodes by rapid thermal-annealing process. Optics Letters. 2008;33(11):1210–1212. Available from: http://ol.osa.org/abstract.cfm?URI=ol-33-11-1210.

[91] Greenwood PDL, Childs DTD, Groom KM, Stevens BJ, Hopkinson M, Hogg RA. Tuning superluminescent diode characteristics for optical coherence tomography systems by utilizing a multicontact device incorporating wavelength-modulated quantum dots. Selected Topics in IEEE Journal of Quantum Electronics. 2009;15(3):757–763.

[92] Hadley GR. Transparent boundary condition for the beam propagation method. IEEE Journal of Quantum Electronics. 1992;28(1):363–370.

[93] Agrawal GP. Fast Fourier transform based beam propagation model for stripe geometry semiconductor lasers: Inclusion of axial effects. Journal of Applied Physics. 1984;56(11):3100–3109.

[94] Xu T, Bardella P, Rossetti M, Montrosset I. Beam propagation method simulation and analysis of quantum dot flared semiconductor optical amplifiers in continuous wave high-saturation regime. Optoelectronics, IET. 2012;6(2):110–116.

[95] Matuschek N, Duelk M. Modeling and simulation of superluminescent light-emitting diodes (SLEDs). Selected Topics in IEEE Journal of Quantum Electronics. 2013;19(5):1–7.

[96] Loeser M, Witzigmann B. Multidimensional electro-opto-thermal modeling of broad-band optical devices. IEEE Journal of Quantum Electronics. 2008;44(6):505–514.

[97] Lin CF, Lee BL. Extremely broadband AlGaAs/GaAs superluminescent diodes. Applied Physics Letters. 1997;71(12):1598–1600.

[98] Wang J, Hamp MJ, Cassidy DT. Design considerations for asymmetric multiple quantum well broad spectral width superluminescent diodes. IEEE Journal of Quantum Electronics. 2008;44(12):1256–1262.

[99] Li LH, Rossetti M, Fiore A, Occhi L, Velez C. Wide emission spectrum from superluminescent diodes with chirped quantum dot multilayers. Electronics Letters. 2005;41(1):41–43.

[100] Gioannini M. Analysis of the optical gain characteristics of semiconductor quantum-dash materials including the band structure modifications due to the wetting layer. IEEE Journal of Quantum Electronics. 2006;42(3):331–340.

[101] Somers A, Kaiser W, Reithmaier JP, Forchel A, Gioaninni M, Montrosset I. Optical gain properties of InAs / InAlGaAs / InP quantum dash structures with a spectral gain bandwidth of more than 300 nm. Applied Physics Letters. 2006;89(6):–. Available from: http://scitation.aip.org/content/aip/journal/apl/89/6/10.1063/1.2266994.

[102] Ozaki N, Takeuchi K, Ohkouchi S, *et al*. Monolithically grown multi-color InAs quantum dots as a spectral-shape-controllable near-infrared broadband light source. Applied Physics Letters. 2013;103(5):–. Available from: http://scitation.aip.org/content/aip/journal/apl/103/5/10.1063/1.4817386.

[103] Ooi BS, McIlvaney K, Street MW, *et al*. Selective quantum-well intermixing in GaAs-AlGaAs structures using impurity-free vacancy diffusion. IEEE Journal of Quantum Electronics. 1997;33(10):1784–1793.

[104] Blazek M, Elsaesser W. Emission state hierarchy governed coherence and intensity noise properties of quantum dot superluminescent diodes. IEEE Journal of Quantum Electronics. 2012;48(12):1578–1582.

[105] Blazek M, Elsäßer W, Hopkinson M, *et al*. Coherence function control of quantum dot superluminescent light emitting diodes by frequency selective optical feedback. Optics Express. 2009;17(16):13365–13372. Available from: http://www.opticsexpress.org/abstract.cfm?URI=oe-17-16-13365.

[106] Chen S, Zhou K, Zhang Z, *et al*. Hybrid quantum well/quantum dot structure for broad spectral bandwidth emitters. Selected Topics in IEEE Journal of Quantum Electronics. 2013;19(4):1900209–1900209.

[107] Park J, Li X. Theoretical and numerical analysis of superluminescent diodes. Journal of Lightwave Technology. 2006;24(6):2473. Available from: http://jlt.osa.org/abstract.cfm?URI=jlt-24-6-2473.

[108] Gioannini M. Numerical modelling of quantum dot superluminescent diodes. In: Lasers and Electro-Optics, 2007 and the International Quantum Electronics Conference. CLEOE-IQEC 2007. European Conference on; Munich, 2007. p. 1–1.

[109] Crowley MT, Patel N, Saiz TA, *et al*. Modelling the spectral emission of multi-section quantum dot superluminescent light-emitting diodes. Semiconductor Science and Technology. 2012;27(6):065011. Available from: http://stacks.iop.org/0268-1242/27/i=6/a=065011.

[110] Ikegami T. Reflectivity of mode at facet and oscillation mode in double-heterostructure injection lasers. IEEE Journal of Quantum Electronics. 1972;8(6):470–476.

[111] Marcuse D. Reflection loss of laser mode from tilted end mirror. Journal of Lightwave Technology. 1989;7(2):336–339.

[112] Sewell P, Reed M, Benson TM, Kendall PC. Full vector analysis of two-dimensional angled and coated optical waveguide facets. IEEE Journal of Quantum Electronics. 1997;33(12):2311–2318.

[113] Farries MC, Buus J, Robbins DJ. Analysis of antireflection coatings on angled facet semiconductor laser amplifiers. Electronics Letters. 1990;26(6): 381–382.

[114] Li X, Jin P, An Q, *et al.* A high-performance quantum dot superluminescent diode with a two-section structure. Nanoscale Research Letters. 2011;6(1): 625. Available from: http://www.nanoscalereslett.com/content/6/1/625.

[115] Rossetti M, Bardella P, Montrosset I. Numerical investigation of power tunability in two-section QD superluminescent diodes. Optical and Quantum Electronics. 2008;40(14–15):1129–1134. Available from: http://dx.doi.org/10.1007/s11082-009-9307-0.

Chapter 3
Q-switched and mode-locked lasers for biophotonic applications

Edik U. Rafailov[1] *and Eugene A. Avrutin*[2]

3.1 Introduction

3.1.1 Ultrashort-pulse lasers

Recent development and progress of ultrafast optical sources made it possible to use them for a broad variety of applications, biomedical and otherwise. These range from body treatment (dentistry, eye surgery, dermatology, tattoo removal, hair removal, photodynamic therapy) [1], displays [2], spectroscopy, LIDAR and photochemistry (cancer detection and treatment) [3,4], optical metrology (precise measurement, navigation, scanners, optical sampling, all-optical clock, fibre-optic sensors) [5] to data storage (CDs, DVDs, Blu-ray disks, magneto-optical disks, holography) [6], material processing [7] and laser cooling [8].

New doors were opened with the generation of femtosecond optical pulses for such areas as material processing, biophotonics, and telecommunication. Due to the ultrashort pulse duration, the peak power can be very high even at modest average power. It is very important for a variety of applications where the reduced thermal stress is essential, for example, micro-machining and photo-ablation of biological tissues. Nano surgery in a living cell is now possible without surface damage of the membrane. High peak power of femtosecond laser is very useful for generating non-linear optical response in biological samples. Several innovative non-linear optical microscopy techniques involving multi-photon fluorescence and second harmonic generation became widely used for biological and biomedical research providing 3D imaging of fine structures even at a cellular level [9]. Femtosecond laser tweezers can be now used in near-field optics as well as for studying fundamental cell biology and cell mechanics [10,11]. As a result of the high peak power such lasers can be used for the non-linear processes such as frequency conversion, optical time-domain multiplexing/demultiplexing, wavelength-division multiplexing and further spectral broadening [12]. Precise timing of the short pulses is used for all-optical clock recovery [13].

[1]Aston Institute of Photonic Technologies (AIPT), School of Engineering and Applied Sciences, Aston University, UK
[2]Department of Electronic Engineering, University of York, UK

The most traditional sources for many of the applications above are solid-state lasers, which are capable of generating femtosecond pulses with high peak power [14,15]. A thin-disk Yb:YAG laser allows producing up to 60 W average power [16] and Kerr-lens-mode-locked (ML) Ti:Sapphire laser enables directly generating the pulses as short as 5 fs [17]. However, solid-state lasers tend to be very expensive, complex and bulky. They are optically pumped and thus a separate pump laser is required often with a cooling system, and the overall wall plug efficiency is limited [18]. There is no direct control over the laser gain medium. The low gain of the crystals defines the required minimum crystal length, which in turn is limiting the pulse repetition rate. Thus, an alternative laser systems free from the limitations above is attractive. One of the ways is to use electrically pumped semiconductors pico-to-femtosecond diode lasers which can be very compact and integrable, whereas their direct electrical pumping makes for high wall plug efficiency.

3.1.2 Progress in the use of diode lasers in biophotonics

In recent years, imaging techniques based on non-linear optical effects have been greatly improved by the progress in technology of ML femtosecond lasers. When biological tissue interacts with an intense ultrashort pulse laser beam, a non-linear response to the applied field strength is observed which is used as the imaging contrast. Multi-photon microscopy allows providing multicolour imaging of a surface as well as allows an image of internal organs to be created [19–24]. When compared to single photon fluorescence microscopy, multi-photon fluorescence microscopy (Figure 3.1) has a number of advantages. As a result of using long wavelength for excitation, the penetration depth is improved due to reduced effect of unwanted Mie and Rayleigh scatterings [25]. The axial resolution of an imaging system is known to be greatly improved by use of a confocal mask [26], which in the case of a single-photon excitation is essential. Two-photon absorption, on the other hand, occurs only at the focal plane of the microscope objective and so the fluorescence is confined in the focal volume (Figure 3.2). Thus three-dimensional high-resolution images can be produced without using a confocal pinhole mask [27,28]. In addition, some fluorophores are only efficiently excited with two photons.

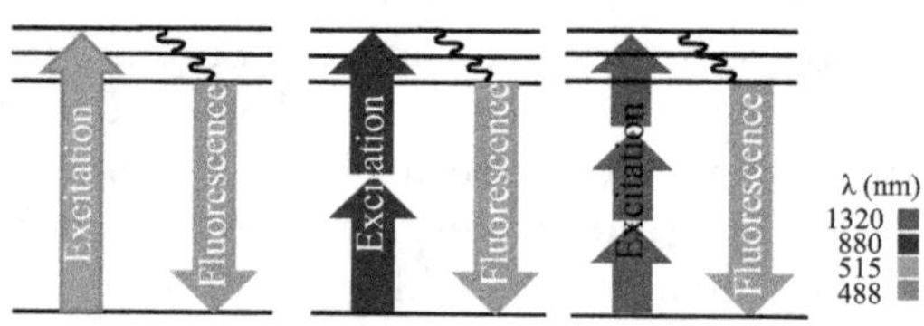

Figure 3.1 Single photon (left) and two-photon (middle) and three-photon (right) excitation diagrams

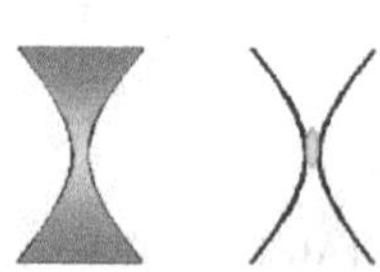

Figure 3.2 One-photon excitation (left) and multi-photon excitation (right) at a focal plane of the microscope objective

The idea of two-photon fluorescence scanning microscopy was proposed by Sheppard *et al.* [29,30]. Two-photon fluorescence microscopy was first demonstrated in 1990 by Denk *et al.* [31]. Three-photon fluorescence was also achieved [32], and not long after that, a microscopic image was recorded using a non-linear crystal [33].

For successful multiphoton imaging, a laser should produce enough output power in order to have high fluorescent intensity I which increases with pulse duration τ, the square of the peak power P_{peak}, and with the reciprocal of optical pulse period T (3.1) [34]:

$$I = k\frac{\tau}{T}P_{\text{peak}}^2 \tag{3.1}$$

where the factor k depends on the two-photon absorption cross section of the fluorophore at the laser wavelength, on the spatial energy distribution in the focus, and on the pulse shape. Knowing the duty cycle relationship $P_{\text{av}}/P_{\text{peak}} = \tau/T$, (3.1) can be rewritten as:

$$I = kP_{\text{av}}P_{\text{peak}} \tag{3.2}$$

where P_{av} is the average power.

The output power required for successful multiphoton imaging is determined from the following considerations. The average power should be high enough for excitation of the fluorophore, while at the same time modest enough not to damage or destroy the biological samples. Due to water absorption, temperature rises and beyond a certain threshold, cells and tissues may degrade because of protein denaturation. For imaging through biological tissue, the wavelength region between 600 nm and 1,300 nm is desirable since this corresponds to a transmission window in the absorption profile as shown in Figure 3.3. At wavelengths higher than 1,300

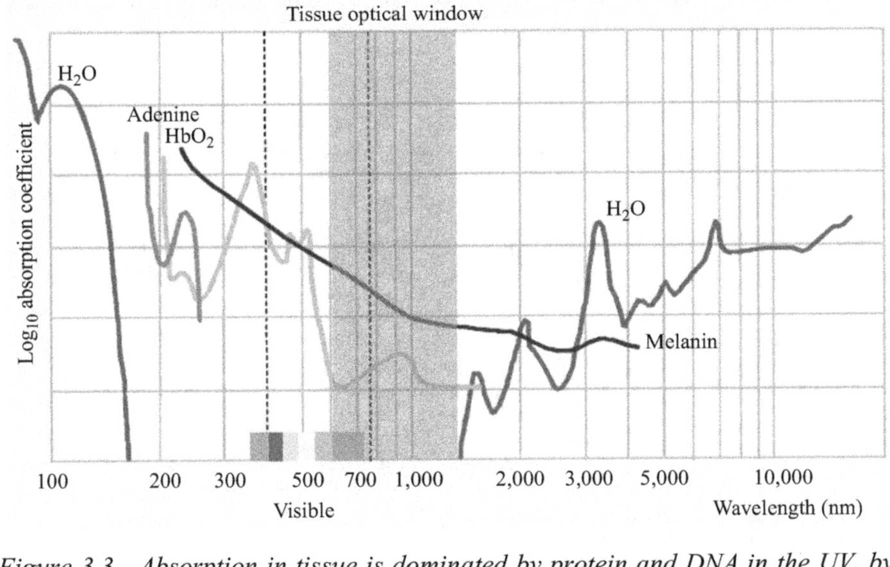

Figure 3.3 Absorption in tissue is dominated by protein and DNA in the UV, by water in the IR, and by haemoglobin and melanin in the visible light

nm, light absorption by water molecules becomes substantial. Several groups have demonstrated high peak power pulse generation with semiconductor lasers [35,36]. For example, Kim *et al.* obtained 1.4 kW peak power at 974 nm from a master oscillator system based on eXtreme Chirped Pulse Amplification (X-CPA) involving colliding pulse external cavity ML semiconductor laser, extensive pulse stretching, multiple amplification stages and strong compression [36]. Schlauch *et al.* demonstrated 622 fs pulses with 513 mW output power from a near 830 nm passively ML tapered laser resulting in world-record peak power of 2.5 kW [37]. At the same time another group has shown the possibility of two-photon fluorescence bioimaging of actin filaments in PtK2 cells using second harmonics generation with a 1.55 μm gain-switched distributed-feedback-Bragg laser diode generating 1 kW peak power after several amplification stages [35].

It has been shown that several tens of watts of peak power can be enough for multiphoton imaging [37,38]. For example, multiphoton imaging of a cell containing Coumarin and mouse kidney was achieved using external-cavity ML laser generating 5 ps pulses at a repetition rate of 500 MHz at a wavelength of 800 nm, with 20 and 32 W peak powers, respectively, at the sample plane. In order to achieve high peak power, a two-stage amplifier configuration was implemented as well [38]. The nature of the laser dynamic regimes involved in these and other experiments will be considered in more detailed below.

3.2 Ultrashort pulse generation in semiconductor lasers: the main techniques and principles

3.2.1 Gain-switching and Q-switching

There are several techniques for producing ultra-short pulses from semiconductors lasers, the three most commonly used ones being gain-switching, Q-switching and ML (see, e.g., [39–41] for some earlier reviews of the techniques).

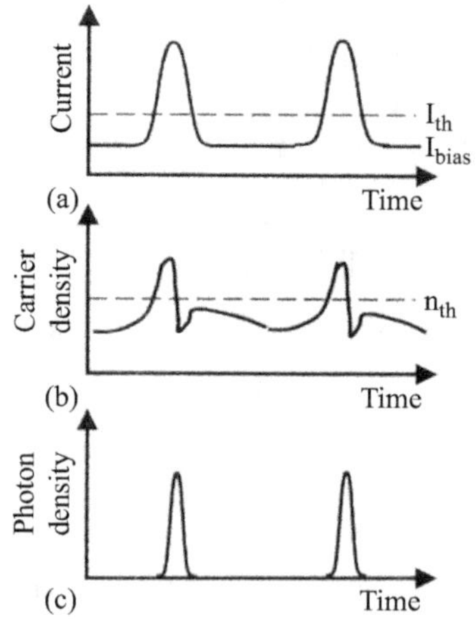

Figure 3.4 Gain-switching laser dynamics: (a) applied current; (b) carrier density; and (c) optical power

Gain-switching (Figure 3.4) is conceptually the easiest technique for generation of short pulses that doesn't involve any external cavity or sophisticated drive electronics [42–44]. The main idea consists of injection of large number of carriers (electrons) by short electrical pulses (shorter, or of the order of, the population inversion relaxation time, which in the context of semiconductor lasers is the carrier lifetime, of the order of 1 ns) [39]. With a correct adjustment of the laser and optical pulse parameters, this results in the emission of a single optical pulse, about an order of magnitude shorter than the electrical pulse. Essentially, this is the first pulse in the relaxation oscillation trail usually seen at the switch-on of a type B laser, but due to the short duration of the pumping pulse, the subsequent relaxation oscillations are not realised – by the time they would occur, the pumping pulse is finished. For a theoretical analysis, a simple rate equation model is sufficient.

A version of gain-switching consists in large-signal harmonic modulation of the laser at a frequency of 0.1–1 GHz, sometimes known as repetitive gain-switching. In this case, each positive period of the sinusoidal modulation is equivalent to a short current pulse, and a stream of short (much shorter than the period of the modulating signal) optical pulses is created with the repetition frequency equal to that of the modulation; see, for example [45].

Q-switching is achieved by changing the cavity loss actively using an intracavity modulator (Figure 3.5) or passively via a saturable absorber or equivalent element (Figure 3.6) which switches the cavity quality factor Q from a low to a high value [46,47]. Both techniques, when applied to solid-state lasers, allows generating pulses with high (Gigawatt) peak power called giant pulses; in semiconductor lasers, the typical peak power can be of the order of hundreds of Watts in an optimised construction and can be amplified further by using extracavity amplifiers (see Section 3.5).

Both active [48] and passive gain-switching have been extensively studied in semiconductor lasers, but passive gain-switching has attracted rather more attention. As in a laser of any type, Passive Q-switching in semiconductor lasers is based on the non-linear absorption behaviour of the *saturable absorber* introduced into a

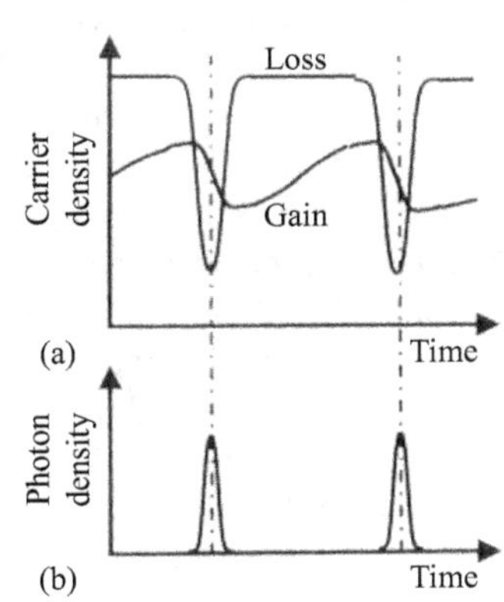

Figure 3.5 Active Q-switching laser time evolution of (a) loss, gain; and (b) optical power

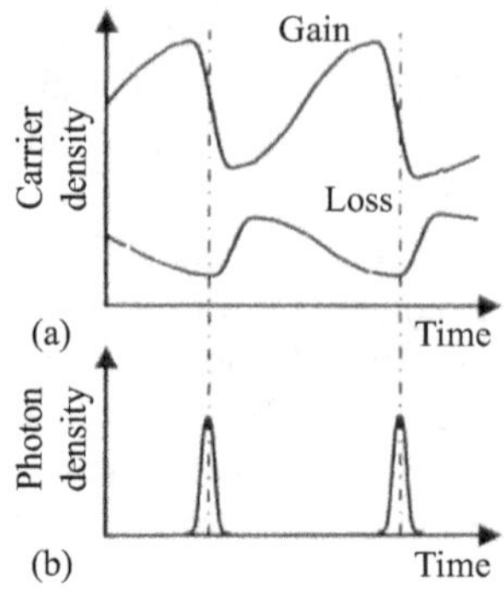

Figure 3.6 Passive Q-switching laser time evolution of (a) loss, gain; and (b) optical power

cavity. In the case of a monolithic semiconductor laser, the saturable absorber is, quite simply, a section of the laser cavity (typically a few per cent to 10%–20% of the laser length), usually electrically isolated from the rest of the laser, for example, by an etched groove, and either left unbiased or reverse biased unlike the rest of the laser which is forward biased. The absorber is saturated (decreases its absorption) with an increased light intensity. This provides a *positive feedback* leading to the damped, decaying relaxation oscillations, which are seen in all class B lasers at switch-on, becoming undamped, self-sustained oscillations. These self-sustained oscillations are variously known as passive Q-switching (usually if they are obtained deliberately), or self-sustained pulsations (the latter term is more frequently used when the regime is an unwanted effect, for example, in ML lasers discussed below).

It can be shown both theoretically and experimentally [41,49] that the range of currents in which Q-switching is maximised, thus maximising the peak pulse power, at a value of the absorber recovery time of a few tens of picoseconds; to reduce the recovery time to such short values, either reverse biasing the absorber section (see below) or ion implantation [50,51] have been used; the former is cheaper and easier to achieve technologically and the latter has the advantage of being potentially applicable to a whole array or even vertical stack of lasers.

A further increase in the energy and peak power of both gain-and Q-switched lasers directly from the laser can indeed be achieved by using lateral arrays [52] or vertical stacks [53] of lasers (the latter may be easier in the case of gain-switching than Q-switching), by using lasers with broad waveguides, with a large ratio of active layer cross section to the confinement factor [54–56], and by employing tapered waveguide constructions [57,52,58,59]. Alternatively, the increase of power can be achieved by means of an external amplifier [60].

With the high peak and average powers thus achieved, both gain and Q-switching pulses can be used for applications involving non-linearities; the use of repetitively gain-switched lasers operating at a wavelength of 980 nm has, for example, been reported for green light generation for second harmonic generation; with a different wavelength, the same technique can conceivably be applied to techniques such as multiphoton imaging.

Finally, it is worth noting that the principles of gain and (passive) Q-switching can be combined: lasers with a saturable absorber can be pumped with single short current pulses for generating picosecond high-power pulses (see, e.g., [51,56]), and repetitively passively Q-switched lasers can be stabilised by external current injection, somewhat akin to repetitive gain-switching though typically requiring lower current amplitude since the laser is already self-pulsating [61]; both principles can in principle be applied in biophotonic applications.

3.2.2 ML techniques in semiconductor lasers: the general principles

ML technique essentially consists, as the name implies, in locking the longitudinal modes of the laser together with a constant (in time) phase relation. In such a regime, when rather than oscillating independently, each mode in the laser cavity operates with a fixed phase relation between it and the other modes, all modes will constructively interfere with one another, producing an intense burst or pulse of light with a constant repetition rate $\Delta\nu$

$$\Delta\nu \approx \frac{c}{2nL} \tag{3.3}$$

where c is the speed of light, L is the length of the cavity, n is the group velocity refractive index averaged over the cavity length (so that nL is the *optical length* of the cavity).

The total oscillating bandwidth $\Delta\nu_L$ ($\Delta\nu_L = N\Delta\nu$, where N is the effective number of modes) is related to the pulse width $\Delta\tau_P$ by the relation

$$\Delta\tau_P = \frac{\beta}{\Delta\nu_L} \tag{3.4}$$

so that the more modes are locked the shorter pulses can be produced. The parameter β is a numerical factor known as the *time-bandwidth product (TBP)*. This is one of the most important figures of merit (FOM) of a mode-locked laser emission. The smaller the magnitude, the shorter the mode-locking pulses for a given spectral width, and thus, assuming that the pulse energy does not depend on the phase relation between modes, the higher the peak pulse power. Hence, for non-linear applications which involve maximising the non-linear signal I given by (3.1) or (3.2), it is desirable to keep the time-bandwidth product as small as possible.

The magnitude of β depends on the shape of the spectral intensity distribution and on the nature of the phase relation between modes. In the ideal case of mode-locking dynamics, the phase difference between the adjacent modes is not just constant in time, but roughly constant from one pair of modes to another:

$$\varphi_l - \varphi_{l-1} \approx \varphi \tag{3.5}$$

where φ is a constant phase difference, and φ_l is the phase of the l-th longitudinal mode. Note that, for the same mode intensity distribution, the actual value of φ, so long as it is a constant $\varphi < \pi$, does not affect the shape and duration of the pulses in the pulse train, it merely shifts the pulse train as a whole in time with respect to some reference time, and thus has almost no physical meaning or significance.

A phase relation that (nearly) exactly satisfies the condition (3.5) corresponds to what is usually referred to as *transform-limited pulse generation* (indeed the pulse train is, mathematically, related to the Fourier transform of the spectrum). In this case, the time-bandwidth product depends only on the pulse shape and is typically less than $\beta < \sim 0.4 - 0.5$ if the spectral width $\Delta\nu_L$ and the pulse duration $\Delta\tau_P$ are both measured at half maximum ($\beta \approx 0.31$ is the value for Gaussian pulses, for which the spectrum shape is also a Gaussian; this is believed to be the smallest value possible and is often taken as a reference against which realistic cases are measured).

In reality, even if the phases of all modes are constant in time so that mode-locking takes place, not all the modes have the same phase (or phase difference), so (3.5) is usually, at best, approximate. This corresponds, in general, to the generation of *chirped* pulses, in which the instantaneous carrier frequency varies during the pulse (which is one of the definitions of chirp in electromagnetic signals). The main physical origin of chirp in ML semiconductor lasers lies in the light intensity dependence of the group refractive index in the cavity (*self-phase modulation*). In the case of chirped pulses, β is greater than in the case of transform-limited pulses (typically between about 0.5 and as much as 8), which is not desirable for most applications, including the biomedical ones, but can be remedied by extracavity chirp compensating schemes, at the cost of a more complex and expensive optical source construction.

Finally, a regime is possible whereby not all the lasing bandwidth is effectively used for generating ultra-short pulses, and so phases of part of the modes drift slowly in time. This is a type of unstable, or weak, mode-locking, which is less advantageous than stable mode-locking for most applications and cannot, in general, be rectified by passive external schemes. It may still be acceptable for some non-linear applications, provided that the instability is weak enough.

Mode-locking is often believed to be the most interesting technique for optical data/telecommunication as well as imaging of the biological samples as it is possible to generate much shorter pulses at much higher repetition rates than using either gain-switching or Q-switching. Mode-locking also tends to have low timing jitter compared to the other two regimes, because each new ML pulse does not emerge from noisy spontaneous emission as in the case of gain- or Q-switching, but is generated purely deterministically by recirculating a pulse in the laser cavity.

Mode-locking a diode laser (Figure 3.7) allows pulses as short as 393 femtoseconds to be generated in a monolithic cavity [62] and down to less than 100 fs in a specialist external cavity.

There are two main techniques for ML: (1) active ML and (2) passive ML (Figures 3.8 and 3.9) which are possible when the laser resonator includes optical modulator or saturable absorber, respectively, see, for example [63,64]. A combination of the two techniques is known as *hybrid* ML which is defined by gain/loss modulation applied to a cavity which also contains a saturable absorber of an equivalent element.

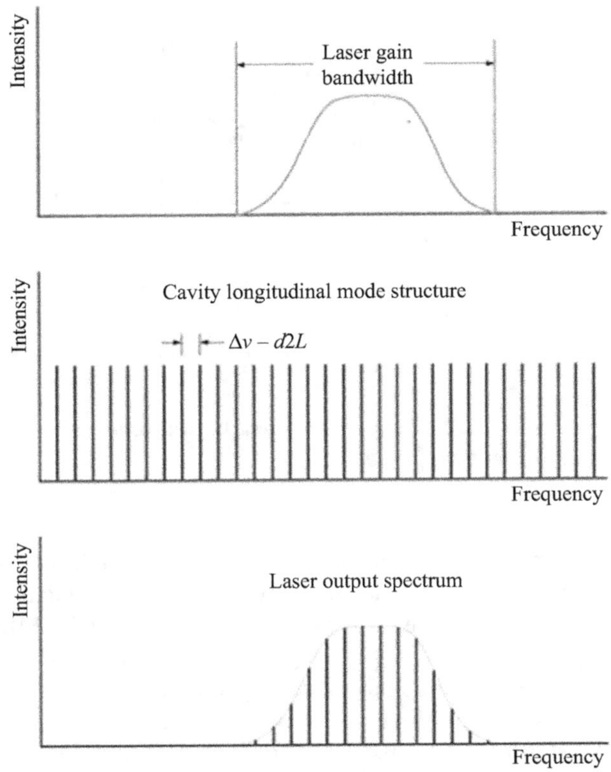

Figure 3.7 Mode-locking spectrum formation

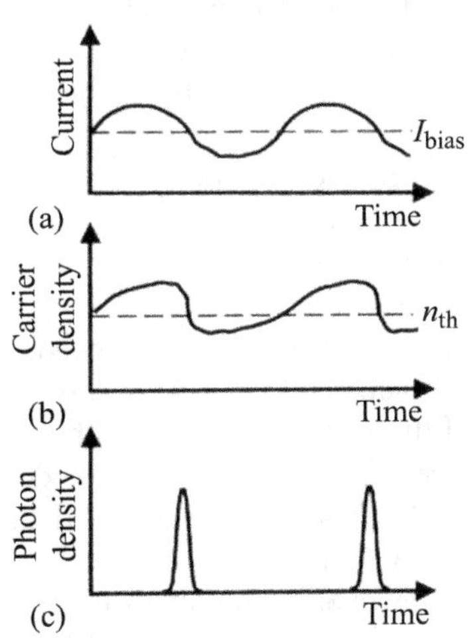

Figure 3.8 Active ML laser dynamics of (a) drive current; (b) carrier density; and (c) photon density or optical power

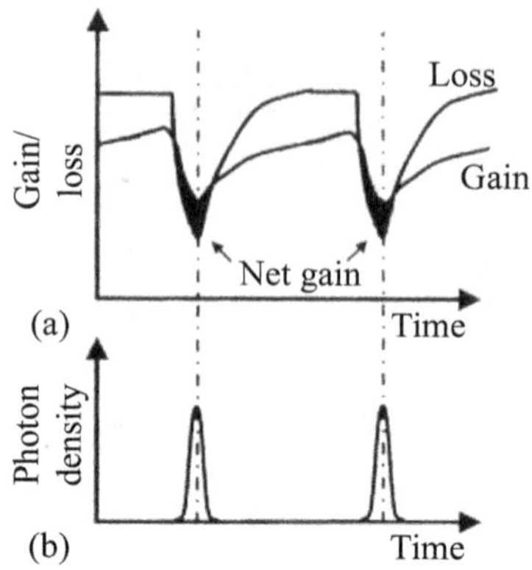

Figure 3.9 Passive ML laser dynamics of (a) loss and gain and (b) photon density or optical power

Active ML can be induced by an amplitude modulator (AM ML; in the case of semiconductor lasers, this is often achieved by a harmonic modulation of pumping applied to a part of the laser cavity), or by a phase modulator (FM ML). We note in passing that only a part of the cavity length has to see the gain modulation for successful mode-locking; harmonically and synchronously modulating the gain over the entire cavity length does not lock the modes together.

A special version of active mode-locking consists in periodic modulation of laser gain near the fundamental cavity frequency with a signal in the form of pulses shorter than the round-trip time; this is known as synchronous mode-locking.

The main idea is achieving locking of the modes by modulating the loss or gain of a laser at a frequency near the intermodal spacing $\Delta\nu$ so that each spectral mode is driven by the modulation sidebands of its neighbours. The modulation of the gain can be achieved by external modulator or by modulating the driving current as shown in Figure 3.9.

Passive ML is typically achieved by introducing a saturable absorber into the laser cavity. In the case of an external cavity laser, this can be a semiconductor saturable absorber mirror (SESAM), whereas in the case of a monolithic construction, a reverse biased section of a waveguide as in a passively Q-switched laser. In any case, the absorption in the absorber needs to be *saturated* (decrease with incident power) faster than the gain of the laser, as in the case of passive Q-switching (hence both of these regimes, or a combination of them as mentioned below, can often be observed from the same laser, depending on the operating point, i.e., the bias applied to the gain element and the absorber). In order for passive mode-locking to be observed in preference to passive Q-switching, the absorber should also have a recovery time much faster than that of the gain, thus forming a narrow net gain window for formation of the pulses (Figure 3.9).

The pulse repetition rate in passive ML is defined, within the first approximation, by the round-trip time of the laser cavity and depends only very weakly on

Table 3.1 A performance summary of the main types of short-pulse generation techniques

Technique	**Pulse width**	**Repetition rate**	**Pulse energy**	**Comments**
Gain-switching	Down to 2 ps	<10 GHz	300 pJ	Good control of repetition rate via electrical drive
Passive Q-switching	1 ps	<20 GHz	3.5 J	Limited control of repetition rate and poor jitter performance
Active Q-switching	5 ps	<10 GHz	1 J	Repetition rate set by electrical drive
Passive mode-locking	200 fs	2 THz	60 pJ	Repetition rate set by cavity and good jitter performance
Active mode-locking	600 fs	<50 GHz	1 pJ	Repetition rate set by cavity and excellent jitter performance

the pumping conditions, while in passive Q-switching the pulse repetition rate is set by amplitudes of driving currents and absorber voltage. A performance summary of the main techniques for ultrashort pulse generation is presented in the Table 3.1.

3.2.3 Saturable absorber properties and their effect on passive ML

Mode-locking theory distinguishes two regimes of saturable absorber operation, depending on the characteristic recovery time of the absorber after the pulse. A *fast* saturable absorber is the one that recovers on a timescale much shorter than the pulse duration, whereas a *slow* saturable absorber recovers on a time scale much longer than the pulse duration. A fast saturable absorber is characterised by the characteristic saturation *power*, whereas a slow saturable absorber is described by the characteristic saturation *energy* as described below. Saturable absorbers in the form of a reverse biased semiconductor section typically used for mode-locking semiconductor lasers are, strictly speaking, neither fast nor exactly slow. The temporal relaxation of their absorption tends to contain, not one, but several time constants. To begin with, there is usually a (weaker) fast (sub-picosecond) element associated with intraband relaxation of carrier energy distribution (in bulk and quantum well (QW) materials) or fast interlevel processes (in quantum dots (QDs), particularly for the case of holes rather than electrons). The most important time constant of the absorber relaxation however is associated with the relaxation of the *total carrier population* in the absorber; the physical origin of this relaxation in reverse biased absorbers is the sweep out of carriers from the absorber by the electric field. Depending on the external bias and the heterostructure parameters [65], this is typically on the scale of a few picoseconds to a few tens of picoseconds, which is either comparable to the pulse duration or longer than it, but not necessarily *much* longer. Thus the saturable absorber is usually treated approximately as a slow one, but its performance in a laser cannot be described or predicted from the saturation energy alone as is the case for a 'genuine' slow absorber; the (dominant) recovery time is also important as discussed below.

In order to achieve stable ML using a slow saturable absorber, several requirements should be met. Firstly, the saturation energy of the absorber E_a^{sat} should be, at the very least, less than that of the gain medium E_g^{sat}; this is traditionally expressed as

$$E_a^{\mathrm{sat}} \equiv \frac{h\nu A_{\mathrm{a}}}{\partial a/\partial N} < \frac{h\nu A_g}{\partial g/\partial N} \equiv E_g^{\mathrm{sat}} \tag{3.6}$$

Here, $h\nu$ is the photon energy (h being Planck's constant and ν the optical frequency), A_{g} and A_{a} are the optical mode cross-sectional areas in the gain and absorber sections of the laser, $\partial a/\partial N$ is the differential loss, $\partial g/\partial N$ is the differential gain near the operating point of the laser (g being the gain and a the absorption, and N denoting the carrier density).

If the modal cross-sectional areas in the gain and absorber sections are the same, as tends to be the case in edge-emitting semiconductor lasers (with the exception of tapered waveguide ones where it is ensured that $A_{\mathrm{g}} > A_{\mathrm{a}}$ to facilitate mode-locking and simultaneously increase the output power) then the condition (3.6) means that the differential loss $\partial a/\partial n$ should be higher than differential gain $\partial g/\partial n$, as shown in Figure 3.10.

It is worth noting that the often-used condition (3.6) can be traced to the early ML theories proposed by H. Haus and G.New in the 1970s and assuming weak gain and loss per round trip; a more modern theory by Vladimirov and Turaev [66] generalises the requirement for successful mode-locking to the case when the gain and loss over the round-trip are significant (as is the case in monolithic semiconductor lasers) in a more stringent form:

$$E_g^{\mathrm{sat}}/E_a^{\mathrm{sat}} > 1/\kappa \tag{3.7}$$

where κ is the dimensionless attenuation of light in the cavity due to outcoupling and internal losses.

The second condition for successful mode-locking concerns the saturable absorber recovery time. Firstly, the recovery time of the absorber should be faster than that of the gain. The absorber's recovery time plays a crucial role in term of

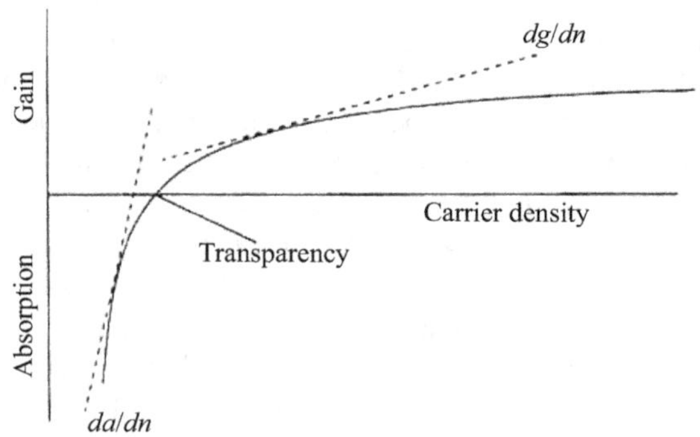

Figure 3.10 Gain/loss dynamics versus carrier concentration for successful ML in laser diodes

ultrashort pulse generation in high repetition rate lasers. It is important for forming narrow net gain as shown in Figure 3.9. If this requirement is met, then the gain overcomes the loss only near the peak of the pulse and the spontaneous emission is not amplified between the pulses. It is also important that the mode-locking period (the inverse of the pulse repetition rate) should be longer than the saturable absorber recovery time.

The saturable absorber's recovery time, whether in the case of QW or QD absorber, is associated primarily with thermionic emission and tunnelling mechanisms [65,67]. At low reverse bias applied to the absorber the recovery time around 62 ps was observed. It was not governed by recombination processes but by the thermal activation of the electrons from the QD ground state (GS) into excited QD states with subsequent emission into the GaAs matrix. As higher reverse bias was applied the decrease in the absorption recovery time down to 700 fs was observed. The tunnelling electron escape mechanism from the QD GS into the GaAs barrier becomes dominant as the barrier height is lowered from E_{b_0} to E_b by electric field and a triangular barrier is formed as shown in Figure 3.11 [65].

Note that all the conditions described above ensure only that mode-locking should be possible in a given laser *at some operating conditions*. The dynamics of the *laser at a given operating point* (usually fully described by, firstly, the current to the gain section, which determines the unsaturated gain, and, secondly, the reverse bias to the absorber section, which determines the sweep out time, the unsaturated absorption, and the absorption cross-section) cannot always be predicted by any simple analytical criterion, and can be quite complex, given the non-linear nature of the mode-locking process. Here, we shall give only the summary of the main dynamic regimes that can be observed in a laser intended for passive mode-locking.

Stable mode-locking, in bulk or QW semiconductor lasers, is typically observed within a finite range of both gain section current and absorber section reverse bias voltage.

At a relatively *low current* and a moderately small bias voltage, the mode-locking pulse trail tends to acquire a *passive Q-switching* envelope, also referred to as *self-sustained pulsations instability* or the *trailing edge instability* (the latter since it arises because of the gain being incompletely depleted at the end of the

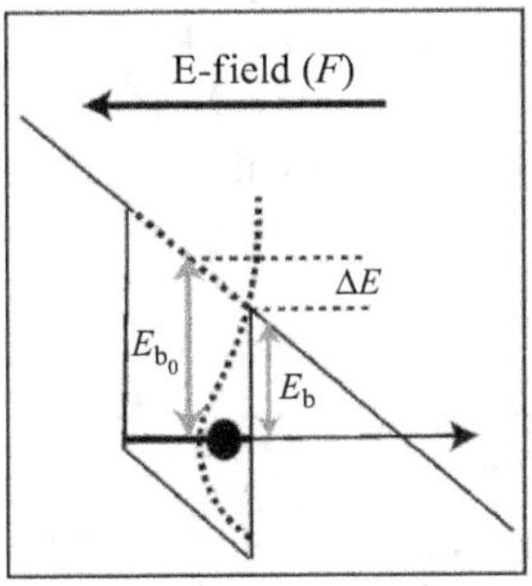

Figure 3.11 Tunnelling electron's escape mechanism with applied electric field

pulse, meaning a positive net gain at the trailing edge). Whilst detrimental for most applications, it may be of some use for applications involving non-linearities such as two-photon spectroscopy, since the combination of passive Q-switching and ML leads to some of the pulses in the train having peak powers several times higher than in the case of stable ML. For values of the reverse bias corresponding to a large ${}^{\partial a}/_{\partial N}$ but a long absorption recovery time, however, the mode-locking substrate of the Q-switching envelope disappears, and the regime becomes pure passive Q-switching with noisy multimode filling inside each Q-switching pulse.

For a high current, significant ${}^{\partial a}/_{\partial N}$ and a short absorption recovery time, the mode-locking pulse train acquires a noisy and chaotic envelope associated with the so-called *leading edge instability* [66] and with a further increase in current the laser dynamics deteriorate into a purely chaotic regime. Further increase in current may once again see stable mode-locking, but at a second harmonic of the repetition frequency, with two pulses per round-trip (in the modal picture of Section 3.1.2, this usually corresponds to every second mode being suppressed in the spectrum). For very long cavities (repetition rates of the order of 0.1–1 ns), regions of stable operation at progressively higher harmonics [66], experimentally, up to the 19th harmonic in an external cavity laser with the fundamental repetition frequency of 60 MHz [68], sometimes separated by areas of unstable behaviour, can be observed. It has been recently shown experimentally (in a vertical external cavity laser construction) and theoretically that stabilisation of fundamental harmonic operation, avoiding higher harmonics, is possible at arbitrarily low repetition rates, provided that a construction with a very large effective value of ${}^{\partial a}/_{\partial N}$ is used and the bias current is suitably low [68]. In such a case, the laser has an S-shaped light-current characteristic, with on-off bistability near threshold; operating within the range of this bistability ensures stable harmonic operation, experimentally down to a repetition rate of 60 MHz [68]. This regime might play some part in future bio-medical applications, potentially replacing gas or solid-state lasers with their low repetition rates, though the peak powers of this regime are by necessity modest and the efficiency limited.

At very high current and /or for very long ($>\sim 1$ ns) or very short (<1 ps) absorber recovery time, the laser is predicted theoretically, and in some cases seen experimentally, to operate in a CW-type regime. In practice, genuine CW operation in a single longitudinal mode is only observed for shorter cavities (fundamental repetition frequency of the order of tens of GHz or higher). For long cavities which have closely spaced, weakly discriminated modes, the single-frequency regime would require currents that are impractically high; these lasers are more likely to display a 'nearly CW' regime with several unlocked modes present, which emerges from unstable mode-locking.

The dynamics of QD mode-locked lasers, which are the main subject of this chapter, are somewhat different from those of QW or bulk lasers. On the one hand, Q-switching instabilities in these lasers are more weakly rendered, making stable mode-locking (fundamental or harmonic) easier to achieve, but on the other hand, an additional complexity arises due to the possibility of mode-locking associated with ground or excited state (ES) transitions, or both, as discussed below.

For more detailed description of mode-locking basics, including the mathematical theory of mode-locking, the reader is referred to our previous monograph [69] or review chapters [41,70], as well as to the chapter by Professor I. Montrosset and co-authors in this edition.

3.3 State-of-the-art ML: QW lasers

GaAs/InGaAs-based semiconductor QW ML lasers have been developed over the past 20 years (see Table 3.2). They have been shown by a number of teams to

Table 3.2 Overview of ML InGaAs QW lasers with following characteristics: τ_p – pulse duration, λ – emission wavelength, $\Delta\lambda$ – full-width half-maximum, f_{rep} – pulse repetition rate, P_{peak} – peak power, TBWP – time bandwidth product, n/a – data are not available.

No. of layers	ML regime	Setup	τ_p (ps)	λ (nm)	$\Delta\lambda$ (nm)	f_{rep} (GHz)	P_{peak} (W)	TBWP	Ref.	Year
4	Active/ Passive/ Hybrid	External	9/ 2.5/ 2.7	840	0.75/ 1.68/ 1.87	5.5	0.02/ 0.07/ 0.06	2.9/ 1.8/ 2.2	[72, 73]	1991
4	Active/ Passive/ Hybrid	Monolithic	13/ 6.5/ 10	840	0.77/ 0.93/ 1.26	5.5	0.013/ 0.028/ 0.018	4.3/ 4.0/ 3.5	[72, 73]	1991
4	Passive/ Hybrid	Monolithic	5.5/ 2.2	840/ 1,580	1.29/ 4.16	11/ 21	0.096/ 0.013	3/ 1.1	[72]	1992
4	Colliding pulse	Monolithic	1.3	1,580	4.99	41	0.015	0.78	[72, 74]	1992
n/a	Hybrid	External+ SOA compression	5.1 (0.46)	830	2.5 (3.2)	0.335	72	5.44 **(0.44)**	[71]	1992
n/a	Hybrid	External+ SOA Compression	**0.2**	830	n/a	0.335	**165**	n/a	[71]	1992
3	Colliding pulse	Monolithic	0.26	1,555	n/a	1,540	0.04	n/a	[75, 76]	1994
3	Passive	Monolithic/ single contact	0.86	1,024	n/a	116	0.06	n/a	[77]	1995
3	Passive	External	3.3	980	n/a	n/a	>2	n/a	[78]	1995
2	Passive	External	3.5	1,550	n/a	9.32	n/a	n/a	[79]	1998
2	Compound cavity	Monolithic	0.24	860	n/a	**2,100**	0.004	n/a	[80]	2001
6	Passive	External	1.7	1,300	n/a	25	~0.047	n/a	[81]	2004
1	Hybrid	External compression	27.1 (3.4)	976	0.6	0.535	0.26	0.64	[82]	2005
1	Passive	External Compression	38.2 (8.8)	976	n/a	0.535	n/a	n/a	[82]	2005
1	Passive	Monolithic	8.3	1,305	n/a	9.48	n/a	n/a	[83]	2007
n/a	Passive	Monolithic	0.7	1,631	5.49	37.9	n/a	0.5	[84]	2009
n/a	Passive	Monolithic	1.8	1,564	4.6	38	n/a	1	[85]	2009
2	Colliding-pulse	Monolithic	1.84/2	830	0.55/0.5	19.75	0.46/ 0.67	**0.44**	[86]	2011

generate ultrashort transform limited pulses (see [70–72] and references therein). For example, hybrid ML 830 nm laser in external cavity configuration in order to get low 335 MHz repetition rate with amplification and compression stages generated 460 fs transform limited pulses resulting in peak power 72 W, which can be improved up to 160 W by using cubic compensator for generation of 200 fs [71]. The highest repetition rate of 2.1 THz was achieved by compound cavity mode-locking 860 nm double QW laser with 2.2 mW average power and 0.24 ps pulses [80]. As shown in Tables 3.1 and 3.2, both InGaAs and GaInAsP QW lasers have been developed for operating in a broad range of repetition frequencies and with potentially powers in the multiwatt range.

Compared to InGaAs structures, InGaAsP lasers (Table 3.3) showed the highest repetition rate of 860 GHz at 1,265 nm using Asymmetric Colliding Pulse harmonic ML technique [87]. For generating higher power, slab coupled lasers were fabricated from the same material [88]. Peak power of 5.8 W was delivered from 4.3 GHz InGaAsP laser incorporating five QW layers with 10 ps pulses at 1,544 nm wavelength [89]. More recently, optical pulses as short as 200 fs were generated with high peak power of 52 W from 830 nm QW laser at 266 MHz repetition rate in external cavity configuration with intracavity dispersion management and grating compressor [90]. An alternative material system such as GaInNAsSb can be used for development of passively ML laser emitting at 1,583 nm [91].

Table 3.3 Overview of ML InGaAsP QW lasers with following characteristics: τ_p – pulse duration, λ – emission wavelength, $\Delta\lambda$ – full-width half-maximum, f_{rep} – pulse repetition rate, P_{peak} – peak power, TBWP – time bandwidth product. n/a – data are not available.

No. of layers	ML regime	Setup	τ_p (ps)	λ (nm)	$\Delta\lambda$ (nm)	f_{rep} (GHz)	P_{peak} (W)	TBWP	Ref.	Material/ Year
6	Hybrid	Monolithic	1.4	1,300	n/a	15.1	n/a	n/a	[92]	1989
5	Passive	Monolithic	1/0.83	1,515/ 1,545	2.68/ 2.89	156/ 80	n/a	0.34/ 0.31	[93]	1992
5	Passive	Monolithic	**0.64**	1,560	4	350	n/a	0.32	93	1992
6	Colliding-pulse harmonics	Monolithic	n/a	~1,265	n/a	**860**	n/a	n/a	[87]	1997
n/a	Passive	Monolithic	1.2	1,584	12.6	~37	n/a	1.8	[94]	1999
8	Passive	Monolithic	0.41	1,550	6.8	133	n/a	0.35	[95]	2002
1/ 2/ 3	Passive	Monolithic	2.8/ 4.5/ 4.9	1,531/ 1,556/ 1,574	1.1/ 0.7/ 1.1	39.63/ 39.46/ 39.49	0.064/ 0.05/ n/a	0.4/ 0.39/ 0.66	[96]	2004
5	Passive	Monolithic Slab coupled	10	1,544	5.7	4.29	**5.8**	7.5	[89]	2006
1	Passive	Monolithic	0.86	1,540	5	21.31	0.5	0.57	[97]	2008
5	Passive	Monolithic Slab coupled	5.8	1,550	~6	4.6	~2	~4.3	[88]	2008
2	Passive	External (compression)	7.4 (**0.2**)	830	1.8 (6.48)	0.266	3.2 (**52**)	5.7 (0.58)	[90]	2010

To summarise, QW-based devices exhibited good performance involving a very broad range of repetition rates (from 60 MHz to 2.1 THz), short pulse (down to 200 fs FWHM) generation using pulse compression technique and high pulse energy (around 60 pJ generation). It is worth noting that the results above have been achieved separately at different wavelengths, from 830 to 1.5 μm. However, certain spectral ranges, particularly between about 1,100 nm (the longest wavelength usually addressed with InGaAs lasers) and 1,280–1,300 nm (the shortest wavelength for which InGaAsP lasers have been developed) have not been addressed with QW lasers [98], partly because of the lack of requirement for traditional communications related applications, partly because of genuine technological difficulty.

QD lasers, on the other hand, can address this wavelength range, which is very useful for biomedical applications. They will be considered in the section below.

3.4 Monolithic ML QD lasers

Emission wavelength of QD-based devices can be in the range of 1,100–1,300 nm, which is important for varieties of applications, including metropolitan area networks and, most importantly here, multiphoton imaging and non-linear spectroscopy. QD-based lasers can offer several advantages such as low threshold current density; temperature-insensitivity and single-frequency generation due to the density of state of QD material is discrete. In addition, ultrafast carrier dynamics of the QD absorber is very useful for generating ultrashort pulses using ML technique. For example, it was shown that is possible generating 393 fs pulses directly from QD-based device [48].

In this section, the latest results achieved from QD lasers will be discussed in more detail, including the generation of high peak power of 17.7 W directly from monolithic tapered gain-guided lasers without use of compression and amplification techniques [99] as well as the generation of optical pulses with a low repetition rate of 191 MHz [100,101] and high peak power of 1.5 W and 25 pJ pulse energy directly from oscillator using external cavity configuration [102]. The main experimental results on mode-locking of QD lasers are summarised in Table 3.4.

Monolithic QD based ML edge-emitting lasers have shown considerable potential as compact and versatile high power ultrashort pulse optical sources [100,103]. The advantages of QDs as active layers for mode-locking are largely due to the distinctive properties of these materials, which in turn rely on quantum confinement of the carriers in all the three dimensions. This alters the carrier kinetics in the dots, leading to relatively slow carrier capture into the dots, which in turn acts to stabilise the laser dynamics and suppress ML instabilities. The reverse-biased saturable absorber sections show somewhat shorter recovery times than those seen in QWs, which also contributes to stable mode-locking. Besides, the discrete energy levels displayed by QD materials can also be exploited to access different, distinct spectral bands in the same device, adding an extra level of functionality to these ultrafast lasers. For instance, it is possible to achieve two-state lasing using the GS and ES transitions in a QD laser [104]. It should be noticed that similar performance

Table 3.4 Overview of ML InGaAsP QD lasers with respect to the following characteristics: τ_p – pulse duration, λ – emission wavelength, $\Delta\lambda$ – full-width half-maximum, f_{rep} – pulse repetition rate, P_{peak} –peak power, TBWP –Time bandwidth product. n/a – data are not available.

No. of layers	ML regime	Setup	τ_p (ps)	λ (nm)	$\Delta\lambda$ (nm)	f_{rep} (GHz)	P_{peak} (W)	TBWP	Ref.	Year
2	Passive	Monolithic	17	1,278	1	7.4	~0.02	3.1	[121]	2001
3	Hybrid	Monolithic	14.2	1,107	0.8	10	0.004	2.8	[122]	2003
3	Passive	Monolithic	5	1,107	0.8	18.3	0.005	0.9	[123]	2003
5	Passive	Monolithic	1.7	1,277	0.86	9.7	0.06	0.27	[124]	2004
5	Passive	Monolithic	7	1,283	~0.4	35	0.006	~0.5	[125]	2004
5	Hybrid	Monolithic	11	1,286	~0.22	20	n/a	0.44	[125]	2004
5	Passive	Monolithic	3	1,281	0.8	50	0.006	0.44	[126]	2004
10	Passive	Monolithic	10	1,291	0.2	18	0.0025	0.36	[127]	2004
10	Passive	Monolithic	3.2	1,260	7.1	5	1.7	4.3	[128]	2005
n/a	Passive	Monolithic	5.7	1,264	4.5	5.2	0.29	4.8	[129]	2005
5	Passive	Monolithic	**0.39**	1,260	14	21	**3**	1	[62,103]	2005
3	Colliding-pulse	Monolithic	7	1,107	0.32	20	0.013	0.54	[130]	2005
10	Passive	External+S OA (Compression)	15 (1.2)	1,268	3.1	4.95	1.22	6 (0.69)	[131]	2005
10	Passive	External+S OA (compression)	14 (0.97)	1,275	6	2.7	n/a	15.5 (1.1)	[114]	2006
10	Passive	External+S OA (compression)	10.8 (1.2)	1,193	4.25	2.7	n/a	9.7 (1.1)	[114]	2006
10	Active	External	24	1,269	2	12.77	n/a	10	[132]	2006
5	Colliding-pulse/harmonic	Monolithic	1.3	1,280	1.4	**238**	0.01	0.33	[133]	2006
10	Hybrid	Monolithic	0.71	1,280	n/a	20	n/a	n/a	[134]	2006
10	Passive	Monolithic	1.5	1,280	6.18	80	n/a	1.7		2006
5	Passive	Monolithic	7	1,190	5.5	21	0.156	8.1	[108]	2006

5	Passive	Monolithic (tapered waveguide)	0.79	1,276	3.6	24	0.5	~**0.5**	[135]	2006
10	Passive	Monolithic	1.8	1,300	~2.2	39.87	n/a	0.72	[136]	2006
10	Passive	Monolithic	1.5	1,300	~6.4	80	n/a	1.7	[136]	2006
15	Hybrid	Monolithic (SOA)	0.7 (0.7)	1,300	8.5	20	0.13	1.1	[137]	2006
15	Hybrid	Monolithic (SOA)	1.8 (1.9)	1,300	2.2	40	0.014	0.7	[137]	2006
15	Passive	Monolithic (SOA)	1.9 (2.2)	1,300	4.2	80	0.011	1.5	[137]	2006
9	Passive/Hybrid	External	n/a	1,205	n/a	0.35-1.5	n/a	n/a	[138]	2007
10	Passive	Monolithic	1.9	1,300	n/a	19.46	0.07	n/a	[139]	2007
6	Passive	Monolithic	6.4	1,216	2.6	7.2	0.224	3.4	[140]	2007
6	Passive/harmonic	Monolithic	6	1,216	2.6	50.7	0.07	3.2	[140]	2007
6	Passive/harmonic	Monolithic	6.6	1,216	1.7	42.4	0.234	2.3	[140]	2007
6	Passive	Monolithic	5.7	1,240	~4.5	4.97	1.04	~0.5	[141]	2007
10	Passive/harmonic	Monolithic (compression)	9.8 (0.96)	1,260	5.5	5	>1	10 (1)	[142]	2008
10	Passive/harmonic	Monolithic (compression)	8.6 (0.58)	1,260	7.2	40	>1	12(0.8)	[100]	2008
5	Passive	External	0.93	1,300	2.3	1(0.31)	0.41	0.385	[143]	2011
10	Passive	Monolithic (tapered waveguide)	0.36	1,280	5.56	17	2.25	0.36	[144]	2009
5	Passive	Monolithic	~1	1,300	~5.6	40	~0.2	~1	[145]	2009
6	Passive	Monolithic	10	1,209	2.8	10.2	~0.11	~5.7	[146]	2009
5	Passive	Monolithic	2	1,300	n/a	10	n/a	n/a	[147]	2009
15	Passive/Hybrid	Monolithic (compression)	1.6 (0.7)	1,310	~1.2 (0.5)	40	n/a (0.006)	1.4 (0.8)	[148]	2010
6	Passive	Monolithic	5.7	1,240	~5	4.97	1.04	~5.5	[149]	2010

(Continues)

Table 3.4 (Continued)

No. of layers	ML regime	Setup	τ_p (ps)	λ (nm)	$\Delta\lambda$ (nm)	f_{rep} (GHz)	P_{peak} (W)	TBWP	Ref.	Year
8	Passive	Monolithic	5	1,240	4.3	4.97	n/a	4.2	[105]	2010
6	Passive	Monolithic	6	1,270	n/a	12.105	n/a	n/a	[111]	2010
6	Passive	Monolithic	3.3	1,207	n/a	12.321	n/a	n/a	[106]	2010
10	Passive	External	13.6	1,274	1.2	1.14	1.5	3.02	[102]	2010
5/10	Passive	Monolithic	3.2/ 3.3	1,250/ 1,260	7.3/8.4	14.65/ 14.57	3.6	4.4/5.2	[150]	2010
10	Passive	External	12	1,268	4	0.281 (0.191)	0.5	~8.9	[100][101]	2011
5	Passive	External	2.2	1,300	5.3	1.1	0.16	2.1	[146]	2011
8	Passive	Monolithic	2.7	n/a	n/a	5	n/a	n/a	[151]	2011
15	Passive	Monolithic	1.2	1,300	n/a	40	n/a	n/a	[152]	2011
10	Passive	Monolithic	1.26	1,260	5.6	10	17.7	1.33	[99]	2011

can be achieved when using QD saturable absorbers to mode-lock non-semiconductor lasers, for example, femtosecond pulses were generated from a $Cr^{4}+$: forsterite ML laser incorporated within a cavity a QD-based SESAM involving not only GS but ES as well [105]. First, optically gain-switched quantum-dot edge-emitting laser generating picosecond pulses from GS and ES states was demonstrated in 2006 [106]. In this context, stable ML operation involving either GS or ES in a two-section QD laser diode for the first time was shown, by controlling the injection current and reverse bias to force laser emission to switch from the GS to the ES transition [107,108]. Self-pulsation regimes in two-state QD laser operating simultaneously on the GS and ES were reported [109,110]. Furthermore, reverse GS/ES transitions were observed by Breuer *et al.* [111–113]. Such versatility of the QD ML diode laser could enable its use as a two-band clock recovery source [114,115], and also potentially for addressing two spectral regions simultaneously in a spectroscopic application. Dual-GS-wavelength ML or continuous-wave (CW) operation was reported by a number of groups with various theoretical explanations proposed [116–119]. In addition, by changing structural parameters (the cavity length and number of QDs in the active region) of the laser GS or ES emission can be achieved [120,127]. In this context, compact cheap electrically pumped QD-based ML lasers are very promising candidates as versatile next generation optical sources which can be successfully applied for time-domain spectroscopy (as well as non-biophotonic applications such as wavelength-division multiplexing and ultrafast optical processing).

A detailed investigation of the influence of bias conditions on the pulse characteristics generated via ES in a two-section InAs/GaAs QD laser has been presented in [153]. A dual-wavelength ML regime was demonstrated, involving both the ground and excited-state transitions with a spectral separation of 83 nm [154], which is the widest spectral separation ever to be observed in a non-vibronic dual-wavelength ML laser.

Besides being an important distinction of QD lasers, ES mode-locking has a number of practical advantages. It shows better tolerance to optical feedback compared to the GS ML regime [155], which is of importance for applications, as the use of an optical isolator can be avoided. This can be attributed to weaker self-phase modulation during ES lasing. Indeed, numerical analysis of chirp showed that a QD laser emitting from the GS gives a chirp considerably higher than from ES, due to a large refractive index variation at the GS lasing wavelength caused by the accumulated carriers in the ES [156]. Furthermore, the use of the ES transition can effectively extend the spectral range of laser emission towards shorter wavelengths.

Therefore, GS/ES mode-locking is interesting both from scientific and practical point of view and will be discussed in more detail below, mainly following the results of [153,154].

By varying the bias conditions, different ML regimes were observed, involving lasing from both GS and ES (Figure 3.12). Ground-state mode-locking (GSML) was observed at a wavelength $\lambda = 1{,}250$ nm and repetition rate of 20.1 GHz, while for excited-state mode-locking (ESML), $\lambda = 1{,}177$ nm and the repetition rate was 19.8 GHz. A map of different operating regimes depending on the operating point (absorber voltage and gain section current) is depicted in Figure 3.12.

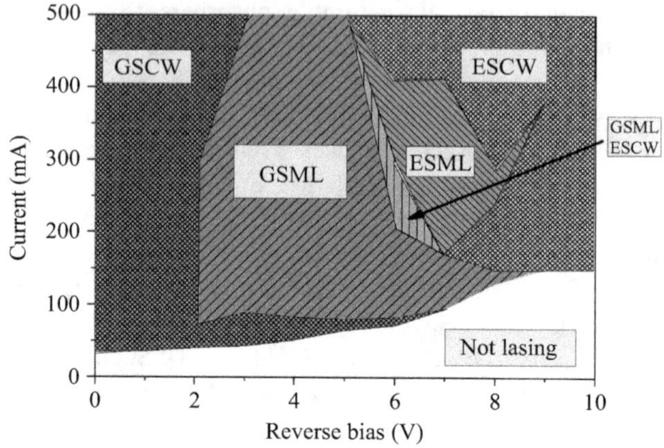

Figure 3.12 Map of the different mode-locking regimes with bias conditions: GSML – ground-state mode-locking; GSML ESCW – ground-state mode-locking in coexistence with excited-state continuous wave emission; ESML – excited-state mode-locking

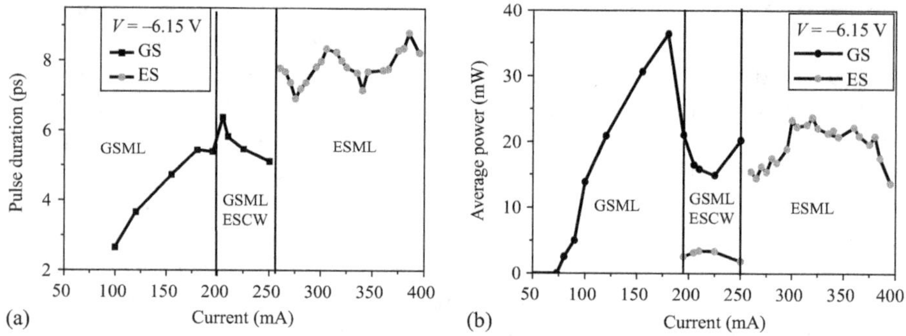

Figure 3.13 Dependence on injected current of (a) pulse duration and (b) average output power for GS and ES mode-locking at 6.15 V reverse bias. GSML – ground-state mode-locking; GSML ESCW – ground-state mode-locking in coexistence with excited-state continuous wave emission; ESML – excited state mode-locking

For quantifying the laser performance, it is instructive to first consider the laser dynamics for a fixed SA voltage bias (say, 6 V) and varying current. For a bias current of up to 200 mA, it was observed (Figure 3.13(a)) that the pulse duration of GSML increased with current; this is not unusual in mode-locked lasers and is sometimes explained by the increase in self-phase modulation effects with optical power, which in combination with the group velocity dispersion in the semiconductor laser leads to a pulse broadening with increasing current [157].

Between 200 and 250 mA, a different GSML regime was also observed, where the coexistence with ES (CW) has the effect of reducing the pulse duration with increasing current. A similar ML regime has been investigated and reported previously [158]. When the current is increased further, only ESML was observed, due to the gain saturation across the GS transition and the higher saturated gain that characterises the ES. As shown in Figure 3.13(a), the pulse duration in the ESML regime varied between 7 and 9 ps, thus exhibiting wider pulse durations than what was obtained for GSML. (It is important to stress that ESML was only achieved at the expense of injecting high drive currents, which usually is associated with longer pulse durations.) The corresponding average power was lower compared to that of the GS due to the increased non-saturated losses (Figure 3.13(b)). The output power changed between 15 and 25 mW with increasing current.

The variations in pulse duration and average power were also investigated with a fixed injection current (300 mA) applied to the gain section, while the reverse bias applied to the saturable absorber was increased, resulting in the observation of the GSML and ESML regimes in succession (Figure 3.14). As expected, the pulse duration of GSML was observed to decrease from 15.5 to 6 ps with reverse bias due to the fact that the absorption recovery becomes increasingly faster (see, e.g., [159,160] and an overview in [41]). Pulses are almost twice shorter compared to Kim *et al.*, where pulse duration changed from 25 to 14 ps with reverse bias [114]. For a higher value of reverse bias, the GSML regime is switched to the ESML regime, whereby the pulse widths vary from 9.3 to 10.3 ps with increasing reverse bias up to 8 V while the average power continues to decrease with reverse bias as is also observed in the GSML regime. From this investigation, it can be concluded that such a device can be very promising for generation of ultrashort pulses using compression techniques not only using GS transitions but ES transitions as well.

The differences in pulse dynamics of GSML and ESML can be partially explained by the different intradot relaxation dynamics of the absorber, for GS or ES transitions. As reported by Piwonski *et al.* [159], the absorption recovery via the GS is mediated mostly by phonon-assisted processes, while the ES absorption

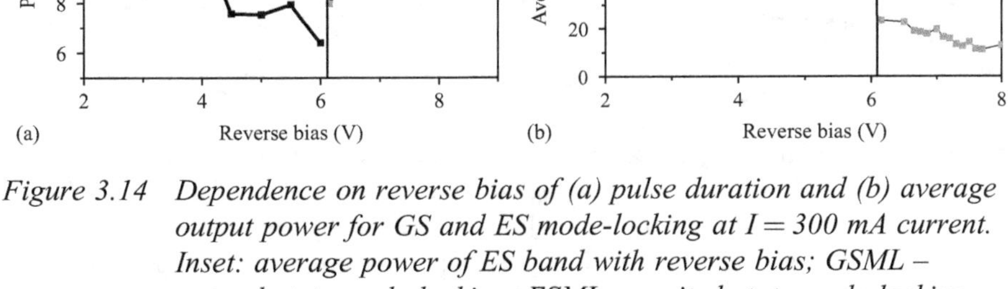

Figure 3.14 Dependence on reverse bias of (a) pulse duration and (b) average output power for GS and ES mode-locking at I = 300 mA current. Inset: average power of ES band with reverse bias; GSML – ground-state mode-locking; ESML – excited-state mode-locking.

recovery is dominated by Auger-type processes. It was also shown [159] that the absorption recovery time reduced more significantly with increasing reverse bias for the GS, while the absorption recovery time for the ES was shown to be somewhat less sensitive to the reverse bias applied, particularly for higher values of reverse bias (5–9 V), which indeed encompasses the range of reverse bias values required to achieve stable pulse generation via ESML [153].

In addition to mode-locking regime involving lasing either at the ground or at the ES transition wavelengths, simultaneous dual-wavelength passive ML regime was also observed from a (different) QD two-section monolithic laser with picosecond pulses generated from both GS ($\lambda = 1{,}263$ nm) and ES ($\lambda = 1{,}180$ nm) transitions simultaneously with the widest spectral separation of 83 nm ever observed in a non-vibronic ultrafast laser [154], though slightly lower than previously reported results from a solid-state laser [161]. The experimental setup and the structure of the device were similar to the one described before except for slightly different wavelength as the result of formation of different size of QDs during the growth.

The regime of simultaneous ML in GS and ES was attained for current levels in the gain section between 330 and 430 mA, and values of reverse bias between 6 and 10 V in the saturable absorber section. A map depicting the different ML regimes is represented in Figure 3.15. It is interesting that ESML was observed both before and after the dual-wavelength ML regime which can be attributed to interplay between the modes due to non-linear absorption.

As can be seen from the Table 3.5, the central emission wavelengths were ~1,180 nm and ~1,263 nm for ES and GS, respectively. The pulses during simultaneous dual-wavelength ML regime can be as short as 5.9 and 7 ps for GS and ES, respectively, depending on bias conditions. As an example, optical spectrum and RF spectrum for the gain section current of 425 mA and the absorber

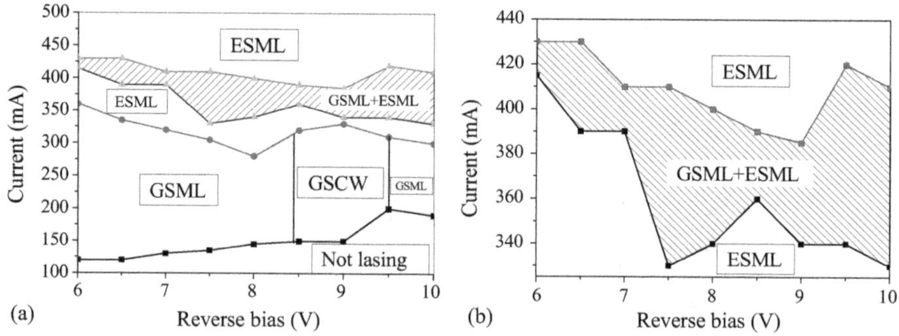

Figure 3.15 (a) Mapping of the different operating regimes observed in 2 mm QD ML laser and (b) mapping of the dual-wavelength ML regime here described (GSML+ESML). Legend: GSML – ground-state mode-locking; ESML – excited-state mode-locking; GSCW – ground-state continuous-wave operation

Table 3.5 Bias conditions: reverse voltage to the absorber (V) and forward current to the gain section (I); pulse duration (τ), average power (P_{av}), peak power (P_{peak}), time bandwidth product (TBWP), full-width half maximum of optical spectra (Δλ), central wavelength (λ), GS – ground state (black) or ES – excited state (red) of dual-mode regime in QD-based ML laser

V (V)	*I* (mA)	*τ* (ps)	P_{av} (mW)	P_{peak} (mW)	TBWP	Δλ (nm)	λ (nm)	State
6	425	5.9	32.1	261.6	6.7	6.0	1,263.40	GS
6	430	7.1	35.5	249.7	9.5	7.1	1,263.74	GS
6.7	400	5.9	17.1	144.1	7.0	6.2	1,263.03	GS
7	394	6.6	11.8	89.9	7.6	6.2	1,263.38	GS
8	375	9.3	12.1	65.2	8.0	4.6	1,263.52	GS
8	390	7.1	12.5	87.9	6.9	5.2	1,265.00	GS
9.2	380	7.3	10.5	71.6	7.3	5.3	1,264.18	GS
9.3	344	9.4	8.3	44.0	8.4	4.7	1,263.28	GS
6	425	8.6	21.4	128.4	9.0	4.9	1,180.55	ES
6	430	8.8	11.8	67.6	9.8	5.2	1,180.70	ES
6.7	400	7.9	21.7	136.7	8.2	4.8	1,180.16	ES
7	394	8.7	24.5	140.9	9.0	4.8	1,179.87	ES
8	375	8.5	29.2	171.9	8.1	4.4	1,179.07	ES
8	390	8.6	33.7	196.6	8.4	4.5	1,179.07	ES
9.2	380	8.6	11.6	67.0	8.0	4.3	1,179.14	ES
9.3	344	7.0	34.8	250.2	6.7	4.5	1,178.49	ES

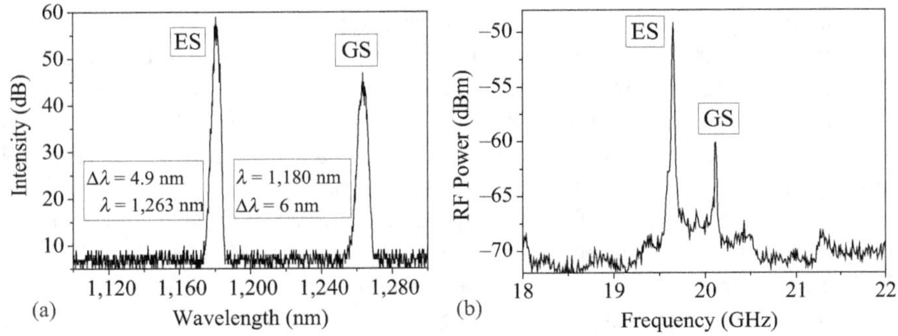

Figure 3.16 (a) Optical spectrum and (b) RF spectrum characteristic of the dual-wavelength ML regime, for an injection current of 425 mA and reverse bias of 6 V

section reverse bias of 6 V are presented (Figure 3.16 (a)) resulting in the spectral separation between GS and ES of 83 nm. As refractive index varies with wavelength, the pulse repetition rates for ES and GS were 0.5 GHz different, namely, 19.6 and 20.1 GHz, respectively (Figure 3.16(b)). The autocorrelation traces obtained for the ES and GS were fitted using Gaussian function and resulting in

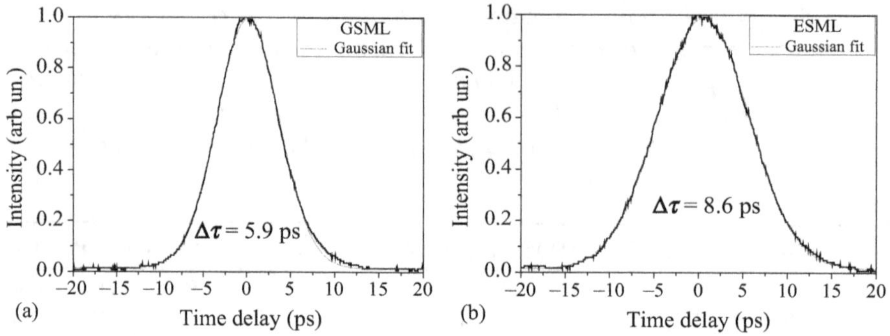

Figure 3.17 Autocorrelation traces for (a) GS mode-locking and (b) ES mode-locking at 6 V reverse bias and 425 mA gain current

pulse durations of 5.9 ps for the GS and 8.6 ps for the ES bands, obtained for a reverse bias of 6 V and an injection current of 425 mA (Figure 3.17(a) and (b)). Knowing the pulse duration, central emission wavelength and full-width half-maximum (FWHM) of optical spectra, the TBWP was estimated for the GS of 6.7 and for the ES of 9. These pulse durations and TBWPs are similar to those previously observed in separate GS/ES ML as discussed above. The average power for the GS was 21.3 mW and for ES 32.1 mW was measured, from which the peak powers of 261 mW and 128 mW, respectively, can be estimated (see Table 3.5).

The dual-wavelength regime of operation from QD-based ML two-section laser was proved by a numerical model based on the delay differential equation approximation [162] taking into account the dynamics of QD material [163]. Dual-wavelength emission is possible for reverse bias higher than 6 V due to the decrease of the absorber recovery time mediated by the tunnelling carrier escape mechanism [148]. The relatively narrow region of dual-wavelength regime is associated with the carrier kinetics between two states (GS, ES) which gives rise to instabilities due to the carrier intraband capture and escape processes between states. Continuing the work of [154], stable two-state ML was later achieved from a two-section ML laser incorporating two repeats of three chirped QD layers [111,112]. It is interesting that reverse-emission-state-transition was observed in those samples, with ML starting with ES emission as 0V reverse bias was applied. Dual-state ML was achieved at lower gain current between 130 and 170 mA compared to the results of [154] (330–430 mA) because loss in the absorber is increased with voltage applied. In other words, the higher absorber-to-gain ratio 1:10 compared to 1:7 leads to the different gain and absorber dynamics in the gain and the saturable absorber sections. The effect of the fast relaxation of ES carriers in the GS, the so-called photon pumping or photon recycling process, results in the reduction of the absorption at GS wavelength at saturable absorber allowing the coexistence of GS and ES ML [111,112].

The behaviour of QD ML was found to be strongly dependent on the laser geometry. In general, the importance of optimised saturable absorber parameters for short pulse generation in a passively ML laser was understood more than 25 years ago [164]. It was found that in multi-section monolithic devices pulse width and peak power show better results for ML lasers with a longer passive section [143]. Sub-picosecond pulses (796 fs) were generated in 20 GHz passive ML quantum-dot laser diode incorporating 5-layer of QD in the active region with absorber-to-gain ratio of 1:4 [165]. The ratio between the gain section and the absorber is known to play vital role for ultrashort pulse generation. It was found [144] that the absorber-to-gain length ratio in QD lasers greater than 1:3 is not desirable as there is not enough gain to overcome the losses in the absorber section while the absorber-to-gain ratio less than 1:14 is not favourable for ML due to insufficient saturable absorption to initiate mode-locking. In this context, the influence of non-linear absorption saturation and the Quantum Confined Stark effect in QD laser on bias conditions were studied [166]. Understanding of ultrafast QD absorber dynamics gives further insight into pulse broadening/shortening mechanisms [167]. To that end, the dependence of pulse duration, average power, peak power on the number of QD layers incorporated in the two-section QD based ML laser was studied experimentally. The experimental setup described above was used. The lasers investigated had the total cavity length of 2 mm and 1.3 mm, with the absorber of 300 μm resulting in the gain-to-absorber ratio of 1:6 and 1:4, respectively, which is very favourable for stable ML as discussed above. All lasers had a ridge waveguide width of 6 μm, the active region incorporated 5, 10 or 15 layers of QDs grown by the Stranski– Krastanow method. All lasers were maintained at 20 °C by a Peltier cooler. Pulse duration variation with bias conditions for 2 mm devices with different number of QD layers (5, 10, 15) is presented in Figure 3.18. The pulse duration varies from 1 to 10 ps depending on the gain section current and the absorber reverse bias. Pulse broadening with increasing gain current for all devices is partly attributed to self-phase modulation effect [145]. The shortest pulse duration of 1 ps from 10 QD layers laser were observed just above lasing threshold with 6 V reverse bias and 122 mA forward current. The central emission wavelength for such bias conditions was around 1,253 nm with the FWHM of 3.67 nm resulting in TBWP of 0.7, which indicates that the pulse was not transform limited. The

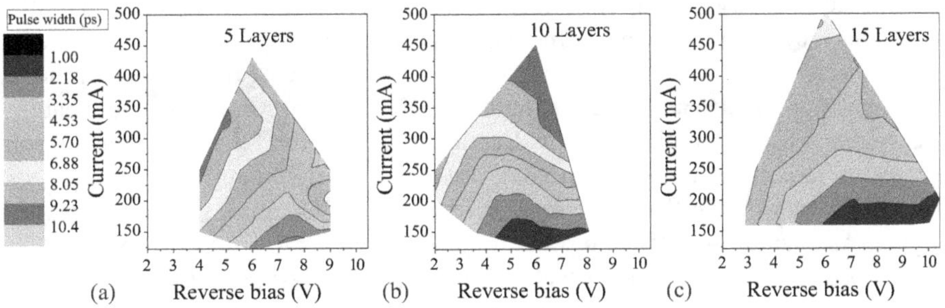

Figure 3.18 Pulse width dynamics with bias conditions for 2 mm laser with (a) 5; (b) 10; and (c) 15 QD layers

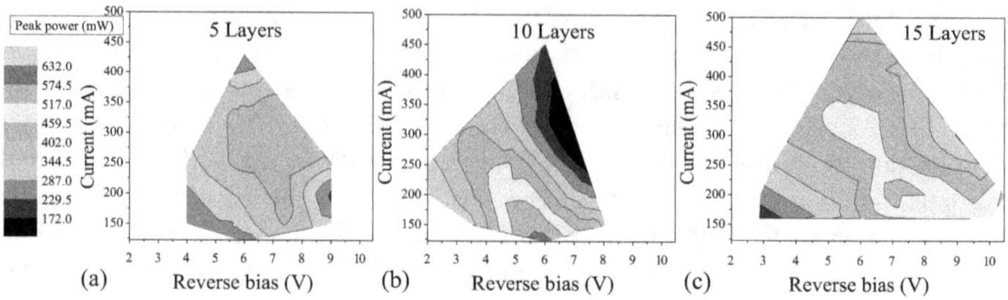

Figure 3.19 Peak power dynamics with bias conditions for 2 mm laser with (a) 5; (b) 10; and (c) 15 QD layers

exponential decrease of pulse duration with increased absorber bias was observed for all devices and explained by exponential dependence of the absorber recovery time on the bias [168]. The range of ML for a 5 QD layer laser is between 4 V and 9 V with current range from 140 to 400 mA (Figure 3.18(a)). As the number of QD layers increased to 10, the range of ML is broadened and shorter pulses can be generated (Figure 3.18(b)). A 15 QD layers laser showed the widest range of ML from 2 to 10 V and driving current from 160 to 500 mA. The threshold current is higher for this laser as more gain is required to overcome optical losses. In Figure 3.19, peak power dynamics are shown. The variation of peak power with reverse bias does not exhibit a clear trend as the increase in average power is somewhat compensated by the pulse broadening. The highest peak power about 600 mW is achieved for the low current region of a 10 QD layers laser due to higher average power than for a 5 QD layers laser and shorter pulse generation. As can be seen in Figure 3.19(b) and (c), peak power with increased number of layers doesn't change much as broadening pulse balances the increased average power.

Furthermore, laser emission involving the ES transition was observed only for 5 layer devices as the lower number of layers allowed for an early saturation of the GS and achievement of the ES laser emission. ES ML was observed for gain current higher than 300 mA and reverse bias higher than 6 V in the absorber for 2mm long laser. In order to induce a transition from GS to ES emission, it is important to saturate the gain associated with the GS transition. This can be achieved by either increasing the forward current in the gain section, and/or by increasing the reverse bias in the saturable absorber, which leads to an increase of the GS non-saturated losses and thus leads to an earlier saturation of the GS gain. Another way of enabling an easier access to ES emission is by decreasing the length of the laser which results in a lower GS gain unable to compensate the total loss in the cavity [127,140]. As an example, it was shown that ES emission from a $\lambda = 1.24$ μm laser incorporating a single InAs quantum-dot layer appears for laser cavity less or equal to 1.5 mm [140]. In addition, a 3 QD layer-laser as well exhibited only ES lasing for the short cavity in the range of 1 to 2 mm long [127]. However, the higher the number of QD layers the weaker the effect of gain saturation [169]. For example, a 10 QD layer InGaAs laser provides enough gain for GS lasing in a broad-area 1 mm long stripe laser [170].

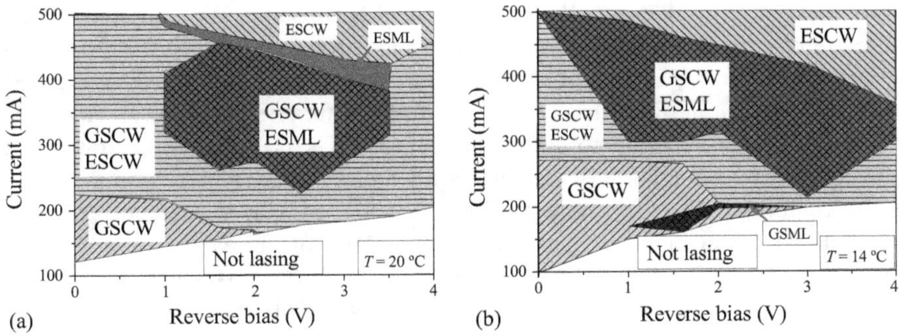

Figure 3.20 Mapping of ML regimes observed for 1.3 mm long 5QD layers device, under an operating temperature of (a) 20 °C and (b) 14 °C. GS/ES: ground-state/excited-state. CW/ML: continuous-wave/mode-locked operation

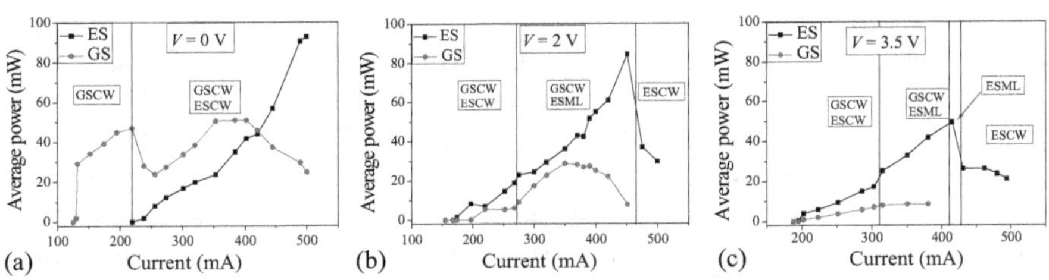

Figure 3.21 Average power dynamics for 1.3 mm long device and 5 QD layers with (a) 0 V; (b) 2 V; and (c) 3.5 V reverse bias. GS/ES: ground-state/excited state. CW/ML: continuous-wave/mode-locked operation

In this context, a shorter device, 1.3 mm long with a 300 μm absorber and a 5 QD layers in the active region, was also investigated. ES transition ML at $\lambda \approx 1{,}188$ nm with or without co-presence of CW GS emission at $\lambda \approx 1{,}274$ nm was observed as shown in Figure 3.20(a) and (b) for operating temperatures of 20 and 14 °C, respectively. At 20 °C, ESML without GSCW was observed only for current values higher than 400 mA (Figure 3.20(a)). As the operating temperature was lowered to 14 °C, a small region of GSML was noticed (Figure 3.20(b)) which is consistent with the fact that GS intensity increases dramatically with the decreasing temperature, while ES intensity does not change significantly with temperature [48,171]. The region of ES ML with GS-CW becomes slightly broader as the temperature decreases as well. Output power for different states was measured at 0, 2 and 3.5 V reverse bias (Figure 3.21(a), (b) and (c)).

As can be seen in the Figure, in a short cavity laser at 0 V reverse bias, both states exist under CW operation but when reverse bias (2–3.5 V) is applied to the absorber section, the ES power is increased while the GS power decreases and

ESML can be achieved due to the saturation of GS emission. With applied voltage, the range of existence of GS CW emission decreases from almost 400 mA (from 125 to 500 mA for $V = 0$ V) to less than 200 mA (from 188 to 380 mA for $V = 3.5$ V). It is explained by the fact that the losses in the saturable absorber are increased with reverse bias and due to finite number of QDs GS gain is saturated with increasing current quicker. As the results ES emission plays more dominant role. As shown in Figure 3.21(b) and (c), ES ML is achieved with coexistence of GS and even without GS at high voltage and at the expense of the gain current. Average power of GS emission for 2 mm long devices initially increases with current then drops when ES emission switches on. In this case of shorter cavity 1.3 mm GS average power for reverse bias of 0 V behaves similar as can be seen in Figure 3.21(b) for 2 mm device and increases in the first place up to 47 mW for current from 125 mA to 240 mA and then the power drops to 28 mW as ES appears. With further increase in current, both GS and ES power increase simultaneously. At current 420 mA, power level of GS and ES is the same and around of 45 mW. Higher values of the gain section current result in a lower GS emission power and a higher ES emission power. As the reverse bias is increased to 3.5 V, the power level of GS reaches 8.6 mW as ES emission becomes more favourable due to GS saturation. The highest average power of 84.5 mW can be attained in ES ML regime at 2 V reverse bias and 450 mA gain current. The average power for ES increases as applied voltage increases up to 2 V and then decreases with voltage while for 2 mm devices average power always decreases with reverse bias for both GS and ES bands as shown in Figure 3.21(b). The increase of average power of ES is associated with gain saturation of GS while the following decrease of average power can be explained by increase of the total loss at expense of reverse bias.

The shortest pulse duration obtained for ES spectral band is 4.6 ps at 2.5 V reverse bias and 230 mA gain current with 16.6 mW average power (Figure 3.22(a)). The FWHM of optical spectra for ES (1,188 nm) was 2.9 nm giving time bandwidth product of 2.83 suggesting that the pulses are not transform limited. Shorter pulses for ES are demonstrated for lower current and the same is observed for GS 2 mm devices. The peak power of ES increases with gain current from 77 mW up to 350 mW (Figure 3.22(b)). The highest peak power of 350 mW is

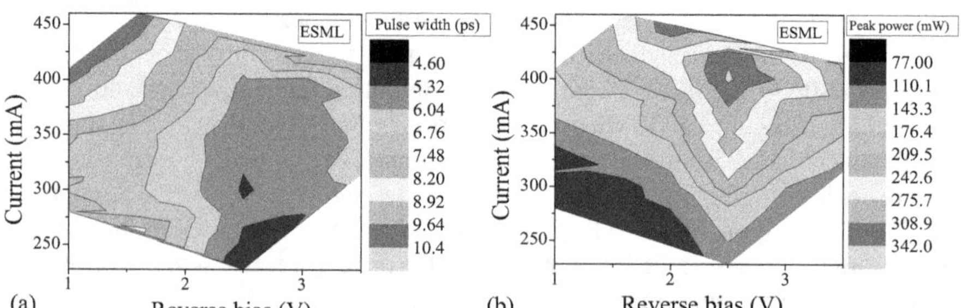

Figure 3.22 (a) Pulse width and (b) peak power evolution with bias conditions for 1.3 mm long device and 5 QD layers

achieved for reverse bias of 2.5 V and driving current of 400 mA. The power level is similar to what was achieved for GS ML regime in 2 mm laser.

To summarise, a 15-layer QD laser showed the widest range of ML from 2 V to 10 V and driving current from 160 mA to 500 mA. The highest peak power of 600 mW is achieved for 10 QD-layer lasers.

Without further amplification, these power levels are still more than an order of magnitude too low for multi-photon imaging applications. We note however that one of the advantages of QDs as an active layer platform is that amplifiers with a high saturation power can be fabricated from the same material, ensuring good spectral overlap of the amplifier gain spectrum with the emission spectrum of the laser, as will be discussed below.

Another property of ML QD lasers which may be very useful for applications in non-linear spectroscopy is the possibility of hysteresis/bistability in mode-locking operation. Such bistable behaviour in the laser dynamics under variation of the absorber bias was demonstrated in two-section QD lasers with and without mode-locking and can be explained by a combination of electric field dependence (Quantum Confined Stark effect) of absorption and the saturation properties in gain and absorber [147,166,172–174]. Recently, it has been reported in two-section passively ML QD lasers with a 7.9 GHz pulse repetition rate, using a device with an active region which comprised of tenfold stack of InGaAs QDs layers. Bistability between 1,166 nm and 1,174 nm was achieved with power suppression ratios of 30 dB. However, the level of output power achieved was rather low, ranging between 0.2–0.8 mW for an operating temperature of 12 °C [172,173]. We demonstrated robust, high power wavelength-bistability between 1,245 nm and 1,295 nm attributing to two different modes for monolithic two-section laser operating at 10 GHz frequency with average powers up to 25 mW for CW and 17 mW for ML regimes. A suppression ratio higher than 40 dB was demonstrated [100,175].

In addition to bistability, QD lasers are very conducive to *tunable* mode-locked operation. Currently available QD growth technology enables a high degree of control over the emission spectrum of QD devices, which can be tailored for different applications, such as a broadly tunable laser [176–180]. For instance, chirped multiple QD external-cavity diode lasers demonstrated impressive 208 nm tunability with near 200 mW maximum output power [181]. Monolithic multi-section devices have only shown continuous sweeping region between 1,029.1 nm to 1,017.4 nm (11.7 nm) and extending to 1,004.3 nm (with a total 24.8 nm sweep range) by applying different current to various sections of the device [180].

Below, a further improvement of this new approach is presented in order to achieve wavelength bistability and tunability in a monolithic multi-section QD laser, by using an active region which incorporates non-identical InAs QD layers which emit at distinct wavelengths. The widest range of wavelength bistability of 54 nm - the largest spectral range reported so far, controllable saturable absorber bias has been demonstrated. Wide wavelength tunability between 1,245 nm and 1,290 nm under ML regime with a 40 dB suppression ratio is also demonstrated. Depending on the range of reverse bias applied, CW or ML regimes are obtained, with average powers up to 31 and 28 mW, respectively. This output power performance represents a further significant improvement when compared with previous results [172,173]. Moreover, electronically-controlled 45 nm tunability range is demonstrated from a monolithic multi-section ML QD laser, allowing for the generation of picosecond pulses electronically tunable between 1,245 nm and 1,290 nm, with a pulse repetition rate of around 10 GHz [182]. This represents a completely new regime of operation of ML laser diodes, which significantly enhances their spectral versatility, while offering the potential for high-speed electronic tuning. This regime however requires a customised laser design.

Namely, we used a multi-section laser (Figure 3.23) with a ridge waveguide width of 6 μm and a total length of 4 mm, giving a pulse repetition rate (fundamental round0-trip frequency) of 10 GHz when ML. The device consists of four 1-mm-long electrically insulated sections; each of these further divided into 300 μm and 700 sub-sections. A reverse bias is applied to the two 300 μm sections placed nearer the back facet, thus forming a distributed saturable absorber with a total length of 600 μm. The gain section is formed by the remaining sections which are forward biased. The output facet (on the gain section side) was strongly antireflection coated, while the back facet (on the absorber side) was high-reflection coated, with reflectivities of approximately 0.1% and 95%, respectively. The QD structure was grown on a GaAs substrate by molecular beam epitaxy. Its active region consists of ten InAs QD layers covered by non-identical InGaAs capping layers, incorporated into $Al_{0.35}Ga_{0.65}As$ cladding layers. The size of the QDs is

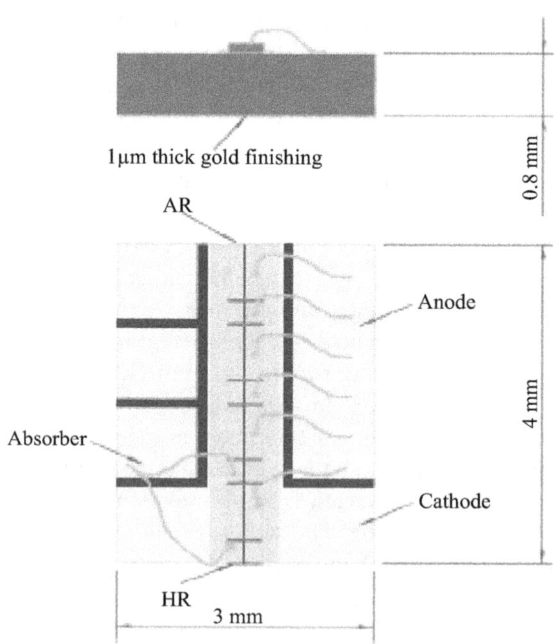

Figure 3.23 Schematic of multi-section QD laser

related to the wavelength emission and can be controlled to some extent by manipulating the thickness of capping layers which leads to variance of the indium segregation into the QDs. As a result, the larger size of QDs, the longer emission wavelength. This structure consists of 3 QD layers with central wavelength emission at 1,211 nm, 3 QD layers at 1,243 nm and finally 4 QD layers at 1,285 nm. The higher number of layers for larger QDs is used to keep the gain spectrum flat as the density of dots is decreased with the increasing QD size as explained in Ref. [176]. The laser was kept at 20 °C by a Peltier cooler. The gain section was pumped with a low-noise current source and the absorber section was connected to a voltage source. The pulse durations were measured by a non-collinear autocorrelator based on second-harmonic generation. The spectral characteristics were measured by a spectrometer and ML performance was further investigated with an RF spectrum analyser in combination with a high-speed 29 GHz photodiode.

Wavelength bistability was observed for fixed currents of 260 mA, 300 mA, and 330 mA applied to the gain section and the reverse bias varied between 0 V and 10 V, in both ascending and descending directions. The threshold current density was 370 A/cm^2. For a fixed current of 260 mA as shown in Figure 3.24(a), the widest spectral separation of 54 nm was obtained between 1,296 and 1,242 nm

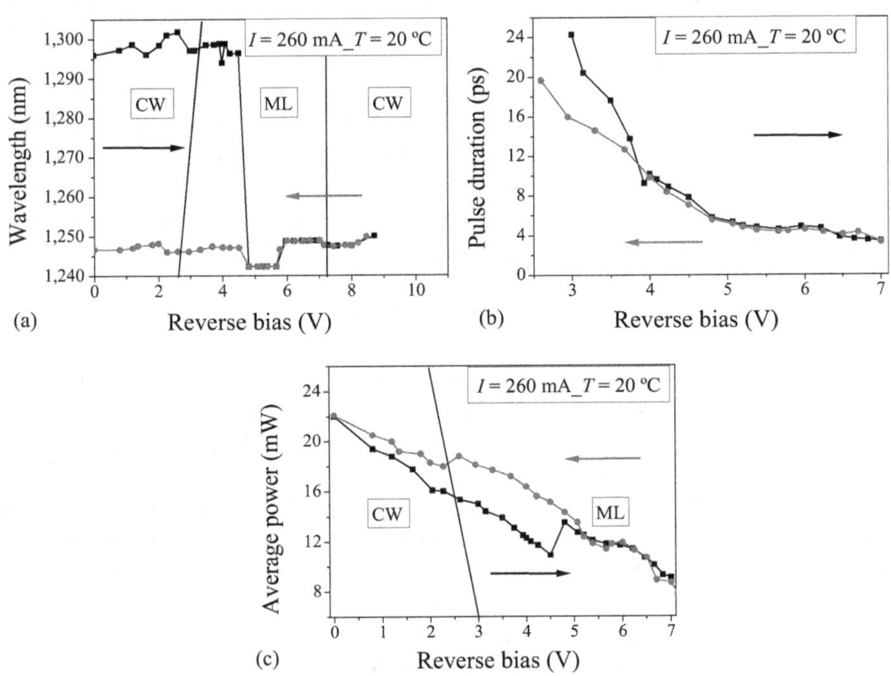

Figure 3.24 (a) Dynamics of the emission wavelength with various values of ascending and descending reverse bias; (b) pulse duration; and (c) output power in the ascending and descending directions for 260 mA fixed gain current

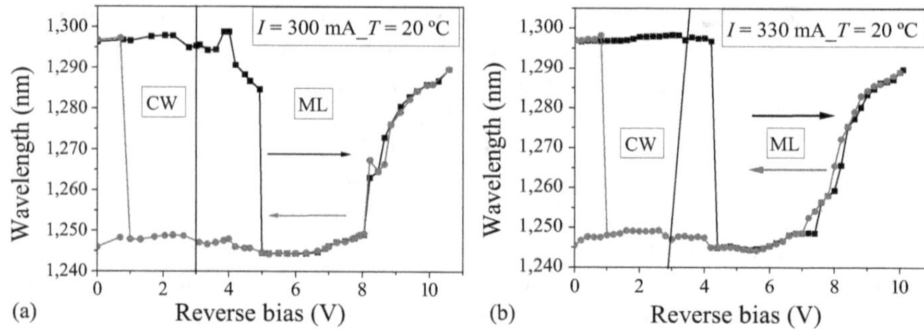

Figure 3.25 Dynamics of the emission wavelength with various values of ascending (black line, squares) and descending (grey line, circles) reverse bias for (a) 300 mA and (b) 330 mA fixed gain current in CW and ML regimes of operation

at −4.6 V reverse bias. The pulse duration changed from 24 ps to 4 ps with reverse bias changing from −2.6 V to −7 V (Figures 3.24 and 3.25). At a constant gain section current of, say, 260 mA, the pulse duration depends on direction of applying voltage which is explained by different operating wavelengths (Figure 3.24(b)). Shorter pulse durations were measured for the 1,245 nm wavelength. A transition in the laser operation mode – between CW and ML – was found for a reverse bias of ~3 V for the ascending voltage variation and of 2.6 V for the descending variation for fixed current of 260 mA. ML wavelength switching was observed at ~4.8 V. The output power hysteresis occurs between the reverse bias values of 0 and 6 V shown in Figure 3.24(c) and may be attributed to the interplay of non-linear saturation of the QD absorption and quantum confined Stark effect in the absorber [147,172–174].

When applying a higher fixed current of 300 and 330 mA, wavelength switching was observed, similar to that observed at 260 mA. At higher reverse bias values, gradually tunable mode-locking frequency shift was obtained (Figure 3.25(a) and (b)). The region of ML in descending direction for 330 mA current is slightly wider then ascending direction as non-linear absorption of the absorber plays role again.

The pulse duration varied from 28 ps to 4 ps with a variation in the absorber voltage from −3 V to −10.6 V (Figure 3.26(a) and (b)). Pulse duration is slightly increased as current increases which is a known effect due to a combination in the increase in self-phase modulation and gain compression, see, for example [183]. The pulse duration (at low reverse bias values) at the gain current of 300 mA is shorter for $\lambda \approx 1{,}285$ nm than $\lambda \approx 1{,}245$ nm, while for higher gain current of 330 mA, the pulse duration is about the same for both wavelengths. Noticeable hysteresis of average power is demonstrated in Figure 3.27(a) and (b) for fixed current of 300 and 330 mA with applied reverse bias associated with switching the laser operation from one wavelength to another. The highest output powers measured for 300 mA fixed current were 31 mW and 28 mW for CW and ML regimes,

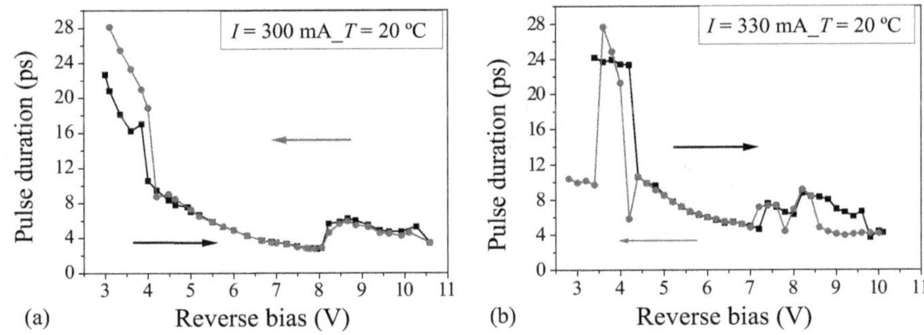

Figure 3.26 Dynamics of the pulse duration with ascending (black line, squares) and descending (grey line, circles) direction of applied reverse bias for (a) 300 mA and (b) 330 mA

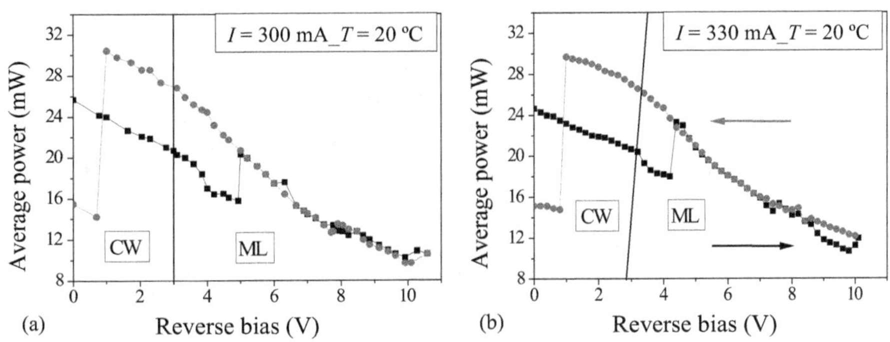

Figure 3.27 Dynamics of the pulse duration with ascending (black line, squares) and descending (grey line, circles) direction of applied reverse bias for (a) 300 mA and (b) 330 mA

respectively. Wavelength tunability region between −7 and −10.6 V is observed for both current values shown.

Within the ML region at high absorber bias, the wavelength can be tuned continuously from 1,245 to 1,290 nm by bias variation. This is likely to be an essentially linear effect, to do with the red shift of the absorption peak of the QD with increasing reverse bias [174,184]. Indeed it was shown that absorption spectra shift to lower photon energies with increasing electric field up to 21% at 1.32 μm at 18 Volt bias for 3 QD layers InAs waveguides [184]. Such changes would favour ML operation towards increasingly longer wavelengths, within the available broadband gain.

To characterise the tunable mode-locking further, Figure 3.28 shows optical spectra (a) along with autocorrelations and rf spectra (b, c). A suppression ratio higher than 40 dB suggests the effectiveness of using the novel QD material for building a tunable electrically controlled optical source (Figure 3.29).

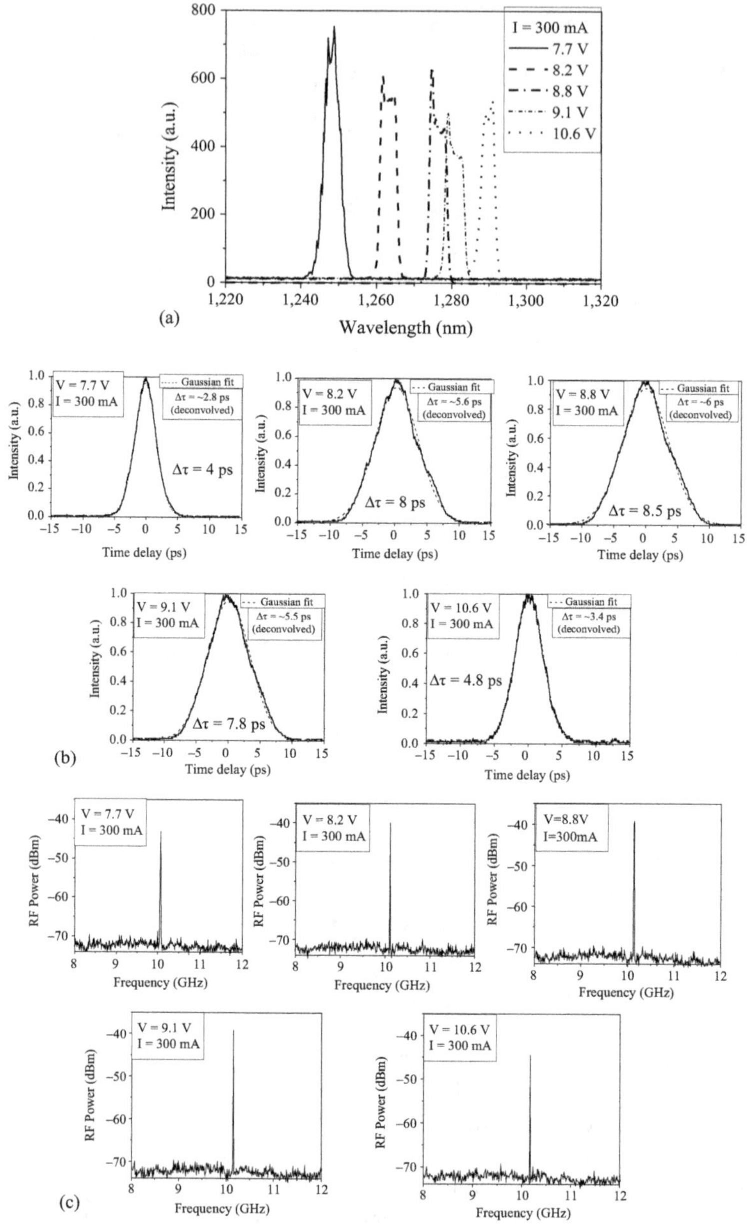

Figure 3.28 (a) ML spectra under wavelength tuning; (b) corresponding autocorrelations; and (c) RF spectra in the ascending direction for a fixed gain section current of 300 mA with applied reverse bias

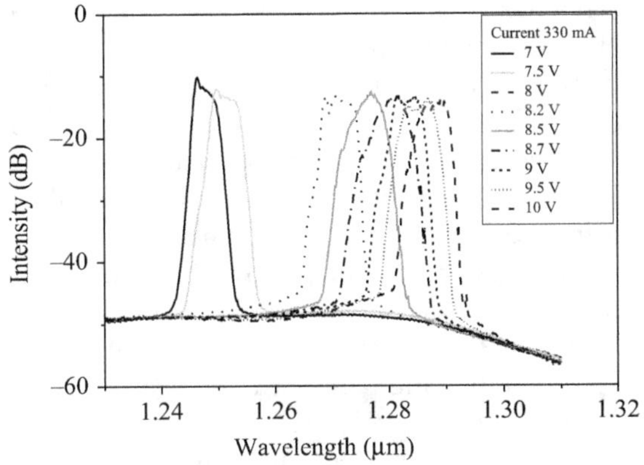

Figure 3.29 Spectral tunability with high suppression ratio more than 40 dB in the descending direction for a fixed current of 330 mA with applied reverse bias

3.5 Short-pulse semiconductor lasers in a master oscillator-power amplifier configuration

As discussed above, monolithic ML semiconductor (particularly quantum dot) lasers can generate stable, nearly transform limited, and if necessary widely tunable picosecond pulses; however, the power levels attained, at least so far, by monolithic diode laser designs are a bit too low to be directly suitable for non-linear biophotonics applications. Therefore, composite laser designs including a pulsed laser (mode-locked or gain-switched) followed by a CW biased amplifier in a discrete master oscillator-power amplifier (MOPA) configurations have attracted considerable attention. Some of these designs, using quantum well semiconductor lasers, have been mentioned in the introductory section [35–38].

Since such laser systems are normally intended for non-linear operation, the major parameter used to characterise them, alongside the peak power, is the calculated as:

$$\mathrm{FOM} = \bar{P}^2 f_{\mathrm{psf}}^2 / \left(F_{\mathrm{rep}} \tau_{\mathrm{p}}\right) \tag{3.8}$$

where $\bar{P}$ is the average power, $f_{\mathrm{psf}} < \sim 1$ is the pulse shape factor needed to recalculate the average power into the peak power, F_{rep} is the ML repetition frequency, and τ_{p} is the pulse duration, usually obtained from deconvolution of an autocorrelation signal. If using the laser-amplifier (sub)system for a non-linear biomedical application, the non-linear signal as described by (3.2) is thus directly proportional to the FOM as described by (3.8).

In the recent years, all-semiconductor MOPA pulsed laser systems using all three major methods of short pulse generation (gain-switching, Q-switching, ML) have been proposed and implemented.

Such systems have demonstrated the peak power and FOM of 100 W and 27 W^2, respectively, at a wavelength of 800 nm [37], 366 W (311 W^2) at 922 nm [185], later extended to 661 W [186], 1.4 kW (271 W^2) at 975 nm [187], 2.5 kW (1,281 W^2) at 830 nm [188] and 6.5 kW (5,233 W^2) at 850 nm [189]. All these systems used an external-cavity configuration and employed, except the one reported in [37], pulse post-compression techniques which allowed the energy of the pulse to be concentrated in sub-picosecond pulses.

Without pulse compression, the peak power and hence the FOM, are by necessity somewhat lower. At wavelengths above 1 μm, 35-ps short pulses with a peak power of 65 W at 1.063 μm have been reported using a gain-switched three-section laser with post-amplification by a two-section SOA with a flared gain section pumped by current pulses, which resulted in a FOM of 0.15 W^2 [190]. A comparable peak power of 50 W at 1.066 μm a and pulse width of 80 ps, corresponding to a FOM of 0.25 W^2, has been reported for active Q-switching by electric modulation of the absorber section of a three-section laser [191]. A peak power of 2.7 W at 1.55 μm and a pulse width of 100 ps leading to a FOM of 0.73 W^2 has been reported in an experiment using a tapered laser with a distributed-feedback (DFB) master oscillator [192]. An increased FOM of 3.12 W^2 has been achieved at 1.06 μm with a peak power of 6.3 W and 74 ps short pulses by gain-switching of a three section laser consisting of a straight DBR section, a straight absorber section and a flared gain section [193]. By gain-switching a single-section straight DFB laser followed by an SOA consisting of a straight section and a flared one, 23-W peak power at 1.063 μm with pulse widths of 40 ps and a FOM of 3.68 W^2 has been realised [194].

Below, we shall consider in most details the latest quantum dot implementation of the master laser-power amplifier type picosecond-pulse emitter [195] which has provided the record performance so far at the wavelength around 1.3 μm, the promising spectral range for deep tissue penetration as discussed in Section 3.1.

The arrangement involved was relatively straightforward (Figure 3.30); possibly the most nontrivial part of the design was the use of tapered waveguide devices for both the laser and the amplifier, both of which had been carefully optimised for high-power short-pulse generation and amplification. The purpose of using the tapered design was different for the laser and amplifier. In the case of the amplifier, the tapered waveguide is well known to lead to a significant increase in the saturation power P_{sat} of the amplifier, which in the MOPA construction *almost entirely* determines the output power achievable – the power at the amplifier input, provided that it reaches a certain minimum value ($P_{min} \sim P_{sat}/G_{sm}$, G_{sm} being the absolute small-signal gain of the amplifier), is almost irrelevant. The tapered laser structure giving high pulsed and average power is still of some use though, as it provides a certain power reserve and allows very broad tolerance to the coupling of light from the laser into the amplifier – in fact, before focusing the laser output into the SOA, the average output power of the laser was attenuated to 6 mW to avoid early gain saturation of the SOA and thus optimise the output.

Both active regions of the laser and amplifier consisted of a 10 layers of InAs/InGaAs QDs. The light field in both of the devices was gain-guided with an

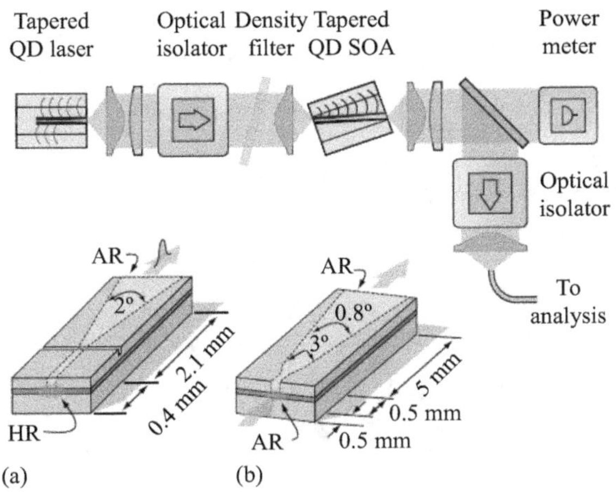

Figure 3.30 Schematic of the master oscillator-power amplifier pulse generator and the geometries of (a) tapered quantum dot laser and (b) tapered quantum dot amplifier

additional weak index-guiding of the light field in the SOA for a better lateral confinement. The laser used was the one studied previously in [99], with a straight saturable absorber 400 μm long and a tapered 2.1 mm long gain section. For the amplification measurements, the laser was operated at a current of 649 mA and at a reverse bias of 6 V which was comfortably within the range of stable mode-locking, without either Q-switched ML (trailing-edge instability) or leading-edge modulation instabilities, and also corresponded to relatively short pulses (with the deconvoluted pulse width of 1.2 ps assuming a Gaussian pulse shape).

In interpreting the results of the measurements, special care was taken to eliminate the amplified spontaneous emission power and evaluate only the power of the amplified ML laser light. The measured amplified pulse parameters are shown in Figure 3.31(a) (TBP and pulse width) and Figure 3.31(b) (peak power and FOM). For calculating the latter, the values $f_{psf} \approx 0.94$ and $F_{rep} \approx 16$ GHz were used.

The minimum TBP of 0.66 at 6.2 A indicated that, although the non-linear amplification in the SOA introduced some self-phase modulation (seen also in the somewhat asymmetric spectra of the amplified pulses), the chirp introduced was relatively weak. In general, the measured TBP correlated with the pulse width. The minimum pulse width amounted to 1.33 ps at 6.2 A, which was only marginally greater than the pulsewidth of the ‘source’ pulses (1.2 ps). The estimated peak power displayed a maximum value of 42 W with a pulse width of 1.37 ps for the highest applied SOA gain current of 7 A. Even though the amplified pulse widths are larger than the pulses of the laser, there was no tendency for the pulse duration to increase with the amplifier current/gain (Figure 3.31(a)) thus giving the potential to a steady peak power increase by increasing the amplification current. Due to such high peak powers and the high repetition rates, the achieved FOM, which is depicted in Figure 3.31(b), reached a value of 38.5 W^2 well exceeding state-of-the-art results for

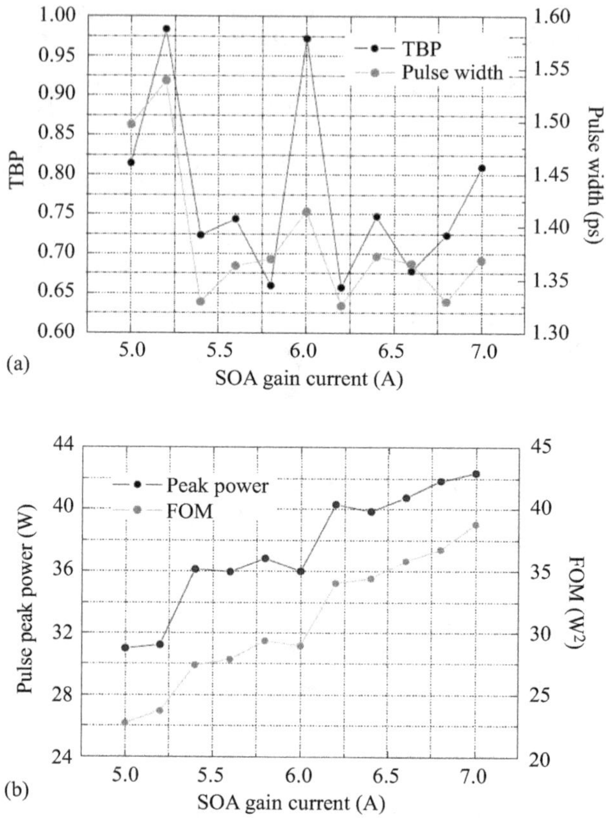

Figure 3.31 Parameters of an amplified pulse as functions of the amplifier current

other ML lasers with post-amplification and without pulse post-compression at wavelengths >1 μm.

The relatively low TBP of 0.80 at the highest applied SOA current of 7 A obviate the use of any postcompression, because a typical dual-grating-based compressor exhibits transmission losses of at least 50%. The scheme was thus very compact, and there may be a possibility of making it even more compact and robust if the laser and amplifier can be integrated monolithically.

3.6 New mode-locking regimes in and for biophotonic applications

All the discussion of mode-locking presented in the previous sections concerned passive mode-locking with a saturable absorber, the fundamental principles of which have been known since the 1960s and 1970s (though the important details pertaining to semiconductor, particularly QD, lasers continue to be a subject of investigation up to now). Recently, more unusual regimes of laser dynamics, similar to mode-locking, have been observed *specifically* in biomedical systems, more precisely in swept-frequency Optical Coherent Tomography (OCT). It involves using a laser whose frequency is swept rapidly, at about 10–100 kHz rate, across a wide wavelength range, about 100 nm, to measure the spectrum of interferences between tissue reflections and a reference reflector, which in turn is used to create a subsurface image of the tissue. The requirements to the laser source for such an operation are as follows: (a) wide tuning range ($\sim$100 nm) for a high ($<$15 μm) depth resolution of imaging; (b) fast tuning (100 kHz) for fast data acquisition (combined with the wide tuning range, this determines the speed of frequency sweeping of up to $\sim$ 1 GHz/ns); and, finally, (c) a narrow dynamic linewidth ($<$ 13 GHz) for a large imaging depth capability (although too narrow a linewidth is to be avoided too, in order to avoid artefacts). Thus, swept-source OCT typically uses external cavity laser diodes frequency-swept by means of a tunable intracavity filter. Originally, ring lasers including a semiconductor amplifier in a long ($\sim$ 1m) fibre ring cavity were used for this purpose [196]. More recently, the development of highly frequency selective, compact reflecting filters [197] has allowed a simpler, more compact Fabry–Pérot construction to be used.

Since such a construction is relatively short (1–10 cm in length), the laser is frequency-swept through its intermodal frequency interval in only a few round-trips and thus operates in a perpetually transient regime. Depending on the combination of the filter bandwidth, the cavity intermodal interval and the sweeping speed, either near-single-mode hopping or quasi-continuous envelope tuning are possible. In particular, during quasi-continuous sweeping from higher to lower optical frequencies (shorter to longer wavelengths), or 'red sweeping' for brevity, with the tuning rate above a certain minimum value, a dynamic regime is observed known as *sliding-frequency mode-locking (SFML)*. During this regime, the laser radiation forms a set of pulses at the repetition rate equal to the laser cavity round-trip frequency or its harmonic, approximately preserving their shape, duration, and peak power from pulse to pulse (as in ordinary mode-locking), but with the central frequency slightly adjusted from pulse to pulse to follow the swept filter frequency. Figure 3.32 shows a profile of these pulses and the corresponding single-pulse spectra simulated with a travelling wave dynamic model [198] (the corresponding experimental measurement can be complicated due to the transient nature of the regime).

SFML is observed in a laser construction that does not include a saturable absorber, nor is any modulation at the round-trip time present, so the mode-locking regime is neither active nor the traditional passive type, but is a version of self-mode-locking; it comes about by the mode interaction through population pulsations and four-wave mixing in the laser. In this, it is very likely to be related to self-mode-locking observed in certain types of nanostructure lasers (see, e.g., [199] and an overview in [41]). However, SFML is observed in lasers that would *not* show the self-mode-locking in the absence of tuning. The SFML behaviour is partly mediated by the fact that the modal interaction in a semiconductor lasers is asymmetric due to self-phase modulation, in that the weak mode on the long-wavelength side of a strong mode acquires additional positive gain due to non-linear modal interaction via population

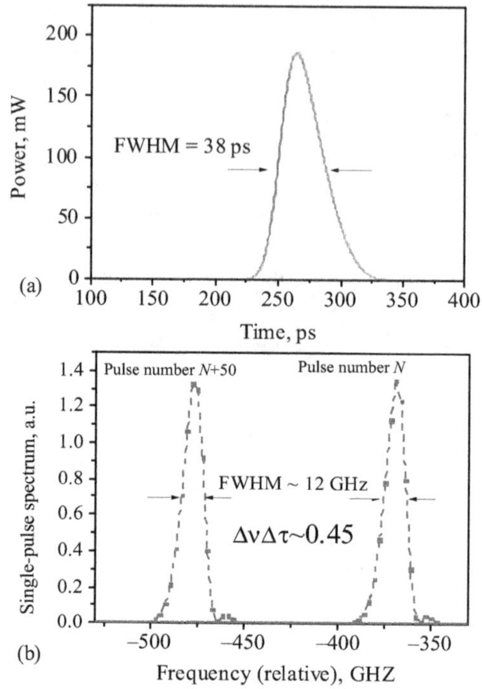

Figure 3.32 Eye diagram (100 pulses) (a) and spectra of two pulses in the stream and (b) for red tuning with $\delta\nu_{filter}$ = 20 GHz and a tuning speed of −6 GHz/ns [198]

pulsations (Bogatov effect). In the process of the 'red sweeping', new modes are being constantly excited on the long-wavelength side of the currently dominant mode; the non-linear modal interaction thus facilitates their emergence in the spectrum, with their phase locked to that of the dominant mode, hence mode-locking.

For the same reason, SFML is not observed under the tuning in the opposite direction ('blue sweeping', from longer to shorter wavelengths, when each new mode that emerges in inhibited, rather than facilitated, by non-linear interaction [198,200]). Perhaps more importantly, the same tuning direction asymmetry is also seen in the power output: the average power under red tuning is greater than under blue tuning. This asymmetry is detrimental to the system operation. It has been shown [198], that the use of a structure with a saturable absorber, combining SFML with traditional passive mode-locking, somewhat reduces this asymmetry; there has also been some suggestions [196] that the use of QD active materials can be useful in that respect as well. Such a QD structure has indeed been demonstrated [189] with the linewidth suitable for OCT applications.

By necessity, SF ML can only be observed over a finite length of time until the laser is tuned out of its gain bandwidth and switches off; therefore it is not a truly stationary process unlike traditional ML.

It also has to be noted also that there is a trade-off between the available tuning speed (which benefits from a broader filter bandwidth) and the instantaneous linewidth (which is, naturally, limited by the filter bandwidth [198,200]). Despite this limitation, systems showing SFML have shown promising performance: spectral tuning ranges of about 120 nm at the wavelengths of both 1,060 and 1,300 nm have been achieved with such systems, enabling OCT with a scan rate of 100 KHz, and a resolution of 8.3 μm and 15 μm in air at 1,060 and 1,300 nm, respectively [197].

The alternative approach to swept-source OCT is *Fourier Domain* mode-locking (FDML) first proposed in [201]. Unlike SFML, FDML does involve periodic modulation of laser parameters at the round-trip frequency, but unlike traditional active mode-locking, it is not the amplitude of laser radiation that is modulated but the spectral properties. The design of a typical FDML laser is, broadly speaking, similar to that of the external ring cavity design in which SFML-like behaviour was first observed [196]. It involves an SOA chip, placed into an external cavity, containing also a fast tunable filter. The filter then scans over the tuning range of the laser at the rate equal, or very close, to the cavity round-trip frequency or one of its harmonics, ensuring that the entire laser spectrum is excited. In order to achieve this resonance between the round-trip and the scanning rate of several kHz, corresponding to a round-trip time in a range of fractions of a *millisecond*, the fibre needs to be a several orders of magnitude longer than one used in the SFML design described in [196], with the length in the range of kilometres rather than metres. In the original demonstration of the principle [201], a fibre cavity with an optical length of 10 km was swept at the fourth harmonic of the round-trip frequency of 29 KHz. Effective sweep rates of up to 290 kHz were demonstrated with a 105 nm tuning range at 1,300 nm centre wavelength. The average output power was 3 mW directly from the laser and 20 mW after post-amplification, which allowed sensitivities of 108 dB to be achieved in an OCT system. The principle, according to the inventors of this regime of operation, should be scalable to a broader range of frequencies with shorter cavities [201]. Indeed in later studies (see, e.g., [202]), the principle was extended to megahertz frequencies.

The properties of the OCT lasers, including the nature of the dynamic regimes and the dynamic linewidth, are highly nontrivial and have been extensively investigated both experimentally and using theoretical studies and numerical methods based on the traditional Haus-type mode-locking model [181,203] and, later, on the more modern large-signal delay-differential model [204,205].

Studies reveal that, in the case of exactly resonant modulation, when the scan rate *exactly* equals the round-trip frequency of the cavity, the tuning in the FDML laser (unlike the SFML one) can be expected to be symmetric, with almost identical dynamics seen both under blue and red tuning. With a finite detuning between the sweeping frequency (rate) and the round-trip, however, the FDML lasers do show an asymmetry between the red and blue frequency sweep direction. Theoretical studies using the delay-differential model [204,205] reveal that the origin of this, as in the case of SFML, is the asymmetric interaction of longitudinal modes of a laser resonator due to self-phase modulation. Experiments and simulations presented in [204,205] also show that, as in the case of SFML, the difference between red and blue tuning in FDML is not just quantitative but qualitative. Namely when the sweeping frequency is somewhat greater than the cavity round-trip frequency, the red sweeping results in a series of discrete frequency jumps while the blue sweeping results in a complex, irregular (chaotic) oscillating output. When the sweeping frequency is somewhat smaller than less than that of cavity round trip, the asymmetry is reversed in a very symmetric fashion. The asymmetric dynamics naturally leads to an asymmetry of the instantaneous spectrum: the mode-switching process has a narrower linewidth than the chaotic oscillation; therefore typically only one direction of sweeping is used for information acquisition in FDML-based OCT The advantage of the FDML laser is however that for this one sweep direction used, the dynamic linewidth of the laser emission can be significantly smaller than the filter bandwidth, since the very large passive resonator involved in FDML provides a high-quality element making for narrow-bandwidth operation. Theoretical calculations [181] predict instantaneous linewidths of the order of a few GHz. Experimental measurement of instantaneous linewidth is not trivial because of the permanently transient nature of the regime, as in SFML. In the original experiments [201], the dynamic linewidths were narrow enough to enable imaging over a 7 mm depth with only a 7.5 dB decrease in sensitivity, implying linewidths of a few GHz. A more direct measurement method, using fast synchronous time gating with an electro-optic modulator, extracting a short temporal fragment ('snapshot') of laser emission, whose spectrum can subsequently be measured with an optical spectrum analyser, has been proposed in [206] giving values of the order of units of GHz. In [181], the measured linewidths and lineshapes were compared with theory, with very good agreement. The lineshape was found to be asymmetric, with a high frequency tail; unlike the case of SFML or ordinary mode-locking, the authors of [181] attributed this spectrum asymmetry to high-order dispersion rather than self-phase modulation. There is no definite trade-off between the tuning speed and the bandwidth in the FDML design, which is one of the reasons it is currently believed to be the best performing laser system for swept-source OCT, though the more compact SFML systems find some use as well.

3.7 Conclusions

We have attempted an overview of fast-pulsating lasers (gain-switched, Q-switched and mode-locked) in various biomedical applications. By necessity, the accents in the chapter have been to an extent coloured by our research experience and interests; apologies are extended to those authors whose work may have been omitted. Different properties of a laser system are important for different applications (e.g., high instantaneous and average power for multiphoton spectroscopy vs. spectral tuning range for OCT), but all the designs above benefit from the properties of semiconductor lasers such as chip compactness, ease of pumping, and high gain per unit length. The use of nanostructures, particularly QDs, has been shown to open new exciting opportunities.

The authors are grateful to Dr. D. Nikitichev for the use of some of the results from his PhD thesis.

References

[1] B. Wardle, *Principles and Applications of Photochemistry*: John Wiley and Sons, Ltd, 2009.

[2] M. S. Brennesholtz and E. H. Stupp, *Projection Displays*, 2nd ed.: John Wiley and Sons, Ltd., 2008.

[3] W. Demtröder, *Laser spectroscopy: basic concepts and instrumentation*, 3rd ed. New York: Springer-Verlag Berlin Heidelberg, 2003.

[4] H. H. Telle, A. G. Ureña, and R. J. Donovan, *Laser Chemistry: Spectroscopy, Dynamics and Applications*. Chichester: John Wiley and Sons Ltd, 2007.

[5] A. Horn, *Ultra-fast Material Metrology*. Darmstadt: Wiley-VCH Verlag GmbH & Co. KGaA, Weinheim, 2009.

[6] K. Curtis, L. Dhar, A. Hill, W. Wilson, and M. Ayres, *Holographic Data Storage: From Theory to Practical Systems*. Chichester: John Wiley & Sons Ltd, 2010.

[7] W. M. Steen and J. Mazumder, *Laser Material Processing*, Fourth ed. New York: Springer-Verlag London Limited, 2010.

[8] H. J. Metcalf and P. Van der Straten, *Laser cooling and Trapping*. New York: Springer-Verlag New York, Inc, 1999.

[9] P. J. Campagnola and L. M. Loew, "Second-harmonic imaging microscopy for visualizing biomolecular arrays in cells, tissues and organisms," *Nat Biotech*, vol. 21, pp. 1356–1360, 2003.

[10] U. K. Tirlapur and K. Konig, "Cell biology: Targeted transfection by femtosecond laser," *Nature*, vol. 418, pp. 290–291, 2002.

[11] D. Day, C. G. Cranfield, and M. Gu, "High-speed fluorescence imaging and intensity profiling of femtosecond-induced calcium transients," *International Journal of Biomedical Imaging*, vol. 2006, pp. 1–6, 2006.

[12] G. A. Keeler, B. E. Nelson, D. Agarwal, *et al.*, "The benefits of ultrashort optical pulses in optically interconnected systems," *Selected Topics in IEEE J. Quantum Electron.*, vol. 9, pp. 477–485, 2003.

[13] S. Arahira, H. Takahashi, K. Nakamura, H. Yaegashi, and Y. Ogawa, "Polarization-, wavelength-, and filter-free all-optical clock recovery in a passively mode-locked laser diode with orthogonally pumped polarization-diversity configuration," *IEEE J. Quantum Electron.*, vol. 45, pp. 476–487, 2009.

[14] D. E. Spence, P. N. Kean, and W. Sibbett, "60-fsec pulse generation from a self-mode-locked Ti:sapphire laser," *Opt. Lett.*, vol. 16, pp. 42–44, 1991.

[15] U. Keller, D. A. B. Miller, G. D. Boyd, T. H. Chiu, J. F. Ferguson, and M. T. Asom, "Solid-state low-loss intracavity saturable absorber for Nd:YLF lasers: an antiresonant semiconductor Fabry-Perot saturable absorber," *Opt. Lett.*, vol. 17, pp. 505–507, 1992.

[16] E. Innerhofer, T. Südmeyer, F. Brunner, *et al.*, "60-W average power in 810-fs pulses from a thin-disk Yb:YAG laser," *Opt. Lett.*, vol. 28, pp. 367–369, 2003.

[17] R. Ell, U. Morgner, F. X. Kärtner, *et al.*, "Generation of 5-fs pulses and octave-spanning spectra directly from a Ti:sapphire laser," *Opt. Lett.*, vol. 26, pp. 373–375, 2001.

[18] C. T. A. Brown, M. A. Cataluna, A. A. Lagatsky, *et al.*, "Compact laser-diode-based femtosecond sources," *New J. Phys.*, vol. 6, 2004.

[19] M. Gu, D. Bird, D. Day, L. Fu, and D. Morrish, *et al.*, *Femtosecond Biophotonics: Core Technology and Applications*. Cambridge: Cambridge University Press, 2010.

[20] A. Zumbusch, G. R. Holtom, and X. S. Xie, "Three-dimensional vibrational imaging by coherent anti-Stokes Raman scattering," *Phys. Rev. Lett.*, vol. 82, p. 4142, 1999.

[21] A. Zoumi, A. T. Yeh, and B. J. Tromberg, "Imaging cells and extracellular matrix in vivo by using second-harmonic generation and two-photon excited fluorescence," *Proceedings of the National Academy of Sciences*, vol. 99, pp. 11014–11019, August 20, 2002.

[22] K. König, "Multiphoton microscopy in life sciences," *J. Microsc.*, vol. 200, pp. 83–104, 2000.

[23] L. Moreaux, O. Sandre, S. Charpak, M. Blanchard-Desce, and J. Mertz, "Coherent scattering in multi-harmonic light microscopy," *Biophys. J.*, vol. 80, pp. 1568–1574, 2001.

[24] W. R. Zipfel, R. M. Williams, and W. W. Webb, "Nonlinear magic: multiphoton microscopy in the biosciences," *Nat Biotech*, vol. 21, pp. 1369–1377, 2003.

[25] X. Deng and M. Gu, "Penetration depth of single-, two-, and three-photon fluorescence microscopic imaging through human cortex structures: Monte Carlo simulation," *Appl. Opt.*, vol. 42, pp. 3321–3329, 2003.

[26] M. Gu and X. S. Gan, "Effect of the detector size and the fluorescence wavelength on the resolution of three- and two-photon confocal microscopy," *Bioimaging*, vol. 4, pp. 129–137, 1996.

[27] M. Gu, "Resolution in three-photon fluorescence scanning microscopy," *Opt. Lett.*, vol. 21, pp. 988–990, 1996.

[28] C. J. R. Sheppard and M. Gu, "Image formation in two-photon fluorescence microscopy," *Optik*, vol. 86, pp. 104–106, 1990.

[29] C. Sheppard, J. Gannaway, and R. Kompfner, "The scanning harmonic optical microscope," *IEEE J. Quantum Electron.*, vol. 13, pp. 912–912, 1977.

[30] C. J. R. Sheppard and R. Kompfner, "Resonant scanning optical microscope," *Appl. Opt.*, vol. 17, pp. 2879–2882, 1978.

[31] W. Denk, J. H. Strickler, and W. W. Webb, "Two-photon laser scanning fluorescence microscopy," *Science*, vol. 248, pp. 73–76, 1990.

[32] G. S. He, J. D. Bhawalkar, P. N. Prasad, and B. A. Reinhardt, "Three-photon-absorption-induced fluorescence and optical limiting effects in an organic compound," *Opt. Lett.*, vol. 20, pp. 1524–1526, 1995.

[33] S. Hell, "Three-photon excitation in fluorescence microscopy," *J. Biomed. Opt.*, vol. 1, p. 71, 1996.

[34] H. Yokoyama, H. Guo, T. Yoda, *et al.*, "Two-photon bioimaging with picosecond optical pulses from a semiconductor laser," *Opt. Express*, vol. 14, pp. 3467–3471, 2006.

[35] K. Kim, S. Lee, and P. J. Delfyett, "1.4 kW high peak power generation from an all semiconductor mode-locked master oscillator power amplifier system based on eXtreme Chirped Pulse Amplification (X-CPA)," *Opt. Express*, vol. 13, pp. 4600–4606, 2005.

[36] T. Schlauch, M. Li, M. R. Hofmann, A. Klehr, G. Erbert, and G. Trankle, "High peak power femtosecond pulses from modelocked semiconductor laser in external cavity," *Electron. Lett.*, vol. 44, pp. 678–679, 2008.

[37] M. Kuramoto, N. Kitajima, H. Guo, Y. Furushima, M. Ikeda, and H. Yokoyama, "Two-photon fluorescence bioimaging with an all-semiconductor laser picosecond pulse source," *Opt. Lett.*, vol. 32, pp. 2726–2728, 2007.

[38] D. H. Yokoyama, A. Sato, H. -C. Guo, K. Sato, M. Mure, and H. Tsubokawa, "Nonlinear-microscopy optical-pulse sources based on mode-locked semiconductor lasers," *Opt. Express*, vol. 16, pp. 17752–17758, 2008.

[39] P. Vasil'ev, I. H. White, and J. Gowar, "Fast phenomena in semiconductor lasers," *Reports on Progress in Physics*, vol. 63, p. 1997, 2000.

[40] P. Vasil'ev, *Ultrafast Laser Diodes: Fundamentals and Applications*. London: Artch House, 1995.

[41] Avrutin, E.A., and Rafailov, E.U., "Ultrafast pulse generation by semiconductor lasers," in *Semiconductor lasers, Fundamentals and Applications*, A. Baranov and E. Tournie, Eds (2013), pp. 149–160.

[42] O. Svelto, "Transient Laser Behavior," in *Principles of Lasers*, edition. New York: Springer, 2010, pp. 319–370.

[43] K. Y. Lau, "Gain switching of semiconductor injection lasers," *Appl. Phys. Lett.*, vol. 52, 1988.

[44] T. Sogawa, Y. Arakawa, M. Tanaka, and H. Sakaki, "Observation of a short optical pulse (< 1.3 ps) from a gain-switched quantum well laser," *Appl. Phys. Lett.*, vol. 53, pp. 1580–1582, 1988.

[45] AuYeung, "Picosecond optical pulse generation at gigahertz rates by direct modulation of a semiconductor laser," *Appl. Phys. Lett.*, vol. 38, pp. 308–310, 1981.

[46] N. G. Basov, V. N. Morozov, V. V. Nikitin, and A. S. Semenov, "Investigation of GaAs laser radiation pulsation," *Soviet Physics Semiconductors*, vol. 1, 1968.

[47] T. Nick, *LASER: The inventor, the Nobel laureate, and the thirty-year patent war.* New York: Simon and Schuster, 2000.

[48] Y. Arakawa, A. Larsson, J. Paslaski, and A. Yariv, "Active Q switching in a GaAs/AlGaAs multiquantum well laser with an intracavity monolithic loss modulator," *Appl. Phys. Lett.*, vol. 48, pp. 561–563, 1986.

[49] P. P. Vasil'ev and I. S. Goldobin, "New method for passive Q switching of an injection laser with an ultrashort absorber recovery time," *Sov. J. Quantum Electron.*, vol. 17, p. 835, 1987.

[50] P. Gavrilovic, N. Stelmakh, J. H. Zarrabi, and D. M. Beyea, "High energy CW Q-switched operation of multicontact semiconductor laser," *Electron. Lett.*, vol. 31, pp. 1154–1155, 1995.

[51] G. B. Venus, A. Gubenko, V. Dashevskii, *et al.*, "Multi-section ion-implantation-induced saturable absorbers for high-power 1.5-μm picosecond laser diodes," 17th IEEE International Semiconductor Laser Conference, Monterrey, CA, 24–28 September 2000, paper ThB3.

[52] I. Y. Khrushchev, D. J. Kitcher, K. A. Williams, I. H. White, F. R. Laughton, and R. V. Penty, "Picosecond Q-switched bow-tie laser diode array," *Electron. Lett.*, vol. 33, pp. 426–428, 1997.

[53] N. Stelmakh, J. Lopez, O. Smolski, and C. Roychoudhuri, "100 W 50 ps gain-switched pulses from vertical-stack laser diode," *Electron. Lett.*, vol. 36, pp. 1022–1024, 2000.

[54] B. S. Ryvkin, E. A. Avrutin, and J. Kostamovaara, "Asymmetric-waveguide laser diode for high-power optical pulse generation by gain switching," *J. Lightwave Technol.*, vol. 27, pp. 2125–2131, 2009; ISSN 2125–2131.

[55] D. V. Kuksenkov, R. V. Roussev, S. Li, W. A. Wood, and C. M. Lynn, Multiple-wavelength synthetic green laser source for speckle reduction, *Proc. SPIE*, vol. 7917, pp. 79170B–1–12, 2011.

[56] B. Lanz, B. S. Ryvkin, E. A. Avrutin, and J. Kostamovaara, "Performance improvement by a saturable absorber in gain-switched asymmetric-waveguide laser diodes," *Opt. Express*, vol. 21, pp. 29780–29791, 2013.

[57] Y. Sheng-Hui, S. Smith, J. Fitz, and C. H. Lee, "Generation of high-power picosecond pulses from a gain-switched two-section quantum-well laser with a laterally tapered energy-storing section," *Photon. Technol. Lett.*, vol. 8, pp. 337–339, 1996.

[58] B. Cakmak, "Passively Q-switched operation of a double contact tapered InGaAs/GaAs diode laser," *Microelectron. Eng.*, vol. 65, pp. 273–278, 2003.

[59] A. Klehr, B. Sumpf, K. H. Hasler, *et al.*, "High-peak-power pulse generation with GHz repetition rate using a Q-switched 1060nm DBR tapered laser," San Francisco, California, USA, 2010, pp. 76161J-12.

[60] J. Chung, N. Kikuchi, and S. Sasaki, "Generation of transform-limited optical pulses using a combination of a gain-switched diode laser and a semiconductor optical amplifier," *Photon. Technol. Lett.*, IEEE, vol. 7, pp. 860–862, 1995.

[61] P. P. Vasil'ev, I. H. White, D. Burns, and W. Sibbett, "High power, low-jitter encoded picosecond pulse generation using an RF-locked self-Q-switched multicontact GaAs/GaAlAs diode laser," *Electron. Lett.*, vol. 29, pp. 1593–1594, 1993.

[62] E. U. Rafailov, M. A. Catalunab, W. Sibbett, *et al.*, "High-power picosecond and femtosecond pulse generation from a two-section mode-locked quantum-dot laser," *Appl. Phys. Lett.*, vol. 87, 2005.

[63] A. Weiner, "Ultrafast optics", Wiley, Hoboken, NJ, 2009, 580 pages.

[64] R. Penty, "Short pulse generation using semiconductor lasers," in *Ultrafast Photonics*, P. Osborne, Ed., ed, 2002.

[65] V. V. Nikolaev and E. A. Avrutin, "Photocarrier escape time in quantum-well light-absorbing devices: Effects of electric field and well parameters," *IEEE Journal of Quantum Electronics*, vol. 39, pp. 1653–1660, 2003.

[66] A. G. Vladimirov and D. Turaev, "Model for passive mode locking in semiconductor lasers," *Phys. Rev. A*, vol. 72, Article Number: 033808, 2005.

[67] K. Y. Lau, "Short-pulse and high-frequency signal generation in semiconductor lasers," *J. Lightwave Technol.*, vol. 7, pp. 400–419, 1989.

[68] M. Marconi, J. Javaloyes, S. Balle, and M. Giudici, "How laser localized structures evolve out of passive mode-locking," *Proc. of the International Conference on Transparent Optical Networks (ICTON-2014)*, Graz, Austria, 7–10 July 2014, paper TUB2.6.

[69] E. U. Rafailov, M. A. Cataluna, and E.A. Avrutin, "Ultrafast lasers based on quantum dot structures," Wiley, NY/Berlin, 2011, 250 pages.

[70] E. A. Avrutin, and E. U. Rafailov, "Advances in mode-locked semiconductor lasers," in: J. Coleman, A.C. Bryce, C. Jagadish (Editors), "Advances in semiconductor lasers," Elsevier, San Diego, CA, Vol. 86, 2012, pp. 93–147 (Semiconductors and Semimetals series).

[71] P. J. Delfyett, L. T. Florez, N. Stoffel, *et al.*, "High-power ultrafast laser diodes," *IEEE J. Quantum Electron.*, vol. 28, pp. 2203–2219, 1992.

[72] D. J. Derickson, R. J. Helkey, A. Mar, J. R. Karin, J. G. Wasserbauer, and J. E. Bowers, "Short pulse generation using multisegment mode-locked semiconductor lasers," *IEEE J. Quantum Electron.*, vol. 28, pp. 2186–2202, 1992.

[73] D. J. Derickson, P. A. Morton, and J. E. Bowers, "Comparison of timing jitter in external and monolithic cavity mode-locked semiconductor lasers," *Appl. Phys. Lett.*, vol. 59, pp. 3372–3374, 1991.

[74] D. J. Derickson, J. G. Wasserbauer, R. J. Helkey, *et al.*, "Comparison of colliding pulse and self-colliding pulse monolithic cavity mode-locked semiconductor lasers," 1992, p. ThB3.

[75] S. Arahira, Y. Matsui, and Y. Ogawa, "Mode-locking at very high repetition rates more than terahertz in passively mode-locked distributed-Bragg-reflector laser diodes," *IEEE J. Quantum Electron.*, vol. 32, pp. 1211–1224, 1996.

[76] S. Arahira, S. Oshiba, Y. Matsui, T. Kunii, and Y. Ogawa, "Terahertz-rate optical pulse generation from a passively mode-locked semiconductor laser diode," *Opt. Lett.*, vol. 19, pp. 834–836, 1994.

[77] K. A. Williams, H. D. Summers, M. K. A. Rahman, *et al.*, "Monolithic passive mode-locking in single contact tapered lasers," in *Lasers and Electro-Optics Society Annual Meeting, 1995. 8th Annual Meeting Conference Proceedings, Volume 1, IEEE*, 1995, pp. 135–136, vol. 2.

[78] A. Mar, R. Helkey, D. B. Young, and J. E. Bowers, "High-power mode-locked semiconductor lasers using flared waveguides," *Appl. Phys. Lett.*, vol. 66, pp. 3558–3560, 1995.

[79] F. Camacho, E. A. Avrutin, A. C. Bryce, and J. H. Marsh, "Passive mode locking in semiconductor lasers with monolithically integrated passive waveguides," *IEE Proc. Optoelectron.*, vol. 145, pp. 43–46, 1998.

[80] D. A. Yanson, M. W. Street, S. D. McDougall, I. G. Thayne, J. H. Marsh, and E. A. Avrutin, "Terahertz repetition frequencies from harmonic mode-locked monolithic compound-cavity laser diodes," *Appl. Phys. Lett.*, vol. 78, pp. 3571–3573, 2001.

[81] F. Robert, A. C. Bryce, J. H. Marsh, A. J. Spring Thorpe, and J. K. White, "Passive mode locking of InAlGaAs 1.3 μm strained quantum wells extended cavity laser fabricated by quantum-well intermixing," *Photon. Technol. Lett., IEEE*, vol. 16, pp. 374–376, 2004.

[82] S. Lee, K. Kim, L. Vaissie, O. V. Smolski, E. G. Johnson, and P. J. Delfyett, "Picosecond pulse generation using a saturable absorber section of grating-coupled surface-emitting laser," *Photon. Technol. Lett., IEEE*, vol. 17, pp. 2676–2678, 2005.

[83] Y. C. Xin, A. Stintz, H. Cao, *et al.*, "Monolithic passively mode-locked lasers using quantum dot or quantum well materials grown on GaAs substrates," *Proc. of SPIE*, vol. 6468, 2007.

[84] J. Javaloyes, L. Hou, P. Stolarz, "Wide range 40-GHz passive mode-locking operation of an AlGaInAs 1.55 μm strained quantum well laser," in *Lasers and Electro-Optics* 67, *2009 and the European Quantum Electronics Conference. CLEO Europe – EQEC 2009. European Conference on*, 2009, pp. 1–1.

[85] A. C. Bryce, P. Stolarz, J. Javaloyes, "Passive mode-locking of AlGaInAs quantum well laser, modelling and experiment," in *LEOS Annual Meeting Conference Proceedings, 2009. LEOS '09. IEEE*, 2009, pp. 515–516.

[86] G. Tandoi, C. N. Ironside, and A. C. Bryce, "Nonabsorbing mirrors for quantum-well colliding pulse mode-locked lasers," *Photon. Technol. Lett., IEEE*, vol. 23, pp. 293–295, 2011.

[87] T. Shimizu, I. Ogura, and H. Yokoyama, "860 GHz rate asymmetric colliding pulse modelocked diode lasers," *Electron. Lett.*, vol. 33, pp. 1868–1869, 1997.

[88] F. R. Ahmad and F. Rana, "Passively mode-locked high-power (210 mW) semiconductor lasers at 1.55 μm wavelength," *Photon. Technol. Lett., IEEE*, vol. 20, pp. 190–192, 2008.

[89] J. J. Plant, J. T. Gopinath, B. Chann, D. J. Ripin, R. K. Huang, and P. W. Juodawlkis, "250 mW, 1.5 μm monolithic passively mode-locked slab coupled optical waveguide laser," *Opt. Lett.*, vol. 31, pp. 223–225, 2006

[90] T. Schlauch, J. C. Balzer, A. Klehr, G. Erbert, G. Tränkle, and M. R. Hofmann, "Femtosecond passively modelocked diode laser with intracavity dispersion management," *Opt. Express*, vol. 18, pp. 24316–24324, 2010.

[91] Y. C. Xin, C.-Y. Lin, Y. Li, *et al.*, "Monolithic 1.55 μm GaInNAsSb quantum well mode-locked lasers," in *Lasers and Electro-Optics, 2007. CLEO 2007. Conference on*, 2007, pp. 1–2.

[92] P. A. Morton, J. E. Bowers, L. Koszi, M. Soler, J. Lopata, and D. Wilt, "Monolithic hybrid mode locked 1.3 μm semiconductor laser arrays," in *Electron Devices Meeting, 1989. IEDM '89. Technical Digest.*, 1989, pp. 865–868.

[93] Y.-K. Chen and M. C. Wu, "Monolithic colliding-pulse mode-locked quantum well lasers," *IEEE J. Quantum Electron.*, vol. 28, pp. 2176–2185, 1992.

[94] D. Kunimatsu, S. Arahira, Y. Kato, and Y. Ogawa, "Passively mode-locked laser diodes with bandgap wavelength detuned saturable absorbers," *Photon. Technol. Lett.*, vol. 11, pp. 1363–1365, 1999.

[95] Y. Nomura, S. Ochi, N. Tomita, *et al.*, "Mode locking in Fabry-Perot semiconductor lasers," *Phys. Rev. A*, vol. 65, p. 043807, 2002.

[96] K. Yvind, D. Larsson, L. J. Christiansen, *et al.*, "Low-jitter and high-power 40-GHz all-active mode-locked lasers," *Photon. Technol. Lett., IEEE*, vol. 16, pp. 975–977, 2004.

[97] K. Merghem, A. Akrout, A. Martinez, *et al.*, "Short pulse generation using a passively mode locked single InGaAsP/InP quantum well laser," *Opt. Express*, vol. 16, pp. 10675–10683, 2008.

[98] P. Crump, S. Patterson, S. Elim, *et al.*, "Extending the wavelength range of single-emitter diode lasers for medical and sensing applications: 12xx-nm quantum dots, 2000-nm wells, >5000-nm cascade lasers," *Proc. SPIE*, vol. 6456, 2007.

[99] D. Nikitichev, Y. Ding, M. A. Cataluna, *et al.*, "High peak power and sub-picosecond Fourier-limited pulse generation from passively mode-locked monolitic two-section gain-guided tapered InGaAs quantum-dot lasers," *J. Laser Phys.*, vol. 22 (4), pp. 715–724, 2012.

[100] M. A. Cataluna, Y. Ding, D. I. Nikitichev, *et al.*, "High-power versatile picosecond pulse generation from mode-locked quantum-dot laser diodes," *IEEE J. Select. Top. Quantum Electron.*, vol. 17(1), pp. 1–9, 2011.

[101] Y. Ding, M. A. Cataluna, D. Nikitichev, I. Krestnikov, D. Livshits, and E. Rafailov, "Broad repetition-rate tunable quantum-dot external-cavity passively mode-locked laser with extremely narrow radio frequency line-width," *Appl. Phys. Express*, vol. 4, article No. 062703, 2011.

[102] Y. Ding, D. I. Nikitichev, I. Krestnikov, D. Livshits, M. A. Cataluna, and E. U. Rafailov, "Quantum-dot external-cavity passively modelocked laser with high peak power and pulse energy," *Electron. Lett.*, vol. 46, pp. 1516–1518, 2010.

[103] E. U. Rafailov, M. A. Cataluna, and W. Sibbett, "Mode-locked quantum-dot lasers," *Nat Photon*, vol. 1, pp. 395–401, 2007.

[104] A. Markus, J. X. Chen, C. Paranthoën, A. Fiore, C. Platz, and O. Gauthier-Lafaye, "Simultaneous two-state lasing in quantum-dot lasers," *Appl. Phys. Lett.*, vol. 82, pp. 1818–1820, 2003.

[105] C. G. Leburn, N. K. Metzger, A. A. Lagatsky, *et al.*, "Femtosecond pulse generation using ground-state and excited-state transitions in quantum-dot SESAMs," OSA Technical Digest Series (CD) (Optical Society of America, 2009, p. MB18.

[106] E. U. Rafailov, A. D. McRobbie, M. A. Cataluna, L. O'Faolain, and W. Sibbett, "Investigation of transition dynamics in a quantum-dot laser optically pumped by femtosecond pulses," *Appl. Phys. Lett.*, vol. 88, 2006.

[107] M. A. Cataluna, W. Sibbett, D. A. Livshits, J. Weimert, A. R. Kovsh, and E. U. Rafailov, "Ground and excited-state modelocking in a two-section quantum-dot laser," in *2005 IEEE LEOS Annual Meeting Conference Proceedings*, ed New York: IEEE, 2005, pp. 870–871.

[108] M. A. Cataluna, W. Sibbett, D. A. Livshits, J. Weimert, A. R. Kovsh, and E. U. Rafailov, "Stable mode locking via ground- or excited-state transitions in a two-section quantum-dot laser," *Appl. Phys. Lett.*, vol. 89, pp. 081124–081123, 2006.

[109] E. A. Viktorov, M. A. Cataluna, L. O'Faolain, *et al.*, "Dynamics of a two-state quantum dot laser with saturable absorber," *Appl. Phys. Lett.*, vol. 90, Article Number: 121113, 2007.

[110] A. Markus, M. Rossetti, V. Calligari, *et al.*, "Two-state switching and dynamics in quantum dot two-section lasers," *J. Appl. Phys.*, vol. 100, 2006.

[111] S. Breuer, M. Rossetti, W. Elsässer, *et al.*, "Reverse-emission-state-transition mode locking of a two-section InAs/InGaAs quantum dot laser," *Appl. Phys. Lett.*, vol. 97, pp. 071118–071113, 2010.

[112] S. Breuer, M. Rossetti, W. Elsässer, *et al.*, "Reverse ground-state excited-state transition dynamics in two-section quantum dot semiconductor lasers: mode-locking and state-switching," *Proceedings of SPIE*, vol. 7720, 2010.

[113] S. Breuer, W. Elsaber, and M. Hopkinson, "State-switched modelocking of two-segment quantum dot laser via self-electro-optical quantum dot absorber," *Electron. Lett.*, vol. 46, pp. 161–162, 2010.

[114] J. Kim, M. T. Choi, and P. J. Delfyett, "Pulse generation and compression via ground and excited states from a grating coupled passively mode-locked quantum dot two-section diode laser," *Appl. Phys. Lett.*, vol. 89, Article Number: 261106, 2006.

[115] J. Kim and P. J. Delfyett, "Interband optical pulse injection locking of quantum dot mode-locked semiconductor laser," *Opt. Express*, vol. 16, pp. 11153–11161, 2008.

[116] J. Liu, Z. G. Lu, S. Raymond, P. J. Poole, P. J. Barrios, and D. Poitras, "Dual-wavelength 92.5 GHz self-mode-locked InP-based quantum dot laser," *Opt. Lett.*, vol. 33, pp. 1702–1704, 2008.

[117] M. G. Thompson, R. V. Penty, and I. H. White, "Regimes of mode-locking in tapered quantum dot laser diodes," in *Semiconductor Laser Conference, 2008. ISLC 2008. IEEE 21st International*, 2008, pp. 27–28.

[118] S. G. Li, Q. Gong, Y. F. Lao, *et al.*, "Two-color quantum dot laser with tunable wavelength gap," *Appl. Phys. Lett.*, vol. 95, Article Number: 251111, 2009.

[119] C. Mesaritakis, C. Simos, H. Simos, S. Mikroulis, I. Krestnikov, and D. Syvridis, "Pulse width narrowing due to dual ground state emission in quantum dot passively mode locked lasers," *Appl. Phys. Lett.*, vol. 96, Article Number: 211110, 2010.

[120] L. F. Lester, A. Stintz, H. Li, *et al.*, "Optical characteristics of 1.24-μm InAs quantum-dot laser diodes," *Photon. Technol. Lett., IEEE*, vol. 11, pp. 931–933, 1999.

[121] X. Huang, A. Stintz, H. Li, L. F. Lester, J. Cheng, and K. J. Malloy, "Passive mode-locking in 1.3 μm two-section InAs quantum dot lasers," *Appl. Phys. Lett.*, vol. 78, pp. 2825–2827, 2001.

[122] M. G. Thompson, C. Marinelli, K. T. Tan, *et al.*, "10 GHz hybrid modelocking of monolithic InGaAs quantum dot lasers," *Electron. Lett.*, vol. 39, pp. 1121–1122, 2003.

[123] K. T. Tan, M. G. Thompson, C. Marinelli, *et al.*, "Investigation of high repetition rate mode-locked quantum dot lasers," in *Lasers and Electro-Optics Society, 2003. LEOS 2003. The 16th Annual Meeting of the IEEE*, 2003, pp. 826–827, vol. 2.

[124] A. E. Gubenko, L. M. Gadjiev, N. D. Il'inskaya, *et al.*, "Mode-locking at 9.7 GHz repetition rate with 1.7 ps pulse duration in two-section QD lasers," in *Semiconductor Laser Conference, 2004. Conference Digest. 2004 IEEE 19th International*, 2004, pp. 51–52.

[125] M. Kuntz, L. F. Lester, D. Chang, C. Langrock, M. M. Fejer, and D. J. Kane, "35 GHz mode-locking of 1.3 μm quantum dot lasers," *Appl. Phys. Lett.*, vol. 85, pp. 843–845, 2004.

[126] M. Kuntz, G. Fiol, M. Lämmlin, *et al.*, "Direct modulation and mode locking of 1.3 μm quantum dot lasers," *New J. Phys.*, vol. 6, p. 181, 2004

[127] M. G. Thompson, K. T. Tan , C. Marinelli, *et al.*, "Transform-limited optical pulses from 18 GHz monolithic modelocked quantum dot lasers operating at ~ 1.3 μm," *Electron. Lett.*, vol. 40, pp. 346–347, 2004.

[128] A. Gubenko, D. Livshits, I. Krestnikov, *et al.*, "High-power monolithic passively modelocked quantum-dot laser," *Electron. Lett.*, vol. 41, pp. 1124–1125, 2005.

[129] L. Zhang, L. S. Cheng, A. L. Gray, *et al.*, "Low Timing Jitter, 5 GHz Optical Pulses from a Monolithic Two-Section Passively Mode-Locked 1250/1310 nm Quantum Dot Laser for High Speed Optical Interconnets," 2005, p. OWM4.

[130] M. G. Thompson, C. Marinelli, X. Zhao, *et al.*, "Colliding-pulse modelocked quantum dot lasers," *Electron. Lett.*, vol. 41, pp. 248–250, 2005.

[131] M. T. Choi, W. Lee, J.-M. Kim, and P. J. Delfyett, "Ultrashort, high-power pulse generation from a master oscillator power amplifier based on external cavity mode locking of a quantum-dot two-section diode laser," *Appl. Phys. Lett.*, vol. 87, pp. 221107–221103, 2005.

[132] M.-T. Choi, J.-M. Kim, W. Lee, and P. J. Delfyett, "Ultralow noise optical pulse generation in an actively mode-locked quantum-dot semiconductor laser," *Appl. Phys. Lett.*, vol. 88, pp. 131106–131103, 2006.

[133] A. R. Rae, M. G. Thompson, R. V. Penty, *et al.*, "Harmonic Mode-Locking of a Quantum-Dot Laser Diode," in *Lasers and Electro-Optics Society, 2006. LEOS 2006. 19th Annual Meeting of the IEEE*, 2006, pp. 874–875.

[134] D. Bimberg, G. Fiol, M. Kuntz *et al.*, "High speed nanophotonic devices based on quantum dots," *Phys. Stat. Sol. (A)*, vol. 203, pp. 3523–3532, 2006.

[135] M. G. Thompson, A. Rae, R. L. Sellin, *et al.*, "Subpicosecond high-power mode locking using flared waveguide monolithic quantum-dot lasers," *Appl. Phys. Lett.*, vol. 88, pp. 133119–133113, 2006.

[136] M. Laemmlin, G. Fiol, M. Kuntz, *et al.*, "Quantum dot based photonic devices at 1.3 μm: Direct modulation, mode-locking, SOAs and VCSELs," *Phys. Stat. Sol. (C)*, vol. 3, pp. 391–394, 2006.

[137] M. Laemmlin, G. Fiol, C. Meuer, *et al.*, "Distortion-free optical amplification of 20–80 GHz modelocked laser pulses at 1.3 μm using quantum dots," *Electron. Lett.*, vol. 42, pp. 697–699, 2006.

[138] A. D. McRobbie, M. A. Cataluna, D. A. Livshits, W. Sibbett, and E. U. Rafailov, "High power all-quantum-dot-based external cavity modelocked laser," *Electron. Lett.*, vol. 43, pp. 812–813, 2007.

[139] F. Kefelian, S. O'Donoghue, M. T. Todaro, J. McInerney, and G. Huyet, "High repetition rate monolithic passively mode-locked semiconductor quantum-dot laser: Investigation of the locking regimes and the RF linewidth," in *Lasers and Electro-Optics, 2007. CLEO 2007. Conference on*, 2007, pp. 1–2.

[140] Y. C. Xin, Y. Li, V. Kovanis, A. L. Gray, L. Zhang, and L. F. Lester, "Reconfigurable quantum dot monolithic multisection passive mode-locked lasers," *Opt. Express*, vol. 15, pp. 7623–7633, 2007.

[141] J. Javaloyes, L. P. Hou, P. Stolarz, *et al.*, "Wide range 40-GHz passive mode-locking operation of an AlGaInAs 1.55 μm strained quantum well laser," in *Lasers and Electro-Optics 2009 and the European Quantum Electronics Conference. CLEO Europe – EQEC 2009. European Conference on*, 2009, pp. 1–1.

[142] A. Gubenko, D. Livshits, S. Mikhrin, I. Krestnikov, and A. Kovsh, "Frequency up-chirping in monolithic passively mode-locked InGaAs quantum dot lasers," in *IEEE Lasers and Electro-Optics Society, 2008. LEOS 2008. 21st Annual Meeting of the*, 2008, pp. 469–470.

[143] M. Xia, M. G. Thompson, R. V. Penty, and I. H. White, "External-cavity mode-locked quantum-dot laser diodes for low repetition rate, sub-picosecond pulse generation," *Selected Topics in IEEE J. Quantum Electron.*, vol. PP, pp. 1–1, 2011.

[144] M. G. Thompson, A. R. Rae, M. Xia, R. V. Penty, and I. H. White, "InGaAs Quantum-dot mode-locked laser diodes," *Selected Topics in IEEE J. Quantum Electron.*, vol. 15, pp. 661–672, 2009.

[145] G. Fiol, C. Meuer, H. Schmeckebier, *et al.*, "Quantum-dot semiconductor mode-locked lasers and amplifiers at 40 GHz," *IEEE J. Quantum Electron.*, vol. 45, pp. 1429–1435, 2009.

[146] C. Y. Lin, Y. -C. Xin, J. H. Kim, C. G. Christodoulou, and L. F. Lester, "Compact optical generation of microwave signals using a monolithic quantum dot passively mode-locked laser," *Photonics Journal, IEEE*, vol. 1, pp. 236–244, 2009.

[147] G. Carpintero, M. G. Thompson, R. V. Penty, and I. H. White, "Low noise performance of passively mode-locked 10-GHz quantum-dot laser diode," *Photon. Technol. Lett., IEEE*, vol. 21, pp. 389–391, 2009.

[148] H. Schmeckebier, G. Fiol, C. Meuer, D. Arsenijević, and D. Bimberg, "Complete pulse characterization of quantum dot mode-locked lasers suitable for optical communication up to 160 Gbit/s," *Opt. Express*, vol. 18, pp. 3415–3425, 2010.

[149] L. Lester, D. J. Kane, N. G. Usechak, *et al.*, "Pulse characteristics of passively mode-locked quantum dot lasers," *Proceedings of SPIE*, vol. 7616, 2010.

[150] D. Nikitichev, Y. Ding, M. Ruiz, *et al.*, "High-power passively mode-locked tapered InAs/GaAs quantum-dot lasers," *Appl. Phys. B*, vol. 103, pp. 609–613, 2011.

[151] Y. Li, Y. Li, C. -Y. Lin, *et al.*, "Pulse characterization of a passively mode-locked quantum dot semiconductor laser using FROG and autocorrelation," 2011, p. JWA89.

[152] M. Radziunas, A. G. Vladimirov, E. A. Viktorov, *et al.*, "Pulse broadening in quantum-dot mode-locked semiconductor lasers: Simulation, analysis, and experiments," *IEEE J. Quantum Electron.*, vol. 47, pp. 935–943, 2011.

[153] D. I. Nikitichev, *et al.*, "Investigation of the pulse dynamics in a mode-locked quantum-dot laser, involving the ground/excited state transitions," presented at the 1st EOS Topical Meeting on Lasers 2009, Capri, Italy, 2009.

[154] M. A. Cataluna, D. I. Nikitichev, S. Mikroulis, *et al.*, "Dual-wavelength mode-locked quantum-dot laser, via ground and excited state transitions: Experimental and theoretical investigation," *Opt. Express*, vol. 18, pp. 12832–12838, 2010.

[155] C. Mesaritakis, C. Simos, H. Simos, *et al.*, "Effect of optical feedback to the ground and excited state emission of a passively mode locked quantum dot laser," *Appl. Phys. Lett.*, vol. 97, pp. 061114–061114–3, 2010.

[156] M. Gioannini and I. Montrosset, "Numerical analysis of the frequency chirp in quantum-dot semiconductor lasers," *IEEE J. Quantum Electron.*, vol. 43, pp. 941–949, 2007.

[157] R. G. M. P. Koumans and R. Van Roijen, "Theory for passive mode-locking in semiconductor laser structures including the effects of self-phase modulation, dispersion, and pulse collisions," *IEEE J. Quantum Electron.*, vol. 32, pp. 478–492, 1996.

[158] M. A. Cataluna, J. Pulka, E. A. Viktorov, G. Huyet, and J. Houlihan, "New mode locking regime in a quantum-dot laser: enhancement by simultaneous CW excited-state emission," 2006, p. CThH3.

[159] T. Piwonski, J. Pulka, E. A. Viktorov, G. Huyet, and J. Houlihan, "Refractive index dynamics of quantum dot based waveguide electro-absorbers," *Appl. Phys. Lett.*, vol. 97, pp. 051107–051103, 2010.

[160] D. B. Malins, A. Gomez-Iglesias, S. J. White, and E. U. Rafailov, "Ultrafast electroabsorption dynamics in an InAs quantum dot saturable absorber at 1.3 μm," *Appl. Phys. Lett.*, vol. 89, pp. 171111–171111–3, 2006.

[161] A. Leitenstorfer, C. Fürst, and A. Laubereau, "Widely tunable two-color mode-locked Ti:sapphire laser with pulse jitter of less than 2 fs," *Opt. Lett.*, vol. 20, pp. 916–918, 1995.

[162] E. A. Viktorov, P. Mandel, A. G. Vladimirov, and U. Bandelow, "Model for mode locking in quantum dot lasers," *Appl. Phys. Lett.*, vol. 88, pp. 201102–201103, 2006.

[163] M. Sugawara, N. Hatori, H. Ebe, *et al.*, "Modeling room-temperature lasing spectra of 1.3-μm self-assembled InAs/GaAs quantum-dot lasers: Homogeneous broadening of optical gain under current injection," *J. Appl. Phys.*, vol. 97, Article Number: 043523, 2005.

[164] K. Y. Lau, "Short-pulse and high-frequency signal generation in semiconductor lasers," *J. Lightwave Technol., Journal of*, vol. 7, pp. 400–419, 1989.

[165] M. G. Thompson, A. R. Rae, R. V. Penty, *et al.*, "Absorber length optimisation for sub-picosecond pulse generation and ultra-low jitter performance in passively mode-locked 1.3μm quantum-dot laser diodes," 2006, p. OThG3.

[166] X. Huang, A. Stintz, H. Li, *et al.*, "Bistable operation of a two-section 1.3 μm InAs quantum dot laser-absorption saturation and the quantum confined Stark effect," *IEEE J. Quantum Electron.*, vol. 37, pp. 414–417, 2001.

[167] E. U. Rafailov, S. J. White, A. A. Lagatsky, *et al.*, "Fast quantum-dot saturable absorber for passive mode-locking of solid-state lasers," *Photon. Technol. Lett., IEEE*, vol. 16, pp. 2439–2441, 2004.

[168] K. A. Williams, M. G. Thompson, and I. H. White, "Long-wavelength monolithic mode-locked diode lasers," *New J. Phys.*, vol. 6, p. 179, 2004.

[169] S. V. Zaitsev, N. Yu. Gordeev, Yu. M. Sherniakov, *et al.*, "Radiation characteristics of injection lasers based on vertically coupled quantum dots," *Superlattices Microstruct.*, vol. 21, pp. 559–564, 1997.

[170] V. M. Ustinov, A. E. Zhukov, A.Y. Egorov, and N. A. Maleev, *Quantum Dot Lasers*. New York: Oxford University Press, 2003.

[171] J. Chen, W. J. Fan, Y. Ding, *et al.*, "Nanostructure model and optical properties of InAs/GaAs quantum dot in vertical cavity surface emitting lasers," *Opto-Electron. Rev.*, vol. 19, 2011.

[172] M. Feng, S. T. Cundiff, R. P. Mirin, and K. L. Silverman, "Wavelength bistability and switching in two-section quantum-dot diode lasers," *IEEE J. Quantum Electron.*, vol. 46, pp. 951–958, 2010.

[173] M. Feng, N. A. Brilliant, S. T. Cundiff, R. P. Mirin, and K. L. Silverman, "Wavelength bistability in two-section mode-locked quantum-dot diode lasers," *Photon. Technol. Lett., IEEE*, vol. 19, pp. 804–806, 2007.

[174] O. Qasaimeh, W. -D. Zhou, J. Phillips, S. Krishna, and P. Bhattacharya, "Bistability and self-pulsation in quantum-dot lasers with intracavity quantum-dot saturable absorbers," *Appl. Phys. Lett.*, vol. 74, pp. 1654–1656, 1999.

[175] D. I. Nikitichev, M. A. Cataluna, Y. Ding, *et al.*, "High-power spectral bistability in a multi-section quantum dot laser under continuous-wave or mode-locked operation," in *CLEO2011*, USA, 2011.

[176] K. A. Fedorova, M. A. Cataluna, I. Krestnikov, D. Livshits, and E. U. Rafailov, "Broadly tunable high-power InAs/GaAs quantum-dot external cavity diode lasers," *Opt. Express*, vol. 18, pp. 19438–19443, 2010.

[177] X. Q. Lv, P. Jin, W. Y. Wang, and Z. G. Wang, "Broadband external cavity tunable quantum dot lasers with low injection current density," *Opt. Express*, vol. 18, pp. 8916–8922, 2010.

[178] X. Q. Lv, P. Jin, and Z. G. Wang, "Broadly tunable grating-coupled external cavity laser with quantum-dot active region," *Photon. Technol. Lett., IEEE*, vol. 22, pp. 1799–1801, 2010.

[179] G. Ortner, C. Ni. Allen, C. Dion, *et al.*, "External cavity InAs/InP quantum dot laser with a tuning range of 166 nm," *Appl. Phys. Lett.*, vol. 88, pp. 121119–121119–3, 2006.

[180] B. J. Stevens, D. T. D. Childs, K. M. Groom, M. Hopkinson, and R. A. Hogg, "All semiconductor swept laser source utilizing quantum dots," *Appl. Phys. Lett.*, vol. 91, Article Number: 121119, 2007.

[181] S. Todor, B. Biedermann, W. Wieser, R. Huber, and C. Jirauschek, "Instantaneous lineshape analysis of Fourier domain mode-locked lasers," *Opt. Express*, vol. 19, pp. 8802–8807, 2011.

[182] D. I. Nikitichev, M. A. Cataluna, Y. Ding, I. Krestnikov, D. Livshits, and E. Rafailov, "Tunable quantum-dot mode-locked monolithic laser," presented at the 2nd EOS Topical Meeting on Lasers (ETML'11), Italy, 2011.

[183] R. G. M. P. Koumans and R. Van Roijen, "Theory for passive mode-locking in semiconductor laser structures including the effects of self-phase modulation, dispersion, and pulse collisions," *IEEE J. Quantum Electron.*, vol. 32, pp. 478–492, 1996.

[184] I. B. Akca, A. Dâna, A. Aydinli, *et al.*, "Electro-optic and electro-absorption characterization of InAs quantum dot waveguides," *Opt. Express*, vol. 16, pp. 3439–3444, 2008.

[185] T. Ulm, F. Harth, H. Fuchs, J. A. L'huillier, and R. Wallenstein, "InGaAs diode laser system generating pulses of 580 fs duration and 366 W peak power," *Appl. Phys. B*, vol. 92, pp. 481–484, 2008.

[186] T. Ulm, A. Klehr, G. Erbert, F. Harth, and J. A. L'huillier, "Femtosecond diode laser MOPA system at 920 nm based on asymmetric colliding pulse mode-locking," *Appl. Phys. B*, vol. 99, pp. 409–414, 2010.

[187] K. Kim, S. Lee, and P. Delfyett, "1.4 kW high peak power generation from an all semiconductor mode-locked master oscillator power amplifier system based on eXtreme Chirped Pulse Amplification(X-CPA)," *Opt. Express*, vol. 13, pp. 4600–4606, 2005.

[188] C. Jördens, T. Schlauch, M. Li, M. R. Hofmann, M. Bieler, and M. Koch, "All-semiconductor laser driven terahertz time-domain spectrometer," *Appl. Phys. B*, vol. 93, pp. 515–520, 2008.

[189] J. C. Balzer, T. Schlauch, A. Klehr, G. Erbert, G. Tränkle, and M. R. Hofmann, "High peak power pulses from dispersion optimised modelocked semiconductor laser," *Electron. Lett.*, vol. 49, pp. 838–839, 2013.

[190] S. Schwertfeger, A. Klehr, T. Hoffmann, A. Liero, H. Wenzel, and G. Erbert, "Generation of sub-100 ps pulses with a peak power of 65 W by gain switching, pulse shortening, and pulse amplification using a semiconductor-based master oscillator–power amplifier system," *Appl. Opt.*, vol. 52, pp. 3364–3367, 2013.

[191] S. Schwertfeger, A. Klehr, T. Hoffmann, A. Liero, H. Wenzel, and G. Erbert, "Picosecond pulses with 50 W peak power and reduced ASE background from an all-semiconductor MOPA system," *Appl. Phys. B*, vol. 103, pp. 603–607, 2011

[192] P. Adamiec, B. Bonilla, A. Consoli, J. M. G. Tijero, S. Aguilera, and I. Esquivias, "High-peak-power pulse generation from a monolithic master oscillator power amplifier at 1.5 μm," *Appl. Opt.*, vol. 51, pp. 7160–7164, 2012.

[193] A. Klehr, B. Sumpf, K. -H. Hasler, J. Fricke, A. Liero, and G. Erbert, "High-power pulse generation in GHz range with 1064-nm DBR tapered laser," *IEEE Photon. Technol. Lett.*, vol. 22, pp. 832–834, 2010.

[194] S. Riecke, S. Schwertfeger, K. Lauritsen, K. Paschke, R. Erdmann, and G. Tränkle, "23 W peak power picosecond pulses from a single-stage all-semiconductor master oscillator power amplifier," *Appl. Phys. B*, vol. 98, pp. 295–299, 2010.

[195] C. Weber, L. Drzewietzki, M. Rossetti, *et al.*, "Picosecond pulse amplification up to a peak power of 42 W by a quantum-dot tapered optical amplifier and a mode-locked laser emitting at 1.26 μm," *Opt. Lett.*, vol. 40, pp. 295–397, 2015.

[196] A. Bilenca, S. H. Yun, G. J. Tearney, and B. E. Bouma, "Numerical study of wavelength-swept semiconductor ring lasers: The role of refractive index nonlinearities in semiconductor optical amplifiers and implications for biomedical imaging applications," *Opt. Lett.*, vol. 31, pp. 760–762, 2006.

[197] M. Kuznetzov, W. Atia, B. Johnson, and D. Flanders, "Compact ultrafast reflective Fabry-Perot tunable lasers for OCT applications," *Proc. SPIE*, 7554, 75541F (2010).

[198] E. A. Avrutin and L. Zhang, "Dynamics of semiconductor lasers under fast intracavity frequency sweeping," in *Proc. 14th Conference on Transparent Optical Networks*, Coventry, UK, 2012, paper Mo.C.4.3., pp. 1–4, IEEE, 2012.

[199] F. Lelarge, B. Dagens, J. Renaudier, *et al.*, "Recent advances on InAs/InP quantum dash based, semiconductor lasers and optical amplifiers operating at 1.55 mu m," *IEEE J. Selected Topics in Quantum Electron.*, vol. 13, pp. 111–124, 2007.

[200] S. Slepneva, B. O'Shaughnessy, B. Kelleher, *et al.*, "Dynamics of a short cavity swept source OCT laser," *Opt. Express*, vol. 22, pp. 18177–18185, 2014.

[201] R. Huber, M. Wojtkowski, and J. G. Fujimoto, "Fourier domain mode locking (FDML): A new laser operating regime and applications for optical coherence tomography," *Opt. Express*, vol. 14, pp. 3225–3226, 2006.

[202] T. Klein, W. Wieser, C. M. Eigenwillig, B. R. Biedermann, and R. Huber, "Megahertz OCT for ultrawide-field retinal imaging with a 1050 nm Fourier domain mode-locked laser," *Opt. Express*, vol. 19, pp. 3044–3062, 2011.

[203] C. Jirauschek, B. Biedermann, and R. Huber, "A theoretical description of Fourier domain mode locked lasers," *Opt. Express*, vol. 17, pp. 24013–24019, 2009.

[204] S. Slepneva, B. Kelleher, B. O'Shaughnessy, S. P. Hegarty, A. G. Vladimirov, and G. Huyet, "Dynamics of Fourier domain mode-locked lasers," *Opt. Express*, vol. 21, pp. 19241–19251, 2013.

[205] B. Kellegher, D. Goulding, S. Slepneva, *et al.*, "Phase and frequency dynamics of Fourier domain mode locked OCT lasers," *Proc. of the SPIE*, vol. 9357, 2015.

[206] B. R. Biedermann, W. Wieser, C. M. Eigenwillig, T. Klein, and R. Huber, "Direct measurement of the instantaneous linewidth of rapidly wavelength-swept lasers," *Opt. Lett.*, vol. 35, pp. 3733–3735, 2010.

Chapter 4
Light-emitting diode technologies

Hoi Wai Choi[1]

4.1 Introduction

Half century since the invention of the visible light-emitting diode (LED), the technology has blossomed from a minuscule light emitter to present-day powerful illuminators, driven by explosive developments in material and processing technologies, including but not limited to advanced epitaxy of complex quantum structures, nanotechnology-enabled devices and sophisticated chip packaging techniques to accommodate the needs of such avant-garde devices. The initial LEDs emit at the red/ infrared spectral regions using the GaAsP alloys. GaP extended the emission range to green, but the elusive blue LED/laser remained a great challenge for a while. That hurdle was eventually overcome in the 1990s when the first candela class blue LEDs were demonstrated using the InGaN/GaN material system; since then the technology took flight and never looked back.

Today LEDs are part of everyday lives, appearing as traffic lights, liquid crystal display (LCD) backlights, desktop lamps, handheld torches, just to name a few. Of course, LEDs are not limited to domestic uses; devices of specific wavelengths are adopted in professional products such as phototherapy, biotechnology, medical imaging, and microscopy owing to their distinct optical emission characteristics not found from any other light sources. The bio-medical related uses typically rely on short-wavelength emitters based on the Al-containing GaN alloys, capable of producing emission down to 200+ nm, owing to the high-energy generated photons which can induce photo-chemical changes to materials. For instance, UV LEDs at 280 nm or below have been used for photo-excitation of fluorescent markers in cells, photo-therapy of skin disorders and killing of germs in liquids.

Although GaN, a compound semiconductor with a wide bandgap, had been studied for decades, several issues including p-type doping and material quality prevented it from flourishing. The long-awaited breakthrough came in around 1996, when the discoveries of S. Nakamura [1] and I. Akasaka [2] made it possible to grow high-quality heterostructures with GaN and its Indium alloys and enabled the effective incorporation and activation of p-dopants, respectively. From then

[1]Department of Electrical and Electronic Engineering, The University of Hong Kong, Hong Kong

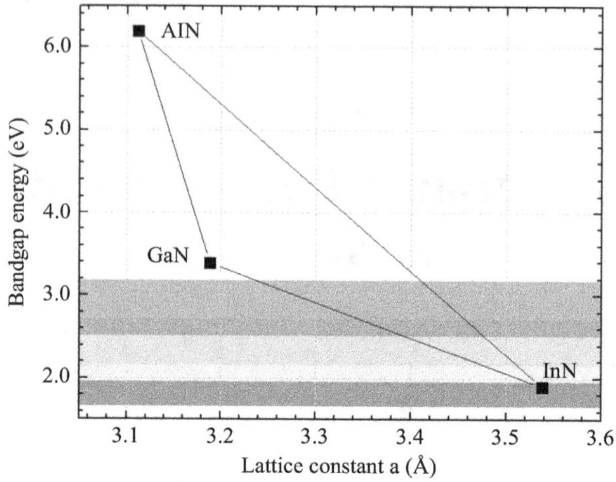

Figure 4.1 Plot of bandgap energies versus lattice constants for III-nitride semiconductors

on, it was possible to grow high quantum efficiency LED structures containing multiquantum wells emitting at wavelengths between ~395 nm (near-ultra-violet) to ~550 nm (green). Replacing In with Al extends the emission range down to the 200 nm. As illustrated in Figure 4.1, the possibility forming ternary ($Al_xGa_{1-x}N$, $In_xGa_{1-x}N$, and $In_xAl_{1-x}N$) and quaternary ($In_xAl_yGa_{1-x-y}N$) alloys provides for a wide spectral range from 1.9 eV (InN) to 6.2 eV (AlN).

The rapid growth of nitride technology can be attributed to intensive engineering efforts in research laboratories in universities and industries around the world. Many problems remain incompletely resolved, including the built-in piezoelectric field, efficiency "droop", the green gap and light extraction among others. In spite of imperfections with the material, there are now many large-scale commercial applications relying on GaN-based devices, including phosphor-coated white LEDs for solid-state lighting, the 405 nm violet laser diode for high-density optical storage and the RGB LEDs for panel displays. Combining a short wavelength LED with suitable phosphors enables generation of white light; the white light LEDs are rapidly replacing compact fluorescent lamps (CFLs) for LCD display backlights and general lighting, the transition being accelerated by rising concerns over energy source depletion.

Among a wide range of domestic and industrial uses, lighting is undoubtedly the major driving force behind the rapid development of nitride LEDs, with market revenues of over US$10 billion in 2012 for packaged LED lamps. However the unique properties of LEDs over other light sources, including compactness, robustness and narrowband emission across the UV-VIS range, make them suitable for a much wide range of niche applications, including those in the high-value chemical and biomedical sectors. For instance, UV-LEDs are used for excitation of fluorescence from biological particles [3] and for killing of bacteria in water purification uses [4]; blue LEDs for curing of dental monomers [5]; red and green LEDs to treat skin disorders [6]; white LEDs for optical communications [7]; and IR LEDs for gas sensing [8].

The scope of this chapter is focused on technologies that targets at improving the efficiencies of LEDs, for lighting applications and beyond. First and foremost, the issue of light confinement is discussed in Section 4.2. The method of chip shaping, an effective means of boosting light extraction, is introduced. In Section 4.3, various strategies for enhancing the functionalities of LEDs are described. Arrays of micrometer scale emitters are demonstrated as efficient light sources, micro-displays as well as for individually addressable excitation sources in Section 4.3.1. A hemispherical lens attached to a circular flip-chip LED is demonstrated as a high-efficiency fiber-coupled light source in Section 4.3.2.1. Combining these technologies, fiber-coupled micro-LED arrays are presented in Section 4.3.2.2.

4.2 Light extraction

It is well known that the performance of an LED is governed by the wall-plug efficiency defined as the ratio of total optical output power from emitter to the electrical input power. It indicates how efficiently the electrical power can be converted into optical power and is mathematically expressed as the product of four factors, namely electrical efficiency, injection efficiency, internal quantum efficiency (IQE), and light extraction efficiency:

$$\eta_{\mathrm{wall}} = \eta_{\mathrm{ele}} \cdot \eta_{\mathrm{EQE}} = \eta_{\mathrm{ele}} \cdot \eta_{\mathrm{inj}} \cdot \eta_{\mathrm{int}} \cdot \eta_{\mathrm{extraction}} \tag{4.1}$$

The electrical efficiency is primarily limited by ohmic losses and driver losses and shows how efficiently energy can be acquired from the power source to drive an LED. The product of the remaining terms is regarded as the external quantum efficiency (EQE) which indicates the portion of photons emitted from the devices due to the injected carriers.

Prior to electron-hole recombination, an injection of electron-hole pair into the active layers is necessitated. The injection efficiency is defined as the proportion of injected electrons being able to reach the active region. The injection efficiency can be expressed as:

$$\eta_{\mathrm{inj}} = \frac{J_{\mathrm{p}}}{J_{\mathrm{p}} + J_{\mathrm{n}} + J_{\mathrm{s}}} \tag{4.2}$$

where J_{p} is the minority carrier hole diffusion current, J_{n} is the minority carrier electron diffusion current, and J_{s} is the space charge recombination current in p-n junction.

During radiative recombination, an electron from the conduction band directly combines with a hole in the valence band to generate a photon. However, non-radiative recombination competes with radiative recombination under practical conditions. Such undesired processes occur via Auger processes or defect levels to produce heat or phonons [9]. The IQE is the fraction of electron-hole pairs in the active region that can recombine radiatively to produce photons:

$$\eta_{\text{int}} = \frac{P_{\text{int}}/(hv)}{I/e} \tag{4.3}$$

where P_{int} is the optical power emitted from the active region.

In an ideal scenario, all photons generated at the active region are expected to escape from the LED and enter into free space, achieving unity extraction efficiency. However, in real life, emitted light may be partially reabsorbed. The light extraction efficiency is typically defined as the portion of light generated in the active region that emits into free space:

$$\eta_{\text{ext}} = \frac{P/(hv)}{P_{\text{int}}/(hv)} \tag{4.4}$$

where P is the optical power emitted from the active region into free space.

The growing demand for blue light LEDs has prompted for devices with maximal EQEs, which is determined by both IQE and light extraction efficiency. With rapid and massive improvements of growth techniques, epitaxial structures and crystal qualities, IQE has been greatly increased to more than 80% [10]. However, the extremely low light extraction efficiencies (<20%) remains one of the major bottlenecks restricting the performance of LEDs, attributed to optical absorption at substrate, current spreading layer, ohmic contacts and bonding wire. The underlying cause is total internal reflections (TIRs).

Owing to high refractive index contrast at semiconductor/air interface, the majority of photons emitted from the active region remains trapped within the semiconductor. Such light-trapping phenomenon is known as TIR, limiting the amount of light that can actually escape the device. According to Snell's law, TIR occurs when light rays strike on the flat-top semiconductor/air interface with incident angle greater than critical angle, as depicted in Figure 4.2:

$$\theta_c = \sin^{-1}\left(\frac{n_{\text{air}}}{n_{\text{sc}}}\right) \tag{4.5}$$

where n_{air} *and* n_{sc} are the refractive indexes of air and semiconductor. Photons emitted at angles beyond the escape cone are likely to be reabsorbed after multiple reflections.

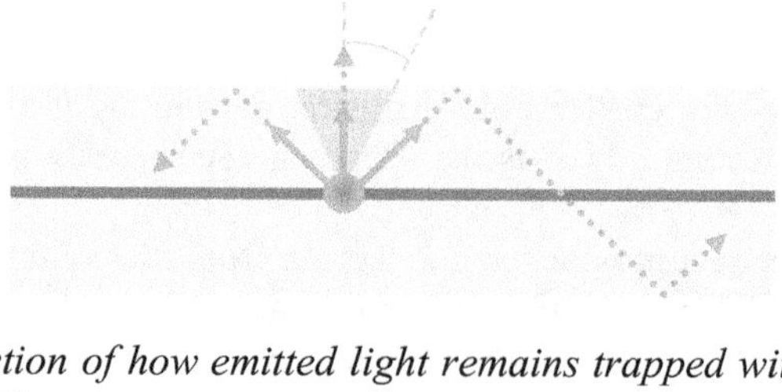

Figure 4.2 Depiction of how emitted light remains trapped within the device via TIR

In particular, the refractive index is about 2.45 for III-nitride semiconductor at visible wavelengths giving an extraction angle (escape angle or escape cone) of about 23.5°. TIR resulting from the narrow escape cone prevents the photons from escaping from the semiconductor. Moreover, the light extraction efficiency from a planar surface can be roughly estimated by:

$$\eta_{\text{ext}} = \int_0^{\theta_c} T_{\text{f}}(\theta)\alpha(\theta, d)\sin\theta d\theta \tag{4.6}$$

where $T_f(\theta)$ is the Fresnel transmission coefficient function and $\alpha(\theta,d)$ is the absorption coefficient function. The flat-top emission surface of a typical LEDs is found to be as low as 4% [11] while the overall extraction efficiency is limited to around 12%. In the following sections, strategies of improving light extraction, as well as functionalities of LEDs are presented.

4.2.1 Chip shaping

With dozens of advanced features loaded into today's LED chip, isn't it strange that the geometries of chips haven't changed much over the decade? The most likely reason for chips to adopt rectangular or square geometries is probably related to the way there are diced. Using the process of wafer sawing whereby only continuous linear cuts are possible, rectangular chips seems the logical way to go. However, geometries with high degrees of symmetries generally do not favor light extraction; nevertheless light extraction efficiency is widely recognized as the key limiting factor for LEDs to realize their full potentials. As discussed in Section 4.1, a light ray incident onto an interface at an angle beyond the critical angle will be completely reflected. Excessive reflection of light at interfaces results in light confinement and subsequent re-absorption, resulting in heat generation. Therefore, improving light extraction is a direct way of combating heat dissipation issues. Surface texturing, in the forms of random roughening or regular and periodic arrays of microstructures and nanostructures (also known as photonic crystals) [12–19] has widely been adopted to minimize optical confinement in the vertical direction. However, little attention has been paid to lateral optical confinement, which is dependent on the dimension and geometry of the LED chip. In the highly symmetrical rectangular or square structure, a light ray that is reflected due to TIR at one interface will most likely be reflected at subsequent interfaces. This can be understood from the fact that all interfaces, or facets, of such chips are either parallel or orthogonal to each other; the incident angle of a light ray remains unchanged after successive reflections unless scattered, so that such confined photons are eventually reabsorbed. The situation will change if the facets are not parallel to each other; this forms the basis of our interests in polygonal, circular and 3D chip geometries. The hypothesis is also supported by optical ray-trace simulations. By examining the percentage of light rays that can be extracted from LEDs of different geometries, their efficiencies can be predicted. The results of our set of simulations on LED chips of circular, triangular, squarish, pentagonal, hexagonal, heptagonal, octagonal and circular (the circle is in fact a polygon with an infinite number of sides) geometries are summarized in Table 4.1.

Table 4.1 Light extraction efficiencies of LED chips of different geometries

Shape	Light extraction efficiency (%)
Triangle	15.07
Square	12.98
Pentagon	15.09
Hexagon	14.39
Heptagon	14.54
Octagon	14.43
Circle	13.98

What stands out from the set of figures is the fact that square LEDs have distinctively lower extraction efficiencies than any other shapes. Considering the fact that LED chips in the market are invariably diced into squares or rectangles, device designers should give serious thoughts about re-designing the chip.

The enabling technology for realizing LEDs of unconventional geometries is laser micromachining. Although lasers have now been widely adopted for chip dicing, they have continued to be used for making linear cuts even though laser beams can easily be steered to carve virtually any shape. In traditional laser-micromachining, a focused beam is directed to the wafer orthogonally. Using the modified set-up which involves the insertion of a laser-turning mirror between the focusing objective and the wafer, the beam is re-directed to strike the wafer at an oblique angle [20]. The setup for laser micromachining consists of a UV laser source, beam focusing optics and an *x*–*y* motorized translation stage. The laser source is a third harmonic ND:YLF diode-pumped solid-state laser manufactured by Spectra Physics. The laser emits at 349 nm, while the pulse repetition rate ranges from single pulse to 5 kHz. After beam expansion and collimation through a beam expander, the laser beam is reflected 90° by a dielectric laser line mirror and focused onto the horizontal machining plane to a very tiny spot several micrometers in diameter with a focusing triplet. The additional feature of this setup, as illustrated in the schematic diagram of Figure 4.3, is the insertion of a UV mirror at an oblique angle within thc optical path between the focusing optics and the machining plane, which serves to deflect the convergent beam to strike the sample at an oblique angle with respect to the horizontal working plane. This oblique incident beam can be used for the cutting of GaN LED chips into virtually any 3D geometry.

Polygonal LEDs of equal emission areas with different number of sides are fabricated from standard large-area (1 mm^2) LED chips which have been fabricated on InGaN/GaN blue LED wafers using standard micro-fabrication techniques, as shown in Figure 4.4. The optical powers of the packaged devices were measured by placing the devices within a 2-inch integrating sphere, and the light from the output port channeled to a calibrated spectrometer via an optical fiber. The results are shown in the light intensity vs. current (L–I) characteristics plot in Figure 4.5,

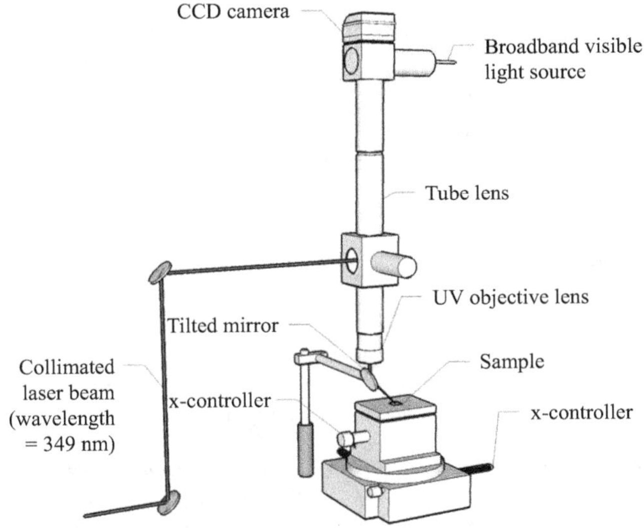

Figure 4.3 Schematic diagram of laser micromachining setup

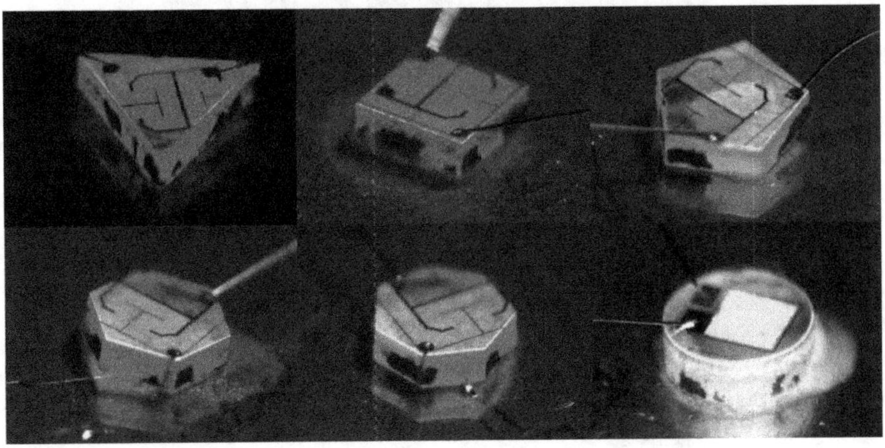

Figure 4.4 LED chips of different geometries processed by laser micromachining

which clearly shows that the rectangular LED emits with the lowest intensity at all currents. Clearly, the advantages of polygonal geometries are evident. Of course, one could argue that the chip packing density, and thus chip count across a wafer, might be compromised. Therefore, we propose the use of triangles and hexagons; such shapes can be closed-packed into any array without sacrificing chip space and thus make economic sense.

Besides 2D LED structures, LED of 3D geometries can also be processed using the laser setup. LEDs of truncated pyramidal (TP) structure, as depicted in

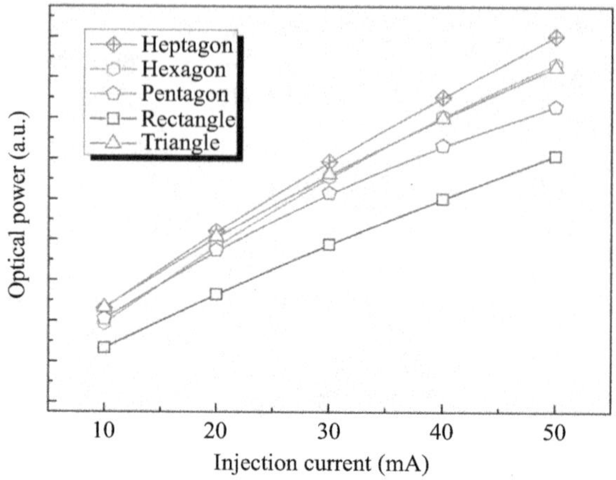

Figure 4.5 Measured light output power versus injection current (L-I) for laser micromachined polygonal LEDs. Reprinted with permission from the Journal of Applied Physics 108, art. 023110. © 2010, AIP Publishing LLC

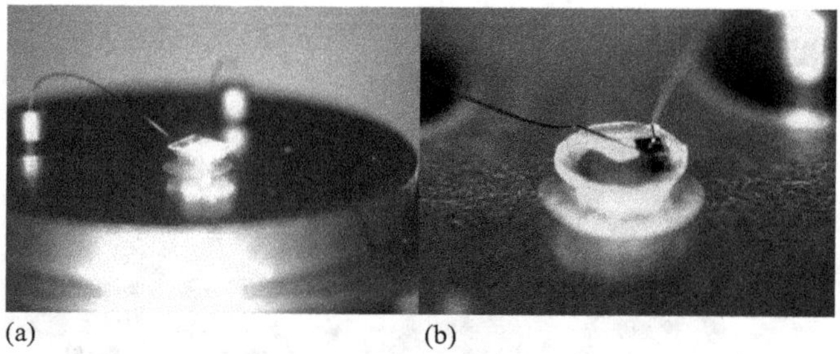

Figure 4.6 LED chips of inverted truncated (a) pyramidal and (b) conical geometries

Figure 4.6(a), are formed simply by applying four successive oblique cuts along the four edges of devices [21]. On average, the light enhancement factor is measured to be ~88% over similar chips with vertical facets, consistent with our theoretical predictions by ray-tracing. The inclined sidewalls serve as reflectors to redirect laterally propagating photons into the escape cone. Devices of TP-geometry also determined to emit with a wider divergence angle.

Combining oblique angle machining and rotary machining, LEDs of truncated conical structures [22] can also be formed, as shown in Figure 4.6(b). Subsequently, a reflective metallic layer is deposited onto the bottom and inclined sidewall surfaces of the chip to form an integrated reflector cup. This structure is particularly beneficial for building white LEDs with homogenous emission. In a typical white LED, an epoxy mixture containing phosphor particles is coated prior to encapsulation. The phosphor mixture is allowed to reflow in order to cover both the chip surface and sidewall facets, resulting in a dome-shape-like nonuniformity, the effects of which are particularly prominent at the edges. Such thickness nonuniformities give rise to nonhomogeneity of color emission at different viewing angles. With the integrated reflector cup design, light emission via sidewalls is suppressed effectively. The majority of the light rays are emitted from the top planar surface, so that the divergence angle of the device can be reduced by 16°. Consequently, only the top planar surface needs to be phosphor coated; planar coating produces significantly improved uniformity that paves the way for superior color homogeneity.

4.3 Device functionalities

4.3.1 Micro-LEDs

The exciting world of InGaN/GaN micro-LEDs began with an optical study of arrays of 10 μm-micro-disks which have been photolithographically defined and dry etched to form pillar structures [23]. The study yielded several important observations regarding these microscale structures, including strong enhancement of quantum efficiencies and recombination lifetimes. In a further study whereby the micro-cavities were shaped into micro-ring geometries, optical resonance modes were observed [24]. The measured mode spacing confirmed the nature of the resonance as whispering gallery (WG) modes associated with the cylindrical symmetry of the microstructures. Following these optical-pumped experiments, electroluminescent devices of micrometer scales were subsequently reported. Electrically injected individual [25] and interconnected micro-LEDs were demonstrated; these were later developed into high extraction efficiency LEDs of different geometries, including arrays of micro-disks [26,27] and micro-rings [28]. In order to minimize the wastage of junction areas, tightly packed micro-hexagonal micro-LEDs have been designed [17]. The effect of dimensions has also been investigated [29]. Electroluminescent micro-LED devices are various geometries are illustrated in Figure 4.7.

Apart from parallel-addressed micro-LEDs, these miniature emitters have also been interconnected in various configurations to form emissive micro-displays. An early demonstration of this idea was reported by Jiang *et al.*, consisting of 10 × 10

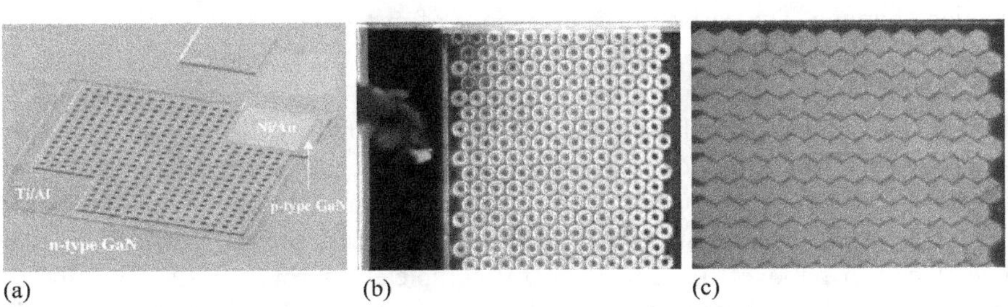

Figure 4.7 Micro-LED arrays of (a) disk; (b) ring; and (c) hexagonal geometries. Figure 4.7(a) reprinted with permission from Applied Physics Letters 77, pp. 3236–8. © 2010, AIP Publishing LLC

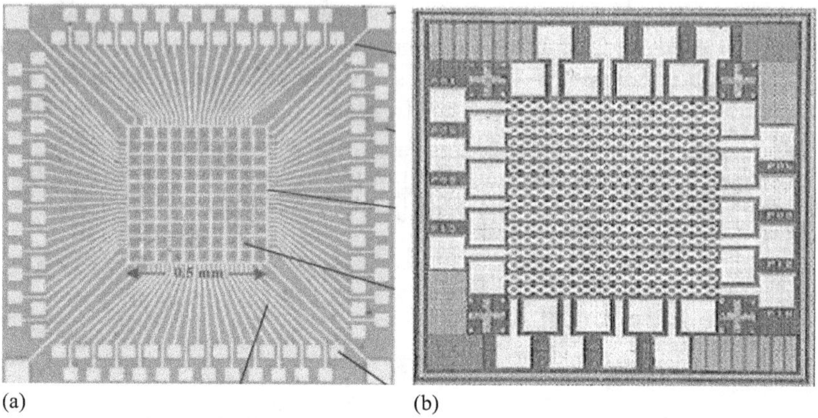

Figure 4.8 (a) Individually addressing and (b) matrix-addressing schemes for micro-LED arrays. Figure 4.8(a) reprinted with permission from Applied Physics Letters 78, pp. 1303–5. © 2001, AIP Publishing LLC

12- μm diameter pixels which are individually addressable as shown in Figure 4.8 (a) [30]. Based on this addressing arrangement, a 100-pixel micro-display requires 100 bonding pads, limiting scalability in terms of array size, prompting for more efficient addressing schemes. Matrix addressing, whereby pixels are interconnected both along a column and a row, significantly reduces the number of pads from $m \times n$ to $m + n$, where m and n are the number of pixels along a column and row, respectively. Having solved one issue brings in new challenges in the processing and fabrication of such devices.

The device is typically sectioned into multiple longitudinal mesa structures by deep etching through the entire depth of the wafer. Individual pixels are formed along these mesas by yet another dry etch process; pixels along each mesa are thus interconnected via the n-type GaN layer. Metallic interconnects are then deposited in the direction orthogonal to the mesas, completing the matrix addressing scheme. Before this can be done, the pixel sidewalls must be properly isolated to avoid junction shortage. The earliest demonstrations of matrix-addressable micro-displays relied on the technique of planarization. The air-gaps between micro-LED pixels were filled with oxide grown by chemical vapor deposition and planarized by chemical mechanical polishing. Micro-displays reported by Jeon *et al.* [31] were fabricated this way. In order to overcome the complexity and mechanical damage induced, the self-aligned contact isolation approach was developed [32,33]. In order to isolate pixels in adjacent columns, the trenches between GaN mesas must be deep-etched to remove the semiconductor completely, terminating at the sapphire interface. GaN LED wafers are typically 3 to 4 μ thick. There are two difficulties associated with subsequent processing of the device. Firstly, metal interconnects must run continuous across the deep trenches and secondly, the metal layer must not create a shortage path between the n-type and p-type GaN layers of the LED

(a) (b)

Figure 4.9 (a) Dry-etched mesas with sloped sidewalls and (b) A 128 × 98 pixel micro-display. Figure 4.9(a) reprinted with permission from the Journal of Vacuum Science & Technology B 23, pp. 99–103.

structure. Based on the processes developed at the University of Strathclyde, mesas with sloped sidewall were etched using a novel recipe; a self-aligned contact lift-off process was developed so that a layer of isolation oxide covers the entire device area except regions where contacts are to be exposed. A large pixel count 128 × 96 elements micro-display was reported, exhibiting superior uniformity, low operation voltage and luminous efficiency, opening the door towards semiconductor emissive micro-displays (Figure 4.9).

Since its inception, there have been tremendous amount of efforts put into the research and development of micro-sized emitter arrays, with variations in dimensions, emission wavelengths, pixel geometries and packaging strategies. As mentioned in an earlier paragraph, the array size of a single device has increased from the earlier proof-of-concept demonstration of 10 × 10 to the 128 × 96 pixel prototype over a relatively short period of time [34]. The dimension of a single pixel in an array also ranges from a few microns to tens of microns. It was discovered that the extraction efficiency scales with the pixel dimension- the smaller the pixel area, the more percentage of light is expected to extracted from the structures. In general, light extraction efficiency is proportional to the ratio between surface area and emission area. In fact, this is the underlying reason for adopting micrometer emitters over conventional large-area emitters. The micro-disk geometry, which has been the most widely adopted choice in many micro-LED works, obviously possess a large surface area to emission area ratio. For the purpose of further enhancing this figure, micro-ring LEDs have been proposed and demonstrated to perform even better than micro-disk LEDs – that is when we compare the optical output assuming equal light generating area. When this is true, the micro-ring LED inevitably occupies more chip-space, an important attribute in commercial sense. A design that optimizes between chip-space and light extraction efficiency is needed. A closed-packed array of hexagons would fit these requirements. An LED with a honeycomb emissive region, offering a filling-factor in excess of 80%, was proven to emit 50% more light that a micro-disk counterpart.

Other geometries were also reported, including the stripe design, although this is adopted for a specific application rather than for efficiency.

The initial micro-LEDs emit at visible colors, mostly at blue and green wavelengths. However, researchers quickly extended the micro-LED concept to ultraviolet wavelength, riding on the tide of rapid developments in AlGaN materials. As a matter of fact, the micro-LED structure was particularly useful to UV emitters. The AlGaN materials have lower IQEs than their InGaN counterparts; the micro-LED design aids with extracting those confined but precious UV photons. On the other hand, the AlGaN-based LEDs suffer from poor current spreading across the n-type conductivity layer due to the presence of an AlGaN/GaN superlattice. Microstructuring alleviates this problem; more on this in a later section. To date, there have been numerous reports of UV micro-emitters, including the 370 nm micro-ring LED [35], a deep-UV emitter at 254 nm with a micro-pixel design [36], and a 64 × 64 pixels 368 nm matrix-addressable array [37] among others.

4.3.1.1 Distinguishing features and performance of micro-sized LEDs

Electrical characteristics

One of the initial questions that needs to be answered regarding dimensional downsizing is its effect on the electrical characteristics of a micro-LED device. A comprehensive study was carried out and reported in ref. [26], whereby interconnected arrays of micro-disk emitters with diameters ranging from 8 μm to 20 μm were investigated. The measured *I–V* curves in Figure 4.10(a) reveal that while the 20 μm micro-LEDs had similar properties as a broad-area LED, the turn-on and 20 mA operating voltage increases as the size of the micro-LEDs are further reduced. When designing micro-LED devices, one should avoid unnecessarily small dimensions, as the p-contact resistance is sensitive to the actual contact area [38]. On the other hand, micro-sectioning of the active region does offer a significant benefit to current spreading across the n-type GaN layer, particularly in UV devices. An AlGaN/GaN superlattice is typically inserted into AlGaN-based UV devices in order to improve the crystalline quality of the upper layers, although the transport properties are usually affected due to the increased resistance. Devices with SLS may suffer from current crowding effects, and this non-uniform emission. The schematic diagram in Figure 4.10(b) illustrates how the micro-LED network alleviates this problem. By separating the emissive active region into multiple micro-scale elements, the SLS layers within each micro-LED elements are no longer interconnected, and thus do not play a role in current spreading [35]. Instead, the role of current spreading is now transferred to the n-type GaN layer underneath, with much lower resistance. The result is a device with much better current spreading, uniform emission across the active region, and an overall increase of optical output.

Optical properties and performance

The major factor limiting the performance of a nitride LED is certainly light extraction efficiency. No matter how high the IQE may be, the large difference in refractive index between GaN and air restricts the fraction of light that can be extraction to

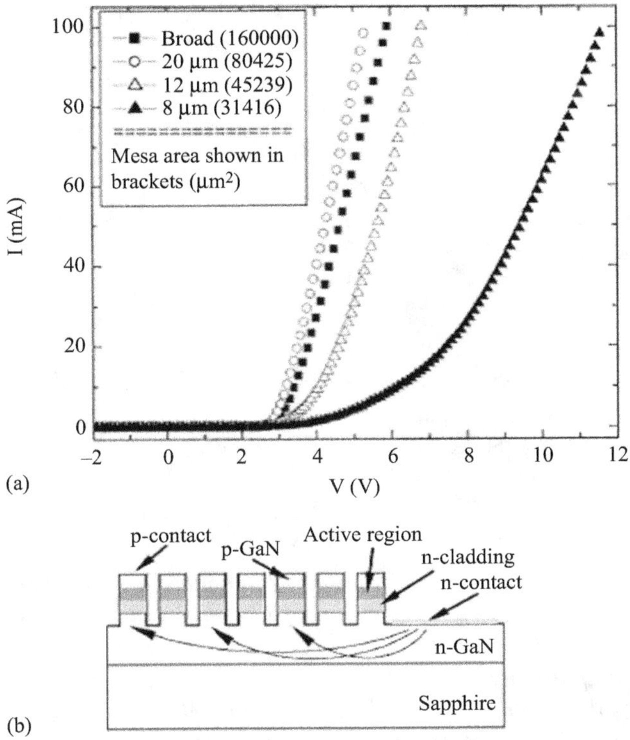

Figure 4.10 (a) I–V characteristics of micro-LED arrays of different disk diameters and (b) schematic diagram illustrating current flow (and current crowding effects) in a UV micro-LED. Figure 4.10(b) reprinted with permission from Applied Physics Letters 86, art. 053504. © 2005, AIP Publishing LLC

below 20% in a typical device geometry. Much of the confined light is eventually reabsorbed by the material. The micro-LED design offers additional pathways for light to escape the device.

An experiment was performed to verify this postulation [39]. An electron beam was focused to a tiny spot to excite cathodoluminescence (CL) at the center of a micro-disk containing quantum wells. The optical signal was detected by a CCD camera. As expected, a spot of light is detected at the point of electron beam incidence, as depicted in Figure 4.11(a). At the same time, a bright ring of light at the periphery of the micro-disk was clearly visible. These photons were generated at the center of the micro-disk; they were channeled along the plane of the disk either directly or via multiple reflections, and escapes via the micro-disk sidewalls. If the dimensions of disk structures were much larger, the light would have been absorbed. The CL spectra also reveal the presence of resonant modes. Based on the modal spacing and their dependence on pillar diameters, the resonance can be

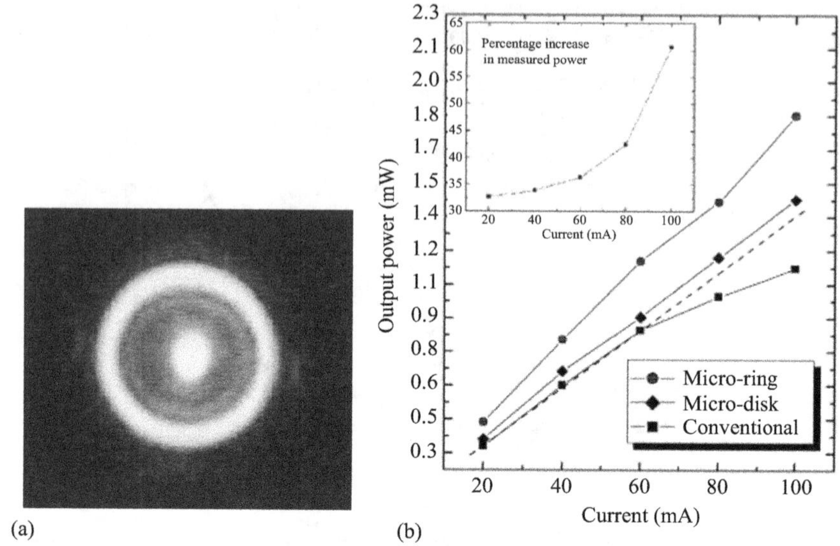

Figure 4.11 *(a) Optical image of a single 12 μm micro-LED element excited by a focused electron beam and (b) L–I plots for micro-ring, micro-disk and conventional LEDs. The percentage increase in measured output power of a micro-ring LED over a conventional LED is shown in the inset. Figure 4.11(a) reprinted with permission from the Journal of Applied Physics 93, pp. 5978–82. © 2003, AIP Publishing LLC*

attributed to WG modes, which serves to reduce the stimulated emission threshold for lasing devices [40].

Enhancement of light extraction efficiency has been proven in electroluminescent devices. The optical performances of micro-LEDs of micro-disk and micro-ring geometries were compared with a conventional broad-area LED of equal emission area, results of which are shown in Figure 4.11(b). It was found that the micro-ring LED had the highest output, as it had the largest sidewall area. The benefit was widened as the injection current increases, which may be due to improved heat-sinking effect due to the microstructured surface. These results prove the feasibility and advantage of adopting micro-LED for high-performance devices.

Transient response of micro-LEDs

The transient EL response of a micro-LED was compared to a broad-area LED [29]. It was found that the turn-on response was of the order of ~30 ps, while the turn-off transient was shorter than a nanosecond, according to the plots in Figure 4.12(a). The significance of the high-speed response of micro-LEDs is that they can potentially replace laser sources as inexpensive short distance optical links in communication networks. More recently, subnanosecond response from matrix-addressable

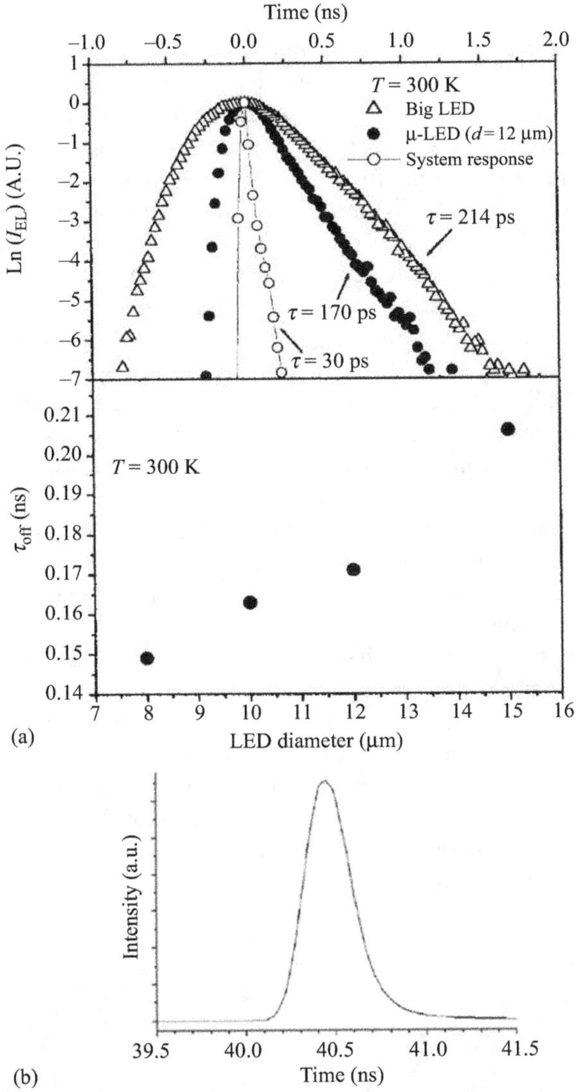

Figure 4.12 *(a) Transient response of a micro-LED compared to a conventional LED, and size dependence of the turn-off time of micro-LEDs and (b) time-correlated single-photon counting (TCSPC) histogram illustrating a 300-ps optical pulse from a micro-LED pixel. Figure 4.12(a) reprinted with permission from Applied Physics Letters 78, pp. 3532–4. © 2001, AIP Publishing LLC*

micro-LED arrays was also reported [41]. A 16 × 16 near-UV emissive array was flip-chip bonded to a custom CMOS driver capable of delivering pulses as short of 300 ps as shown in Figure 4.12(b), opening up the possibility of performing multi-channel time-resolved measurements using micro-LEDs as excitation sources.

Emission characteristics

Just like their large-area counterparts, micro-LEDs emits with a Lambertian distribution as reported in [40] and [42]. The divergent nature of micro-LED emission means that additional optics is needed to manipulate the beams. As GaN LEDs are typically grown on transparent sapphire, it is possible to extract light through the substrate in a flip-chip configuration. Intuitively, optical components can be fabricated directly onto the sapphire surface, creating a monolithic integrated system. This approach was demonstrated in [43], whereby an array of microlenses was formed on the polished sapphire face of a GaN LED by a combination of photoresist reflow and dry etching. A schematic diagram showing the device structure is illustrated in Figure 4.13(a). The resultant device was measured to emit was a narrower divergence cone.

For individually addressable micro-LED arrays, a different approach was adopted to modulate emission. Jeon *et al.* reported on a 64 × 64 370 nm array with integrated polymer microlenses [44]. A layer of optically transparent polymer was spin-coated onto the pixels. The microlenses were then patterned and shaped by photolithography. A schematic diagram of this configuration is shown in Figure 4.13(b). Confocal microscopy indicates that the emission profiles from individual pixels were nearly collimated with narrow divergence. This resolves the crosstalk issue and thus raises the resolution of the devices.

Another way to alleviate the issue of crosstalk, and to facilitate optical signal delivery, is via the use of fiber optics. Instead of using the emissive arrays directly, the signals can first be coupled to an optical fiber. Two different types of fibers have been used for this purpose. In the paper by Xu *et al.* [45], a 30,000 channel imaging fiber was adopted to transmit the optical signals from one end to the other. In a separate report, a specially designed fiber matched to the micro-array was used, demonstrating effective pixel-to-pixel coupling without loss of resolution [42]. The fiber-coupling solution minimizes crosstalk between pixels due to the inherent

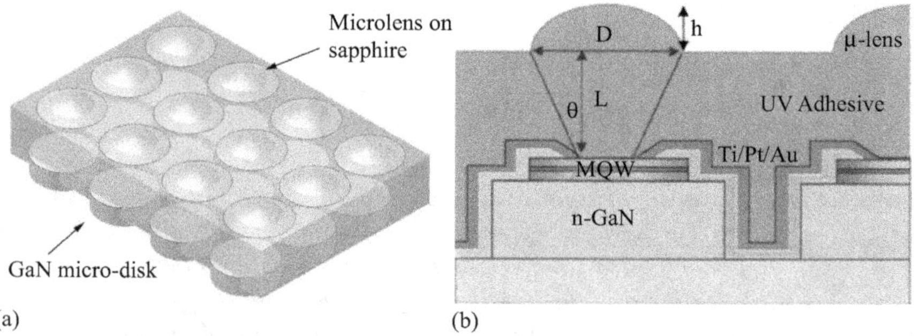

Figure 4.13 Figure 4.13 Schematic diagrams showing (a) a micro-lens array fabricated on the polished sapphire face of a flip-chip micro-LED and (b) a micro-LED device integrated with a polymer lens. Figure 4.13(b) reprinted with permission from Applied Physics Letters 86, art. 221105. © 2005, AIP Publishing LLC

narrow acceptance and divergence cones of fiber optics, although the LED-fiber coupling efficiency needs further optimization. This topic will be discussed further in Section 4.3.2.1.

4.3.2 Fiber-coupling

4.3.2.1 Fiber-coupled LEDs

Fiber-coupled light sources are commonly adopted in a wide range of applications including microscopy, spectroscopy and optical communications, using different types of light sources. Coupled to broadband tungsten halogen or deuterium bulbs, the light guided by fiber can be used for microscopy illumination, ultraviolet curing, fluorescence excitation [46] or biological photo-stimulation [47] among others. However, due to the large dimensions of such light sources compared to a fiber, coupling efficiencies are invariably lower than 10%. On the other hand, the narrow divergence and high directionality of laser beams makes them particularly suitable for efficient coupling to fibers [48]; fiber-coupled laser diodes are thus widely used in optical communication systems [49]. LEDs are miniature light emitters that are rapidly replacing conventional light sources, characterized by high quantum efficiency, quick response time, robustness and long lifetimes among its long list of attributes, and have become popular choices for panel displays, LCD backlights and gradually general lighting. Being light emitters of small footprint that emit with relatively wide divergence, how suitable are LEDs for fiber coupling?

Fiber (butt) coupling efficiency, η, depends mainly on (i) relative dimensions of light source and fiber; (ii) divergence of light source (iii) numerical aperture of fiber and (iv) distance between light source and fiber. Based on condition (i), LEDs, with dimensions similar to those of fiber, are ideal coupling sources. Nevertheless, the omni-directional emissivity of LEDs means that only light emitted from one emission surface of the chip fall within the acceptance cone of the fiber. With traditional light sources, the use of parabolic reflections and focusing optics can improve η. For a light source as small as an LED, the best strategy is probably to minimize the distance between the LED and fiber. Based on a simple electromagnetic wave (EM) model, η from a 1 mm $\times$ 1mm LED with 120° FWHM into a 2-mm-diameter fiber with NA $= 0.4$ is $\sim$ 15% at 0.1 mm separation distance, and η falls off quickly with increasing gap. Reducing divergence through the use of optics, whilst maintaining minimal separation, is thus the key to boosting η. However, the inclusion of external optics incurs additional free-space and interface losses which cannot be neglected. A high-efficiency fiber-coupled LED structure is proposed, achieved through geometrical shaping of the chip and integration of hemispherical optics via liquid capillary bonding. The chip is designed to promote emission from the top surface; the hemispherical lens helps to reduce the emission divergence, while liquid capillary bonding minimizes Fresnel losses between chip and optics and enhance transmission of evanescent waves along the interface. Overall, beam coupling to fiber can be maximized. Of course, there are other methods for improving fiber coupling efficiencies, including the use of lensed fibers. Fibers with concave termination, fabricated by pressing the heated planar cut end of a fiber onto a hemispherical mold

[50], may also be used to collect more light rays and boost η. However, such fibers may not be able to capture as much light as the scheme proposed here.

The LED chips used in this work are fabricated by standard micro-fabrication processes on GaN/sapphire wafers containing InGaN/GaN quantum wells, emitting at a central wavelength of 470 nm. After device fabrication, the wafer is lapped down to $\sim$200 μm and the LED chips are diced and shaped into truncated-conic structures (TCs) by ultraviolet (UV) nanosecond laser micromachining. Chips of circular geometry naturally emit circular beams, enabling maximum coupling with optical fibers, which are invariably circular. Additionally, by coating a metallic layer onto the inclined sidewall, an integrated reflector is formed which suppresses lateral emissions effectively. Such 3D structures are formed with a modified laser micromachining setup as we discussed in the Section 4.2.1, by deflecting the laser beam with a turning mirror so that it is incident onto the wafer at an angle of $\sim$25° to the normal, and rotating the LED chip about its central axis. The shaped chip with top and bottom diameters of 1.7 mm and 1.3 mm was then flip-chip bonded onto a ceramic package, referred to as a TC-LED, which is shown in Figure 4.14(a). Without bond wires, the entire sapphire surface is exposed, on which a 2 mm diameter BK-7 hemispherical lens is attached via liquid capillarity [51]. The hemispherical lens was dipped in acetone and brought into contact with the sapphire surface; a force is then applied to press them together. As the solvent evaporates, the chip and lens are pulled towards each other via liquid surface tension. The drying process can be monitored by observing the evolution of interference fringes. When the interface is free of fringes, the bonding process is complete and a tight bond has been established. The assembly, referenced to as HL-LED, is shown in Figure 4.14(b). For reference, an identical TC-LED is prepared, without attachment of a hemispherical lens. Optical transmission across a capillary-bonded interface is investigated both experimentally and numerically. Optical characteristics of the prepared LEDs are measured. Finally, the beams from the LEDs are coupled to 2 mm-diameter plastic optical fiber (POFs) with an NA of 0.5, and the coupling efficiency evaluated.

The proposed LED assembly is expected to excel due to three major factors: (a) chip geometry; (b) hemispherical lens; and (c) liquid capillary bonding.

Figure 4.14 Microphotographs showing (a) truncated-conical TC-LED; (b) hemispherical lens attached HL-LED; and (c) HL-LED coupled to a POF

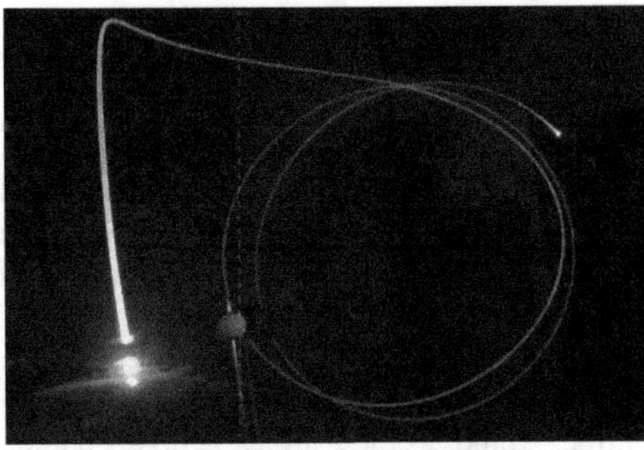

Figure 4.15 Photograph of an LED-fiber coupling system: the illumination LED source coupled to a 10-feet POF

The combination of (a) and (b) is meant to narrow down the divergence of the emitted beam [52]. From two-dimensional far-field radiation patterns of the LEDs measured using a photo-goniometric setup, the FWHM of the HL-LED is effectively reduced to ~120° compared to ~70° for the TC-LED. The lens-integrated device also demonstrates an overall increase in light extraction efficiency. The total emission optical powers, as measured by a radiometric-calibrated spectrometer via a 2-inch integrating sphere, are 3.50 and 3.09 mW for the HL-LED and TC-LED, respectively. This can be attributed to the hemispherical lens, which serves as both an index matching medium and focusing optics. The focusing ability is demonstrated from the emission intensity in the vertical direction, which has been enhanced by 57.5%. The reduced divergence and enhanced optical power in the normal direction suggest that more light can be coupled to a fiber, since more emitted light rays fall within the acceptance cone of a fiber. To determine the fiber coupling efficiencies, light emitted from the LEDs is coupled to a 10 feet long POF with an NA of 0.5, as illustrated in Figure 4.14(c). The fiber is mounted onto a 3-axis micro-positioner for precise alignment so as to capture as much light as possible from the device. The entire system is shown in the photograph of Figure 4.15. The coupling efficiency is determined by comparing the optical power at the end of the fiber with that emitted by the LED, neglecting attenuation along the fiber. The maximum coupling efficiency achieved with the HL-LED is 53.8%, compared to 30.2% from the TC-LED, representing a significant increase of 78%, signifying the effectiveness of the proposed approach for assembling a fiber-coupled light source.

4.3.2.2 Fiber-coupled micro-LED arrays

Apart from the application as micro-displays, micro-LED arrays have also been used for biological imaging and detection [53], fluorescence detection [54] and maskless lithography [55]. Most of the reported two-dimensional micro-LED arrays are designed based on the matrix-addressing method, a strategy which reduces the overall number of bonding pads. However, the complexity of the device fabrication process is increased; so is the requirement on the driving circuitry [56]. To overcome these difficulties, we demonstrate a scheme of generating a two-dimensional micro-display from a linear LED array via a specially designed fiber bundle whereby fibers are linearly arranged on one end, while having a 2D arrangement at the other. Such a 1D to 2D conversion idea is derived from round to keyed linear fiber bundles used for coupling light through the narrow slit at the entrance of a spectrograph.

The fiber bundle enables multiple optical signals to be simultaneously transmitted away from the source, eliminating the need for bulky and lossy optics as with free-space transmission [47]. At the output end, the signals can be projected for imaging purposes. Distinguishing from the use of a 30,000-channel multi-core imaging fiber in the demonstration by Xu *et al.* [57], our fiber-coupling system is further optimized, requiring just as many fiber strands as there are LED pixels without loss of resolution. Apart from imaging purposes, this fiber-coupled LED system can be adapted for multi-channel data transmission [58], a task which has traditionally been fulfilled by laser sources, or for photoexcitation of biological materials [57]. The divergent nature of LED emission results in inevitable crosstalk between adjacent pixels.

The linear micro-LED arrays are designed with 47 pixels arranged across two rows or 64 pixels in a single row fabricated on an MOCVD-grown InGaN/GaN or AlGaN/GaN LED wafer emitting at central wavelengths of 470 and 370 nm, respectively. Photographs of the blue and near-ultraviolet arrays are shown in Figure 4.16(a) and (b), respectively.

The fiber bundle was designed such that the two ends are bi-linear and 2D terminated, respectively. The diameter of each strand of fiber core is about 90 μm. At the input (entrance) end, the fibers are aligned such that the center-to-center spacing and arrangement is identical to that of the blue bilinear micro-LED array. This arrangement allows light from each emitting pixel to effectively couple light into its corresponding fiber. The fibers are twisted along the cable to give a 6×8 two-dimensional arrangement at the output (exit) end, at a resolution of ~1,000 dots per inch (dpi). A three-dimensional schematic diagram of the coupling setup, indicating the conversion from a linear-arranged input to a two-dimensional output, is shown in Figure 4.17.

The device is driven by an FPGA-constructed circuit which enables alphanumeric characters to be displayed. The fiber bundle is mounted on a 3-axis micro-manipulator for precise alignment to the micro-array. Optimal alignment is achieved when

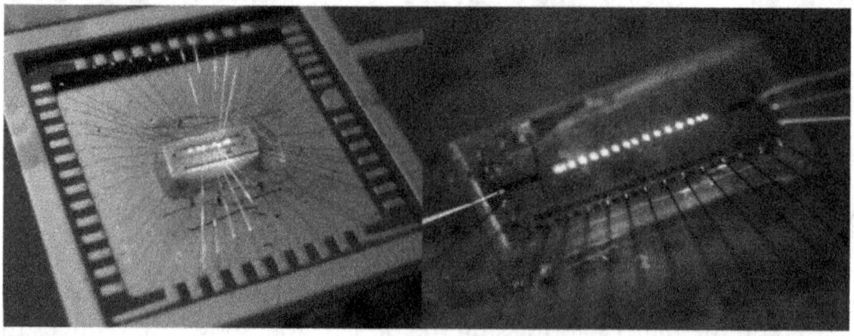

(a) (b)

Figure 4.16 (a) Blue and (b) near-ultraviolet linear micro-LED arrays

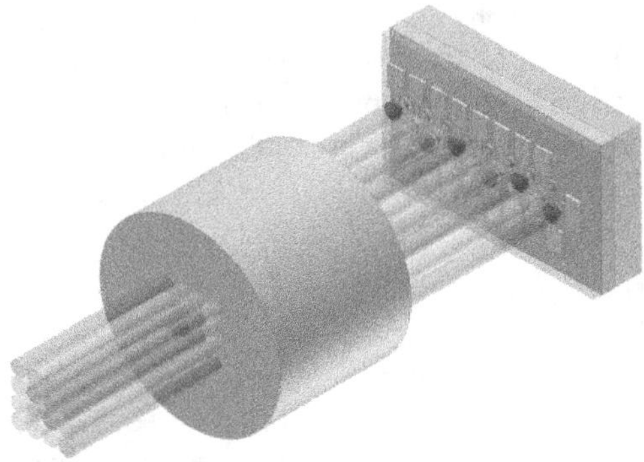

Figure 4.17 A 3D schematic diagram showing the fiber-coupling system, highlighting the pixel-to-pixel nature of coupling between the linearly arrayed LED pixels and the fiber strands, together with fibers arranged in a 6 × 8 array at the output (exit) end

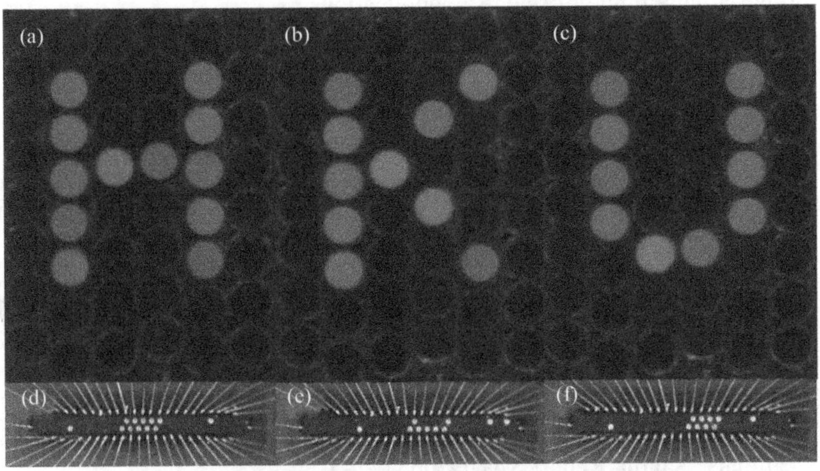

Figure 4.18 (a)–(c) Optical microscope images from the exit end of the fiber, displaying the letters "HKU". The corresponding illuminated pixels on the micro-LED array are shown in (d)–(f).

maximum optical signal is detected at the output end. The measured turn-on voltage of the device is ~3.8 V, while the injection current at 3.3 V is 1.58 mA for each 20-μm pixel. The LEDs are operated at this voltage since they are driven by the digital output of an FPGA board. At this current, each pixel emits 26.9 μW of monochromatic light (measured with a calibrated Si photodetector placed in proximity to the device). Emission at microwatts level is due to the micrometer dimensions of the pixels. With the dual n-pad design, the pixels emit with high uniformity across a row. The functionality of the fiber-coupled micro-LED array as a micro-display is demonstrated by displaying illuminated pixels at the 2D end of the fiber bundle. Figure 4.18

shows optical microphotographs captured from the 6 × 8 fiber array, displaying a sequence of letters "HKU", as well as the corresponding illuminated pixels on bi-linear micro-LED array. The output appears sharp and of good contrast, displaying clear and legible patterns with negligible crosstalk, attributed to the narrow acceptance cone of the fiber.

4.4 Conclusions

LED technology is inherently superior in terms of energy efficiency (with wall plug efficiency exceeding 40%), reliability (lifetime exceeding 50,000 h), robustness and environmental friendliness (involving no hazardous materials). Compared to the CFL which delivers 60 lumens/W of energy consumed, the most advanced white light LEDs are capable of providing over 100 lumens of optical power with the same amount of energy consumed. With further breakthroughs in material epitaxy, chip design, device processing and die packaging, there is plenty of room for boosting energy efficiencies to even higher levels, benefitting not just general lighting but a wide range of scientific applications.

References

[1] S. Nakamura, S. Pearton, and G. Fasol, The Blue Laser Diode, Springer (2000).

[2] H. Amano, M. Kito, K. Hiramatsu, and I. Akasaki, "p-type conduction in Mg-doped GaN Treated with low-energy electron beam irradiation (LEEBI)", *Jpn. J. Appl. Phys.* **28** L2112 (1989).

[3] M. Gherasimova, J. Han, Y.K. Song, A.V. Nurmikko, Y.L. Pan, and R. K. Chang, "VY LEDs for fluorescence detection of biological particles: from materials to applications", *Proc. of SPIE* 6134, Light-Emitting Diodes: Research, Manufacturing, and Applications X, 61340H (2006).

[4] M.H. Crawford, M.A. Banas, M.P. Ross, *et al.*, "Final LDRD report: ultra-violet water purifications systems for rural environments and mobile applications", Sandia Report SAND2005–7245.

[5] Y. Nomura, W. Teshima, N. Tanaka, Y. Yoshida, Y. Nahara, and M. Okazaki, "Thermal analysis of dental resins cured with blue light-emitting diodes (LEDs)", *J. Biomed Mater Res* **63**, 209 (2002).

[6] C. Fritsch, H. Stege, G. Saalmann, G. Goerz, T. Ruzicka, and J. Krutmann, "Green light is effective and less painful than red light in photodynamic therapy of facial solar keratosis", *Photodermatol. Photoimmunol. Photomed.* **13**, 181 (1997).

[7] H.L. Minh, D. O'Brien, G. Faulkner, *et al.*, "100-Mb/s NRZ visible light communications using a postequalized white LED", *IEEE Photon Tech. Lett.* **21**, 1063 (2009).

[8] S.D. Smith, A. Vass, P. Bramley, J.G. Crowder, and C.H. Wang, "Comparison of IR LED gas sensors with thermal source products," *IEE Proc.: Optoelectron.* **144**, 266 (1997).

[9] O. Heikkila, J. Oksanen, and J. Tulkki, "The challenge of unity wall plug efficiency: The effects of internal heating on the efficiency of light emitting diodes," *J. Appl. Phys.* **107**, 033105 (2010).

[10] T. Akasaka, H. Gotoh, T. Saito, and T. Makimoto, "High luminescent efficiency of InGaN multiple quantum wells grown on InGaN underlying layers," *Appl. Phys. Lett.* **85**, 3089 (2004).

[11] S.D. Lester, F.A. Ponce, M.G. Craford, and D.A. Steigerwald, "High dislocation densities in high-efficiency GaN-based light-emitting-diodes," *Appl. Phys. Lett.* **66**, 1249 (1995).

[12] M.S. Minsky, M. White, and E.L. Hu, "Room-temperature photoenhanced wet etching of GaN," *Appl. Phys. Lett.*, **68**, 1531 (1996).

[13] Y. Gao, T. Fujii, R. Sharma, *et al.*, "Roughening hexagonal surface morphology on laser lift-off (LLO) N-face GaN with simple photo-enhanced chemical wet etching," *Jpn. J. Appl. Phys.*, **43**, L637 (2004).

[14] C.C. Yang, R.H. Horng, C.E. Lee, *et al.*, "Improvement in extraction efficiency of GaN-based light-emitting diodes with textured surface layer by natural lithography," *Jpn. J. Appl. Phys.* **44**, 2525 (2005).

[15] S.X. Jin, J. Li, J.Y. Lin, and H.X. Jiang, "InGaN/GaN quantum well interconnected microdisk light emitting diodes," *Appl. Phys. Lett.*, **77**, 3236 (2000).

[16] H.W. Choi, M.D. Dawson, P.R. Edwards, and R.W. Martin, "High extraction efficiency InGaN micro-ring light-emitting diodes," *Appl. Phys. Lett.* **83**, 4483 (2003).

[17] H.W. Choi and S. J. Chua, "Honeycomb GaN micro-light-emitting diodes," *J. Vac. Sci. Tech. B*, **24**, 800 (2006).

[18] J.J. Wierer, M.R. Krames, J.E. Epler, *et al.*, "InGaN/GaN quantum-well heterostructure light-emitting diodes employing photonic crystal structures," *Appl. Phys. Lett.* **84**, 3885 (2004).

[19] D.H. Kim, C.O. Cho, Y.G. Roh, *et al.*, "Enhanced light extraction from GaN-based light-emitting diodes with holographically generated two-dimensional photonic crystal patterns," *Appl. Phys. Lett.* **87**, 203508 (2005).

[20] X.H. Wang, P.T. Lai, and H.W. Choi, "Laser-micromachining of optical microstructures with inclined sidewall profile", *J. Vac. Sci. Tech. B* **27**, 1048 (2009).

[21] W.Y. Fu, K.N. Hui, X.H. Wang, K.K.Y. Wong, P.T. Lai, and H.W. Choi, "Geometrical shaping of InGaN light-emitting diodes by laser micromachining", *IEEE Photon. Tech Lett.* **21**, 1078 (2009).

[22] L. Zhu, X.H. Wang, P.T. Lai, and H.W. Choi, "Angularly-uniform white light-emitting diodes using and integrated reflector cup", *IEEE Photon. Tech Lett.* **22**, 513 (2010).

[23] R.A. Mair, K.C. Zeng, J.Y. Lin, *et al.*, "Optical properties of GaN/AlGaN multiple quantum well microdisks", *Appl. Phys. Lett.* **71**, 2898 (1997).

[24] K.C. Zeng, L. Dai, J.Y. Lin, and H.X. Jiang, "Optical resonance modes in InGaN/GaN multiple-quantum-well microring cavities", *Appl. Phys. Lett.* **75**, 2563 (1999).

[25] S.X. Jin, J. Li, J.Z. Li, J.Y. Lin, and H.X. Jiang, "GaN microdisk light emitting diodes", *Appl. Phys. Lett.* **76**, 631 (2000).
[26] H.W. Choi, C.W. Jeon, M.D. Dawson, P.R. Edwards, and M.D. Dawson, "Fabrication and performance of parallel-addressed InGaN micro-LED arrays", *IEEE Photon. Tech Lett.* **15**, 510 (2003).
[27] S.X. Jin, J. Li, J.Y. Lin, and H.X. Jiang, "InGaN/GaN quantum well interconnected microdisk light emitting diodes", *Appl. Phys. Lett.* **77**, 3236 (2000).
[28] H.W. Choi, C.W. Jeon, and M.D. Dawson, "InGaN microring light-emitting diodes", *IEEE Photon. Tech Lett.* **16**, 33 (2004).
[29] S.X. Jin, J. Shakya, J.Y. Lin, and H.X. Jiang, "Size dependence of III-nitride microdisk light-emitting diode characteristics", *Appl. Phys. Lett.* **78**, 3532 (2001).
[30] H.X. Jiang, S.X. Jin, J. Li, J. Shakya, and J.Y. Lin, "III-nitride blue microdisplays", *Appl. Phys. Lett.* **78**, 1303 (2001).
[31] C.W. Jeon, H.W. Choi, and M.D. Dawson, "A novel fabrication method for a 64 × 64 matrix-addressable GaN-based micro-LED array", *IEEE Photon. Tech Lett.* **15**, 1516 (2003).
[32] H.W. Choi, C.W. Jeon, and M.D. Dawson, "Tapered sidewall dry etching process for GaN and its applications in device fabrication", *J. Vac. Sci. Tech. B* **23**, 99 (2005).
[33] H.W. Choi, C.W. Jeon, and M.D. Dawson, "Fabrication of matrix-addressable micro-LED arrays based on a novel etch technique", *J. Crystal Growth* **268**, 527 (2004).
[34] H.W. Choi, C.W. Jeon, and M.D. Dawson, "High-resolution 128 x 96 nitride microdisplay", *IEEE Elect. Dev. Lett.* **25**, 277 (2004).
[35] H.W. Choi and M.D. Dawson, "Improved current spreading in 370 nm AlGaN microring light emitting diodes", *Appl. Phys. Lett.* **86**, 053504 (2005).
[36] S. Wu, V. Adivarahan, M. Shatalov, A. Chitnis, W.-H. Sun, and M.A. Khan, "Micro-pixel design milliwatt power 254 nm emission light emitting diodes", *Jap. J. Appl. Phys.* **43**, L1035 (2004).
[37] C.W. Jeon, H.W. Choi, E. Gu, and M.D. Dawson, "High-density matrix-addressable AlInGaN-based 368-nm microarray light-emitting diodes", *IEEE Photon. Tech Lett.* **16**, 2421 (2004).
[38] N. Blanc, P. Gueret, P. Buchmann, K. Datwyler, and P. Vettiger, "Conductance statistics of small-area ohmic contacts on GaAs", *Appl. Phys. Lett.* **56**, 2216 (1990).
[39] H.W. Choi, C.W. Jeon, M.D. Dawson, P.R. Edwards, R.W. Martin, and S. Tripathy, "Mechanism of enhanced light output efficiency in InGaN-based microlight emitting diodes", *J. Appl. Phys.* **93**, 5978 (2003).
[40] S. Chang, N.B. Rex, R.K. Chang, G. Chong, and L.J. Guido, "Stimulated emission and lasing in whispering-gallery modes of GaN microdisk cavities", *Appl. Phys. Lett.* **75**, 166 (1999).
[41] J.D. McKendry, B.R. Rae, Z. Gong, *et al.*, "Individually addressable AlInGaN micro-LED arrays with CMOS control and subnanosecond output pulses", *IEEE Photon. Tech Lett.* **21**, 811 (2009).
[42] L. Zhu, C.W. Ng, N. Wong, K.K.Y. Wong, P.T. Lai, and H.W. Choi, "Pixel-to-pixel fiber-coupled emissive micro-light-emitting diode arrays", *IEEE Photonics* **1**, 1 (2009).

[43] H.W. Choi, C. Liu, E. Gu, *et al.*, "GaN micro-light-emitting diode arrays with monolithically integrated sapphire microlenses", *Appl. Phys. Lett.* **84**, 2253 (2004).

[44] C.W. Jeon, E. Gu, and M.D. Dawson, "Mask-free photolithographic exposure using a matrix-addressable micropixellated AlInGaN ultraviolet light-emitting diode", *Appl. Phys. Lett.* **86**, 221105 (2005).

[45] H. Xu, J. Zhang, K.M. Davitt, Y.-K. Song, and A.V. Nurmikko, "Application of blue–green and ultraviolet micro-LEDs to biological imaging and detection", *J. Phys. D* **41**, 094013 (2008).

[46] B.A. Flusberg, E.D. Cocker, W. Piyawattanametha, J.C. Jung, E.L.M. Cheung, and M.J. Schnitzer, "Fiber-optic fluorescence imaging", *Nature Methods* **2**, 941 (2005).

[47] L. Campagnola, H. Wang, and M.J. Zylka, "Fiber-coupled light-emitting diode for localized photostimulation of neurons expressing channel-rhodopsin-2", *J. Neurosci. Methods* **169**, 27 (2008).

[48] F. Dorsch, P. Hennig, and M. Nickel, "High-brightness fiber-coupled diode laser module", *Proc. SPIE* **3285**, 192 (1998).

[49] C. DeCusatis, "Optical data communication: fundamentals and future direction", *Opt. Eng.* **37**, 3082 (1998).

[50] H. Sakata and A. Imada, "Lensed plastic optical fiber employing concave end filled with high-index resin", *J. Light. Tech* **20**, 4 (2002).

[51] E. Yablonovitch, D.M. Huang, T.J. Gmitter, L.T. Florez, and J.P. Harbison, "Van der Waals bonding of GaAs epitaxial liftoff films onto arbitrary substrates," *Appl. Phys. Lett.* **56**, 2419 (1990).

[52] C. Griffin, E. Gu, H.W. Choi, *et al.*, "Beam divergence measurements of InGaN/GaN microarray light-emitting diodes using confocal microscopy", *Appl. Phys. Lett.*, **86**, 041111 (2005).

[53] H. Xu, J. Zhang, K.M. Davitt, Y.-K. Song, and A.V. Nurmikko, "Application of blue–green and ultraviolet micro-LEDs to biological imaging and detection", *J. Phys. D: Appl. Phys* **41**, 094013 (2008).

[54] Y.-L. Pan, V. Boutou, R.K. Chang, I. Ozden, K. Davitt, and A.V. Nurmikko, "Application of light-emitting diodes for aerosol fluorescence detection", *Opt. Lett.*, **28**, 1707 (2003).

[55] C.W. Jeon, E. Gu, and M.D. Dawson, "Polymer microlens arrays applicable to AlInGaN ultraviolet micro-light-emitting diodes", *Appl. Phys. Lett.* **86**, 221105 (2005).

[56] Z. Gong, H.X. Zhang, E. Gu, C. Griffin, and M.D. Dawson, "Micro-pixellated flip-chip InGaN light emitting diodes integration with CMOSS", *IEEE Trans. On Elect. Dev.* **54**, 10 (2007).

[57] H. Xu, M.D. Krisina, W. Dong, *et al.*, "Combining multicore imaging fiber with matrix addressable blue/green LED arrays for spatiotemporal photonic excitation at cellular level", *IEEE 2008 IEEE J. Sel. Top. Quantum Electron.* **14**, 167 (2008).

[58] H.-W. Tsao, W.-M. Cheng, and S.-L. Tsao, "Simulation of 4×4 light coupling to a single-mode fiber", *Microwave and Opt. Tech. Lett.* **46**, 103 (2005).

Chapter 5
GaN-based blue vertical-cavity surface-emitting lasers

Yu-Pin Lan[1], Ying-Yu Lai[2], Yung-Chi Wu[3], Tien-Chang Lu[3], Hao-Chung Kuo[3], and Shing-Chung Wang[3]

In this chapter, we provide a brief review of the optical and electrical pumped GaN-based vertical-cavity surface-emitting lasers from the fabrication technology to output characteristics analysis. Three types of vertical cavity surface-emitting laser (VCSEL) structures are discussed in the following paragraph including their pros and cons among design and process difficulties. First type is fully epitaxial VCSEL structure, second one is the hybrid microcavity consists of a bottom epitaxial distributed Bragg reflector (DBR) and a top dielectric DBR, and third one is vertical resonator composes dielectric top and bottom DBR. We also present several examples of GaN-based VCSELs achieve laser action under optical and electrical pumping at room temperature.

5.1 Introduction

5.1.1 Wide-bandgap III-N materials

Wide-bandgap nitride materials have attracted great attention over past decade due to their promising potential for the applications of optoelectronic devices such as flat panel display, competing storage technologies, automobiles, solid state lighting, and biotechnology, and so on [1–4]. The III-N materials are synthesized mainly using the three kinds of elements of group III, gallium (Ga), aluminum (Al), and indium (In), to form the binary and ternary compounds such as GaN, InN, AlN, $In_xGa_{1-x}N$, $Al_xIn_{1-x}N$, and $Al_xGa_{1-x}N$, etc. The bandgap of these materials cover a very wide range from 0.9 to 6.1 eV (Figure 5.1), which implies the large band offset in heterostructure could be achieved in this material system. The large band offset is very useful to confine carrier for the high-speed and high power electronic devices and light

[1]Institute of Light and Energy Photonics, College of Photonics, National Chiao Tung University, Taiwan
[2]Materials Science and Engineering, National Tsing-Hua University, Taiwan
[3]Institute of EO Engineering & Department of Photonics, National Chiao Tung University, Taiwan

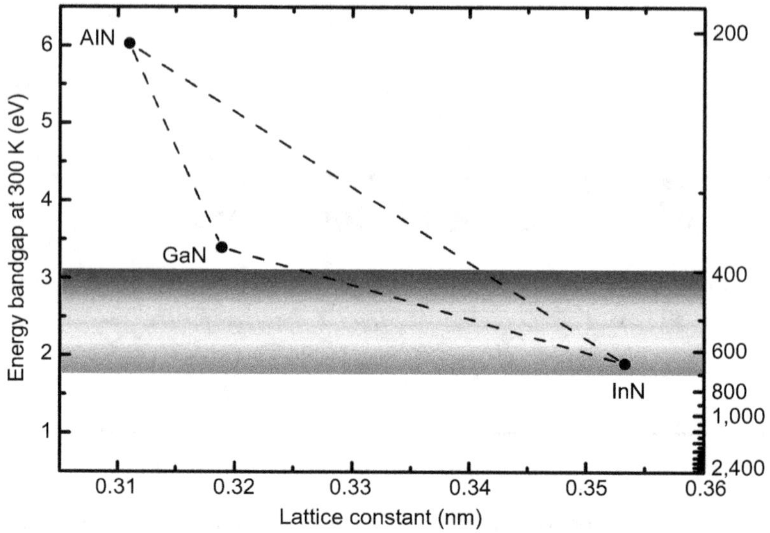

Figure 5.1 The band-gap diagram of III-V group semiconductor materials

emitting devices [5]. In addition, the wide-range direct bandgap among the visible region provides a large potential for full-color display and blue light-emitting diodes (LED)-based solid-state lighting. Furthermore, nitride materials still have lots of advantages such as high bond energy (~2.3eV), high saturation velocity ($\sim 2.7 \times 10^4$ cm/s), high breakdown field ($\sim 2 \times 10^6$ V/cm), and strong excitonic energy (>50 meV) [6,7]. Although wurtzite nitride compounds have some unique properties such as piezoelectric field and spontaneous polarization which is harmful to the efficiency of light emitting devices, the material system still is a very promising candidate for the future optoelectronic applications due to their superior material properties.

5.1.2 Nitride-based VCSELs

5.1.2.1 Vertical cavity surface-emitting lasers

Although the optimization of the edge-emitting laser keeps going, some properties of this kind of laser are unfavorable. One of those properties is its elliptic beam shape. On one hand, the coupling efficiency would be low as the elliptic beam is coupled into optical fiber (typically in the form of circular core). On the other hand, for the application of storage, the elliptic beam not only makes each writing pixel larger but also raises expenses for correcting light shape. Usually, this kind of laser shows slightly large divergence angle to be over ten degree. This also is disadvantageous to the projection. Furthermore, the side-emitting laser devices also makes the testing of devices a tough task. The wafer should be cut into several stripes (several laser devices on one strip) before the testing. For a commercial product, the complicated testing would result in a poor producing efficiency and be disadvantageous. Therefore, in order to have a superior laser device, Iga [8] demonstrated a new kind of laser diodes, VCSEL, in 1977. VCSEL is a vertical-emitting-type laser. It is formed by sandwiching

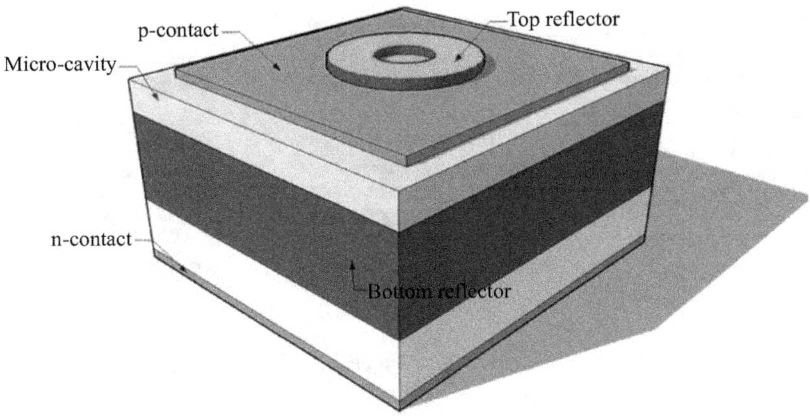

Figure 5.2 The schematic diagram of a vertical-cavity surface-emitting laser diode

a few-lambda cavity in a pair of reflectors, usually in the form of DBR, with a very high reflectivity (>99.9%) as shown in Figure 5.2. In contrast to EELs, photons in the cavity of VCSEL are vertically in resonance and emit outside perpendicularly to the surface of the structure. This laser diode can have many advantageous properties than conventional edge-emitting laser, such as circular beam shape, lower divergence angle, two-dimensional laser array possible, efficient testing, low threshold, and so on. Owing to these superior performances, VCSELs had become very attractive and started to be applied to the commercial products at long wavelength range.

In fact, short-wavelength VCSELs are also very promising for the applications of storage, display, and projection. In particular, the use of two-dimensional arrays of blue VCSELs could further reduce the read-out time in high-density optical storage and increase the scan speed in high-resolution laser printing technology. In recent years, III-nitride EELs have been commercialized and applied to many practical products. But the developments of the VCSEL are still limited in the III-nitride system due to their difficult fabrication process [9–20].

5.1.2.2 III-N-based VCSEL specification

To begin with, the epitaxy technique of III-N material system has significantly improved by Prof. Nakamura at the UCSB in early 1990s, but is still not mature enough as the III-As system. The major obstacle is the high quality epitaxial DBRs. Thus there should be some alternative ways have to be involved in to achieve the VCSEL structure among the III-N system. In view of this, the deposited dielectric type DBR is a promising choice. Therefore, the III-N VCSEL could be roughly divided into three major schemes from the cavity composition: the fully epitaxial mirror type, the hybrid mirror type, and the fully dielectric mirror type, as shown in Figure 5.3.

The top DBR, gain medium, and the bottom DBR are monolithically grown by epitaxy in fully epitaxial mirror type VCSEL scheme. Although such cavity has the

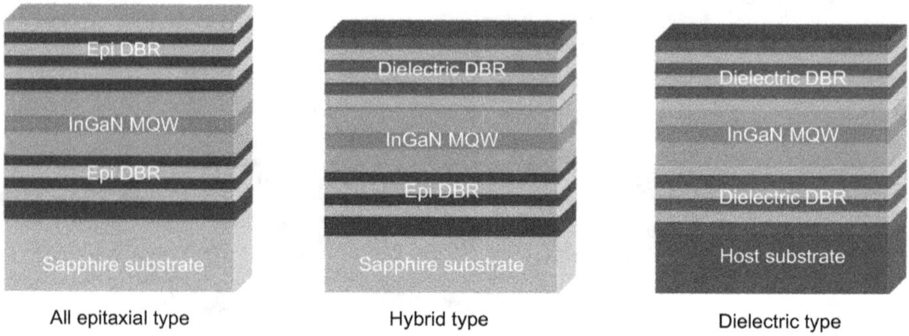

Figure 5.3 The schematic diagram of three nitride-based VCSELs structures

simplest fabrication process since the DBR and the gain medium are already grown by epitaxy, it still suffers from some key issue and is not suitable for practical device application. The most drawback is the poor crystal quality of large numbers of epilayers, small index contrast among nitride material and corresponding nitride-DBR. Furthermore, the top DBR is directly grown on gain medium and prohibits the possibility to insert a transparent conductive layer (TCL) for high-resistance p-GaN layer. The aforementioned innate limitations of fully epitaxial type VSEL result in the high threshold density and only optically pumped devices were reported. As for the hybrid mirror type VCSEL, the *p-i-n* diode gain medium is monolithically grown on the bottom nitride DBR, and the top DBR is deposited onto the gain medium using thin film deposition technique. The dielectric type DBR is not restricted to the lattice constant limitations as the epi-DBR and provides a higher reflectance and wider stop bandwidth by choosing two large index contrast materials. Such structures also makes the insertion of TCL between the p-GaN and the top DBR possible and promises an efficient current injection scheme. The third case, fully dielectric mirror type VCSEL, whose top and bottom DBR are all made of dielectric materials, possess the most complicated fabrication processes among these three types of VCSEL including laser lift off (LLO), metal bonding, lapping, and polishing. But the good DBR performance and the efficient vertical current injection scheme still make this type of VCSEL a very popular way toward the practical applications.

This article is organized as follows. Section 5.2 introduces the concept of distributed Bragg reflection, the fabricated technology and characteristics of epitaxial DBR and dielectric DBR based on nitride materials. Three kinds of GaN-based VCSELs have described in Section 5.3. Finally, we briefly reported a possible method for improving GaN-based VCSEL device.

5.2 Distributed Bragg reflectors

DBRs served as high reflecting mirror in numerous optoelectronic and photonic devices. It is a periodic structure formed by stacking several pairs of two $1/4$-λ-thick layers with different refractive index. Consider a DBR consisting of m

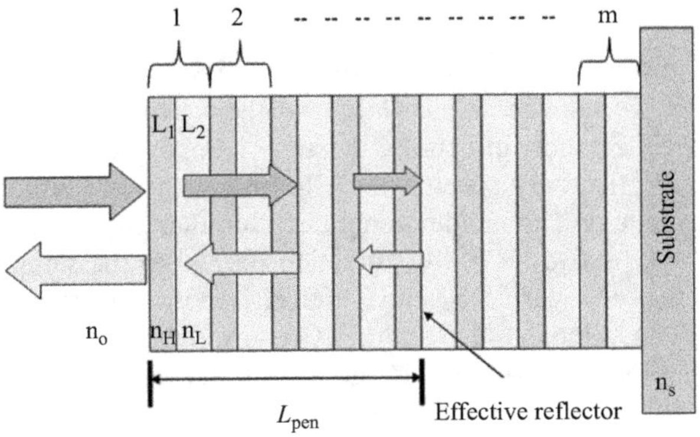

Figure 5.4 A schematic diagram of DBRs

pairs of two dielectric, lossless materials with high- and low- refractive index n_H and n_L, as shown in Figure 5.4. The thickness of the two layers is assumed to be a quarter wave, that is, $L_1 = \lambda_B/4n_H$ and $L_2 = \lambda_B/4n_L$, where the λ_B is the Bragg wavelength.

Multiple reflections at the interface of the DBR and constructive interference of the multiple reflected waves increase the reflectivity with increasing number of pairs. The reflectivity has a maximum at the Bragg wavelength λ_B. The reflectivity of a DBR with m quarter wave pairs at the Bragg wavelength is given by

$$R = \left(\frac{1 - \frac{n_s}{n_o}\left(\frac{n_L}{n_H}\right)^{2p}}{1 + \frac{n_s}{n_o}\left(\frac{n_L}{n_H}\right)^{2p}} \right) \tag{5.1}$$

where the n_o and n_s are the refractive index of incident medium and substrate.

The high-reflectivity or stop-band of a DBR depends on the difference in refractive index of the two constituent materials, Δn ($n_H - n_L$). The spectral width of the stop-band is given by

$$\Delta\lambda_{stopband} = \frac{2\lambda_B \Delta n}{\pi n_{eff}} \tag{5.2}$$

where n_{eff} is the effective refractive index of the mirror. It can be calculated by requiring the same optical path length normal to the layers for the DBR and the effective medium. The effective refractive index is then given by

$$n_{eff} = 2\left(\frac{1}{n_H} + \frac{1}{n_L}\right)^{-1}. \tag{5.3}$$

5.2.1 Nitride-based DBRs

5.2.1.1 The simulation of nitride-based DBRs [21]

In the following, we take the AlN, GaN, and their ternary alloy for example to discuss how the reflectance influent by the alloy composition and the pair number of the epi-DBR. Reflectivity spectra of DBR structures were simulated using the transfer matrix method. The incident angle of illumination and wavelength of the reference light were set to be 0° (the direction normal to the sample surface) and 410 nm, respectively. Figure 5.5 shows the reflectivity spectra of three different nitride-based DBRs, AlN/GaN DBR, $Al_{0.25}Ga_{0.75}N$/GaN DBR, and $Al_{0.35}Ga_{0.65}N$/GaN DBR. The AlN/GaN type has a high reflectivity larger than 99% at 450 nm with fewer pair number of 25. Other two DBRs with low aluminum contents show a large required pairs (>50 pairs) and narrow stop-band (<20 nm). Compared to those nitride reflectors, the AlN/GaN DBR obviously reveals relatively wide stop-band and high reflectivity with relatively few pairs since the index contrast between AlN and GaN is larger than that of AlGaN and GaN. The refractive index of GaN and AlN at wavelength of 410 nm, used as the parameters in the simulation, were $n_{GaN} = 2.45$ and $n_{AlN} = 2.05$. Figure 5.6 shows the reflectance spectrum of 5, 10, 15, and 25 pairs of AlN/GaN DBR. The reflectivity value at center wavelength rapidly rises with the increasing of used pairs. As the pairs of DBR were 25 pairs, AlN/GaN DBR could theoretically achieve a high reflectivity of 99% at 450 nm and a wide stop-band about 46 nm. The superiority of the AlN/GaN DBR also could be further confirmed using the simulation.

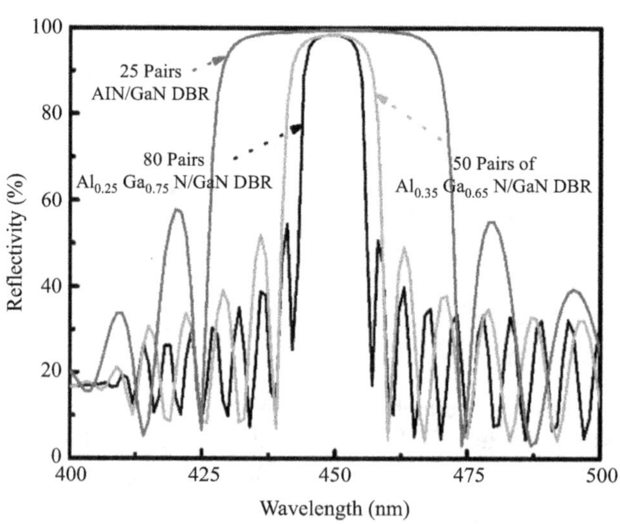

Figure 5.5 Simulated reflectivity spectra of three different nitride-based DBRs with high reflectivity

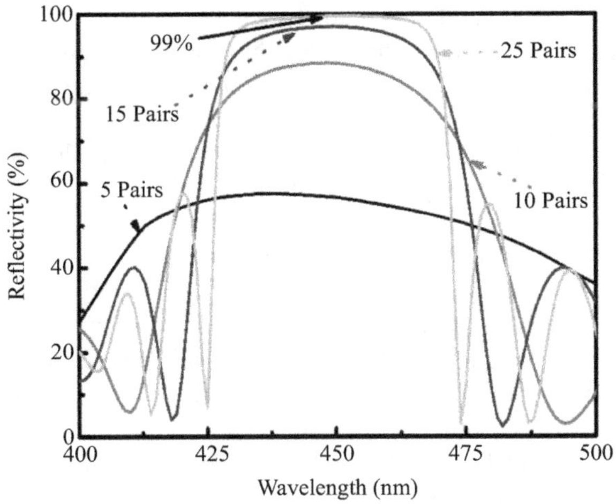

Figure 5.6 Simulated reflectivity spectra of 5, 10, 15, and 25 pairs of AlN/GaN DBRs

5.2.1.2 The fabrication of nitride-based DBRs

Highly reflective DBRs have played a very important role in the operation of VCSELs. High-quality and high-reflectivity DBRs are necessary to achieve threshold condition due to the relatively short gain region of a VCSEL [22,23]. In general, there are several kinds of material systems used in nitride-based DBRs, including AlN/GaN [24–28], AlGaN/GaN [29–32], AlInN/GaN [33–36], and AlGaN/AlN [37,38]. Among these material choices, the AlN/GaN-based DBRs offer the highest refractive index contrast among the III-nitride compounds and provide highly reflective structures together with a large stop-band width. However, the large lattice mismatch between AlN and GaN is up to 2.4%, which generally results in tensile strain and formation of cracks. These cracks tend to grow into V-shaped grooves and seriously affect the reflectivity of the DBR due to scattering, diffraction, and absorption. There is some way to solve this problem such as insertion of AlN/GaN superlattices (SLs). Another way to achieve high-quality nitride-based DBR is using the AlInN and GaN. The AlInN could be lattice matched to the GaN by tuning the Al and In composition. Thus the AlInN/GaN DBR could achieve a large number of pair number without cracks. But the drawback is also obvious to be low index contrast compared with than of AlN/GaN type. In the following, we introduce two approaches to suppress crack formation, one is inserting AlN/GaN SL in AlN/GaN DBR structure and the other is the lattice-matched AlInN/GaN-based DBRs.

5.2.1.2.1 AlN/GaN SL inserted AlN/GaN DBRs [39]

The $Al_xGa_{1-x}N/Al_yGa_{1-y}N$ system is usually used in the DBR structure due to their smaller lattice mismatch. Nevertheless, the refractive index contrast decreases with increasing Al composition in GaN or Ga composition in AlN, which leads to a reduced stop-band width and the requirement of increased number of pairs to achieve high reflectivity. In order to reduce the crack problems encountered in the AlN/GaN DBRs, insertion of the AlN/GaN SL into each of the five DBR periods is a proper way to release the strain induced by the lattice mismatch. The thicknesses of AlN and GaN layers are ~3–5 nm in SL. Figure 5.7 shows cross-sectional transmission electron microscopy (TEM) images of the SL-inserted DBR structure. The lighter layers represent AlN layers while the darker layers represent GaN layers. The interfaces between AlN and GaN layers are sharp and abrupt in low-magnification TEM image. No cracks can be observed in the TEM image. However, detailed observations by this TEM image reveal that the V-shaped defects in the AlN layers are always observable at the GaN-on-AlN interfaces and are filled in with GaN. These V-shaped defects have been reported earlier due to various origins such as stacking mismatch boundaries and surface undulation [40]. The GaN/AlN SL insertion layers were followed by one more AlN layer to identify the change from the AlN layer to the GaN layer. The cross section of one set of 5.5 pairs of GaN/AlN SL insertion layers is shown in Figure 5.7, the interface between GaN and AlN is sharp and abrupt. The GaN/AlN SL insertion layers were followed by one more AlN layer to identify the change from the AlN layer to the GaN layer. Here, a set of GaN/AlN SL insertion layers can be seen as a quasi-alloy of an $Al_xGa_{1-x}N$ layer for a low-refractive-index quarter-wave layer in the DBR structure. The effect of the SL insertion layers on the structural characteristics of the DBR structures can be investigated by the measurement of X-ray asymmetrical reciprocal space map (RSM). The $(10\bar{1}5)$ RSMs of the AlN/GaN-based DBRs without and with SL insertion layers are shown in Figure 5.8, respectively. The perpendicular axis represents the reciprocal lattice *c* and the parallel axis represents reciprocal lattice *a*. Compared with the RSM patterns of non-SL and SL DBRs in Figure 5.8, the diffraction patterns around spots labeled A and B are due to the DBR layers for the high aluminum composition. The diffraction spot labeled C is due to the AlN/GaN SL insertion layers, as shown in Figure 5.8. Both RSM patterns in non-SL and SL DBRs reveal that the DBR layers are under tensile strain that is partially relaxed because the diffraction peaks in the $(10\bar{1}5)$ RSMs do not align in

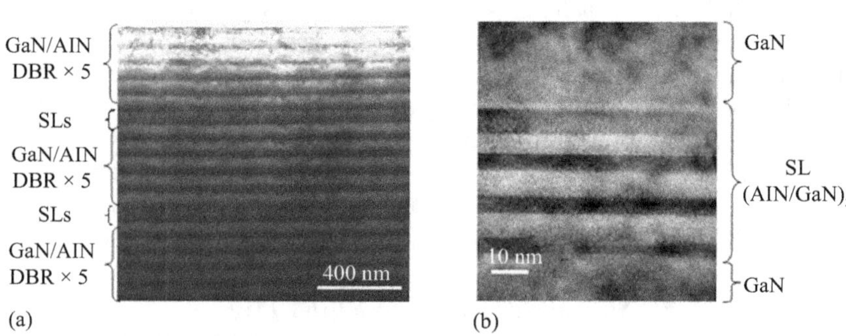

Figure 5.7 Cross-sectional TEM images of (a) the DBR sample and (b) one set of 5.5 pairs of GaN/AlN SL insertion layers [39]

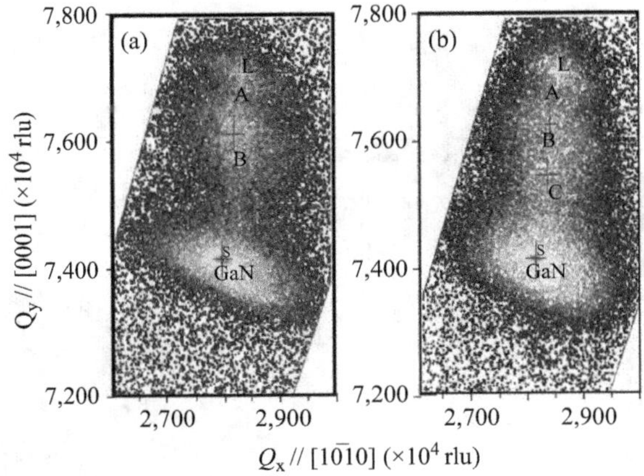

Figure 5.8 Reciprocal space maps: (a) non-SL samples and (b) SL DBR samples [39]

a vertical straight line. The relaxation could be due to the creation of the misfit dislocation [40]. The relaxation process of AlN/GaN SL layers ensures relatively better coherency, i.e., GaN and AlN SL layers are fully strained against each other. Therefore, the SLs behave like effective bulk layers that have an in-plane lattice constant between bulk GaN and AlN-based DBR layers, as shown in Figure 5.8 [41]. The subsequent growth of five-pair AlN/GaN-based DBR could follow the AlN/GaN SLs, which will make the DBR layers suffer relatively smaller strain as compared with DBR layers grown on bulk GaN layers. Consequently, the insertion of the SL layers during the growth of the DBR layers could act as a strain buffers between DBRs and the underlying GaN bulk layer because the in-plane lattice constants of the SL layers are close to those of the AlN layers in the DBRs. Figure 5.9 shows the AFM image of the 29-pair AlN/GaN DBR testing sample. The surface is lumpy and the average roughness measured was about 3.5 nm over a 5×5 μm^2 surface area with crack-free surface morphology. The reflectivity spectrum of the AlN/GaN DBR is shown in Figure 5.10. The peak reflectivity of 99% at 416 nm with flat-topped stop-band of 25 nm was measured.

5.2.1.2.2 Lattice-matched AlInN/GaN-based DBRs

From Figure 5.11, the lattice constant of AlInN could be tuned to match the GaN at appropriate In composition. It seems a very promising way but is limited by difficult epitaxy technique in reality. In 2003, Carlin and Ilegems demonstrated high-reflectivity near-lattice-matched AlInN/GaN-based DBRs by tuning the In composition to ~ 16% [33]. They used an AIXTRON 200/4 RF-S metalorganic vapor phase epitaxy system to grow low-temperature GaN nucleation layer followed by 1-μm-thick GaN buffer layer on 2 in. *c*-plane sapphire substrates. The advantage of using near lattice matched AlInN as the low-index material is shown in Figure 5.11, where the maximum index contrast achievable with AlGaN/GaN in the early stage is

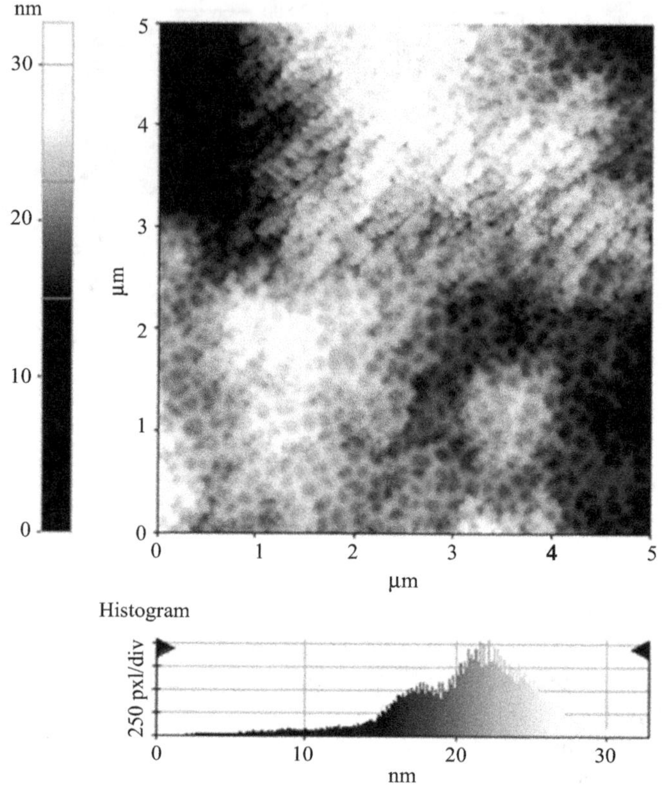

Figure 5.9 AFM image of the DBR sample

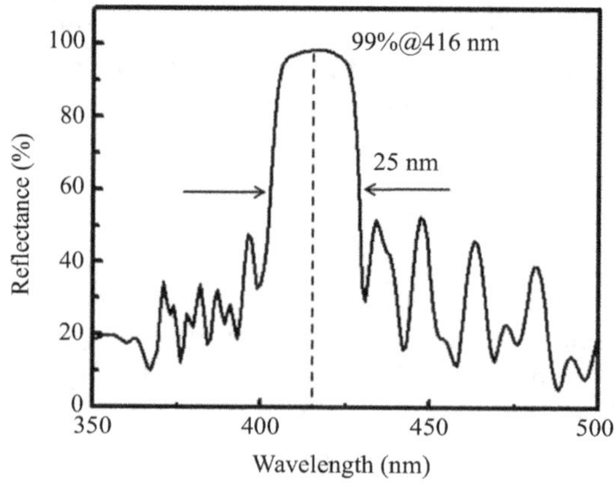

Figure 5.10 The reflectivity spectrum of the AlN/GaN DBR

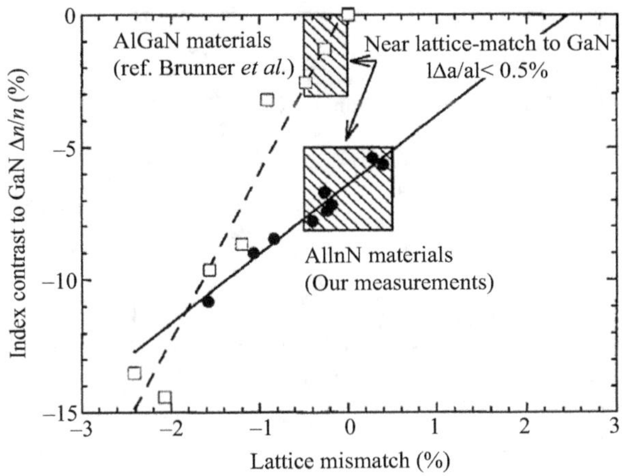

Figure 5.11 Index contrast versus lattice mismatch to GaN: comparison between AlInN/GaN and AlGaN/GaN materials systems [33]

about 3%, while up to 8% is obtained with AlInN/GaN within ±0.5% lattice mismatch that is sufficient to avoid relaxation in blue DBR applications. Figure 5.12 shows the reflectivity spectra of the 20-pair DBRs, a peak reflectivity of over 90% and a 35 nm stop-band width at 515 nm have been exhibited. Although the AlInN/GaN has a great advantage of lattice matching, the difficult epitaxy requirement still restricts its development on practical usage.

5.2.2 Dielectric DBRs

5.2.2.1 The simulation of dielectric DBRs

Compared to the epi-nitride-DBR, dielectric mirror has the advantage of the large refractive index contrast between two different dielectric materials so it needs only a few pairs of DBR to form high reflectivity mirror. In most dielectric DBRs, SiO_2 is widely used as the low refractive index material due to some of its advantaged characteristics such as relative low refractive index than many other dielectric materials. Furthermore, it is easy and cheap to get, hard to decompose, and high transparent window from the wavelength of 180 nm to 8 μm. As for the high refractive index material, Ta_2O_5 is a proper selection owing to benefits of low absorption and high transparency in visible to IR ray. The refractive index of SiO_2 and Ta_2O_5 at wavelength of 450 nm, used as the parameters in the simulation, are $n_{SiO_2} = 1.463$ and $n_{Ta_2O_5} = 2.15$. Figure 5.13 shows the reflectance spectrum of 3, 5, and 8 pairs of Ta_2O_5/SiO_2 DBR. The 8 pairs of Ta_2O_5/SiO_2 mirror can have a high reflectivity of 99% and the wide stop-band about 128 nm. Therefore, we use at least 8 pairs of Ta_2O_5/SiO_2 DBR as the top mirror in the following experiments. And it needs only 8 pairs of Ta_2O_5/SiO_2 DBR to achieve 99% reflectance, which is two times fewer than that of the largest index contrast case in nitride-based case, the AlN/GaN. It presents the high potential of dielectric type DBR on the progress of the nitride VCSELs.

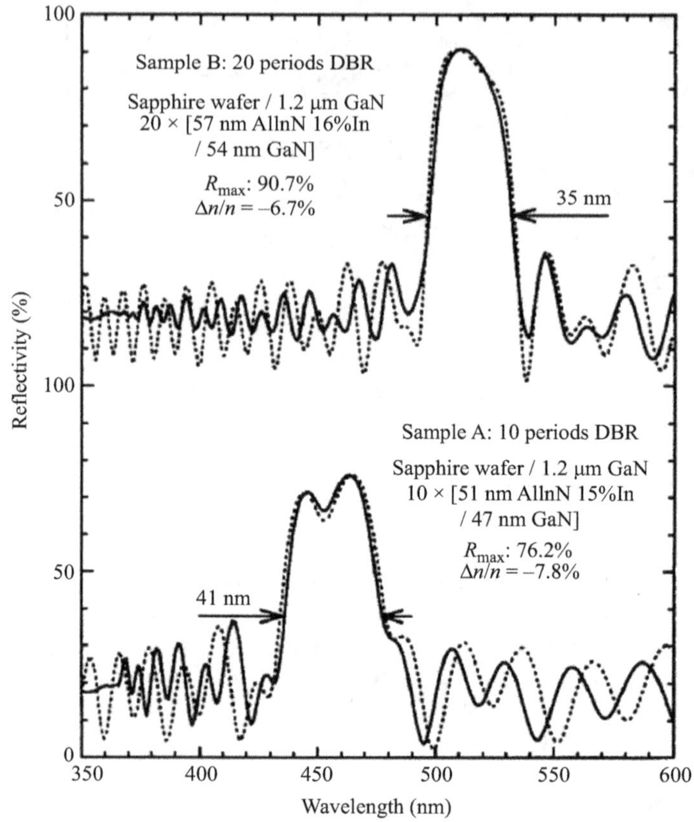

Figure 5.12 Experimental (full lines) and simulated (dotted lines) reflectivity spectra of AlInN/GaN DBRs [33]

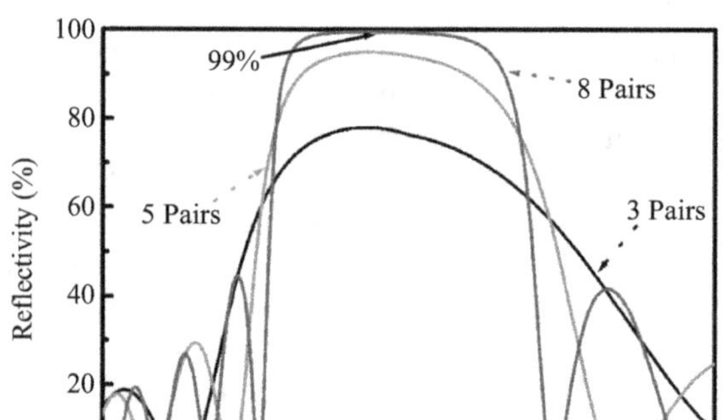

Figure 5.13 Simulated reflectivity spectra of 3, 5, and 8 pairs of Ta_2O_5/SiO_2 DBRs

5.2.2.2 Deposition of dielectric mirrors

There are many deposition techniques such as plasma-enhanced chemical vapor deposition (PECVD), sputter, and e-gun deposition and so on. Here, we present an example of dielectric Ta_2O_5/SiO_2 DBR which is deposited using the electron gun evaporation. The dielectric mirror was coated onto as-grown GaN sample surface in an O_2 ambient at the controlled temperature below 170°C. The reflectivity spectrum of the dielectric mirror was measured as shown in Figure 5.14. The Ta_2O_5/SiO_2 DBR shows a very high reflectivity as high as 99% centered at 451 nm with a wide stop-band of about 130 nm.

5.3 GaN-based VCSELs

5.3.1 *Fully epitaxial nitride-based VCSEL*

Here we take the study reported by Redwing *et al.* as an example. They are the first group to demonstrated optical pumped fully epitaxial GaN-based VCSEL at room temperature [42]. According to the report the fully epitaxial VCSEL structure consists of a 10 μm GaN active region sandwiched between 30-period $Al_{0.12}Ga_{0.88}N/Al_{0.4}Ga_{0.6}N$-based DBRs with the reflectivity values about 84%–93% from theoretical prediction. The fabricated approaches were utilized metalorganic chemical vapor phase epitaxy (MOVPE) to grow a thin AlN buffer layer of 15 nm, followed by growth of a 2 μm $Al_{0.12}Ga_{0.88}N$ spacer layer and the epilayers for the VCSEL structure. They used a pulse nitrogen laser with wavelength of 337 nm, pulse width of 0.8 ns, and pulse energy of 0.94–6.3 μJ/pulse, focus spot of ~70 μm on a cleaved edge of the sample, shown schematically in Figure 5.15.

The surface luminescence emission with half divergent angle of ~22° was collected into monochrometer by lens. Figure 5.16 showed the emission spectra at three different pump intensities. The lasing threshold was estimated as $I_{th} = 2.0$ MW/cm^2.

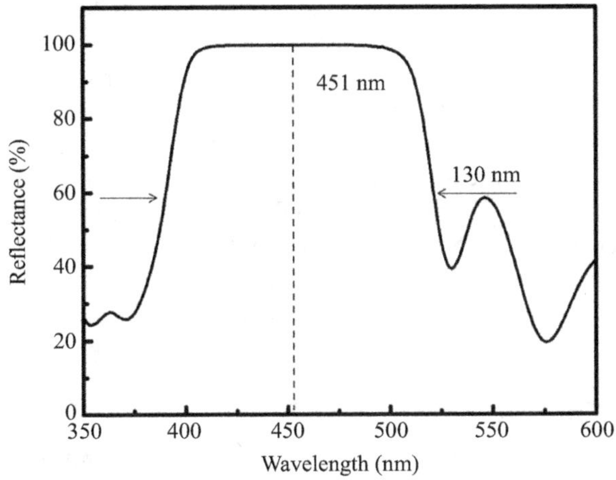

Figure 5.14 The reflectivity spectrum of the Ta_2O_5/SiO_2 DBR

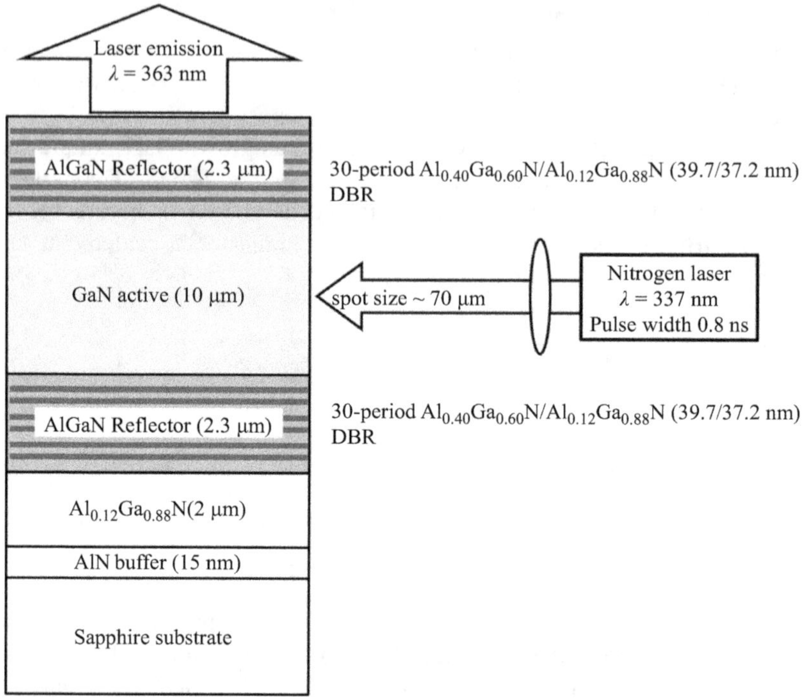

Figure 5.15 The schematic of the VCSEL structure and simply measured map [42]

Below lasing threshold ($\sim 0.65 I_{th}$), a single photoluminescence (PL) spectrum with center peak of 368 nm and full width at half-maximum (FWHM) of ~8 nm (70 meV) was observed. A peak wavelength of 363.5 and linewidth 0.3 nm at input intensity of 2.2 MW/cm^2 ($1.1 I_{th}$) was obvious. At highest powers (2.6 MW/cm^2~$1.3 I_{th}$), additional peaks at 362.1 and 364.8 nm were noted, equally separated from center peak of 363.5 nm by ~1.3 nm. The spectral feature indicated multipass amplification in the cavity considered as Fabry–Perot modes. The FWHM of three modes was measured from 0.2 to 0.5 nm. The results presented above represent the observation of laser action in a GaN-based vertical-cavity structure.

Summaries of the optically pumped GaN-based VCSEL, a narrow laser mode at center wavelength of 363.5 nm and pumping threshold of ~2 MW/cm^2 operating at room temperature was realized. Although the fully epitaxial approach was demonstrated, how to maintain the crystal quality among such thick epitaxial layer and apply to the electrically driven scheme are still challenged.

5.3.2 Hybrid nitride-based VCSEL

The major requirement for the cavity structure of hybrid nitride-based VCSEL, combines an epitaxially grown DBR and a dielectric type DBR, is to grow high-reflectivity and high-quality nitride-based DBRs. The related technologies have

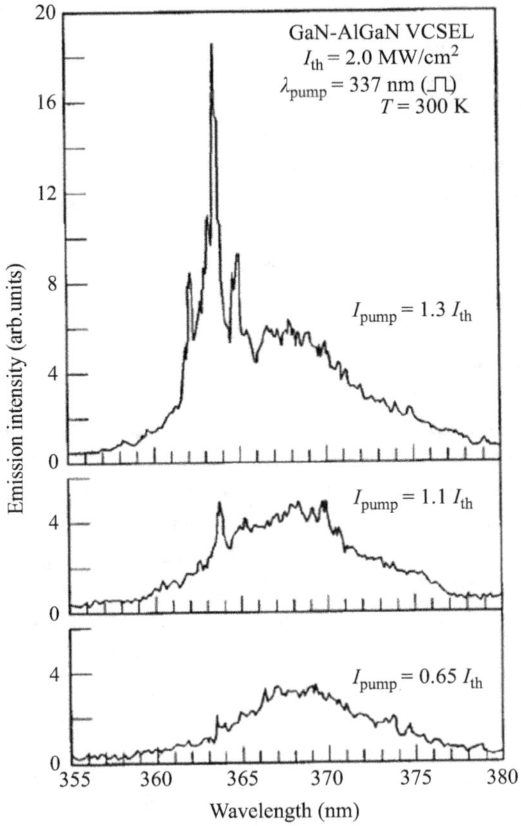

Figure 5.16 Room-temperature surface emission spectra for the GaN-AlGaN VCSEL structure obtained at pump intensities of 1.3 MW/cm^2 (lower trace), 2.2 MW/cm^2 (middle trace), and 2.6 MW/cm^2 (upper trace). These intensities correspond to 0.65 I_{th}, 1.1 I_{th}, and 1.3 I_{th} [42].

been described previous unit. Here we mention our research on hybrid DBR nitride-based VCSEL and perform characteristics of electrically pumped GaN-based hybrid VCSELs. Also, a similar result reported by researchers at ÉCOLE POLYTECHNIQUE FÉDÉRALE DE LAUSANNE (EPFL), in which the bottom DBR is achieved by the AlInN/GaN material system is presented.

5.3.2.1 The growth and fabrication of hybrid DBR VCSEL

The hybrid DBR GaN-based VCSELs were grown in a low-pressure high-speed rotating disk metal–organic chemical vapor deposition (MOCVD) system (EMCORE D75). Two-inch-diameter (0 0 0 1)-oriented sapphire substrates were used for the growth of AlN/GaN-based DBR and cavity. During the growth, trimethylindium (TMIn), trimethylgallium (TMGa), trimethylaluminum (TMAl), and ammonia (NH_3) were used as the In, Ga, Al, and N sources, respectively. The growths of the DBR and micro-cavity are described as follows. First, the substrate was thermally cleaned in the hydrogen ambient for 5 min at 1,100 °C, and then a 30 nm-thick GaN nucleation layer was grown at 500 °C. The growth temperature was raised up to 1,100 °C for the growth of a 2 μm-thick GaN buffer layer. The subsequent epitaxial structure consisted of a 29-pair of quarter-wave AlN/GaN DBR with six AlN/GaN SL insertion layers that are used to reduce tensile strain between the AlN and GaN to avoid crack problem. Each of the five DBR periods inserted one SL for the first 20 pairs of DBRs. Then the SL was inserted into each of the three DBR periods for the remaining nine pairs of DBRs so as to reduce the tensile strain. A 7λ-thick cavity ($\lambda = 410$ nm) including a 860 nm-thick Si-doped n-GaN layer, 10 pairs $In_{0.2}Ga_{0.8}N$/GaN (2.5 nm/12.5 nm) multiple quantum wells (MQWs), a 24 nm-thick AlGaN layer as the electron blocking layer, a 110 nm-thick Mg-doped p-GaN layer, and a 2 nm-thick p^+ InGaN layer as the contact layer. The AlGaN electron blocking layer was served to reduce the electron overflow to the p-GaN layer. A 0.2 μm-thick SiN_x layer was deposited by PECVD as a current confined layer. The current injection aperture of VCSEL devices was about 10μm in diameter. Then, a 30 nm-thick indium-tin-oxide (ITO) layer was deposited as the current spreading layer by an e-gun due to the poor conductivity of the p-GaN layer and annealed at 600 °C for 10 min by rapid thermal annealing. Here, the ITO layer consisted of a 10 nm amorphous ITO layer and a 20 nm crystalline RT ITO layer was employed by an ohmic contact. The 2 nm-thick p^+ InGaN layer on the p-GaN surface can further reduce the series resistance between the thin ITO layer and the p-GaN layer with a slight increase of absorption. Then, the p-contact and n-contact were patterned by a lift-off procedure and deposited with Ni/Au of about 20 nm/150 nm and Ti/Al/Ni/Au of about 20 nm/150 nm/20 nm/150 nm by the e-gun system, respectively. Finally, 10 pairs Ta_2O_5/SiO_2 of the top dielectric DBR were deposited by the ion-assisted e-gun system to complete the whole GaN-based VCSEL devices. The scheme of device is shown in Figure 5.17. Both of 29-pair epitaxial-grown AlN/GaN and the

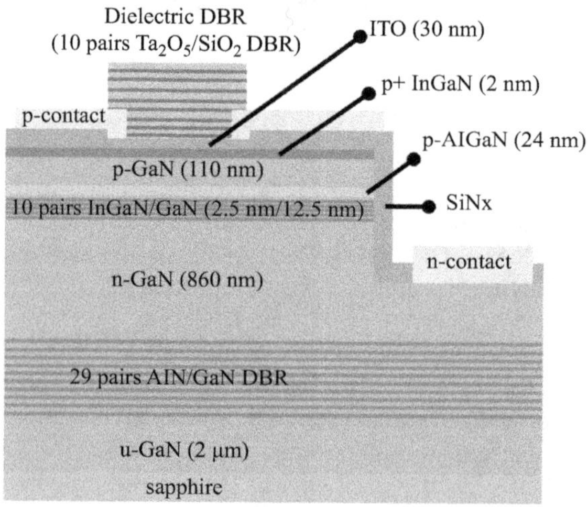

Figure 5.17 The schematic diagram of the GaN-based VCSEL structure

10-pair Ta_2O_5/SiO_2 dielectric DBR were measured higher reflectivity of over 90% at the peak wavelength at 410 nm [43].

5.3.2.2 Characteristics of electrically pumped GaN-based VCSELs

The GaN-based VCSEL devices with current injection apertures of about 10 μm in diameter were tested by using a CW current source (Keithely 238). The emission light was collected by a 100 μm diameter multimode fiber and fed into the spectrometer using a grating of 1,800 g/mm with a spectral resolution of about 0.15 nm. The output from the spectrometer was detected by a CCD to record the emission spectrum. The L-I-V was shown in Figure 5.18, measured at environment temperature of 300 K. A clear lasing transition from spontaneous emission to stimulated emission can be observed, the laser threshold current is around 9.7 mA corresponding to the current density about 12.4 kA/cm^2. The relative low threshold at room temperature operation could be due in part to the successful prevention of carrier overflow by using the electron blocking layer on top of the MQWs and the lower internal absorption loss of the thinner ITO layer. The turn-on voltage is about 4.3 V indicating the good electrical contact of the 30 nm ITO transparent layer and the 2 nm-thick InGaN layers. The output laser intensity from the sample increased linearly with current injection beyond the threshold current. However, the laser intensity started to roll over at higher injection current beyond 15 mA due to the thermal effect.

Figure 5.19 shows the emission spectra of our GaN-based VCSEL devices at current injection of $0.6I_{th}$, $1I_{th}$, and $1.2I_{th}$, respectively. The laser emission wavelength was measured to be 412 nm with a linewidth of about 0.5 nm. The inset of Figure 5.19 shows the CCD image of a lasing spot size of about 2 μm in diameter. Finally, using the angular-resolved measurement system, the laser intensity at different angles emitted from the GaN-based VCSELs was collected by using a 600 μm

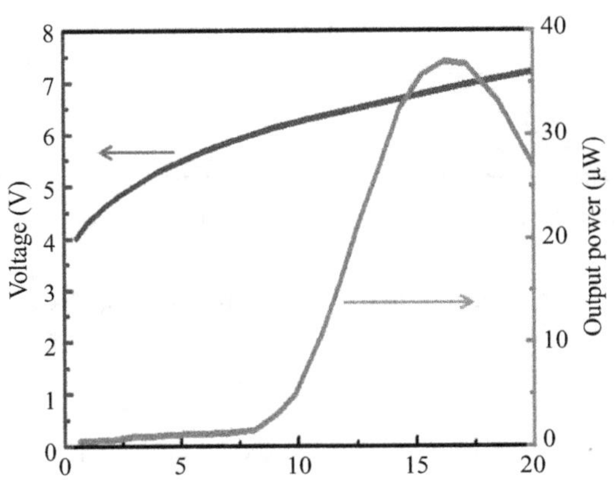

Figure 5.18 The output power and the voltage as a function of injection current at 300 K

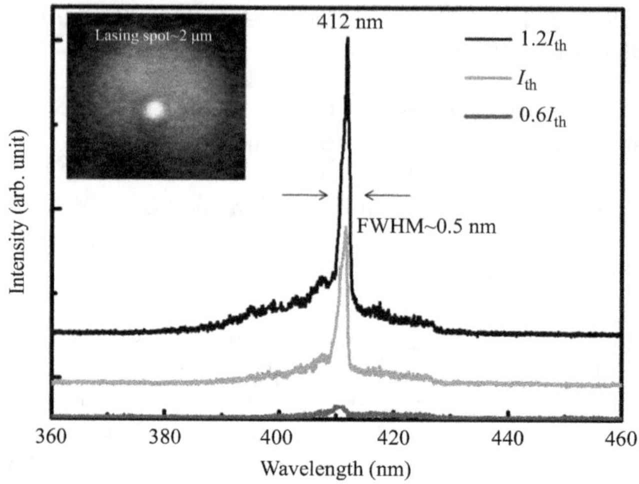

Figure 5.19 The emission spectra of our GaN-based VCSEL devices at current injection of $0.6I_{th}$, $1I_{th}$ and $1.2I_{th}$, respectively. The insert image shows the lasing spot size of about 2 μm in diameter.

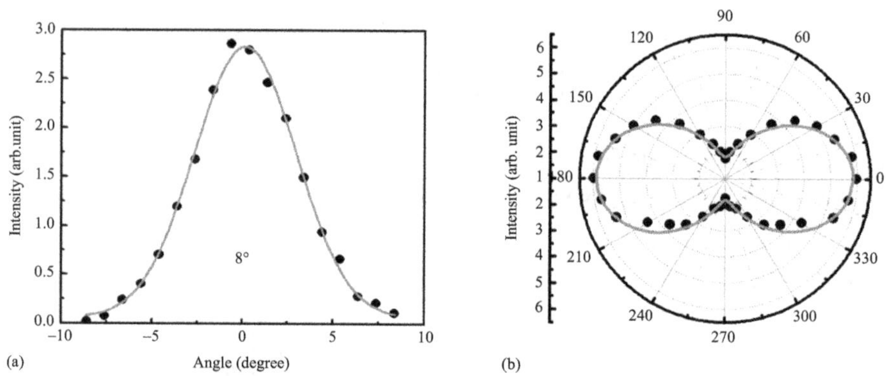

Figure 5.20 (a) The measurement data at different angles and the solid curve is the fitting curve and (b) the DOP of the laser beam [43]

fiber. The measured data at different angles and the fitting curve marked solid line are shown in Figure 5.20. The laser beam divergence angle is about 8°. Figure 5.20 shows the degree of polarization (DOP) of the laser beam. The DOP value was estimated to be about 55°.

In this report, a CW RT operation of GaN-based VCSELs with hybrid mirrors has been demonstrated. The laser has a threshold current at 9.7 mA corresponding to 12.4kA/cm^2. The laser emission wavelength is 412 nm with a linewidth of about 0.5 nm. The DOP, and divergence angle of the laser are measured to be 55% and 8°, respectively.

Except for the AlN/GaN, another material system, AlInN/GaN, which provide not lattice mismatching issue, is also promising to be applied for demonstrating GaN VCSEL. In 2012, the research group lead by Prof. N. Grandjean has demonstrated a VCSEL by using 41.5-pair AlInN/GaN bottom DBR at room temperature under pulsed electrical injection [44]. The VCSEL exhibit a relatively high threshold current density of 140 kA/cm^2, which may be ascribed to absorption losses of the current spreading layer. In view of this, they performed a theoretical calculation and shown that the lasing threshold could be further reduced to ten times lower by adjusting the layer thickness and position of the current spreading layer.

5.3.3 Dielectric DBR nitride-based VCSEL

A dielectric DBR VCSEL has top and bottom dielectric DBR to form a resonated cavity. Since one of the dielectric DBR has to be deposited on one side of the GaN microcavity with the InGaN/GaN MQWs after the GaN thin films were detached from the sapphire substrate by the laser lift-off technique [10]. As a result, the VCSEL fabrication process would become complicated and the precise control of the GaN microcavity thickness would be difficult. Furthermore, the thickness of the GaN microcavity should keep as thick as possible to avoid the damage of the InGaN/GaN MQWs during the laser lift-off process. Such the thick cavity length could increase the threshold condition and reduce the microcavity effect. Here we reported two cases of dielectric DBR GaN-based VCSELs driven by optical pump and electric injection.

5.3.3.1 Optically pumped dielectric DBR VCSEL [10]

The schematic fabrication steps for dielectric DBR VCSEL are shown in Figure 5.21. The layers structure of the GaN-based cavity, grown on a (0001)-oriented sapphire substrate by MOCVD is fabricated by steps: a 30-nm nucleation layer, a 4-μm GaN bulk layer, MQWs consisting of 10 periods of 5-nm GaN barriers and 3-nm $In_{0.1}Ga_{0.9}N$ wells, and a 200-nm GaN cap layer. The peak emission wavelength of the MQWs for the as-grown sample was obtained to be 416 nm. Then, the dielectric DBR consisting of six pairs of SiO_2 and TiO_2 was evaporated on the top of GaN-based cavity. The stop-band center of the DBR was tuned to 450 nm. The reflectivity of the SiO_2/TiO_2 DBR at 414 nm is obtained to be 99.5%. Next, because the adhesion between the dielectric material and GaN surface is weaker than that between the adhesive and GaN surface, an array of disk-like SiO_2/TiO_2 DBR mesas with the diameter of 60 μm was formed by standard photolithography and chemical wet etching process. The DBR mesas were formed in order to increase the adhesion area between the adhesive and GaN to prevent the epitaxial structure separating from the silica substrate during the laser lift-off process. The wafer was then mounted onto a silica substrate using the adhesive. A KrF excimer laser radiation at 248 nm was guided into the sample from back side of the sapphire to separate the sapphire from the epitaxial layers. After the laser lift-off process, the sample was dipped into the H_2SO_4 solution to remove the residual Ga on the exposed GaN buffer layers. In the next step, the sample was lapped and polished by diamond powders to smooth the GaN surface since

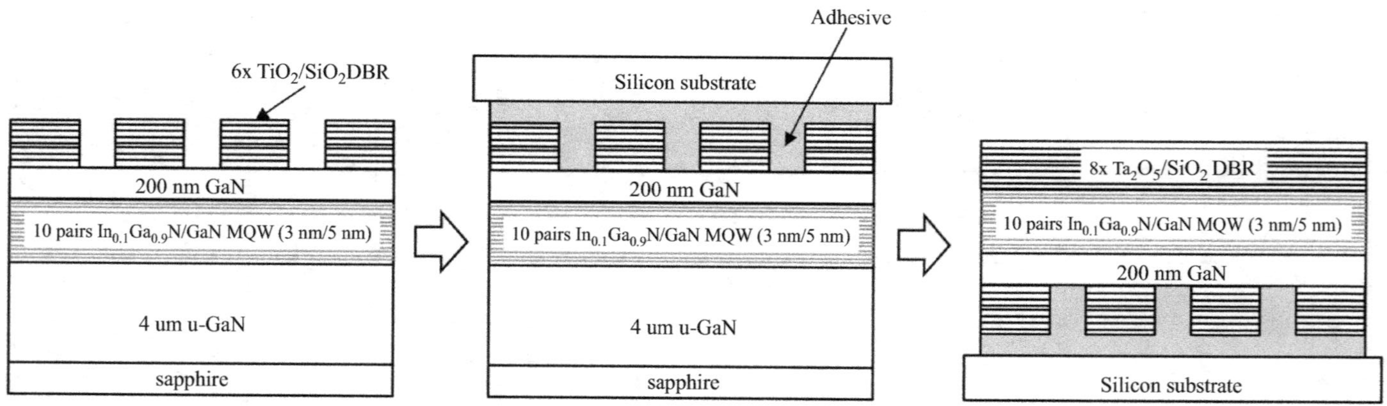

Figure 5.21 Schematic process steps of the dielectric DBR VCSELs incorporating with two dielectric DBRs fabricated by the laser lift-off technique

the laser lift-off process left a roughened surface. The mean surface roughness of the polished GaN surface measured by the AFM is about 1 nm over a scanned area of 20 × 20 μm^2. However, to prevent the possible damage of the quality in MQWs during the lapping process, the 4-μm GaN bulk layer was preserved. Finally, the second DBR consisting of eight pairs of SiO_2 and Ta_2O_5 was deposited on the top of the polished GaN surface. The reflectivity of the SiO_2/Ta_2O_5 DBR at 414 nm is 97%. The stop-band center of the DBR was also tuned to 450 nm. Then, the laser emission characteristics of the completed dielectric DBR VCSELs were investigated by the optically pumped scheme.

Figure 5.22 shows the laser emission intensity from the dielectric DBR VCSEL as a function of the pumping energy at room temperature conditions. A clear evidence of threshold condition occurred at the pumping energy of $E_{th} = 270$ nJ corresponding to an energy density of 21.5 mJ/cm^2. The output laser intensity from the sample increased linearly with the pumping energy level beyond the threshold energy. The pumping light with the emission wavelength of 355 nm has experienced a 60% transmission through the SiO_2/Ta_2O_5 DBR layers and undergone a 98% absorption in the thick GaN layer. Figure 5.23 shows the evolution of the VCSEL emission spectrum with the pumping energy at room temperature. When the pumping energy is below the threshold, the spontaneous emission spectrum shows multiple cavity modes. The mode spacing is about 7 nm corresponding to a cavity length of 4.3 μm, which is nearly equal to the thickness of the epitaxial cavity. The linewidth of a single cavity mode is 0.8 nm as shown in the inset of Figure 5.23. The cavity quality factor (Q factor) estimated from the linewidth is about 518. Considering the optical absorption of GaN layer, an estimated effective cavity reflectivity based on this Q factor is about 97%, which is close to the cavity reflectivity formed by the two dielectric DBRs. This result indicates the laser cavity structure was nearly intact after the laser lift-off process. As the pumping energy

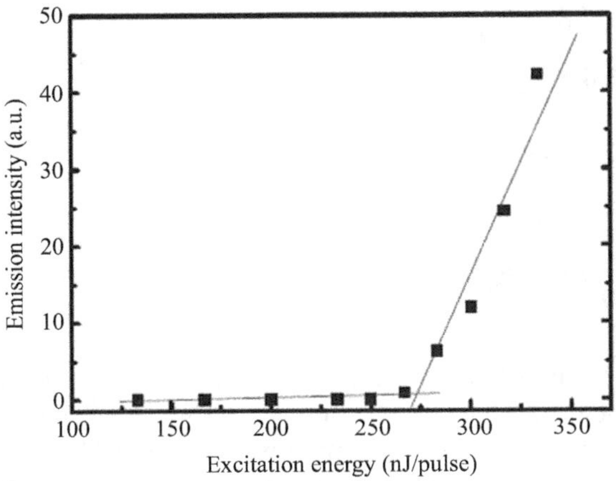

Figure 5.22 The laser emission intensity as a function of the pumping energy at room temperature conditions for the dielectric DBR VCSEL [10]

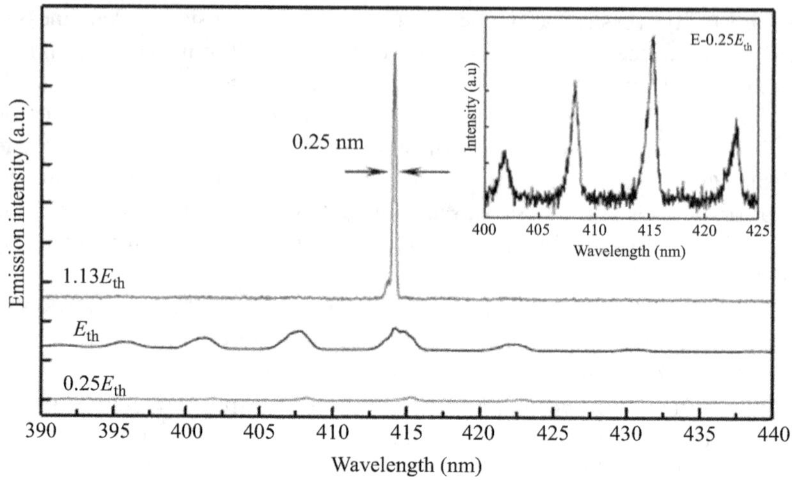

Figure 5.23 Emission spectra from the dielectric DBR VCSEL at various pumping energy. The lasing emission wavelength is 414 nm with a linewidth of 0.25 nm. The inset shows the rescaled emission spectrum under pumping power of 0.25E_{th} [10]

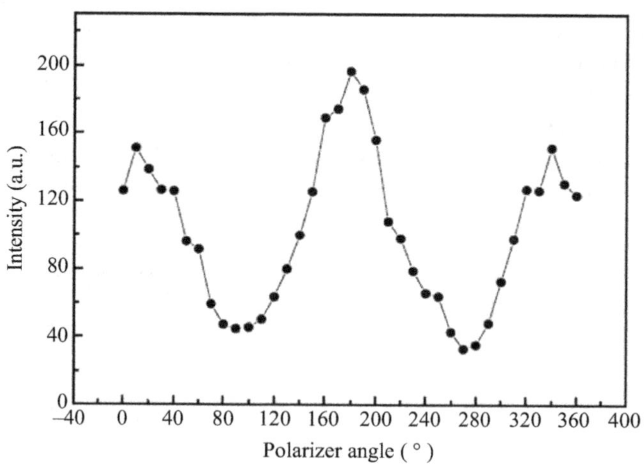

Figure 5.24 The polarization characteristic of the dielectric DBR VCSEL. The DOP is calculated to be 70% [10]

increased above the threshold, a dominant laser emission line appeared at 414 nm with a narrow linewidth of about 0.25 nm. The lasing wavelength is located at one of the cavity modes near the peak emission wavelength of the InGaN MQWs. The laser emission polarization contrast between two orthogonal directions was measured as shown in Figure 5.24. The difference between the two angles of minimum intensities is 180°, and a DOP of about 70% is estimated.

5.3.3.2 Electrically pumped dielectric DBR nitride-based VCSEL

In 2008, Nichia Corporation firstly demonstrated a room-temperature continuous-wave lasing of a blue GaN-based VCSEL with a sapphire-based epitaxial p-i-n diode structure, top and bottom dielectric DBR and a vertical current injection scheme [13]. In 2011, they reported a room-temperature lasing under CW operation for the blue GaN-based VCSEL with a pair of dielectric DBRs and a vertical current injection configuration using GaN substrates for better epitaxial quality [44]. They grew epitaxial layer on c-plane GaN substrates by MOCVD. Five InGaN/GaN MQWs constructed the active layer. A thin ITO layer was formed as a p-type ohmic contact and a current spreading layer [45]. The VCSELs were fabricated by conventional lithography, sputtering, inductively coupled plasma reactive ion etching (ICP-RIE), bonding to a silicon substrate, and chemical mechanical polishing. Measurements were performed at room temperature on a wafer under CW operation for the blue GaN-based VCSEL, which had a current aperture diameter of about 8 μm.

The current–light output (I–L) and current–voltage (*I–V*) characteristics of the blue GaN-based VCSEL are shown in Figure 5.25. The light intensity increases abruptly due to the transition from spontaneous emission to laser emission when the injection currents exceed approximately 1.5 mA. The threshold voltage is approximately 3.3 V. The threshold current densities were estimated to be 3.0 kA/cm^2 by assuming uniform current injection over the current aperture. The maximum output power of the blue GaN-based VCSEL was 0.70 mW at 11 mA. Figure 5.25 shows emission spectra for three different driving currents of $0.8I_{th}$, near I_{th}, and $1.2I_{th}$, respectively. The intensities of the 451 peaks increased dramatically above the threshold current is clearly revealed. Line width narrowing was not observed since the measurement resolution was about 1 nm.

In summary, we give two examples to briefly introduce optically pumped and electrically pumped dielectric DBR GaN-based VCSELs at RT. The optically pumped GaN-based VCSELs had a Q factor of 518, a threshold pumping energy of

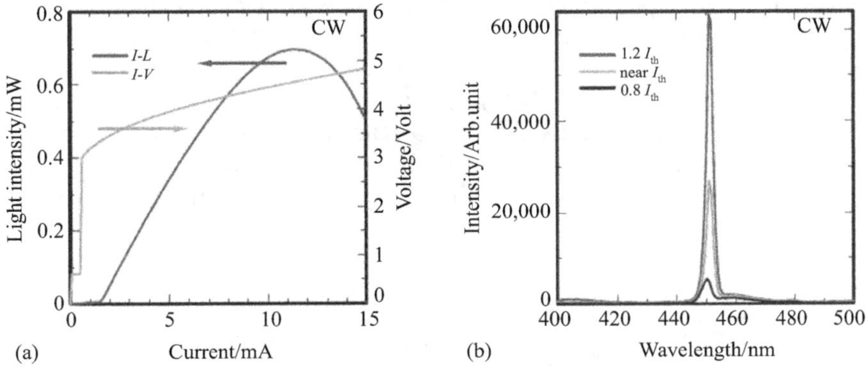

Figure 5.25 (a) I–L and I–V characteristics of GaN-based blue VCSEL under CW operation and (b) emission spectra of GaN-based VCSELs for three different driving currents of $0.88I_{th}$, near I_{th}, and $1.2I_{th}$, respectively [44]

270 nJ, and the laser emits blue–violet wavelength light at 414 nm with a narrow linewidth of 0.25 nm. The laser emission has a DOP of 70%. The laser emission patterns clearly indicate a vertical lasing action of the VCSEL. The current drive blue GaN-based VCSEL had a current aperture diameter of about 8 μm, a threshold current of 1.5mA (3.0 kA/cm^2) and a threshold voltage of 3.3 V. The emission peak was 451 nm and had a maximum output power of 0.7 mW.

5.4 Recent progress and future perspectives

Apart from the aforementioned reports, several interesting reports on the dielectric DBR-based GaN VCSELs by not using traditional approaches such as LLO and bonding, have been demonstrated [46–49]. Following shows two types of electrically pumped GaN VCSEL and corresponding performance.

Traditionally, fabrication of dielectric DBR-based GaN VCSEL containing the LLO and wafer bonding process would cause unwanted damages to the gain medium and difficult thickness controlling capability. To avoid such issues, Prof. S. Nakamura at the UCSB developed a new approach to combine two dielectric DBR into GaN VCSEL [46]. The VCSEL structure is shown in Figure 5.26. They utilize photo-electrochemical (PEC) to undercut the $In_{0:12}Ga_{0:88}N$ sacrificial layer with a precise thickness control. The fabricated VCSEL can be electrically injected at room-temperature and exhibit a single longitudinal mode lasing at 411.9 nm. Due to the non-polar GaN they used, the polarization direction of all VCSELs is locked along the $[1\bar{2}10]$ a-direction of GaN crystal structure.

In 2015, Sony Corporation reports a dielectric DBR GaN VCSEL structure by utilizing the lateral overgrowth on dielectric DBR array capped GaN substrate [47]. Through the epitaxial lateral overgrowth, the GaN gain medium can be sandwiched between two dielectric DBR, as shown in Figure 5.27. Without the complicated process of traditional VCSELs devices, such type of device can provide mW level output power, which is the highest value so far.

To achieve high-performance RT CW GaN-based VCSELs, several important issues including low electrical conductivity of p-type GaN, growing high reflectivity DBR and high material quality should be improved. To improve the lateral current spreading of p-type GaN layer, the ITO film has been widely used as a transparent conductor due to its high transparency (~90%) in the visible spectrum and its low electrical resistivity (< 10^3 Ω·cm). Nevertheless, we mentioned an approach to improve both of current injection and optical confinement of VCSEL

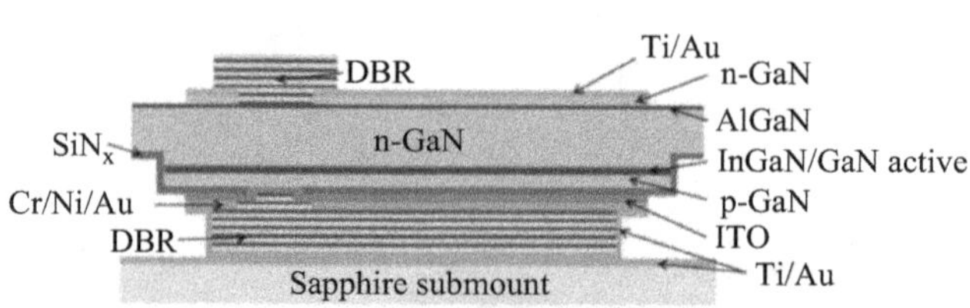

Figure 5.26 VCSEL structure fabricated by PEC technique [46]

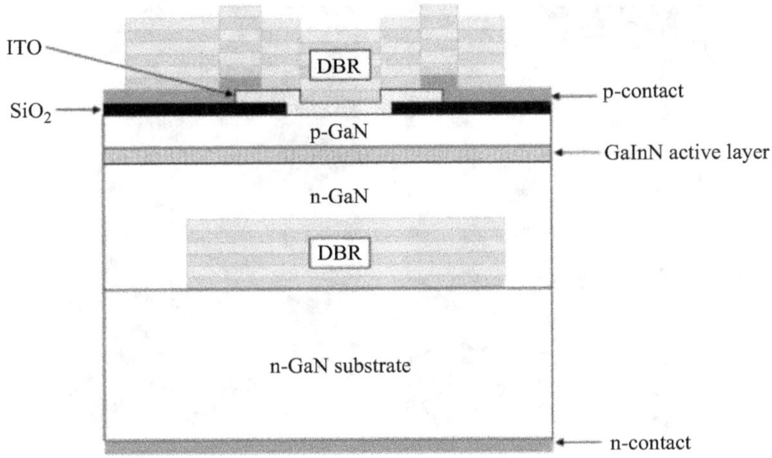

Figure 5.27 VCSEL structure fabricated by lateral overgrowth [47]

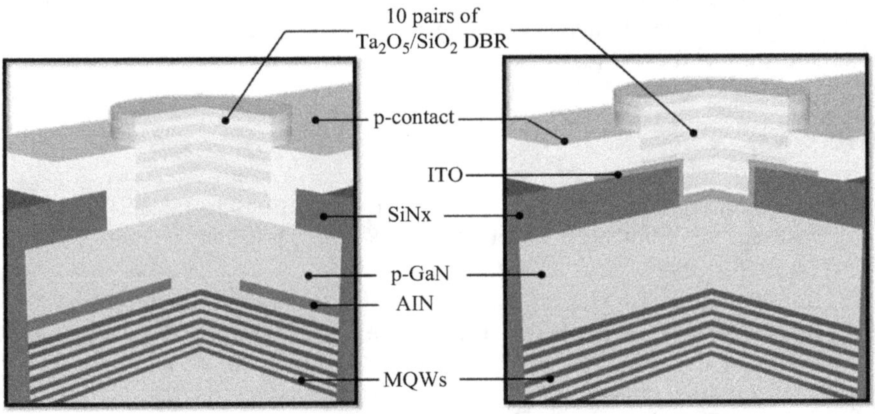

Figure 5.28 The diagram of MCLED with a buried AlN current aperture shows in the left side and right side is without a buried AlN current aperture [49]

structures by inserting a thin AlN layer on the InGaN/GaN MQW as a current blocking layer and optical confinement layer. We realized the concept on a GaN-based high-Q microcavity light-emitting diode (MCLED), the simple scheme is shown in Figure 5.28 [49]. Comparison of MCLED with and without a buried AlN current aperture in the wafer test, shown in Figure 5.29, the MCLED with a buried AlN current aperture shows a concentrated emission light, not only confirms the current is uniformity injection into the current aperture but also shows the AlN plays a role of current blocking layer successfully.

We still have a lot of work ongoing to improve the optical and electrical characteristics of GaN-based VCSELs. One is the heat dissipation problem due to

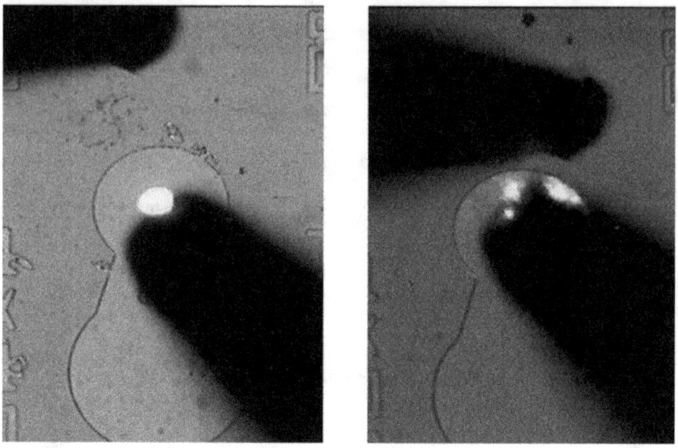

Figure 5.29 The wafer test of MCLED with and without a buried AlN current aperture

worse thermal conductivity of p-type GaN and sapphire substrate. Another one is to get better optical confinement according to large refractive index different. The research results will continually be developed in the future.

References

[1] S. Nakamura, "The roles of structural imperfections in InGaN-Based blue light-emitting diodes and laser diodes," *Science*, vol. 281, pp. 956–961, Aug 14 1998.

[2] S. Nakamura, T. Mukai, and M. Senoh, "Candela-class high-brightness InGaN/AlGaN double-heterostructure blue-light-emitting diodes," *Applied Physics Letters*, vol. 64, pp. 1687–1689, Mar 28 1994.

[3] S. Nakamura, M. Senoh, N. Iwasa, and S. Nagahama, "High-power InGaN single-quantum-well-structure blue and violet light-emitting-diodes," *Applied Physics Letters*, vol. 67, pp. 1868–1870, Sep 25 1995.

[4] S. Nakamura, M. Senoh, S. Nagahama, *et al.*, "Room-temperature continuous-wave operation of InGaN multi-quantum-well-structure laser diodes with a long lifetime," *Applied Physics Letters*, vol. 70, pp. 868–870, Feb 17 1997.

[5] Y. Arakawa, "Progress in GaN-based quantum dots for optoelectronics applications," *IEEE Journal of Selected Topics in Quantum Electronics*, vol. 8, pp. 823–832, Dec 10 2002.

[6] S. N. Mohammad and H. Morkoc, "Progress and prospects of group-III nitride semiconductors," *Progress in Quantum Electronics*, vol. 20, pp. 361–525, 1996.

[7] H. Morkoç, *Nitride semiconductors and devices*. Berlin; New York: Springer, 1999.

[8] K. Iga, F. Koyama, and S. Kinoshita, "Surface emitting semiconductor-lasers," *IEEE Journal of Quantum Electronics*, vol. 24, pp. 1845–1855, Sep 1988.

[9] J. F. Carlin, J. Dorsaz, E. Feltin, *et al.*, "Crack-free fully epitaxial nitride microcavity using highly reflective AlInN/GaN Bragg mirrors," *Applied Physics Letters*, vol. 86, pp. 031107, Jan 17 2005.

[10] J. T. Chu, T. C. Lu, H. H. Yao, *et al.*, "Room-temperature operation of optically pumped blue-violet GaN-based vertical-cavity surface-emitting lasers fabricated by laser lift-off," *Japanese Journal of Applied Physics Part 1-Regular Papers Brief Communications & Review Papers*, vol. 45, pp. 2556–2560, Apr 7 2006.

[11] J. T. Chu, T. C. Lu, M. You, *et al.*, "Emission characteristics of optically pumped GaN-based vertical-cavity surface-emitting lasers," *Applied Physics Letters*, vol. 89, pp. 121112, Sep 18 2006.

[12] E. Feltin, R. Butte, J. F. Carlin, J. Dorsaz, N. Grandjean, and M. Ilegems, "Lattice-matched distributed Bragg reflectors for nitride-based vertical cavity surface emitting lasers," *Electronics Letters*, vol. 41, pp. 94–95, Jan 20 2005.

[13] Y. Higuchi, K. Omae, H. Matsumura, and T. Mukai, "Room-temperature CW lasing of a GaN-based vertical-cavity surface-emitting laser by current injection," *Applied Physics Express*, vol. 1, pp. 121102, Dec 5 2008.

[14] K. Iga, "Possibility green/blue/UV surface emitting-lasers Kenichi IGA," *Blue Laser and Light Emitting Diodes*, pp. 263–266, Mar 1996.

[15] C. C. Kao, T. C. Lu, H. W. Huang, *et al.*, "The lasing characteristics of GaN-based vertical-cavity surface-emitting laser with AlN-GaN and Ta2O5-SiO2 distributed Bragg reflectors," *IEEE Photonics Technology Letters*, vol. 18, pp. 877–879, Mar–Apr 2006.

[16] C. C. Kao, Y. C. Peng, H. H. Yao, *et al.*, "Fabrication and performance of blue GaN-based vertical-cavity surface emitting laser employing AlN/GaN and Ta2O5/SiO2 distributed Bragg reflector," *Applied Physics Letters*, vol. 87, pp. 081105, Aug 22 2005.

[17] T. Someya, R. Werner, A. Forchel, M. Catalano, R. Cingolani, and Y. Arakawa, "Room temperature lasing at blue wavelengths in gallium nitride microcavities," *Science*, vol. 285, pp. 1905–1906, Sep 17 1999.

[18] Y. K. Song, H. Zhou, M. Diagne, *et al.*, "A quasicontinuous wave, optically pumped violet vertical cavity surface emitting laser," *Applied Physics Letters*, vol. 76, pp. 1662–1664, Mar 27 2000.

[19] T. Tawara, H. Gotoh, T. Akasaka, N. Kobayashi, and T. Saitoh, "Low-threshold lasing of InGaN vertical-cavity surface-emitting lasers with dielectric distributed Bragg reflectors," *Applied Physics Letters*, vol. 83, pp. 830–832, Aug 4 2003.

[20] H. L. Zhou, M. Diagne, E. Makarona, *et al.*, "Near ultraviolet optically pumped vertical cavity laser," *Electronics Letters*, vol. 36, pp. 1777–1779, Oct 12 2000.

[21] G. Webb-Wood, U. Ozgur, H. O. Everitt, F. Yun, and H. Morkoc, "Measurement of $Al_xGa_{1-x}N$ refractive indices," *Physica Status Solidi A–Applied Research*, vol. 188, pp. 793–797, Nov 23 2001.

[22] W. W. Chow, K. D. Choquette, M. H. Crawford, K. L. Lear, and G. R. Hadley, "Design, fabrication, and performance of infrared and visible vertical-cavity surface-emitting lasers," *IEEE Journal of Quantum Electronics*, vol. 33, pp. 1810–1824, Oct 1997.

[23] K. Iga, "Surface-emitting laser – Its birth and generation of new optoelectronics field," *IEEE Journal of Selected Topics in Quantum Electronics*, vol. 6, pp. 1201–1215, Nov–Dec 2000.

[24] H. M. Ng, T. D. Moustakas, and S. N. G. Chu, "High reflectivity and broad bandwidth AlN/GaN distributed Bragg reflectors grown by molecular-beam epitaxy," *Applied Physics Letters*, vol. 76, pp. 2818–2820, May 15 2000.

[25] T. Shirasawa, N. Mochida, A. Inoue, *et al.*, "Interface control of GaN/AlGaN quantum well structures in MOVPE growth," *Journal of Crystal Growth*, vol. 189, pp. 124–127, Jun 15 1998.

[26] T. Someya and Y. Arakawa, "Highly reflective GaN/Al0.34Ga0.66N quarter-wave reflectors grown by metal organic chemical vapor deposition," *Applied Physics Letters*, vol. 73, pp. 3653–3655, Dec 21 1998.

[27] C. M. Wu, B. P. Zhang, J. Z. Shang, *et al.*, "High-reflectivity AlN/GaN distributed Bragg reflectors grown on sapphire substrates by MOCVD," *Semiconductor Science and Technology*, vol. 26, May 11 2011.

[28] H. H. Yao, C. F. Lin, H. C. Kuo, and S. C. Wang, "MOCVD growth of AlN/GaN DBR structures under various ambient conditions," *Journal of Crystal Growth*, vol. 262, pp. 151–156, Feb 15 2004.

[29] H. S. Cheong, T. V. Cuong, H. G. Kim, *et al.*, "Suppression of cracks and V-shaped defects, and improvement of reflectivity of GaN/AlGaN distributed Bragg reflectors by insertion of multiple interlayers," *Physica Status Solidi A–Applied Research*, vol. 201, pp. 2799–2802, Sep 10 2004.

[30] H. Dartsch, S. Figge, T. Aschenbrenner, A. Pretorius, A. Rosenauer, and D. Hommel, "Strain compensated AlGaN/GaN-Bragg-reflectors with high Al content grown by MOVPE," *Journal of Crystal Growth*, vol. 310, pp. 4923–4926, Nov 15 2008.

[31] N. Nakada, H. Ishikawa, T. Egawa, and T. Jimbo, "Suppression of crack generation in GaN/AIGaN distributed Bragg reflector on sapphire by the insertion of GaN/AIGaN superlattice grown by metal-organic chemical vapor deposition," *Japanese Journal of Applied Physics Part 2-Letters*, vol. 42, pp. L144–L146, Feb 15 2003.

[32] A. Pretorius, T. Aschenbrenner, H. Dartsch, S. Figge, D. Hommel, and A. Rosenauer, "Transmission electron microscopical investigation of AlGaN/GaN distributed Bragg reflectors," in *Physica Status Solidi C: Current Topics in Solid State Physics, Vol 6, Suppl 2*, R. Butte, Ed., ed, 2009, pp. S680–S683.

[33] J. F. Carlin and M. Ilegems, "High-quality AlInN for high index contrast Bragg mirrors lattice matched to GaN," *Applied Physics Letters*, vol. 83, pp. 668–670, Jul 28 2003.

[34] G. Cosendey, J.-F. Carlin, N. A. K. Kaufmann, R. Butte, and N. Grandjean, "Strain compensation in AlInN/GaN multilayers on GaN substrates: Application to the realization of defect-free Bragg reflectors," *Applied Physics Letters*, vol. 98, May 2 2011.

[35] H. Dartsch, C. Tessarek, T. Aschenbrenner, *et al.*, "Electroluminescence from InGaN quantum dots in a fully monolithic GaN/AlInN cavity," *Journal of Crystal Growth*, vol. 320, pp. 28–31, Apr 1 2011.

[36] Z. Gacevic, S. Fernandez-Garrido, D. Hosseini, S. Estrade, F. Peiro, and E. Calleja, "InAlN/GaN Bragg reflectors grown by plasma-assisted molecular beam epitaxy," *Journal of Applied Physics*, vol. 108, Dec 1 2010.

[37] C.-C. Chen, C.-H. Chiu, Y.-C. Yang, *et al.*, "Tunable light emission from GaN-based photonic crystal with ultraviolet AlN/AlGaN distributed Bragg reflector," *Japanese Journal of Applied Physics*, vol. 50, pp. 04DG09, Apr 20 2011.

[38] B. Liu, R. Zhang, J. G. Zheng, *et al.*, "Composition pulling effect and strain relief mechanism in AlGaN/AlN distributed Bragg reflectors," *Applied Physics Letters*, vol. 98, Jun 27 2011.

[39] T. C. Lu, J. R. Chen, S. W. Chen, *et al.*, "Development of GaN-based vertical-cavity surface-emitting lasers," *IEEE Journal of Selected Topics in Quantum Electronics*, vol. 15, pp. 850–860, May/Jun 2009.

[40] H. K. Cho, J. Y. Lee, and G. M. Yang, "Characterization of pit formation in III-nitrides grown by metalorganic chemical vapor deposition," *Applied Physics Letters*, vol. 80, pp. 1370–1372, Feb 25 2002.

[41] S. Einfeldt, H. Heinke, V. Kirchner, and D. Hommel, "Strain relaxation in AlGaN/GaN superlattices grown on GaN," *Journal of Applied Physics*, vol. 89, pp. 2160–2167, Feb 15 2001.

[42] J. M. Redwing, D. A. S. Loeber, N. G. Anderson, M. A. Tischler, and J. S. Flynn, "An optically pumped GaN-AlGaN vertical cavity surface emitting laser," *Applied Physics Letters*, vol. 69, pp. 1–3, Jul 1 1996.

[43] T.-C. Lu, S.-W. Chen, T.-T. Wu, *et al.*, "Continuous wave operation of current injected GaN vertical cavity surface emitting lasers at room temperature," *Applied Physics Letters*, vol. 97, pp. 071114, Aug 18 2010.

[44] D. Kasahara, D. Morita, T. Kosugi, *et al.*, "Demonstration of Blue and Green GaN-Based Vertical-Cavity Surface-Emitting Lasers by Current Injection at Room Temperature," *Applied Physics Express*, vol. 4, pp. 072103, Jun 29 2011.

[45] G. Cosendey, A. Castiglia, G. Rossbach, J.-F. Carlin, and N. Grandjean, "Blue monolithic AlInN-based vertical cavity surface emitting laser diode on free-standing GaN substrate," *Applied Physics Letters*, vol. 101, pp. 151113, Oct 10 2012.

[46] C. Holder, J. S. Speck, S. P. DenBaars, S. Nakamura, and D. Feezell, "Demonstration of Nonpolar GaN-Based Vertical-Cavity Surface-Emitting Lasers," *Applied Physics Express*, vol. 5, pp. 092104, Sep 5 2012.

[47] S. Izumi, N. Fuutagawa, T. Hamaguchi, M. Murayama, M. Kuramoto, and H. Narui, "Room-temperature continuous-wave operation of GaN-based vertical-cavity surface-emitting lasers fabricated using epitaxial lateral overgrowth," *Applied Physics Express*, vol. 8, pp. 062702, May 21 2015.

[48] Y. K. Song, M. Diagne, H. Zhou, A. V. Nurmikko, R. P. Schneider, and T. Takeuchi, "Resonant-cavity InGaN quantum-well blue light-emitting diodes," *Applied Physics Letters*, vol. 77, pp. 1744–1746, Sep 18 2000.

[49] B. S. Cheng, Y. L. Wu, T. C. Lu, *et al.*, "High Q microcavity light emitting diodes with buried AlN current apertures," *Applied Physics Letters*, vol. 99, pp. 041101, Jul 25 2011.

Chapter 6
Quantum cascade lasers

Manijeh Razeghi[1]

Quantum cascade lasers (QCLs) and associated devices are compact, room temperature (RT) sources capable of producing electromagnetic radiation in the mid-infrared and terahertz (THz) spectral regions. This article will discuss the basic operation of QCLs, the evolution of performance in the past several years, and an overview of material growth and characterization, and QCL device fabrication. This article will also review recent research on mid-infrared QCLs that has resulted in record high wallplug efficiency (WPE), high continuous wave (CW) output power, single mode operation, wide wavelength tunability, and RT THz generation.

6.1 Introduction

The word "LASER" is an acronym for "Light Amplification by Stimulated Emission of Radiation". To qualify as a laser, a device should include a gain medium which amplifies light by stimulated emission of radiation, while at the same time provides optical feedback in the form of a resonator. Since the invention of the laser in 1960 [1], solid, liquid, and gas gain media have been developed for this purpose. The diode laser, which is a special class of the solid-state laser and has a semiconductor active medium, has been widely used as a compact source at RT. Some of the common materials for semiconductor lasers are GaInN, generating blue light for Blue-Ray disc optical storage, GaInP, emitting red light for laser pointers and DVD players, AlGaAs, acting as light source in compact disc systems, and GaInAsP, which serves both high power pumping applications in the 0.78–1 μm wavelength range and telecommunication applications at wavelengths from 1.3 to 1.5 μm. Much of the development of this material system has been detailed in previous work [2].

The conventional semiconductor laser undergoes interband transitions, where electrons in the conduction band combine with holes in the valence band in order to emit photons (light). As the wavelength of emitted photons depends on the energy difference between the conduction band and the valence band, large changes in

[1]Department of Electrical Engineering and Computer Science, Northwestern University, USA

wavelength can be obtained only through a change in the semiconductor bandgap, which requires changing the material composition or moving to a new material system.

In contrast to the interband laser, the QCL [3] is a semiconductor laser where light emission takes place by intersubband transitions between energy levels inside artificially created quantum wells. The choice of material is based on the quantum well depth required to reach the shortest wavelength (highest energy) desired. Once the material is chosen, however, the energy of intersubband transitions can be tuned over a very wide range (infrared to THz) simply by changing the thickness of the wells. In addition, intersubband lasers are free from Auger recombination which limits high temperature operation of conventional lasers in the mid-infrared (mid-IR) and THz spectral regions. Another benefit of this technology is that multiple emitting stages can be cascaded in series to achieve a high output power for various applications.

All of these features make the QCL ideally suited to a variety of applications. Perhaps the main initial driving force for QCL development was for chemical spectroscopy. Rotational and/or vibrational resonances in molecules give rise to distinct, characteristics absorption features in the mid-IR and THz spectral regions. The use of lasers to probe these features offers a fast and extremely sensitive modality for the identification and quantification of almost any chemical. Chemicals of interest to detect in very low concentrations include toxic industrial chemicals, nerve agents, and explosives. In the medical field, mid-IR and THz lasers can be used to monitor breath for abnormal cell metabolism byproducts.

Another application of these lasers is for free space communication. In the near- infrared, there has been a lot of research into free-space optical links to complete the "last-mile" connectivity problem for high-speed Internet. These systems have good bandwidth and are cheaper to install than fiber-optic cables. One downside to these systems is their sensitivity to weather. Scattering by rain and fog severely attenuates the signal, which reduces the bandwidth, distance, and/or availability of the data link. At longer wavelengths, it has been proven that the scattering by fog is dramatically reduced.

While the above applications can be performed at moderate laser powers, the recent development of high-power and RT QCLs have significantly expanded or enhanced the application space. Standoff spectroscopy and chemical imaging are currently in development, by which chemicals can be imaged and identified at a safe distance. Infrared countermeasure development, for the protection of military and civilian aircraft, is also an important topic, as the use of QCLs is projected to reduce the size, weight, and cost of these systems significantly. In the medical field, it is even feasible to project the use of high-power QCLs to target and remove specific tissues based on their infrared absorption signatures.

Finally, recent demonstrations of various kinds of THz sources have opened up new applications related to homeland security screening, non-destructive package inspection, and astronomy. THz QCLs have been demonstrated with significant power output at cryogenic temperatures over a wide range. However, there is a noticeable lack of compact, RT THz sources with sufficient power for these applications.

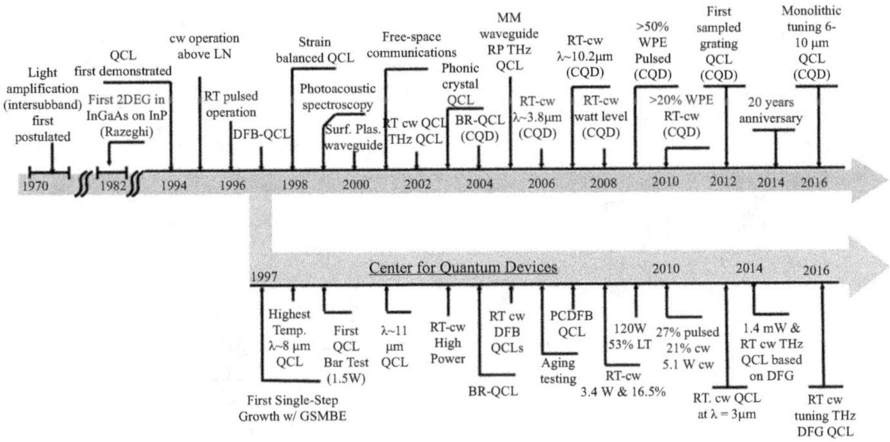

Figure 6.1 *Time line of some important milestones in the development of QCL. 2DEG = two-dimensional electron gas. CW = continuous wave; LN = liquid nitrogen; LT = low temperature; RT = room temperature; DFB = distributed feedback; BR = buried ridge; MM = metal-metal; RP = resonant phonon*

6.2 Summary of quantum cascade laser development

Although the theoretical prediction of an intersubband laser happened in the 1970s (Figure 6.1), it took many design and technological achievements before the first $\lambda = 4.2$ μm QCL could be demonstrated (at 10 K) at Bell Laboratories in 1994 [3]. This demonstration sparked a new wave of innovation, which quickly led to many improvements, including new wavelengths (including cryogenic THz operation) and RT mid-IR operation.

The Center for Quantum Devices (CQD) at Northwestern University began working on the QCL in 1997 [4], and has continuously pushed the envelope for high power QCL development. Many power records have been set and broken, including the shortest and longest wavelength RT, CW QCLs at 3 and 11.5 μm [5,6]. In addition, we have demonstrated the world's highest power conversion efficiency, or WPE, QCLs, with 53% pulsed mode WPE [7] at cryogenic temperature and 27% WPE at RT [8] demonstrated. Narrow linewidth, single mode source technology has also been significantly enhanced, with special attention paid to power scaling of this technology. Most recently, this has been expanded to the development of electrically tunable sources [9] and RT THz emission based on difference frequency generation within a dual-wavelength QCL waveguide [10].

6.3 QCL Basics

The QCL is comprised of a series of quantum wells, often referred to as a superlattice (SL), which is designed to have discrete energy levels due to quantum

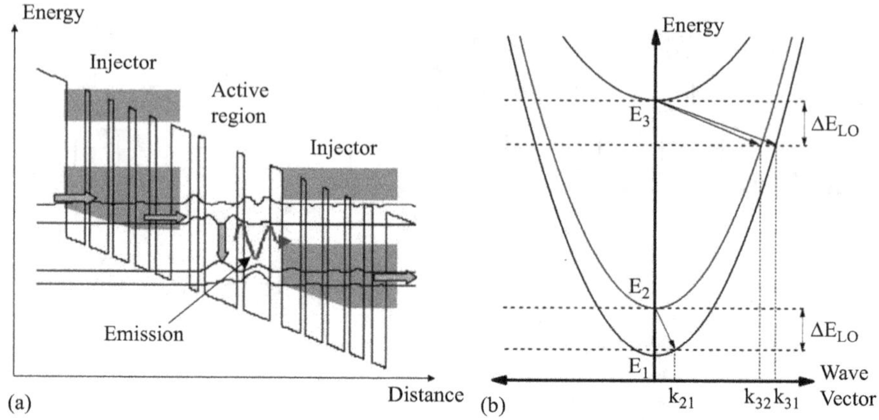

Figure 6.2 *(a) Energy diagram of a typical QCL showing a generalized electron conduction path, relevant active region wavefunctions, and the location of photon emission and (b) dispersion diagram of the upper laser level, lower laser level, and the ground state of a simple QCL active region*

confinement. Functionally, one emitting stage can be broadly divided into two parts, as shown in Figure 6.2(a). The active region is where light emission takes place. In this region, the electrically injected electrons in the upper laser level lose energy in the form of photons during their transition to a lower laser level. The electrons in the lower laser level are quickly depopulated by highly efficient polar longitudinal optical (LO) phonon emission and resonant tunneling into the subsequent injector region. The injector region serves as an energy filter, which collects low energy electrons and can re-inject them into another active region. This occurs under a suitable bias, which provides the necessary potential energy to the electron. Many energy states can present in the injector, often forming what is called an SL "mini-band". High-energy electrons are blocked from entering the injector by means of destructive interference in the form of a "minigap".

The lifetime of the electron levels are generally determined by LO phonon scattering [11], which originates from lattice vibrations of polar semiconductors. Energy dispersion curves of the upper laser level, lower laser level and ground state of a QCL active region with three subbands are shown in Figure 6.2(b). From the theory of LO phonon emission, if the energy separation between the lower laser level (6.2) and ground level (6.1) is designed to be in resonance with the LO phonon energy, the resultant small momentum transfer will lead to a very high scattering rate, thus making the lower laser level depopulation very fast. As long as the momentum transfer is greater from level 3 (giving level 3 a longer lifetime), this will ensure a population inversion between levels 3 and 2 and a net gain for the system at this energy (E3–E2).

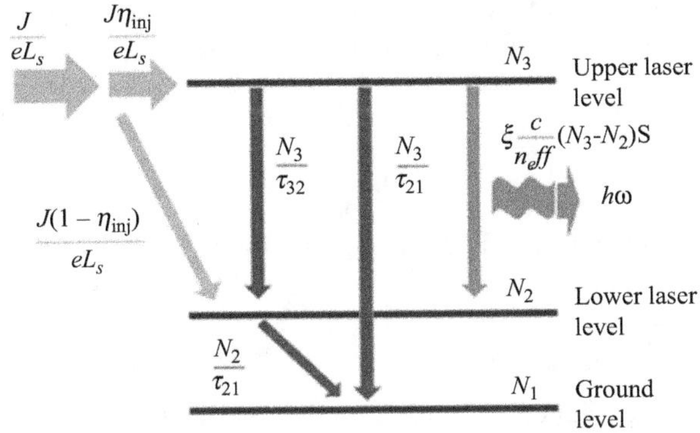

Figure 6.3 General rate equation model for the QCL

6.4 Rate equations and figures of merit

A simple three-level model is often used to describe the QCL operation. A schematic of the model and the general rate equations are

$$\frac{dN_3}{dt} = \frac{J\eta_{\text{inj}}}{eL_s} - \frac{N_3}{\tau_3} - S\xi\frac{c}{n_{\text{eff}}}(N_3 - N_2) \tag{6.1}$$

$$\frac{dN_2}{dt} = \frac{J\left(1-\eta_{\text{inj}}\right)}{eL_s} + \frac{N_3}{\tau_{32}} - \frac{N_2}{\tau_{21}} + S\xi\frac{c}{n_{\text{eff}}}(N_3 - N_2) \tag{6.2}$$

$$\frac{dS}{dt} = \frac{c}{n_{\text{eff}}}[S\xi(N_3 - N_2) - \alpha_{\text{total}}S] \tag{6.3}$$

where J denotes the current density, L_s is the length of one emitting stage (active plus injector regions), and η_{inj} is the injection efficiency into the upper state. N_3, N_2 are the upper and lower state population, respectively (Figure 6.3). $1/\tau_3$, $1/\tau_2$, $1/\tau_{21}$ are the scattering rate out of upper laser level, lower laser level and in between these levels, respectively. ξ is a gain coefficient which is given by

$$\xi = \frac{4\pi e^2}{\lambda n_{\text{eff}}\varepsilon_0}\frac{|z|^2}{\gamma} \tag{6.4}$$

where n_{eff} is the effective modal refractive index, z is the dipole strength of the radiative transition, and γ is the linewidth of the transition. The photon density is denoted by S, which is zero below threshold. A steady-state solution is obtained by setting the left side of the equations to zero.

After algebraic manipulation, below threshold, the population difference between upper laser level and lower is given by

$$\Delta N = N_3 - N_2 = \frac{J\eta_{\text{inj}}}{eL_s}\left(\frac{1}{\eta_{\text{inj}} - \eta_i} - \frac{1}{\eta_{\text{inj}}}\right)\tau_2 \tag{6.5}$$

where the internal quantum efficiency is given by

$$\eta_{\text{inj}} = \frac{\tau_{32} - \tau_2}{\tau_{32} + \tau_{32}\tau_2/\tau_3 - \tau_2} \tag{6.6}$$

Naturally, the population difference increases with an increase in pump current until the optical gain, represented by

$$g = \xi(N_3 - N_2)\Gamma \tag{6.7}$$

compensates for the total optical loss in the laser waveguide/ cavity, where Γ is the optical confinement factor.

At this point, the laser reaches threshold, and the threshold current density is given by

$$J_{\text{th}} = \frac{eL_s\alpha_{\text{total}}}{\xi\Gamma} \Big/ \left\{\left[\tau_3\left(1 - \frac{\tau_2}{\tau_{32}}\right) + \tau_2\right]\eta_{\text{inj}} - \tau_2\right\} \tag{6.8}$$

where α_{total} is the total optical loss.

Above threshold, the photon density S is nonzero and increases with the stimulated emission term. Under steady-state conditions,

$$\text{S} = \frac{\Gamma\eta_i}{eL_S \frac{c}{n_{\text{eff}}}\alpha_{\text{total}}}(J - J_{\text{th}}) \tag{6.9}$$

and the output power is given by

$$\text{P} = \hbar\omega\left(\frac{NAL_s}{\Gamma}\right)\left(\frac{c}{n_{\text{eff}}}\right)S\alpha_{\text{m}} \tag{6.10}$$

where N is the number of emitting stages and A is the cross-sectional area of the QCL core. The total waveguide loss is denoted by

$$\alpha_{\text{total}} = \alpha_{\text{m}} + \alpha_{\text{w}} \tag{6.11}$$

where α_{m} and α_{w} represent the mirror and waveguide loss, respectively. The slope efficiency is given by

$$\eta_s = \frac{dP}{dI} = \frac{\text{N}\hbar\omega}{e}\eta_i\frac{\alpha_{\text{m}}}{\alpha_{\text{m}} + \alpha_\omega} \tag{6.12}$$

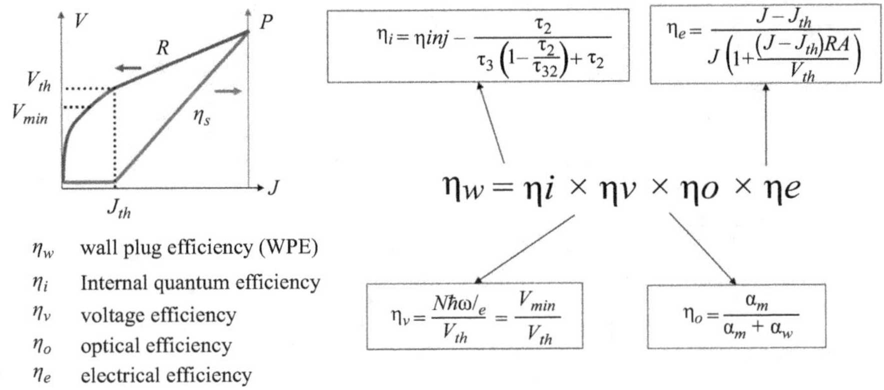

Figure 6.4 Breakdown of wallplug efficiency into sub-efficiencies

In general, the best laser operation is achieved by maximizing the WPE, which is simply the optical power output divided by the electrical power input. The WPE can be broken down into several components, or sub-efficiencies, which depend both on the optical and electrical properties of the laser. A diagram showing this breakdown is given in Figure 6.4. New terms introduced for this analysis include the threshold voltage, V_{th}, which is the voltage that corresponds to the threshold current density, J_{th}. In addition, there is the resistance-area product, RA, which is the slope of the *I*–*V* curve and takes into account nonideal electrical transport in the QCL above threshold. As can be seen from this model, general guidelines used in laser design include maximizing injection efficiency and upper state lifetime while minimizing parasitic resistance, voltage drops, and waveguide loss. In addition, to allow for good CW operation at high temperature, we also must make the process temperature insensitive, as described in a recent publication [12].

6.5 The QCL waveguide

The QCL waveguide has evolved a lot since the first demonstration. The waveguide serves three purposes. The first is to provide adequate confinement of the waveguide optical mode in order to maximize overlap with the gain medium in the waveguide core. This requires a material with a lower refractive index than the core to serve as a cladding. For InP-based QCLs, both bulk InP and AlInAs have been used for this purpose. As the top metal contact supports a guided surface plasmon mode which can introduce significant loss, a thick cladding is required to isolate the waveguide core. Improved separation can also be achieved at a reduced cladding layer thickness by doping the contact layer highly to make it highly conductive. This creates an anomalous reduction in the real part of the refractive index near the plasma frequency which helps minimize evanescent penetration through this layer.

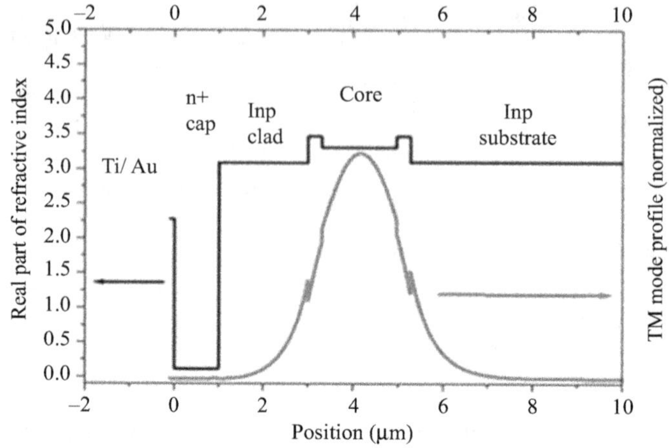

Figure 6.5 Plot of the real component of the refractive index profile and the corresponding fundamental TM mode profile for a QCL waveguide structure

The second purpose of the waveguide is efficient removal of heat from the laser core. Though both AlInAs and InP can be used as cladding material, InP is a binary compound which has ~10 times higher thermal conductivity. High-power CW QCLs tend to use InP exclusively as cladding material. The third use of the waveguide is to conduct electrical current to the core region. Besides surface plasmon loss, another prime contributor to waveguide loss is free carrier absorption. An increase in cladding doping reduces device resistance (lower RA product) but increases the free carrier absorption (part of α_w). So, to have a balance between free carrier loss and voltage drop, the cladding doping profile is often not constant, and may be optimized based on the shape of the optical mode within the waveguide. A plot of the real component of the refractive index profile and the corresponding fundamental TM mode profile for a typical QCL waveguide structure is shown in Figure 6.5.

6.6 Choosing a material system

As mentioned previously, the QCL emitting wavelength is not dependent on the semiconductor band gap, but rather the quantum well depth (conduction band offset) that is achievable with a given pair of materials. This allows the use of well-developed, commercially available materials to access previously hard to reach portions of the electromagnetic spectrum. Though $Ga_xIn_{1-x}As/Al_yIn_{1-y}As$ grown on InP was the first system in which a QCL was demonstrated, it was a fortuitous choice. When $Ga_xIn_{1-x}As$ and $Al_yIn_{1-y}As$ are lattice matched to InP substrate ($x = 0.47$, $y = 0.48$), the relative conduction band offset between them is ~520 meV. At cryogenic temperatures, this allows emission at $\lambda > 4$ μm. However, at RT, significantly higher energy barriers are required to have good injection efficiency, which limits "efficient" operation to much longer wavelengths.

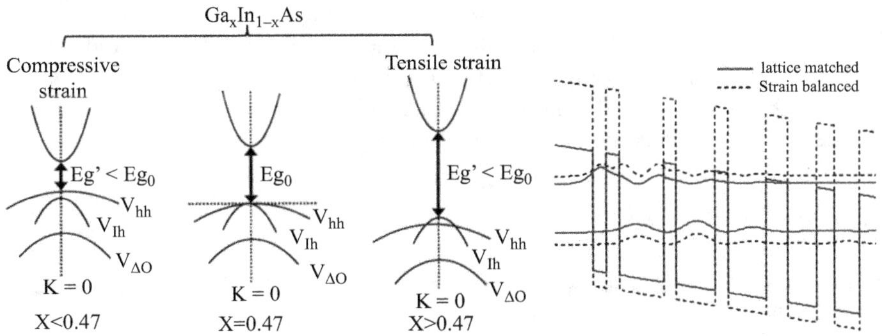

Figure 6.6 *(a) Shift in band structure of strained $Ga_xIn_{1-x}As$ grown on InP as composition is changed. The band gap is depicted by arrows. Eg0 is the unstrained bandgap. The valence bands are shown below the bandgap, and the conduction band is shown above. (b) Conduction band profiles and the corresponding electron wavefunctions of lattice-matched $Ga_{0.47}In_{0.53}As/Al_{0.48}In_{0.52}As$ and strain-balanced $Ga_{0.33}In_{0.67}As/Al_{0.65}In_{0.35}As$ structures*

For short wavelengths and high temperature emission, the conduction band offset needs to be increased. Several near lattice-matched systems, including $Ga_{0.47}In_{0.53}As/AlAs_{0.56}Sb_{0.44}$ on InP or InAs/AlSb on InAs have been attempted with larger band offsets, but require the use of more exotic materials and specialized epitaxial growth equipment. An alternative way to achieve a higher band offset is through strain balancing in the $Ga_xIn_{1-x}As/Al_yIn_{1-y}As$ system.

Strain balancing is achieved by incorporating tensile strain in the barriers and compressive strain in the quantum wells of the SL in order to increase the relative separation between their conduction band edges. For the $Ga_xIn_{1-x}As/Al_yIn_{1-y}As$ system, this is achieved when $x < 0.47$ and $y > 0.48$. Normally, if these materials are grown individually on InP, the significant mismatch in lattice constants leads to the buildup of strain energy that is released through dislocation formation after some critical thickness. This in turn leads to rough surfaces, and poor optical and transport properties.

However, if the layer is thinner than the critical thickness, it will hold the strain energy and assume the lattice constant of the substrate in the horizontal direction. This is referred to as pseudomorphic growth. By alternating thin layers of alternating tensile and compressive strain, the total strain energy can be kept below the critical value for dislocation formation indefinitely.

The exact location of the conduction and valence band edges can be estimated using Model Solid theory [13]. This theory describes how the biaxial deformation of the crystal lattice shifts the average valence band and conduction band values due to the hydrostatic component of strain. The top of the valance bands are then split by the shear component of the strain. The shifting band structure behavior as a function of composition for $Ga_xIn_{1-x}As$ is shown in Figure 6.6(a). When tensilely strained $Al_yIn_{1-y}As$ ($y > 0.48$) and compressively strained $Ga_xIn_{1-x}As$ ($x < 0.47$)

are combined, it leads to a net increase in the quantum well depth, up to a maximum value of ~1.2 eV. A diagram showing the effect on band edges and electron wavefunctions is shown in Figure 6.6(b). For band structure calculations of strain-balanced QCLs, the shifted band energies as well as the strain-dependent changes in effective masses must be taken into account.

6.7 Material growth and characterization

Strain-balanced material growth has become a core focus of our QCL research at the CQD. The material growth of $Ga_xIn_{1-x}As/Al_yIn_{1-y}As$ has been done by both gas-source molecular beam epitaxy (GSMBE) and metalorganic chemical vapor deposition (MOCVD). Both techniques have shown very precise control over layer composition, thickness, and interface quality. However, at present, our most highly strained growth (for $\lambda < 3.5$ μm) has only been attempted with GSMBE. High quality and repeatable laser wafers are the result of careful optimization of growth conditions and pre-growth calibration. The structural, optical, and electrical properties of each growth run are analyzed to ensure excellent quality control.

Surface morphology is a sensitive indicator of instability or dislocation formation during epitaxial growth. The surface morphology can be investigated by atomic force microscopy (AFM). A 10 × 10 μmAFM scan of a typical QCL grown by GSMBE is shown in Figure 6.7(a). A root mean square (rms) surface roughness of ~0.25 nm is measured, which is consistent with the normal monolayer fluctuation in GSMBE growth.

Another structural technique routinely employed is X-ray diffraction. X-ray diffraction is one of the most powerful tools to determine layer thickness and

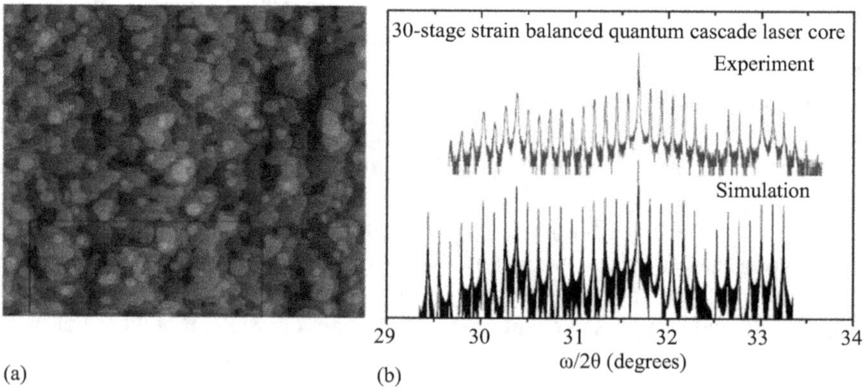

Figure 6.7 (a) A 10 × 10 micron AFM scan of the as-grown QCL showing an rms roughness of 0.25 nm and (b) comparison of experimental and simulated X-ray diffraction spectra for a strain-balanced QCL core

composition of a periodic structure like the QCL. Many diffraction peaks in addition to the substrate peak show up in the diffraction spectrum due to interference between incident and reflected X-rays [2]. SL period, net mismatch, and growth rates can be extracted by simulating and fitting the spectrum using commercial software like X'Pert Epitaxy©. For specific calibration structures, we can also extract layer composition, which must be tightly controlled to achieve the desired emission wavelength. Figure 6.7(b) shows an example diffraction spectrum and simulation for the core region of a strain-balanced QCL after etching off the top cladding layer. A good match between the experimental and simulated diffraction envelopes indicate an accurate layer thickness and compositional control. The low background and sharp secondary SL peaks in the experimental curve also indicate good homogeneity, flatness and interface quality.

The optical properties of a semiconductor can be studied quickly and non-destructively by photoluminescence (PL) [2]. The band-to-band optical emission from a QCL core can be compared to band structure simulations to estimate thickness and composition of quantum wells. When correlated with X-ray diffraction data, the overall uncertainty of calibration can be reduced significantly. The PL intensity also indicates the radiative recombination efficiency of the excited carriers and will clearly show evidence defect levels, impurities, and surface states. An experimental PL emission curve and a simulated band diagram for a QCL core is shown in Figure 6.8.

In the QCL, well-defined concentrations of electrons are made available via n-type doping with Si. Control of the Si (and electron) concentration is critical for ensuring the correct operating current density range, resistance, and waveguide loss for a QCL. Though bulk layers are calibrated using the Hall effect, more complex doping profiles can be obtained through electrochemical capacitance-voltage (ECV) measurements [2]. With this technique, the carrier concentration can be measured and the semiconductor can be etched in a cycle. A typical ECV average doping profile of a QCL core along with the InP buffer region and substrate is shown in Figure 6.9.

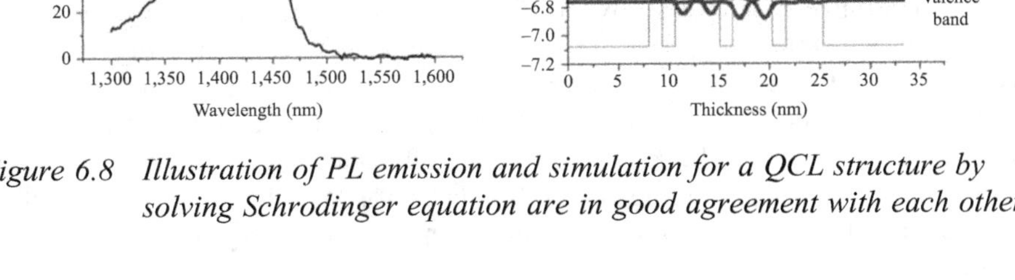

Figure 6.8 Illustration of PL emission and simulation for a QCL structure by solving Schrodinger equation are in good agreement with each other

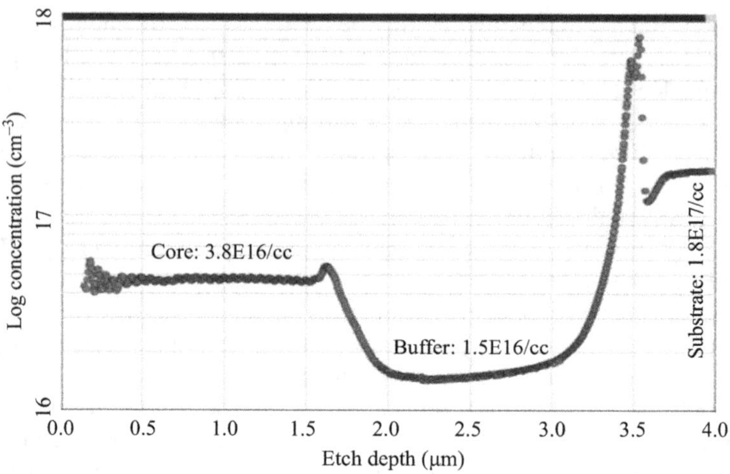

Figure 6.9 ECV doping profile of a QCL structure

6.8 Laser fabrication

Once the QCL wafer is grown, it goes through a number of processing steps by which a laser diode is created. The most popular laser geometry is the double channel ridge [14]. Ridges are made narrow (9–20 µm) for CW operation, while high peak power application utilize wider ridges (100–400 µm). This type of geometry is chosen for its ease of fabrication as well as strong optical and electrical confinement.

The ridge is defined by photolithography and subsequent (often chemical) etching. This is followed by depositing an insulator like SiO_2. A small window is opened on the top of the center ridge before depositing Ti/Au as a top electrical contact. In most cases, this is followed by a thick (2–5 µm) gold layer to help remove waste heat. The back side of the wafer is then thinned to ~100 µm and polished. A bottom contact is finally made by depositing AuGe/Ni/Au (Figure 6.10).

A variation on this process is the buried ridge waveguide, which requires one additional epitaxial growth step. After double-channel formation, semi-insulating InP is grown selectively by MOCVD in the channel regions. This effectively planarizes the wafer and allows significantly narrower ridges (down to ~3 µm) to be fabricated. These narrow, buried ridges exhibit very good thermal conductance and low optical losses, which can be critical for high efficiency CW operation.

Optical mirrors are created by cleaving the wafer perpendicular to laser cavity. To maximize output from the front facet, the back facet is often high reflectance (HR) coated with dielectric and possibly metal layers. This reduces mirror loss and the threshold current density. The front facet's reflectivity (and total mirror loss) is then adjusted via a partial HR (PHR) or anti-reflective (AR) coating to maximize laser efficiency. For packaging, individual lasers or laser bars are bonded either epilayer-side up (small heat transfer) or epilayer-side down (large heat transfer) to a heat sink for testing.

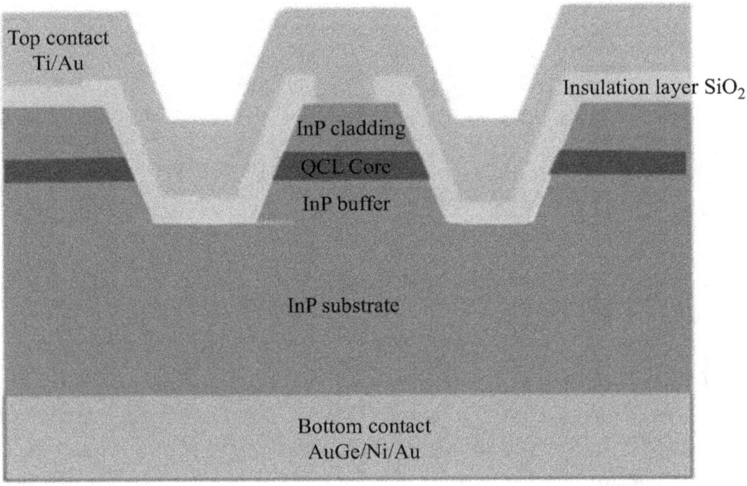

Figure 6.10 Schematic cross section of a double-channel QCL

6.9 Development of high-efficiency lasers

Once the design and production tools for QCLs have been developed, how do we best apply them to realizing the full potential of this technology? One early criticism of the QCL was that it suffered from a low WPE, which is one of the most important figures of merit for many applications. A QCL with a higher WPE will convert more input electrical power into useful electromagnetic radiation. This will simplify thermal packaging and minimize heat loss.

However, raising efficiency is a complex process, dependent on many parameters. In order to simplify the challenge and provide a basis for more thorough optimization, the problem was first analyzed in the low-temperature (cryogenic) limit. In conventional QCLs, the injector acts as a reservoir of electrons from which electrons are selectively injected with the correct energy into the upper laser level of the next stage. Generally the injector is a chirped SL with a built-in energy difference between the ground state of the injector and the lower laser level. If this difference is too small, carriers can be re-excited from the ground state back into the lower laser level, decreasing laser gain and increasing the threshold current. However, at sufficiently low temperatures, the backfilling is negligible, and the extra voltage drop (voltage defect) in the injector is unnecessary. The voltage defect can be reduced by using a single well as an injector instead of an SL injector. In addition to better voltage efficiency, the shorter-emitting stage length leads to a reduced resistance and an increased differential gain. With a low loss waveguide, this allows for shorter cavities and higher internal efficiencies. Altogether, a large increase in WPE is obtained, with a maximum demonstrated value of 53% at 40 K [7]. Though not well suited for RT operation, this design/demonstration was an

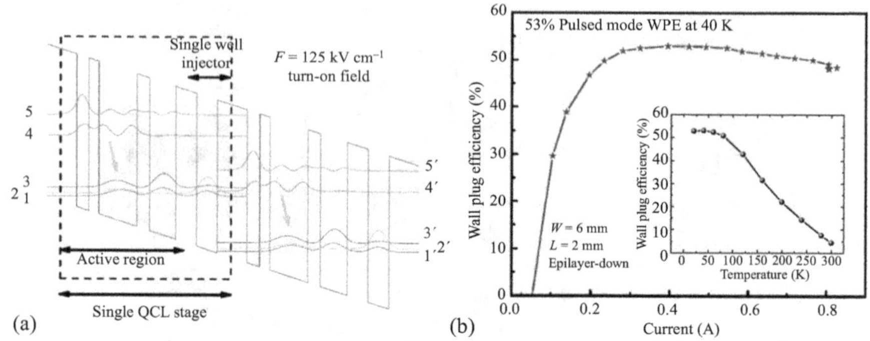

Figure 6.11 (a) Conduction band edge and wave function diagram for a single well injector QCL and (b) WPE vs. current characteristics of a pulsed laser operating at 40 K. Inset of (b) shows maximum WPE vs. heat sink temperature

important milestone in QCL development. Band diagram and performance data are illustrated in Figure 6.11.

The next challenge was to analyze WPE limitations with temperature effects included [12]. Different active region designs had been explored in the course of developing high power QCLs. One limitation of the QCL efficiency is related to interface scattering in the active region, which gets worse as the barrier height is increased. Therefore, in contrast to a conventional deep well design, the active region was modified with the insertion of near lattice-matched (NLM) material as the first well and barrier, giving rise to a smaller conduction band offset on both interfaces of this barrier. Due to the large overlap the upper laser level has with these interfaces, this should lead to an increase in the upper laser state lifetime, τ_3. A secondary benefit of the NLM insertion resulted in the increase of the energy difference between the upper laser level and the next highest energy level from 80 meV, in a conventional three well active region, to 90 meV in the new design. This suppresses thermally assisted carrier leakage and increases injection efficiency.

In terms of the injector design, a similar goal is to reduce electron escape from the upper laser level and maximize injection efficiency. To address this, AlAs inserts were included as a part of the barriers in the injector region to reduce electron escape to the continuum by raising the escape activation energy. Also, unlike designs for low temperature optimization, a high temperature design must suppress thermal back-filling of the lower laser level. A voltage defect, i.e., the energy difference between the lowest laser level and ground state of the injector, of ~180 meV was set, which balances optimization of voltage efficiency and reduction of thermal backfilling.

To accomplish all of the objectives, a QCL core having five different material compositions was designed for high-power, CW operation [8]. The entire structure is strain balanced. A band diagram for this structure is shown in Figure 6.12(a). A 40-period QCL core instead of a standard 30-period core was also used to increase the mode confinement factor and increase the slope efficiency and output power.

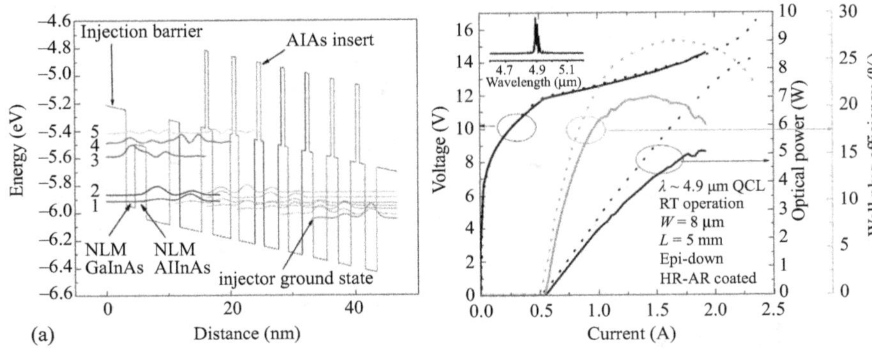

Figure 6.12 *(a) Conduction band edge and wave function diagram and (b) P–I–V and WPE vs. current curve of a QCL with ridge width of 8 μm and cavity length of 5 mm, bonded epi-down under CW operation (solid lines) and pulsed operation (dash dotted line) at RT. The inset of (b) shows the emission spectrum at 4.9 μm in CW mode*

Pulsed testing of this wafer yielded a maximum WPE of 27% for double channel waveguides. Buried ridge processing was used for CW operation and a maximum output power of 5.1 W and WPE of 21% were obtained in CW mode at RT. This efficiency and power have not been matched by any other QCL to date. The power–current–voltage (*P–I–V*) and WPE curves for a QCL under CW operation at RT are shown in Figure 6.12(b).

6.10 Developing short-wavelength, high-power QCLs

As discussed earlier, the shortest wavelength achievable in a QCL is limited by the quantum well depth. Though the high-efficiency lasers are based on strain-balanced SLs, there is not enough depth to realize lasers below 4 μm with good performance. Nevertheless, this spectral region is vitally important, especially for spectroscopy, as many hydrocarbons have their fundamental modes of vibration in the 3–4 μm wavelength range.

To reach shorter wavelengths, we have to gradually increase the quantum well depth and maintain good material and interface quality. Toward this end, the first step was to grow a QCL based on a more highly strained $Ga_{0.24}In_{0.76}As/Al_{0.76}In_{0.24}As$ design, to accommodate a large optical transition energy at shorter wavelength [15]. A three-well active region, or single phonon resonance (SPR) structure, was used for the active region. This is in contrast to double phonon resonance or bound to continuum designs as the upper laser level is at a lower absolute energy. This creates better confinement of electrons and increases the activation energy of electron escape into the continuum.

The wafer was processed into a buried ridge geometry with ridge width of 8.3 μm and a cavity length of 4 mm. Emission was observed at $\lambda \sim 3.76$ μm, as

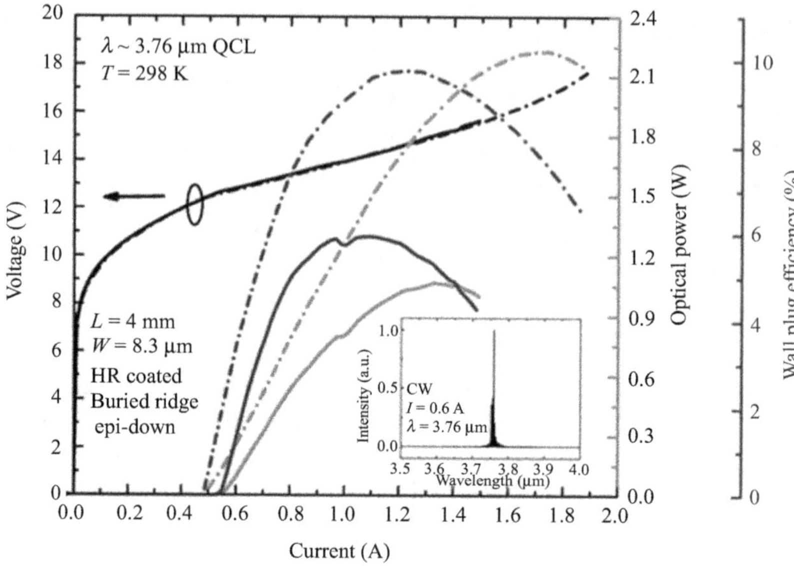

Figure 6.13 Pulsed (dash dotted) and CW (solid line) P–I–V curves for a HR coated 8.3 μm wide and 4 mm long, buried ridge heterostructure QCL at RT. Corresponding WPE is also plotted. Inset shows spectrum in CW operation at 0.6 A

predicted by the band structure simulation. The maximum RT, CW output power was 1.1 W, with a threshold current density of 1.67 kA/cm^2 and slope efficiency near threshold of 2.16 W/A. A maximum RT, WPE of 6% and 10% were obtained in CW and pulsed operation, respectively. Figure 6.13 shows the *P–I–V* curves and corresponding WPE.

While an important first step, the goal is to reach much shorter wavelengths with QCL technology. At these shorter wavelengths, however, additional issues besides the conduction band offset start to become a problem. The main physical issue is the interaction of indirect conduction band valleys (X and L) with the extremely high lying direct valley Γ states used as the upper laser level of the QCL. This interaction leads to possible carrier trapping and leakage paths, which can seriously degrade device performance.

If we review possible III-V material systems with large band offsets, we see that both GaN-based and InAs-based materials exhibit the largest energy separation of Γ and L eigenstates, which make them attractive for short-wavelength QCLs. However, GaN-based QCLs have not been demonstrated at any wavelength due to extreme challenges with respect to growth and tunneling-based transport. InAs/AlSb structure shave been attempted, with fairly robust operation up to 400 K down to $\lambda \sim 3$ μm, but show a very high threshold current density (~4 kA/cm^2 at RT), making them unsuitable for CW operation [16].

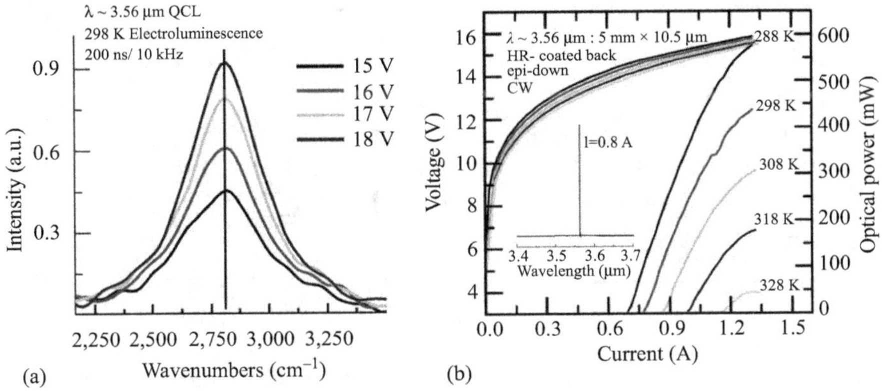

Figure 6.14 *(a) RT, pulsed mode EL spectrum of $\lambda \sim 3.56$ μm QCL as a function of voltage and (b) P–I–V curve of a HR coated 10.5 μm wide and 5 mm long QCL emitting at 3.56 μm, at RT. The inset shows the CW, RT emission spectrum*

Our own effort has concentrated on extending the use of strain-balanced $Ga_xIn_{1-x}As/Al_yIn_{1-y}As$ SLs to achieve shorter wavelengths. In the limit that x goes to 0, we should achieve a similar Γ–L separation that is present in the InAs/AlSb system, but maintain an InP-compatible structure with simple interfaces. Two QCLs with cores based on a $Ga_{0.2}In_{0.8}As/Al_{0.82}In_{0.18}As$ SL were grown with slight differences in their layer thicknesses [17]. The design wavelengths were near $\lambda =$ 3.5 μm. No dislocations were found in the 30-stage core regions even at this high strain. To verify the material quality of the SL, an electroluminescence (EL) sample was prepared from the longer wavelength wafer and tested with 200 ns pulses at 10 kHz frequency. The curves are shown in Figure 6.14(a), and the linewidth is found to be 47.4 meV, which is very promising.

The two wafers were processed into double-channel lasers with widths of 10.5 and 8.6 μm, and they emitted at $\lambda \sim 3.56$ μm and $\lambda \sim 3.39$ μm QCL, respectively. CW threshold current densities of 1.35 and 1.2 kA/cm^2, respectively, were demonstrated near RT. Maximum RT, CW powers of 437 mW and 403 mW were obtained at $\lambda \sim 3.56$ μm and $\lambda \sim 3.39$ μm, respectively. Figure 6.14(b) shows the *P–I–V* curve for the longer wavelength laser in CW operation at different heat sink temperatures. The inset shows the emission spectrum. Though not the first demonstration of CW emission at this wavelength, we were able to bring output power up by an order of magnitude compared to other groups. These results remain the best to date.

Our next goal was to achieve similar results at $3 < \lambda < 3.2$ μm. At the time, only pulsed mode had been demonstrated by another group down to $\lambda \sim 3$ μm [16]. A higher conduction band offset was achieved by increasing the Al composition in the $Al_yIn_{1-y}As$ barrier to $y = 85\%$ and $y = 89\%$ for emission at 3.2 and 3 μm, respectively [5]. The Ga composition inside the $Ga_xIn_{1-x}As$ well was $x = 23\%$ and $x = 21\%$, respectively. However, the extremely high tensile strain level in AlInAs

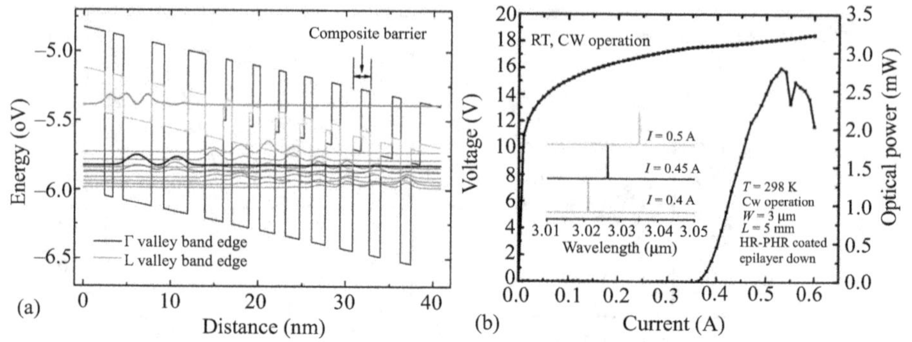

Figure 6.15 (a) Conduction band edge and wave function diagram for a 3 μm QCL and (b) CW P–I–V curve of a HR-PHR coated 3 μm wide and 5 mm long QCL emitting at 3.02 μm, at RT. The inset shows the CW, RT emission spectrum

can cause relaxation in the core region if the global strain is not zero or if the thickness of an individual layer is larger than the critical thickness. In order to neutralize the net strain per period, composite barriers were used in the injector region instead of incorporating more indium inside $Ga_xIn_{1-x}As$, as 2D growth of highly strained GaInAs is very difficult. In this approach, the coupling of the wavefunctions in the injectors was ensured by choosing the proper thicknesses of the composite barriers, while the addition of tensile strain was minimized by tuning the composition of the lesser strained $Al_yIn_{1-y}As$ layers. A conduction band diagram showing the layer structure and electron wavefunctions for a $\lambda = 3$ μm QCL is shown in Figure 6.15(a).

For RT, CW operation, a combination of narrow, buried ridge processing and epi- down bonding were necessary. As short-wavelength QCLs have a small differential gain, the front facet is PHR coated and the back side is HR coated in order to minimize the threshold current density. The PHR has a reflectivity of 85% and reduces the mirror loss from 1.4 cm^{-1} to only 0.21 cm^{-1}, as compared to a device with only an HR coating on the back. A $\lambda \sim 3.2$ μm QCL having a ridge width and cavity length of 7 μm and 5 mm, respectively, gave a maximum CW output power of 20 mW at RT. A $\lambda \sim 3$μm QCL having a ridge width and cavity length of 3 μm and 5 mm, respectively, gave a maximum CW output power of 2.8 mW at RT and is shown in Figure 6.15(b). The emission spectrum at a driving current of 0.4 A peaks at a wavelength of 3.02 μm. This $\lambda = 3.02$ μm emission is the shortest-wavelength demonstration of CW operation at RT for any QCL to date.

6.11 Power scaling of quantum cascade lasers

High-power, CW operation at RT is possible only with narrow ridge widths. However, in some applications where very high peak power is required, the QCL power can be scaled up by increasing the ridge width, thus increasing the laser gain

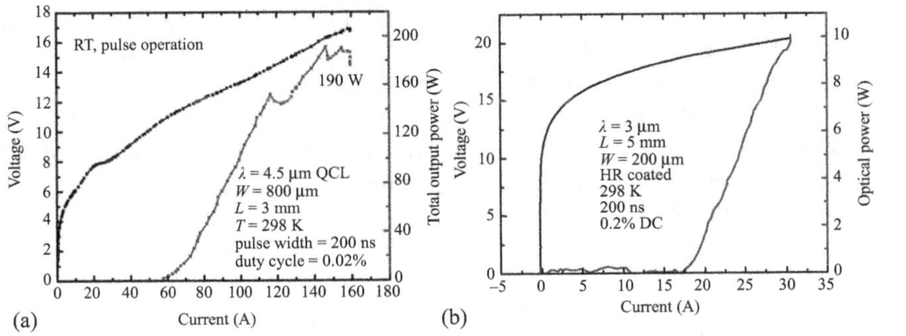

Figure 6.16 (a) RT, P–I–V curve of a $\lambda = 4.5$ µm BA QCL having a ridge width of 800 µm and cavity length 3 mm, showing a peak power output of 190 W and (b) RT, P–I–V curve of a $\lambda = 3$ µm BA QCL having a ridge width of 200 µm and cavity length 5 mm, showing a peak power output of 10 W

volume. This trend was observed for scaling of power from 50 to 400 µm [18]. As observed, both the threshold current and roll over current increase linearly with ridge width. However, WPE decreases with ridge width from 16% at a width of 50 µm to 12% at a width of 200 µm and then saturates. This behavior may be attributed to heating which increases with ridge width and then saturates when the ridge width approaches the thickness of the substrate. A broad area (BA) QCL emitting at $\lambda \sim 4.4$ µm with a ridge width of 400 µm and cavity length of 4 mm exhibited peak power up to 116 W. The lasing far field consisted of distinct dual lobes at angles around $\pm 38°$, which suggest that the near field is composed of periodic local maxima. This also indicates that the near field is stable and there is no random formation of filaments, which would create many narrow, bright spots in the near field. This is, of course, expected, as the linewidth enhancement factor, which is one of the main reasons for filament formation, is much less than in conventional interband lasers. To test if even more peak power can be achieved, another BA QCL with a ridge width of 800 µm and cavity length of 3mm was processed from a separate $\lambda \sim 4.5$ µm QCL and was also tested in pulsed mode at RT. A peak power of 190 W was observed, as shown in Figure 6.16.

Similar power scaling efforts were explored at longer and shorter wavelengths.

At $\lambda \sim 10$ µm, a peak power of 25 W at RT was obtained for a cavity width of 95 µm [19]. Performance was robust, with 15 W still available up to 100 °C. More recently, a $\lambda \sim 3$ µm BA QCL, having ridge width and cavity length of 200 µm and 5 mm, respectively, showed a maximum pulsed output power of 10 W at RT (Figure 6.16(b)).

6.12 Spectral and spatial filtering of quantum cascade lasers

Though the power of a QCL can be improved by using wider ridges, beam quality and spectral quality are not easily controlled quantities. For applications such as chemical spectroscopy, a narrow emission linewidth is necessary to pinpoint specific absorption features, and a single-lobed, diffraction-limited output is necessary for efficient power transfer into a well-defined spot.

Spectral filtering in a narrow waveguide can be accomplished though one-dimensional periodic distributed feedback (DFB) gratings integrated into the QCL waveguide. These gratings exhibit a Bragg condition, with maximum reflectivity at a wavelength given by

$$\Lambda_{\mathrm{B}} = 2n_{\mathrm{eff}}\Lambda \tag{6.13}$$

where n_{eff} is the effective index of the waveguide dominant transverse mode (ideally TM00) and Λ is the spatial period of the grating. The strength of the grating is related to the amplitude of the permittivity induced at the Bragg frequency and is characterized by the splitting between allowed propagating longitudinal modes in terms of frequency and/or threshold gain. When implemented correctly, the grating enhances feedback, lowering the threshold gain, at only one wavelength.

Though we demonstrated many DFB QCLs capable of CW operation at RT, power was typically limited to ~100 mW [20–22]. For standoff spectroscopy or photoacoustic spectroscopy, significantly higher powers are needed. Toward this end, we utilized the recently developed high-power CW multimode lasers described above to radically enhance output power of single wavelength emission (single longitudinal mode) QCLs.

To achieve a single wavelength emission, the DFB coupling needs to be strong enough so that only a single longitudinal laser cavity mode is supported. For long cavities, this can be achieved with a surface plasmon grating [23]. Due to the strong surface plasmon interaction at the metal-semiconductor interface, even a shallow grating depth of 120 nm in the cap layer will generate a coupling coefficient (κ) of 1.37 cm^{-1} and modal loss discrimination of 0.4 cm^{-1}. The small value of κ ensures high output coupling efficiency, and the modal loss discrimination ensures only one grating mode to lase. To further suppress natural cavity longitudinal (Fabry–Perot) modes, the back facet was HR coated, while the front facet was AR coated.

Figure 6.17 shows a *P–I–V* curve at RT for an 11 μm wide and 5 mm long DFB QCL. Power of 2.4 W of RT, CW operation at 10% WPE is obtained. The right inset shows a single-lobed far field characteristic (near diffraction-limited). The left inset shows the tuning of the lasing spectrum from 2,087 to 2,084 cm^{-1} as a function of current between 1 and 1.7 A. Single-mode emission with a side mode suppression ratio (SMSR), which is the relative strength of side peak or noise in the emission spectrum, of 30 dB over the entire current range is observed.

For peak power scaling of single mode lasers, spectral and spatial filtering requires selective transverse and longitudinal feedback as the laser cavity width is increased. This can be achieved using photonic crystal distributed feedback (PCDFB) [24], by which a two-dimensional diffraction grating (as opposed to the 1-D DFB described above) is placed within the laser resonator. Matching lateral and longitudinal filtering strength to the cavity geometry is necessary to fully suppress higher order optical modes.

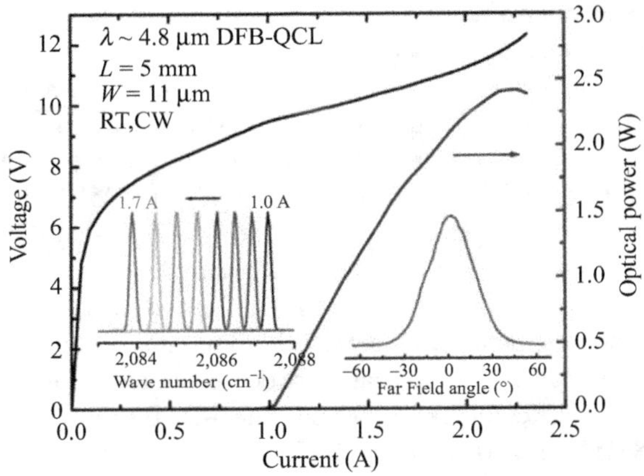

Figure 6.17 P–I–V curve of DFB QCL showing a peak CW, RT power of 2.4 W. The insets show single mode far field and tuning of laser emission wave number with current

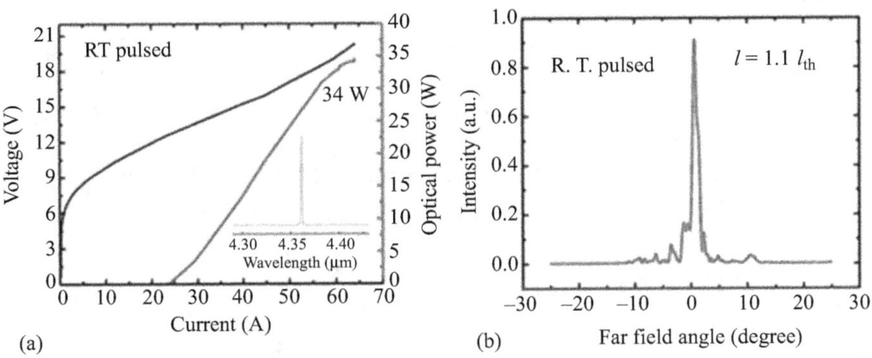

Figure 6.18 (a) RT, P–I–V curve of a 3 mm long, 400 μm wide PCDFB QCL, showing a peak output power of 34 W and (b) parallel far field intensity spectra measured at a driving current of $I = 1.1I_{th}$

In our work, PCDFB features are defined on a 100 nm GaInAs layer grown 100 nm above the laser waveguide core. A second epitaxial growth step by MOCVD was necessary to bury the grating within the upper InP waveguide cladding. The PCDFB structure implemented was a rectangular lattice with first and second-order feedback in transverse and longitudinal directions, respectively. The aspect ratio of the structure was $\Lambda_t/\Lambda_l = 3$ and the cavity tilt angle was 14°. The targeted wavelength was designed to coincide with the QCL gain peak.

From the *P–I–V* curve shown in Figure 6.18(a), a peak power of 34 W is obtained from a device 3 mm long and 400 μm wide, while the slope efficiency and threshold current are 560 mW/A and 2 kA/cm^2, respectively. The inset shows single mode spectral emission at a wavelength of 4.36 μm. The far field characteristic is shown in Figure 6.18(b) at a current of $I = 1.1I_{th}$. The M^2 beam quality parameter is calculated to be 2.5.

Though the PCDFB can provide a single mode emitting spectra and good beam quality, the design is complex and requires strong coupling in the form of buried grating features. To simplify the fabrication process, the surface grating-based angled grating DFB (β-DFB) QCL was developed near $\lambda = 10$ μm [25]. This laser is based on an angled cavity to provide transverse mode filtering and a 1-D diffraction grating to provide longitudinal DFB for narrow linewidth spectral emission. The cavity length (L), ridge width (W) and tilt (θ) were 2.3 mm, 200 μm and 18.4°, respectively. First-order gratings were implemented to a depth of 200 nm on the top contact layer.

The spectrum and far field for this laser are demonstrated in Figure 6.19(a) and (b), respectively. For three different grating periods, the β-DFB shows a single mode output spectrum with a SMSR of ~30 dB. The β-DFB also shows a single lobe in the far field, in the direction normal to the surface. The threshold current density of the β-DFB QCL is 7.5 kA/cm^2. The peak output power per facet and the WPE are 3.13 W and 0.35%, respectively, for this first demonstration. Most recently, by splitting the output power of a DFB seed laser into a 16-element laser amplifier array through a tree-array multi-mode interferometer, output power up to 20 W, 30 W, and 51 W has been achieved at 3.8 μm, 4.8 μm, and 8.0 μm, respectively.

The above schemes (PCDFB, β-DFB) were used to obtain spatial beam quality control, but only operate in pulsed mode due to the wide laser geometry. At the same time, the emission is intrinsically elliptical. From the theory of diffraction, it is known that the far field divergence is inversely proportional to the dimension of the emitting aperture in the near field. Though the parallel emission from the cavity has a reduced divergence as the width increases, the waveguide height is always small and

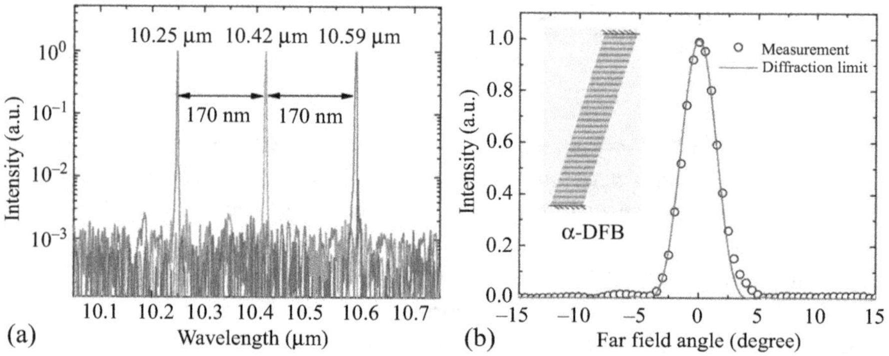

Figure 6.19 (a) Room temperature single mode performances of the β-DFB QCL. Testing is performed at room temperature in pulsed mode. (a) Emission spectra of three β-DFB devices with grating periods of 1.60, 1.63, and 1.66 μm. Single mode operation is obtained for all three devices with a wavelength spacing of 170 nm for adjacent grating periods. The SMSR is 30 dB. (b) Parallel far field of the β-DFB device emitting at 10.42 μm, in comparison to the calculated diffraction limit. The inset shows the cavity geometry, where the red lines represent the orientation of the DFB gratings

there tends to be strong vertical (perpendicular) divergence from the exit facet. A more symmetric aperture would, in turn, lead to a more symmetric far-field.

Achieving good beam quality and CW operation at RT at the same time would be extremely useful. A surface-emitting QCL, having an intrinsically large emitting aperture, is ideal for a low divergence and highly directional beam. However, CW operation requires a narrow ridge as well, which led us to develop high power, CW ring lasers.

A schematic diagram of a ring QCL is shown in the inset of Figure 6.20(a). As shown, epilayer-down bonding was utilized for efficient heat removal, and light is collected from the substrate side. 20 μm long second-order azimuthal diffraction grating lines were written in a ring with a radius of 400 μm, with a grating period of $\Lambda = 1.54$ μm and duty cycle of 60%. This pattern was then transferred into a 300 nm $Ga_{0.47}In_{0.53}As$ grating layer. After that, a 3.32 μm low-doped InP top cladding and highly doped InP top contact were regrown by MOCVD. The wafer was then processed into a ring shaped double channel geometry with a ridge width of 9 μm [26]. After polishing the substrate, AuGe/Ni/Au is evaporated on the substrate side as a bottom contact, with the exception of an emitting aperture of 600 μm in radius centered with respect to the ring cavity. The device was bonded epilayer-down on AlN submounts with indium solder and tested.

A maximum CW power of 0.51 W at 0.84 A, with a maximum WPE of 6.25% at 0.64 A is obtained at RT as shown in Figure 6.20(a). The device exhibits a single mode output spectrum up to a current of 0.64 A, above which it shows multimode behavior. The kink in the power-current curve marks the transition from single mode to multimode operation. The far-field of the device, at a current of 0.5 A, is illustrated in Figure 6.20(b). The far field consists of concentric rings, where the brightest ring spans an angle of 12°±1°. The far field characteristic indicates that the device operates in a very high-order mode. More recent efforts have shown that low

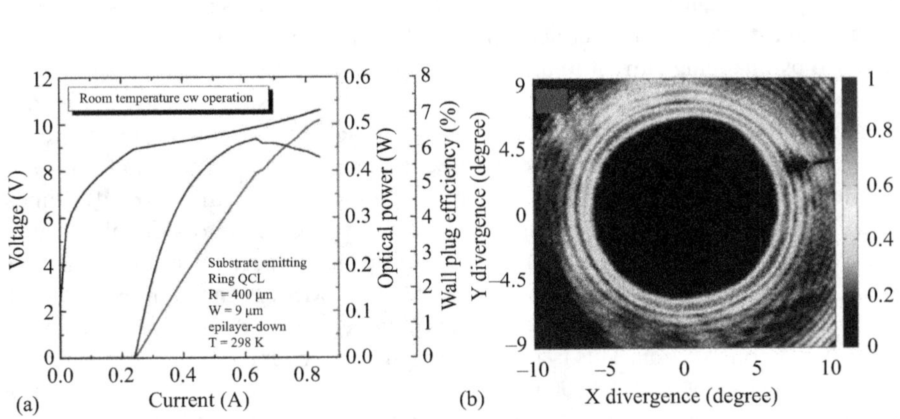

Figure 6.20 (a) Continuous wave P–I–V and WPE curves at 298 K for a ring QCL having a ridge width of 9 μm and radius length 400 μm, showing a peak CW power output of 0.51 W. Inset shows schematic of ring epilayer-down ring laser and (b) far field image measured at a driving current of 0.5 A

order spatial modes can be exhibited from these devices as well, with excellent beam quality. Building on this effort, it may be possible to realize significant power scaling using two-dimensional arrays of ring QCLs while maintaining good beam quality.

6.13 Monolithic wavelength tuning with quantum cascade lasers

In many applications such as spectroscopy, along with high power and high SMSR, a broad tunability is also necessary. This allows a single laser source to scan over multiple absorption features to improve accuracy and chemical versatility. However, the conventional DFB QCL has a tuning range of only ~5 cm^{-1}, which typically limits detection to a single type of molecule. External cavity QCLs, on the other hand, are capable of a wide tunability, but the requirement of precision optics, susceptibility of this system to mechanical vibrations, and limited scan speed due to large grating mass limits the technological potential.

A dual-section laser based on sampled grating distributed feedback (SGDFB), with a design based on devices proven at telecom wavelengths (~1.5 μm), was alternatively explored for monolithic, electrically controlled tuning [9]. In this approach, sampled gratings with two different periods Z_1 and Z_2 are etched on two different sections of the laser waveguide, with two different contacts. Sampled gratings produce a comb-like reflectivity spectrum, with multiple peaks whose spacing is inversely related to the sampling period. As the two different sampling periods Z_1 and Z_2 are combined within a single waveguide, they produce two reflectivity frequency combs with different spacings. Where these separate combs overlap creates individual supermodes, which have strong feedback within the dual-section laser cavity.

As designed, the current injected into the front and back section, If and Ib, are independent of each other. As the current in one of the sections is increased, the refractive index in that section changes, causing a shift in that section's reflectivity spectrum. The frequency comb overlap between the two sections then shifts to another wavelength, giving rise to Vernier-type step tuning. Continuous tuning can also be obtained by changing current in both sections simultaneously. The comb spacing for each section is designed to be less than ~5 cm^{-1}, which is the spectral tuning of a single DFB QCL, so that continuous tuning can be achieved over the entire range.

A SGDFB QCL with front and rear section lengths of ~1.6 mm and ~1.4 mm was fabricated. Both sections use 30-period gratings, which are sampled 3 times, where the grating period is 753 nm. The wafer is processed into a double-channel ridge waveguide with ridge width of 10 μm and cleaved into a device cavity length of 3 mm. The tuning characteristic, output power and SMSR is shown in Figure 6.21. Single mode tuning of more than 50 cm^{-1} is obtained above threshold by increasing the current in one section. This is a 5–10 times increase over the electrical tuning of a simple DFB laser. More than 100 mW of CW power, with a mean SMSR of 24 dB is obtained at RT over the tuning range. Though development is only a beginning, this result shows significant potential for even wider tuning and even higher power output. Further development has led to broadband monolithic tuning between 6.1 and 9.2 μm by photonic integration of an eight-element SGDFB turnable QCL array and an on-chip beam combiner.

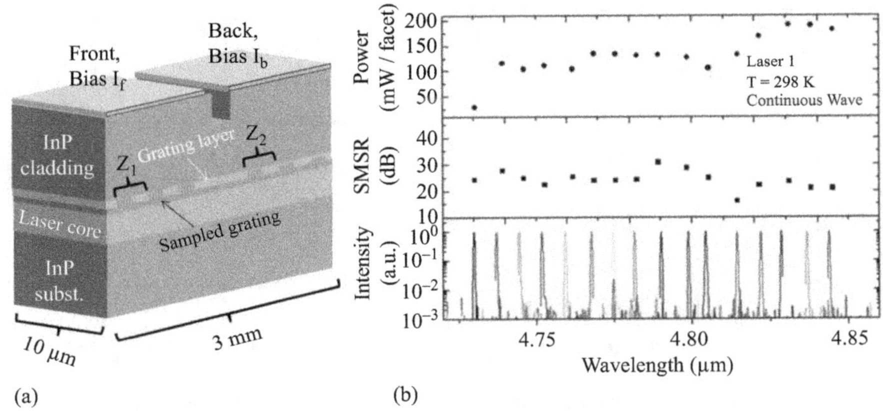

Figure 6.21 *(a) Schematic diagram of the SGDFB geometry and (b) the bottom plot shows the single-mode emission spectra for an electrically tuned, SGDFB QCL at RT. Top-most figure shows continuous wave output power, the middle one shows SMSR as a function of emission wavelength*

6.14 Room temperature THz emission based on difference frequency generation in mid-IR QCLs

As mentioned, the THz region of the electromagnetic spectrum (0.1–10 THz) lacks compact, high power, RT sources. Traditional THz QCLs operate at cryogenic temperatures, but there are significant physical challenges related to increasing the operating temperature. On the other hand, THz sources based on intracavity difference frequency generation (DFG) in mid-IR QCLs are developing rapidly in the recent years, and become the only semiconductor light source that emits over 1-6 THz range with optical power up to mW at RT [27]. When a QCL active region is designed with strong coupling between the lower lasing levels and injector levels, which results in a large nonlinear susceptibility $\chi^{(2)}$, THz emission can be generated within the cavity [28]. Thus, this type of THz source is free from the temperature limitation suffered by the THz QCLs based on direct optical transition, and ideally its working temperature is only limited by the mid-IR QCLs, and can be tuned over a broad wavelength range with broadband heterogeneous active region design [29]. It not only shares the common features of the mid-IR QCLs which are mass reproducible, RT operation, low cost, compact size, and high efficiency, but also carries the potential of delivering THz emission with high power in a wide frequency range.

Figure 6.22(a) shows a lattice-matched $Ga_{0.47}In_{0.53}As/Al_{0.48}In_{0.52}As$ SPR structure at a wavelength of $\lambda_1 \sim 9.0$ μm designed for THz DFG QCLs based on the single-phonon-resonance (SPR) design. The band structure is shown at the high fields when the maximum mid-IR powers are expected. The state 1 is separated by about one optical-phonon energy from state 2, which helps to depopulate state 2

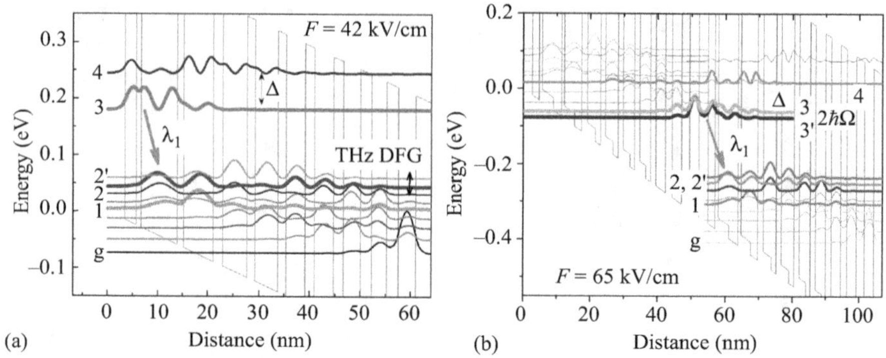

Figure 6.22 *(a) Band structure of a SPR structure at $\lambda \sim 9$ μm with large nonlinear DFG susceptibility based on lattice-matched Ga0.47$In_{0.53}As/Al_{0.48}In_{0.52}As$ material system and (b) band structure of a strongly coupled SPR structure at $\lambda \sim 7.8$ μm based on strain-balanced $Al_{0.63}In_{0.37}As/Ga_{0.35}In_{0.65}As/Ga_{0.47}In_{0.53}As$ material system*

with ultrafast optical-phonon scattering. Compared with the bound-to-continuum (BTC) structure in Ref. [30], the lifetime of the upper lasing state 3 caused by phonon scattering is increased from 0.39 ps for the BTC design to 0.46 ps for the SPR design, while the lifetime of the lower lasing state 2 is maintained at 0.1 ps. The major contributions to the THz DFG nonlinearity stem from the adjacent states 2' that are within 10–20 meV of the lower lasing state 2. The electron density in the states involved in DFG can be estimated with a simplified rate equation model by applying the lasing threshold condition ($G_{gain} = (\alpha_m + \alpha_w)/\Gamma$) at which the clamping occurs. G_{gain}, α_m, α_w, and Γ are gain, mirror loss, waveguide loss, and confinement factor, respectively. Given the waveguide loss of 8 cm^{-1}, mirror loss of 2.3 cm^{-1} for a 3-mm long HR-coated cavity, and confinement factor of 0.4, the nonlinear susceptibility is estimated to be $|\chi^{(2)}| = 3.97 \times 10^4$ pm/V at 4 THz.

The limited conduction band offset (~0.5 eV) for the lattice-matched design results in an energy spacing of ~240 meV between the upper laser level and the continuum states located above the barriers, which is not able to confine the electrons in the upper laser level effectively and induces significant thermally activated carrier leakage into the continuum. This leads to less efficient performance of the devices compared to the state-of-the-art shorter-wavelength counterparts. To further improve the THz performance, a nonlinear SPR active region based on the strain-balanced $Al_{0.63}In_{0.37}As/Ga_{0.35}In_{0.65}As/Ga_{0.47}In_{0.53}As$ material system is designed [31]. The band structure is shown in Figure 6.22(b). The inserted $Ga_{0.47}In_{0.53}As$ layers are used to balance the material strain within one stage. The conduction band offset is enhanced to ~0.74 eV. In the present strain-balanced design, the increased conduction-band offset and interface roughness increases the broadening the oscillation linewidth to ~15–20 meV, compared to that of ~10 meV for the lattice-matched active region design [28]. This allows for a stronger coupling design with an energy splitting of 2 $\hbar\Omega$ = 16.5 meV between the

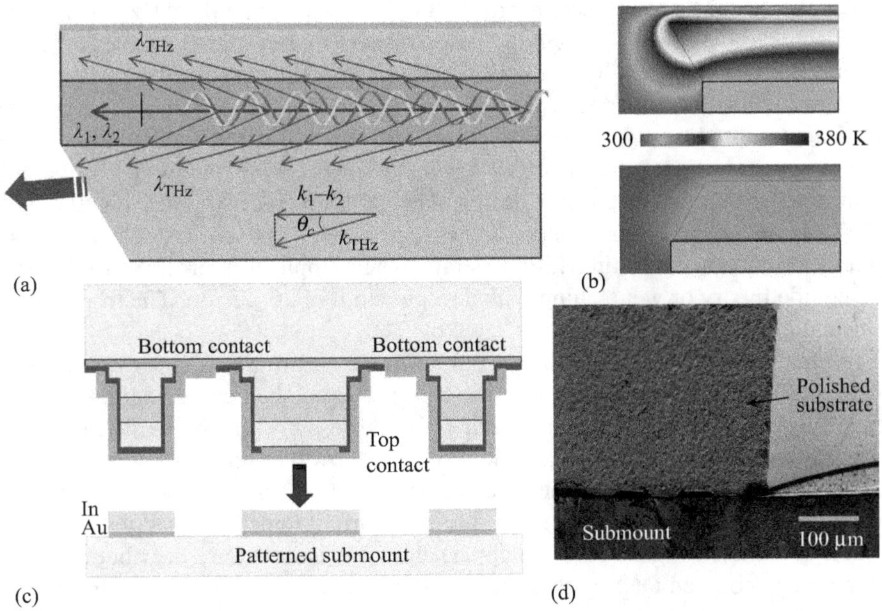

Figure 6.23 (a) Schematic of Čerenkov phase matching in a mid-IR quantum cascade lasers; (b) the thermal distributions of the epi-up and epi-down mounted Čerenkov devices; and (c–d) schematic of the epi-down mounting scheme and the SEM image of an epi-down mounted device

injector and upper lasing level. The optical nonlinearity is engineered by adjusting the injector barrier thickness so that the lower laser state 2 couples with its neighboring levels 2' with an energy spacing of 14–17 meV (3.4–4.1 THz). Given a threshold gain $g_{th} = 5\ \text{cm}^{-1}$ for CW operation, a total nonlinear susceptibility of $|\chi^{(2)}| = 2.0 \times 10^4$ is obtained from the present design.

To achieve a high mid-IR-to-THz power conversion efficiency, the phase mismatching should be close to zero for an effective nonlinear interaction. In a waveguide with modal phase-matching scheme, due to a faster frequency dependent effective index in the THz range with respect to the mid-IR index ($n_{\text{mid-IR}}$), the modal phase matching is only satisfied within a relatively narrow frequency range for a certain waveguide [32]. In contrast to this narrow-band phase-matching scheme, the Čerenkov scheme [33] has been used as an effective broad-band phase-matching method for THz generation in the externally pumped optical rectification or DFG setups [34]. Čerenkov nonlinear emission occurs when the nonlinear polarization wave propagates at a higher phase velocity compared to that of the nonlinear radiation, as shown in Figure 6.23(a). For THz DFG in QCLs, the waveguide for Čerenkov phase-matching scheme can be designed so that the THz refractive index (n_{THz}) in the QCL substrate is higher than the mid-IR group effective refractive index ($n_{\text{mid-IR}}$). In this case, the fundamental mid-IR wave propagates faster than the second-order DFG THz wave, and the THz wave travels at an angle $\theta_C = \cos^{-1}(n_{\text{mid-IR}}/n_{\text{THz}})$ with respect to the mid-IR wave, θ_C is the Čerenkov angle. This can be achieved by replacing the n-doped InP substrate with a semi-insulating InP substrate where n_{THz} (~3.47–3.71) larger than $n_{\text{mid-IR}}$ (~3.35) is fulfilled automatically in the 1–6 THz range. This broad-band Čerenkov phase-matching scheme is used to take full advantage of the wide bandwidth of the THz nonlinear medium based on SPR design. The QCL epilayer structure together with a 200-nm InGaAs bottom contact layer is grown by gas-source molecular beam epitaxy on a semi-insulating InP substrate. The sample was processed into ridge waveguide by dry or wet etching with a ridge width of 20–25 μm. The front facet is polished into 30° to facilitate the Čerenkov THz output.

Since the THz light is coupled out from the substrate with a facet-polished angle of 20–30° for Čerenkov phase-matching scheme, the ideal scenario is to use a thick substrate to couple out as much light as possible. In the experiments, the 350 μm thick substrate is normally not polished, however, thick substrate will lead to inefficient heat removal in the epilayer-up mounting scheme, and a less efficient current injection scheme due to the single-side current injection through a thin layer of bottom contact. Figure 6.23(b) (upper) show the temperature distributions of an epilayer-up mounted QCL with an asymmetric contact pattern near the front facet. Clearly, the epilayer-up mounted Čerenkov device suffers severely from thermal and electrical issues, as the heat near the front facet is not effectively dissipated and the non-uniform electrical distribution across the active region is observed. To address these two issues, double-sided current injection and epilayer-down mounting on a patterned submount are proposed [10], as shown in Figure 6.23(c). The heat dissipation is improved significantly as shown in Figure 6.23(c) (lower), and the electrical distribution is much more uniform with electrical field variation less than 0.5 kV/cm. Figure 6.23(d) is a SEM image of an epi-down mounted Čerenkov THz device on a diamond submount.

Ever since the first demonstration, THz sources based on DFG QCLs have strived for high power operation at RT [35]. Through the above systematic optimizations to the nonlinear active region, DFB waveguide, phase-matching scheme, and device bonding technique, the THz output power experiences an exponential increase in recent years, as depicted in Figure 6.24(a). Currently, the highest THz power up to 1.9 mW with a conversion efficiency of 0.8 mW/W^2 is achieved from a HR coated, 3-mm long, 23-μm wide composited DFB laser epilayer-down mounted a patterned diamond submount, as shown in Figure 6.24(b). The measured THz power is not corrected with any collection efficiency. The maximum THz power WPE for this device reaches to 0.7×10^{-5} [27].

Spectral emission is single mode at 3.5 THz with SMSR as high as 30 dB as shown in Figure 6.24(c). The far field testing in Figure 6.24(d) indicates that the device exhibits a good beam profile with divergence angles of 12.5° in the vertical direction and 36° in the lateral direction with a dual-peak distribution. Ideally, the far field pattern of the Čerenkov THz emission inside of the QCL waveguide should be a conical shape with the cone angle equal to the Čerenkov angle because of the much longer emitting wavelength respecting to the width of the QCL

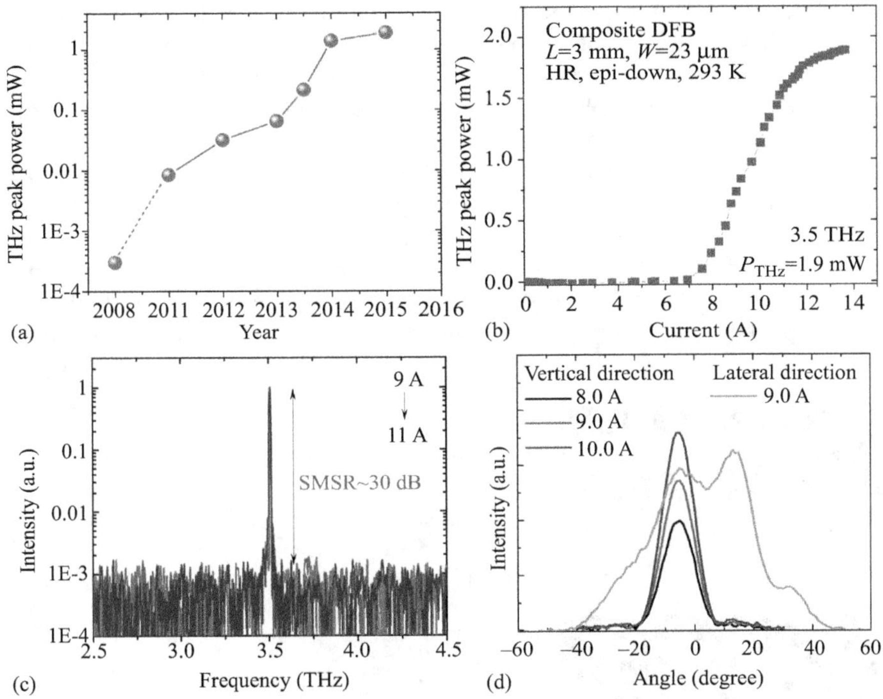

Figure 6.24 (a) Recent demonstrated THz power records of the THz sources based on DFG QCLs at room temperature as a function of year. THz power (b), spectra (c), and far field (d) as functions of currents for a high power THz sources based on DFG QCLs

waveguide [35]. This indicates that only part of this radiation cone is able to be coupled out through the polished substrate facet limited by the Brewster's angle $\theta_B \approx \pm 16.5°$, which is inferred from $\theta_B = \sin^{-1}(n_{air}/n_{InP})$ with $n_{air} = 1$, and $n_{InP} = 3.5$. Considering another 30% reflection to both the mid-IR and THz emissions, the outcoupling efficiency of the emitted THz power with respect to the generated power is as low as 15%. Given the maximum recorded 1.9 mW output power, the total generated THz power in that device is estimated to be ~12 mW. Therefore, there is a great potential for THz power enhancement when an efficient THz outcoupler is applied to the THz waveguide. By enlarging the outcoupling aperture with a Si prism to allow for more THz light to be coupled out from the substrate or extract the THz light from the entire cavity via diffraction grating [36], even higher THz power and efficiency in a wider spectral range in pulsed and CW modes can be obtained. This is of special importance to high THz frequency generation (>4.5 THz) where the phonon absorption in the semi-insulating InP substrate plays a leading role to the THz absorption prior to outcoupling. On the other hand, the power of a mid-IR QCL has been scaled up to 200 W [37], with a similar strategy, the THz power can be further enhanced to tens mW by scaling the device area.

While a lot of effort has been devoted to improving the THz power and frequency tuning range of the THz sources based on DFG in QCLs, the capability of CW operation at RT is of utter importance to wide application and commercialization. Nevertheless, RT CW operation has been difficult to be achieved for the following reasons. First, the design priority has been to achieve a large nonlinear susceptibility for THz generation. However, this may compromise the mid-IR performance. Second, in order to obtain a large population inversion, the doping of the structure is much higher than a typical mid-IR QCL, which results in a high threshold current density, precluding RT CW operation. These issues are recently addressed by utilizing a low-loss buried-ridge waveguide design and highly dissipative epi-down mounting scheme (Figure 6.25(a)), room temperature-continuous

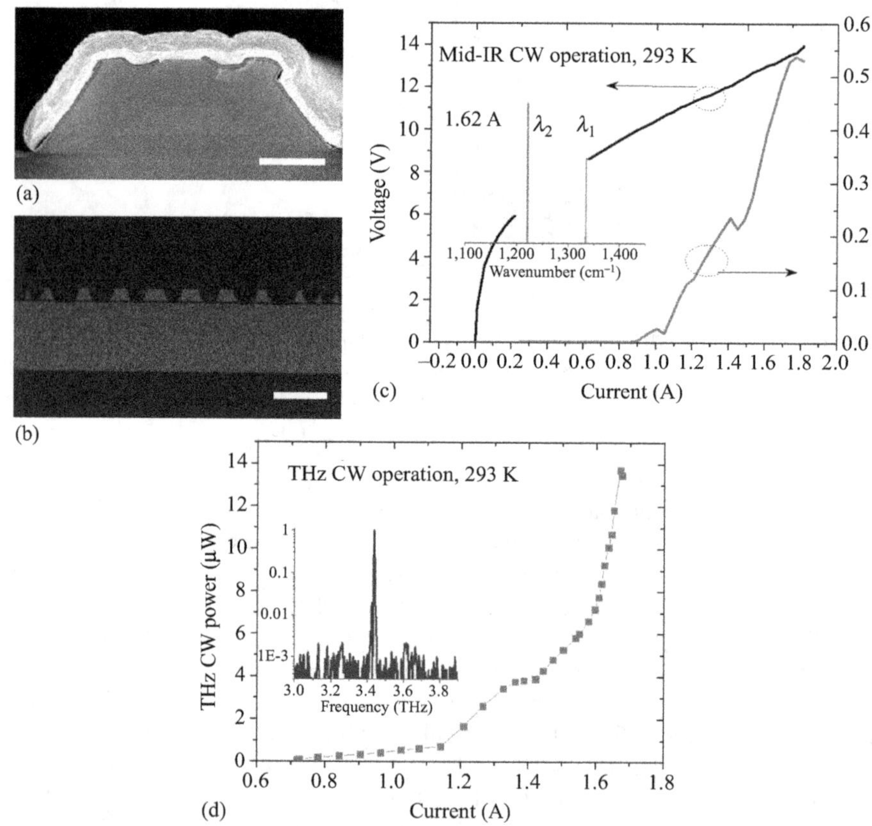

Figure 6.25 SEM images of a buried ridge (a) and buried composite DFB grating waveguide (b). The white bars in (a) and (b) correspond to 10 and 2 μm, respectively. (c) power-current-voltage (P–I–V) of a composite DFB THz device at room temperature CW operation. Inset: CW lasing mid-IR spectrum at 1.62 A. (d) THz CW power as a function of current. Inset: CW THz spectrum at 1.62 A

wave (RT-CW) operation at 3.6 THz was demonstrated with a continuous power of 3 μW [35].

To further enhance the THz CW power, a new strong-coupled strain-balanced QCL design for efficient THz generation-based intracavity DFG with a epi-down mounting scheme. A single-phonon resonance active region made from a composite strain-balanced $Al_{0.63}In_{0.37}As$/ $Ga_{0.35}In_{0.65}As/Ga_{0.47}In_{0.53}As$ active region is designed with strong coupling of 2 $\hbar\Omega = 16.5$ meV. The FWHM of 380 cm^{-1} is even broader than the previous dual-core lattice-matched active region with a FWHM of ~330 cm^{-1} in the 9–11 μm wavelength range [28], which makes this homogenous active region able to provide a sufficiently wide gain to support the 1–5 THz emission.

In CW operations, a 4-mm long composite DFB device emits up to 0.53 W with threshold current densities of 1.87 kA/cm^2, as shown in Figure 6.25(b). The inset is the CW lasing spectrum at 1.62 A. Stable dual-wavelength operation at $\lambda_1 = 7.46$ μm and $\lambda_2 = 8.15$ μm with a frequency spacing of 3.4 THz is observed. For a FP device with identical geometry, a maximum power of 3 W and threshold current density of 1.95 kA/cm^2 in pulsed mode operation, and a maximum power of 1.1 W and a threshold current density of 2.2 kA/cm^2 in CW operation are demonstrated, respectively. The maximum WPEs are 10% and 4% for pulsed mode and CW operations, respectively. Figure 6.25(c) is the THz power of the composite DFB QCL in CW operations 293 K. Maximum CW power of 14 μW with conversion efficiencies of $\eta = 0.35$ mW/W^2, and THz WPEs of 0.8×10^{-6} is obtained. The superlinear increase in CW THz power in Figure 6.25(c) is due to the rapid power increase in λ_1 and dramatic power balancing between λ_1 and λ_2 in the current range of 1.45–1.68 A. The THz CW WPE is about one order of magnitude higher than the previous demonstration [35]. The increased THz power and WPE are mainly attributed to the enhanced mid-IR power and efficiency of the strong-coupled strain-balanced active region design [31].

A compact, high-power, RT THz source emitting in a wide frequency range will greatly benefit THz system development for applications in spectroscopy, communication, sensing, and imaging. The THz source based on DFG in QCL has been designed with a broad gain spectrum and a giant nonlinear susceptibility over a broad THz frequency range [28]. To achieve monolithic tuning which has a wider application, composite DFB arrays were demonstrated. The device consists of an array of composite DFB devices with different grating period designs emitting at different THz frequencies, as shown in Figure 6.26(a).

Single-mode operations from 1.0 to 4.6 THz are demonstrated from a composite DFB array as shown in Figure 6.26(b), with a frequency step-tuning range ($\Delta\nu$) of 3.6 THz and a mean SMSR above 30 dB. This range represents 1.28 times that of the central frequency. For each of the device, the THz spectral position is pretty stable with respect to the temperature or current changes. The average current tuning rate is about 1 GHz/A and the linewidth is about 6–8.6 GHz, which is mainly limited by the resolution of the Fourier-transform infrared spectrometer (0.125 cm^{-1}). Much wider, continuous tuning can be achieved by employing the dual-section sampled grating design [9,38] for each emitting element.

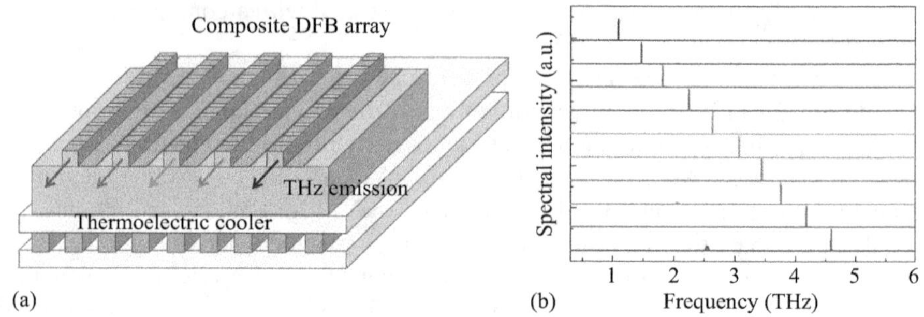

Figure 6.26 (a) Schematic of composite DFB array and (b) THz emitting spectra from 1.0 to 4.6 THz

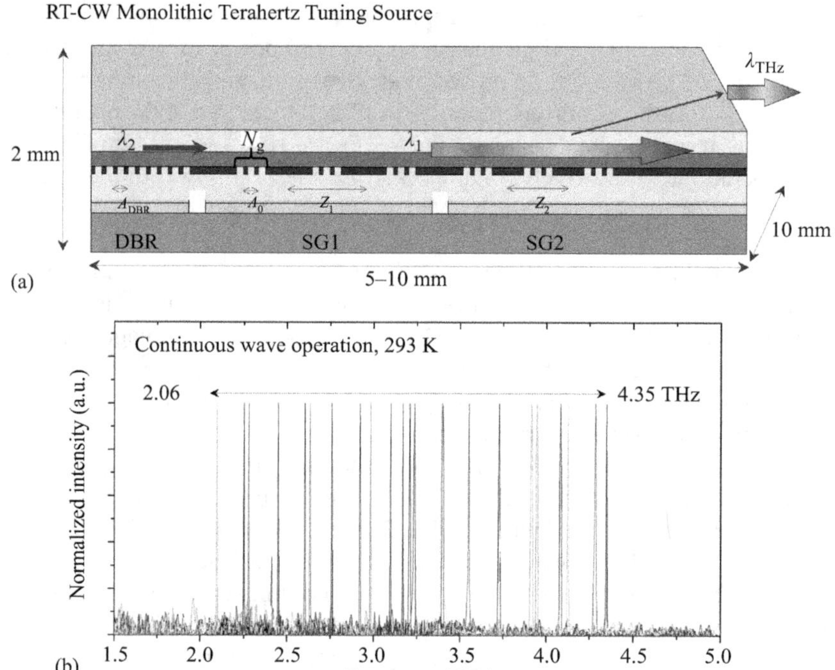

Figure 6.27 (a) Schematic of SGDFB-DBR waveguide design for THz tuning and (b) combined tuning of 2.06–4.35 THz of the two epi-down mounted THz SGDFB-DBR devices

To achieve wide THz tuning from a single THz emitter, we recently demonstrated a monolithic tunable THz source based on multi-section sampled-grating distributed feedback-distributed Bragg reflector (SGDFB-DBR) design [39], as shown in Figure 6.27(a). Both sampled-grating (SG) sections are sampled with a very short grating section ($\Lambda_0 = 1.22$ μm, $N_g = 15$) for 7 times. Here, N_g is the grating number in one short section. The sampling periods $Z_1 = 207$ μm and $Z_2 =$ 185 μm were used for the front and back SG sections, respectively. To enhance the power performance, the front SG section was further elongated with a 1.5-mm unpatterned section for power amplification. Laser bars with 6.3-mm cavity length were cleaved, containing one 1-mm DBR section ($\Lambda_{DBR} = 1.329$ μm) on the back, one 2-mm SG section (SG1) in the middle, and one 3.3-mm SG section plus the amplifier (SG2) in the front [31].

These three-section devices are then epi-down mounted on patterned diamond submounts for CW measurements. The measured THz tuning spectra are presented in Figure 6.27(b). A wide frequency tuning from 2.06 to 3.17 THz with a step of 185 GHz by changing the DC current on SG1, and 2.09 to 3.1 THz with a step of 168 GHz by changing the DC current on SG2, are achieved, respectively. Another device with a slightly different grating period designs of $\Lambda_0 = 1.196$ μm and $\Lambda_{DBR} = 1.36$ μm targeting a larger frequency spacing, exhibits a full tuning range of 3.2 to 4.35 THz, as shown in Figure 6.27(b). The combined tuning range for the two devices is 2.29 THz. The SMSR ranges from 12 to 25 dB in the tuning range measured with an uncooled far-infrared DTGS detector. In the tuned THz frequencies, the THz output power ranges from 0.6 μW at 2.06 THz to 4.2 μW at 3.42 THz. The conversion efficiency peaks around 3.0–3.5 THz with $\eta_{max} = 0.36$ mW/W^2 at 3.2 THz.

6.15 Conclusion

In summary, InP-based QCL technology has demonstrated its versatility as a source over a wide spectral range. However, many challenges remain which limit the performance of QCLs. For example, non-idealities such as intervalley scattering, carrier escape to the continuum, heat removal from the core region, and interface scattering still affect the RT performance of QCLs, especially at short wavelengths. Some of these can be lessened or removed by better gain medium and waveguide design strategies, while the quality of strained material growth can still be improved to realize better interfaces and better compositional control. Advances in power scaling of QCLs, spatial and spectral filtering, and monolithic tuning are also being explored for practical applications. In addition, by integrating a mid-IR laser and nonlinearity into the same device, RT THz emitters are now being developed. With many of these areas showing continual improvements, it is likely that many more exciting ideas and applications will be developed over the next several years.

References

[1] T. H. Maiman, "Stimulated optical radiation in ruby," *Nature*, vol. 187, pp. 493–494, 08/06/print 1960.

[2] M. Razeghi, *The MOCVD Challenge: A survey of GaInAsP-InP and GaInAsP- GaAs for photonic and electronic device applications, Second Edition (Electronic Materials and Devices)*: CRC Press, 2010.

[3] J. Faist, F. Capasso, D. L. Sivco, C. Sirtori, A. L. Hutchinson, and A. Y. Cho, "Quantum cascade laser," *Science*, vol. 264, pp. 553–556, 1994.

[4] S. Slivken, C. Jelen, A. Rybaltowski, J. Diaz, and M. Razeghi, "Gas-source molecular beam epitaxy growth of an 8.5 μm quantum cascade laser," *Applied Physics Letters*, vol. 71, pp. 2593–2595, 1997.

[5] N. Bandyopadhyay, Y. Bai, S. Tsao, S. Nida, S. Slivken, and M. Razeghi, "Room temperature continuous wave operation of lambda [bold ~] 3–3.2 μm quantum cascade lasers," *Applied Physics Letters*, vol. 101, p. 241110, 2012.

[6] S. Slivken, A. Evans, J. Nguyen, *et al.*, "Overview of quantum cascade laser research at the center for quantum devices," *Proceedings of SPIE*, vol. 6900, pp. 69000B1-8, 2008.

[7] Y. Bai, S. Slivken, S. Kuboya, S. R. Darvish, and M. Razeghi, "Quantum cascade lasers that emit more light than heat," *Nature Photonics*, vol. 4, pp. 99–102, 2010.

[8] Y. Bai, N. Bandyopadhyay, S. Tsao, S. Slivken, and M. Razeghi, "Room temperature quantum cascade lasers with 27% wall plug efficiency,"*Applied Physics Letters*, vol. 98, p. 181102, 2011.

[9] S. Slivken, N. Bandyopadhyay, S. Tsao, *et al.*, "Sampled grating, distributed feedback quantum cascade lasers with broad tunability and continuous operation at room temperature," *Applied Physics Letters*, vol. 100, pp. 261112–4, 2012.

[10] Q. Y. Lu, N. Bandyopadhyay, S. Slivken, Y. Bai, and M. Razeghi, "Room temperature terahertz quantum cascade laser sources with 215 μW output power through epilayer-down mounting," *Applied Physics Letters*, vol. 103, pp. 011101–4, 2013.

[11] J. H. Smet, C. G. Fonstad, and Q. Hu, "Intrawell and interwell intersubband transitions in multiple quantum wells for far-infrared sources," *Journal of Applied Physics*, vol. 79, pp. 9305–9320, 1996.

[12] Y. Bai, N. Bandyopadhyay, S. Tsao, E. Selcuk, S. Slivken, and M. Razeghi, "Highly temperature insensitive quantum cascade lasers," *Applied Physics Letters*, vol. 97, p. 251104, 2010.

[13] C. G. Van de Walle, "Band lineups and deformation potentials in the model-solid theory", *Physical Review B*, vol. 39, pp. 1871–1883, 1989.

[14] Y. Bai, S. Slivken, S. R. Darvish, and M. Razeghi, "Room temperature continuous wave operation of quantum cascade lasers with 12.5% wall plug efficiency," *Applied Physics Letters*, vol. 93, pp. 021103–3, 2008.

[15] N. Bandyopadhyay, Y. Bai, B. Gokden, *et al.*, "Watt level performance of quantum cascade lasers in room temperature continuous wave operation at lambda ~ 3.76 μm," *Applied Physics Letters*, vol. 97, pp. 131117–3, 2010.

[16] P. Laffaille, J. C. Moreno, R. Teissier, M. Bahriz, and A. N. Baranov, "High temperature operation of short wavelength InAs-based quantum cascade lasers," *AIP Advances*, vol. 2, p. 022119, 2012.

[17] N. Bandyopadhyay, S. Slivken, Y. Bai, and M. Razeghi, "High power, continuous wave, room temperature operation of lambda ~ 3.4 μm and

lambda ~ 3.55 μm InP-based quantum cascade lasers," *Applied Physics Letters*, vol. 100, pp. 212104–4, 2012.

[18] Y. Bai, S. Slivken, S. R. Darvish, A. Haddadi, B. Gokden, and M. Razeghi, "High power broad area quantum cascade lasers," *Applied Physics Letters*, vol. 95, pp. 221104–3, 2009.

[19] M. Razeghi, S. Slivken, Y. B. Bai, B. Gokden, and S. R. Darvish, "High power quantum cascade lasers," *New Journal of Physics*, vol. 11, p. 125017, 2009.

[20] S. R. Darvish, W. Zhang, A. Evans, J. S. Yu, S. Slivken, and M. Razeghi, "High-power, continuous-wave operation of distributed-feedback quantum-cascade lasers at $\lambda \sim 7.8$ μm," *Applied Physics Letters*, vol. 89, p. 251119, 2006.

[21] S. R. Darvish, S. Slivken, A. Evans, J. S. Yu, and M. Razeghi, "Room-temperature, high-power, and continuous-wave operation of distributed-feedback quantum-cascade lasers at $\lambda \sim 9.6$ μm," *Applied Physics Letters*, vol. 88, p. 201114, 2006.

[22] J. S. Yu, S. Slivken, S. R. Darvish, A. Evans, B. Gokden, and M. Razeghi, "High-power, room-temperature, and continuous-wave operation of distributed-feedback quantum-cascade lasers at $\lambda \sim 4.8$ μm," *Applied Physics Letters*, vol. 87, p. 041104, 2005.

[23] Q. Y. Lu, Y. Bai, N. Bandyopadhyay, S. Slivken, and M. Razeghi, "2.4 W room temperature continuous wave operation of distributed feedback quantum cascade lasers," *Applied Physics Letters*, vol. 98, pp. 181106–3, 2011.

[24] B. Gokden, Y. Bai, N. Bandyopadhyay, S. Slivken, and M. Razeghi, "Broad area photonic crystal distributed feedback quantum cascade lasers emitting 34 W at lambda ~ 4.36 μm," *Applied Physics Letters*, vol. 97, pp. 131112–3, 2010.

[25] Y. Bai, S. Slivken, Q. Y. Lu, N. Bandyopadhyay, and M. Razeghi, "Angled cavity broad area quantum cascade lasers," *Applied Physics Letters*, vol. 101, pp. 081106–5, 2012.

[26] Y. Bai, S. Tsao, N. Bandyopadhyay, *et al.*, "High power, continuous wave, quantum cascade ring laser," *Applied Physics Letters*, vol. 99, pp. 261104–3, 2011.

[27] M. Razeghi, Q. Y. Lu, N. Bandyopadhyay, *et al.*, Quantum cascade lasers: from tool to product. *Optics Express*, vol. 23, p. 8462, 2015.

[28] Q. Y. Lu, N. Bandyopadhyay, S. Slivken, Y. Bai, and M. Razeghi, "Widely tuned room temperature terahertz quantum cascade laser sources based on difference-frequency generation," *Applied Physics Letters*, vol. 101, pp. 251121–4, 2012.

[29] N. Bandyopadhyay, Y. Bai, S. Slivken, and M. Razeghi, "High power operation of $\lambda \sim 5.2$–11 μm strain balanced quantum cascade lasers based on the same material composition," *Applied Physics Letters*, vol. 105, p. 071106, 2014.

[30] Q. Y. Lu, N. Bandyopadhyay, S. Slivken, Y. Bai, and M. Razeghi, "Room temperature single-mode terahertz sources based on intracavity

difference- frequency generation in quantum cascade lasers," *Applied Physics Letters*, vol. 99, p. 131106, 2011.

[31] Q. Y. Lu, D. H. Wu, S. Sengupta, S. Slivken, and M. Razeghi, "Room temperature continuous wave, monolithic tunable THz sources based on highly efficient mid-infrared quantum cascade lasers," *Scientific Reports*, vol. 6, p. 23595, 2016.

[32] Q. Y. Lu, N. Bandyopadhyay, S. Slivken, Y. Bai, and M. Razeghi, "High performance terahertz quantum cascade laser sources based on intracavity difference frequency generation," *Optics Express*, vol. 21, pp. 968–973, 2013.

[33] P. K. Tien, R. Ulrich, and R. J. Martin, "Optical second harmonic generation in form of coherent Cerenkov radiation from a thin-film waveguide," *Applied Physics Letters*, vol. 17, p. 447, 1970.

[34] M. Theuer, G. Torosyan, C. Rau, *et al.* Efficient generation of Cherenkov-type terahertz radiation from a lithium niobate crystal with a silicon prism output coupler. *Applied Physics Letters*, vol. 88, p. 071122, 2006.

[35] Q. Y. Lu, N. Bandyopadhyay, S. Slivken, Y. Bai, and M. Razeghi, "Continuous operation of a monolithic semiconductor terahertz source at room temperature," *Applied Physics Letters*, vol. 104, p. 221105, 2014.

[36] M. Razeghi, Q. Y. Lu, N. Bandyopadhyay, and S. Slivken, Recent development of high power, widely tunable THz quantum cascade laser sources based on difference-frequency generation. *Proc. of SPIE*, vol. 9585, pp. 958502–1, 2015.

[37] D. Heydari, Y. Bai, N. Bandyopadhyay, S. Slivken, and M. Razeghi, "High brightness angled cavity quantum cascade lasers," *Applied Physics Letters*, vol. 106, p. 091105, 2015.

[38] S. Slivken, N. Bandyopadhyay, Y. Bai, Q. Y. Lu, and M. Razeghi, "Extended electrical tuning of quantum cascade lasers with digital concatenated gratings," *Applied Physics Letters*, vol. 103, p. 231110, 2013.

[39] Q. Y. Lu, S. Slivken, N. Bandyopadhyay, Y. Bai, and M. Razeghi, "Widely tunable room temperature semiconductor terahertz source," *Applied Physics Letters*, vol. 105, p. 201102, 2014.

Chapter 7

Diode laser systems based on nonlinear frequency conversion

Ole Bjarlin Jensen[1], *Anders Kragh Hansen*[1], *Christian Pedersen*[1], *Peter E. Andersen*[1], *and Paul Michael Petersen*[1]

The availability of diode lasers is mainly limited by the semiconductor material. As described in the preceding chapters, diode lasers can be manufactured from the ultraviolet spectral range to the mid-infrared. Depending on the wavelength range, the output power can range from few mW to several tens of watts from single-emitter diode lasers. However, some spectral regions are still not easily accessible using direct emitting diode lasers. This is especially true for the green-yellow spectral range, where only very recently the first direct emitting green diode lasers have appeared. This spectral range is of high importance for several biophotonics applications as some of the important chromophores have absorption peaks here. One particular chromophore of high importance is hemoglobin which has absorption peaks located in the green-yellow range at 542 nm and 577 nm. The absorption properties of hemoglobin are utilized in applications like photocoagulation in treatment of retinal detachments or dermatological treatments of vascular lesion like port wine stains or hemangiomas.

Traditionally this wavelength range has been accessed by different laser systems like Argon-ion lasers, frequency-doubled solid-state lasers, copper vapor lasers or dye lasers. All these laser systems have different drawbacks like low efficiency, high cost, bulky size and/or toxic laser materials. Diode lasers in this spectral range are currently not powerful enough to perform vascular treatments. Diode lasers in the near infrared spectral range can on the other hand be very powerful and furthermore the electrical to optical efficiency of these lasers is very high approaching 75%. An obvious method to reach high power outputs in the green–yellow spectral range with diode lasers is frequency conversion of high power near infrared diode lasers.

Nonlinear frequency conversion can also be used to generate light in other spectral regions, where diode lasers are not available. An example could be in the mid-infrared range where the semiconductor materials are not well developed.

[1]Department of Photonics Engineering, Technical University of Denmark, Denmark

By using difference frequency generation (DFG) between two lasers, light can be generated at long wavelengths. By proper choice of the two interacting lasers, almost the entire infrared spectrum can be covered. If the two interacting lasers are diode lasers, the tunability enables generation of tunable infrared light. Infrared light is extensively used within spectroscopy and many applications depend on light sources based on DFG.

As described below, the conversion efficiency of nonlinear frequency conversion depends strongly on the properties of the fundamental light. Three properties of the fundamental laser light are of paramount importance. First, the power level of the fundamental laser sources is important as the conversion efficiency scales with the intensity of the interacting waves. Secondly, the spectral properties of the laser sources are important in order to achieve phase matching in the nonlinear materials. Finally the spatial properties of the laser sources are important to achieve a high intensity throughout the length of the nonlinear material.

In this chapter, we will give a short introduction to nonlinear frequency conversion with second-order nonlinear optics to provide the basic background to the topic and highlight the important parameters to consider when designing a light source based on frequency conversion. In the following section, the different commonly used implementations of nonlinear frequency conversion will be introduced and some examples of demonstrated laser sources will be given. Finally, we will provide an outlook for the development of frequency converted diode laser light sources and briefly discuss some of the applications.

7.1 Nonlinear frequency conversion using second-order nonlinear optics

In this section, we review the principles of nonlinear frequency conversion based on second-order nonlinear effects. One of the reasons that nonlinear optics is attractive for the development of new light sources is that the nonlinear interactions in materials can be used to generate new frequencies. In the following, the basic theory for nonlinear frequency conversion is reviewed and in the succeeding sections, some important applications such as frequency doubling and sum frequency generation will be discussed.

Nonlinear optical effects can be described in terms of the complex dielectric susceptibilities. If we want to understand the various nonlinear optical effects it is useful to consider the expansion of the polarization in the material P in powers of the electric field [1]:

$$P(\mathbf{E}) = \varepsilon_0\left(\chi^{(1)}\mathbf{E} + \chi^{(2)}\mathbf{E}^2 + \chi^{(3)}\mathbf{E}^3 + \cdots\right) \tag{7.1}$$

where $\mathbf{E}$ denotes the total electric field. $\mathbf{E}$ is often composed of different waves with different frequencies, polarizations, and $\mathbf{k}$ vectors. The first term $\chi^{(1)}\mathbf{E}$ in (7.1) corresponds to the linear optical properties, such as the refractive index and the absorption. The second term $\chi^{(2)}\mathbf{E}^2$ corresponds to the second-order nonlinear effects. These include second harmonic generation (SHG), electro-optic linear Kerr effects, and sum frequency generation (SFG) and DFG. The third-order term $\chi^{(3)}\mathbf{E}^3$ is responsible for effects such as third harmonic generation, two photon generation and four-wave mixing. The polarization in (7.1) may be written as $\mathbf{P}(\mathbf{E}) = \varepsilon_0\chi^{(1)}\mathbf{E} + \mathbf{P}^{nl}$ where $\mathbf{P}^{nl}$ is the nonlinear polarization. The nonlinear response of the induced polarization acts back on material and this interaction may be described by the Maxwell equations and the material equations. The Maxwell equations describe how the electric field $\mathbf{E}$ and the magnetic field $\mathbf{H}$ develop:

$$\nabla \times \mathbf{H} = \mathbf{J} + \frac{\partial \mathbf{D}}{\partial t} \tag{7.2}$$

$$\nabla \times \mathbf{E} = -\mu_0 \frac{\partial \mathbf{H}}{\partial t} \tag{7.3}$$

$$\nabla \cdot \mathbf{D} = \rho \tag{7.4}$$

$$\nabla \cdot \mathbf{B} = 0 \tag{7.5}$$

where $\mathbf{D}$ and $\mathbf{B}$ are the electric and magnetic fluxes, $\mathbf{J}$ is the current density, and ρ is the charge density. The material equations are given by

$$\mathbf{D} = \varepsilon_0\mathbf{E} + \mathbf{P} \tag{7.6}$$

$$\mathbf{J} = \sigma\mathbf{E} \tag{7.7}$$

$$\mathbf{B} = \mu_0\mathbf{H} + \mathbf{M} \tag{7.8}$$

with σ as the conductivity of the material. Taking the curl of (7.2) and using the identity $\nabla \times \mathbf{E} = \nabla\,(\nabla \cdot \mathbf{E}) - \nabla^2\mathbf{E}$ lead to the nonlinear wave equation:

$$\nabla^2\mathbf{E} = \mu_0\sigma\frac{\partial \mathbf{E}}{\partial t} + \mu_0\varepsilon_0\varepsilon\frac{\partial^2 \mathbf{E}}{\partial t^2} + \mu_0\frac{\partial^2 \mathbf{P}^{NL}}{\partial t^2} \tag{7.9}$$

where we also have used (7.3), (7.6), and (7.7). Equation (7.9) is the fundamental equation that governs the temporal and spatial evolution of the interacting light beams in the nonlinear medium.

We now consider second-order nonlinear effects where two fields with angular frequencies ω_1 and ω_2 are incident on the nonlinear medium as shown in Figure 7.1. For simplicity we use a scalar approximation. The angular frequency is given in terms of the wavelength, λ, as $\omega = 2\pi c/\lambda$.

The nonlinear polarization generates new frequencies at $2\omega_1$, $2\omega_2$, $\omega_2 \pm \omega_1$ and 0. In a given experiment usually only one frequency is generated and the specific frequency will depend on a phase matching condition described below.

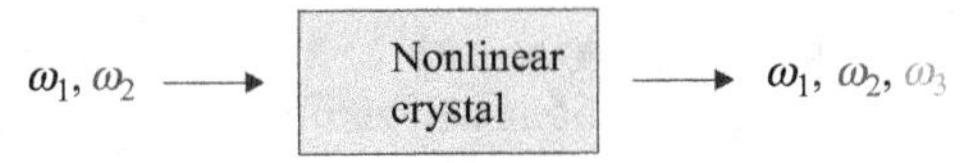

Figure 7.1 Three-wave interaction in a nonlinear crystal

Therefore, in the theoretical analysis we need only to include three beams. The electric fields of these 3 beams are in a plane wave approximation given by:

$$E_i(z,t) = \tfrac{1}{2}A_i e^{i(\omega_i t - k_i z)} + c.c.\; i = 1,2,3 \tag{7.10}$$

where $\mathbf{A}_i$ is the complex amplitude, $\mathbf{k}_i$ is the wave vector, and c.c. denotes the complex conjugate. Inserting these three beams together with the nonlinear polarization $P = \varepsilon_0\chi^{(2)}E^2$ in (7.9) lead to the following 3 coupled wave equations:

$$\frac{dA_1}{dz} = \frac{\mu_0\sigma_1 c}{2n_1}A_1 - \frac{i\omega_1 a d_{\text{eff}}}{n_1 c}A_3 A_2^* e^{-i\Delta kz} \tag{7.11}$$

$$\frac{dA_2}{dz} = \frac{\mu_0\sigma_2 c}{2n_2}A_2 - \frac{i\omega_2 a d_{\text{eff}}}{n_2 c}A_3 A_1^* e^{-i\Delta kz} \tag{7.12}$$

$$\frac{dA_3}{dz} = \frac{\mu_0\sigma_3 c}{2n_3}A_3 - \frac{i\omega_3 a d_{\text{eff}}}{n_3 c}A_1 A_2 e^{i\Delta kz} \tag{7.13}$$

where we have used the slowly varying envelope approximation and inserted the effective nonlinear coefficient d_{eff} in exchange of the nonlinear susceptibility $\chi^{(2)}$ according to [2,3]. In (7.11–7.13), $\Delta k = k_1 + k_2 - k_3$, $(\mu_0/\varepsilon)^{1/2} = \mu C/n_i$ and $a = \tfrac{1}{2}(2 - \delta_{\omega_1\omega_2})$, where $\delta_{\omega_1\omega_2} = 1$ for $\omega_1 = \omega_2$ and $\delta_{\omega_1\omega_2} = 0$ for $\omega_1 \neq \omega_2$.

In the case of SHG, we have $\omega_1 = \omega_2 = \omega_3$, $\omega_3 = 2\omega$, $A_1 = A_2 = A(\omega)$ and $a = \tfrac{1}{2}$. The fundamental equation (7.13) reduces to

$$\frac{dA(2\omega)}{dz} = -\frac{i\omega d_{\text{eff}}}{n(2\omega)c}A^2(\omega)e^{i\Delta kz} \tag{7.14}$$

where absorption has been neglected. If we assume that $A(\omega)$ remains undepleted, thus constant throughout the crystal, we obtain the second harmonic intensity:

$$I(2\omega) = \frac{2\omega^2 d_{\text{eff}}^2}{n^2(\omega)n(2\omega)\varepsilon_0 c^3}L^2 I(\omega)^2\left(\frac{\sin(\Delta kL/2)}{\Delta kL/2}\right)^2 \tag{7.15}$$

where L is the length of the nonlinear crystal. Franken *et al.* [4] reported in 1961 the first experimental observation of SHG. They focused a ruby laser with a wavelength of 694 nm into a quartz crystal and observed generation of light at 347 nm.

We can in a similar fashion to SHG treat the more general sum frequency generation case where two fields of different angular frequency ω_1 and ω_2 are incident on the nonlinear crystal (see Figure 7.1) and generate a field with angular frequency $\omega_3 = \omega_1 + \omega_2$. The result is:

$$I(\omega_3) = \frac{2\omega_3^2 d_{\text{eff}}^2}{n(\omega_1)n(\omega_2)n(\omega_3)\varepsilon_0 c^3}L^2 I_1(\omega_1)I_2(\omega_2)\left(\frac{\sin(\Delta kL/2)}{\Delta kL/2}\right)^2 \tag{7.16}$$

Note that ω_3 is used in (7.16) giving rise to a factor 4 improvement compared to the SHG case (7.15) for equal incident wavelengths. This apparent inconsistency is explained from Figure 7.1. The general SFG case relates to two fields of intensity $I_1(\omega_1)$ and $I_2(\omega_2)$, respectively, incident on the nonlinear crystal whereas in the

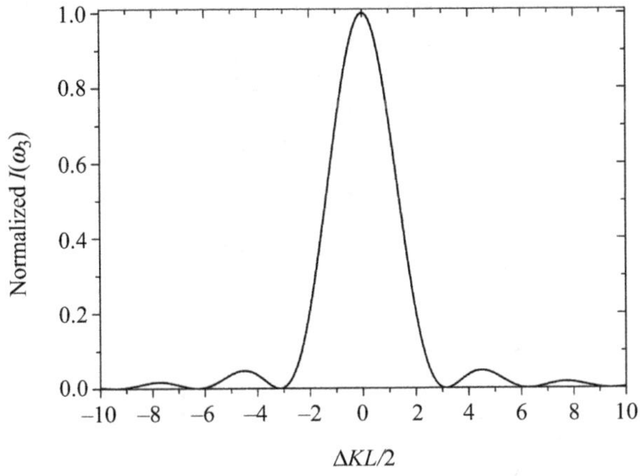

Figure 7.2 The second harmonic intensity versus the wave vector mismatch ΔkL/2

SHG case these 2 fields are merged to only one field of intensity $I(\omega)$. The SHG equation thus can be derived from the SFG equation with $I_1(\omega_1) = ½\ I(\omega)$ and $I_2(\omega_2) = ½\ I(\omega)$ inserted. The same result is obtained using $\alpha = 1$ in (7.11–7.13).

In Figure 7.2, we notice that the maximum frequency conversion is obtained when the condition $\Delta k = 0$ is satisfied. This condition is called phase matching and leads to efficient nonlinear frequency conversion since the pump wave and the nonlinear frequency converted wave are in phase. When phase matched condition is not met the transfer of energy from the fundamental field to the nonlinearly generated field diminishes quickly as a function of crystal length L as seen from Figure 7.2. Investigating phase matching in further detail, we explicitly express the angular frequency dependency of Δk:

$$\Delta \mathrm{k} = \frac{n(\omega_3)\omega_3}{c} - \frac{n(\omega_1)\omega_1}{c} - \frac{n(\omega_2)\omega_2}{c} \tag{7.17}$$

where $n(\omega_i)$ denotes the refractive index of the medium at angular frequency ω_i.

From inspection of (7.17), it is easily seen that generally $\Delta k \neq 0$ due to dispersion. It is therefore an important engineering task to establish phase matching for efficient interaction.

Phase matching has been obtained in numerous ways [5–7]. Here we will only illustrate two commonly used methods, angle tuning using birefringent nonlinear crystals and quasi-phase-matching (QPM) using periodically poled crystals.

Angle tuning for obtaining phase matched three-wave interaction has been the primary method since the early days of nonlinear optics [5]. This method relies on the continuous change of the refractive index with angle of incidence for an extra ordinary polarized beam propagating in an anisotropic crystal. To illustrate this method we consider SHG inside a uniaxial $LiNbO_3$ crystal. We assume that the fundamental beam (ω) propagates with ordinary (o) polarization at an internal

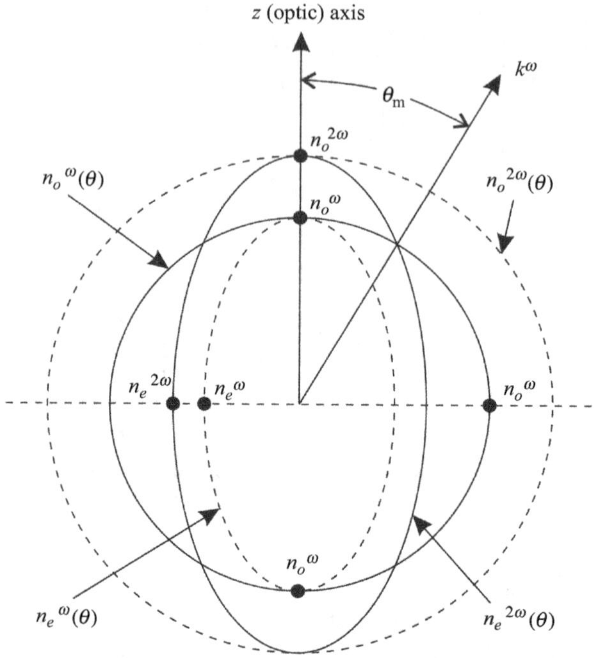

Figure 7.3 The index ellipsoid for SHG using a positive uniaxial crystal

angle θ with respect to the optic axis generating an extraordinary (e) polarized SHG beam (2ω). This setup is labeled an eoo interaction since two ordinary polarized beams generate an extraordinary polarized beam. The polarization state of a field is also indicated by applying a subscript to the corresponding refractive index. Since only the refractive index of the extraordinary beam is θ dependent, (7.16) yields

$$\Delta k(\theta) = \frac{n_e(2\omega,\theta)2\omega}{c} - \frac{n_o(\omega)\omega}{c} - \frac{n_o(\omega)\omega}{c} = \frac{2\omega}{c}[n_e(2\omega,\theta) - n_o(\omega)] \tag{7.18}$$

$n_e(2\omega,\theta)$ can be computed from the index ellipsoid:

$$\frac{1}{n_e^2(2\omega,\theta)} = \frac{\cos(\theta)}{n_o^2(2\omega)} + \frac{\sin(\theta)}{n_e^2(2\omega)} \tag{7.19}$$

where $n_o^2(2\omega), n_e^2(2\omega)$ and $n_o(\omega)$ are calculated from the crystal's Sellmeier equations [8].

The Sellmeier equation is typically determined from experiments or tables. Since θ is a free parameter it offers a tuning mechanism for phase matching. The index ellipsoid and the phase matching condition is shown in Figure 7.3.

One major drawback to angle tuning is that the Poynting vector (the direction of energy transfer) is nonparallel to the k vector for beams with extraordinary polarization in a birefringent medium when propagating at an angle to the optic axis.

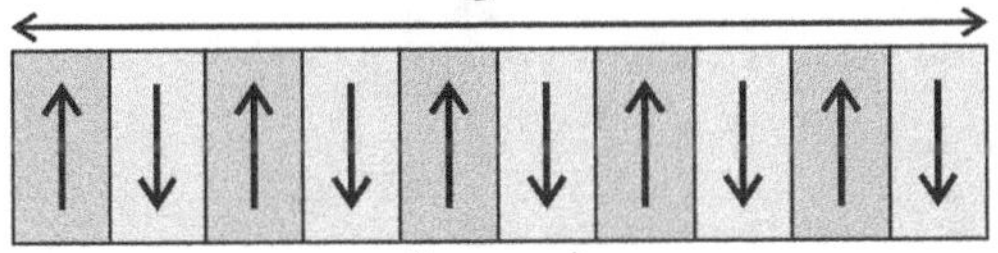

Figure 7.4 The periodically poled material with poling period Λ

This means that any fields with extraordinary polarization involved in the process will tend to separate from those with ordinary polarization, reducing the overlap between the interacting beams and thus reducing the conversion efficiency. Furthermore, walk-off will degrade the spatial beam properties of the fields involved because of interference between the field contributions generated at different propagation distances in the crystal.

A recent and widespread phase matching technique that leads to efficient nonlinear conversion is QPM [6]. In this technique the sign of the nonlinear second-order susceptibility, d is reversed periodically along the axis of propagation. For $LiNbO_3$ the polarization reversal of the crystals domains can be accomplished by applying a high voltage to a periodic set of electrodes attached to the crystal c-axis face. When the applied voltage exceeds the so-called coercive field, typically 25 kV/mm for $LiNbO_3$, the polarization direction of the crystals domains flips and the crystal enters a new and stable state, but with periodically reversed domains, see Figure 7.4. For periodically poled $LiNbO_3$ crystals the poling may also be obtained with proton exchange [9] or electron beam writing [10].

With reference to (7.11) it follows that such periodic modulation of the second-order susceptibility d can be interpreted as a π phase shift in the Δkz term. From Figure 7.4, we can conclude that quasi-phase-matching is obtained when $\Delta k\Lambda = 2\pi$. For the case of SHG, Λ can be explicitly written as: $\Lambda = 2\pi/(k_2 - 2k_1) = \lambda/(2(n_e(2\omega) - (n_e(\omega)))$, where we note that all the fields are polarized along the optic axis ($\theta = 0$).

The superiority of QPM over angle tuning relates to the ability to obtain phase matching through synthesis (i.e., the poling period). For the case of $LiNbO_3$ all three fields can be in the same polarization state exploiting the largest nonlinear second-order coefficient of the crystal. The effective nonlinear coefficient in QPM crystals is reduced to $d_{\text{eff}} = 2d/\pi$ for first-order QPM. At present $LiNbO_3$, $LiTaO_3$, and KTP are the most prominent crystals that can be poled. Most other nonlinear crystals need to be phase matched through angle tuning, temperature tuning or other means.

Finally, we introduce phase matching tolerances of QPM. This term refers to the reduction in output intensity due to phase mismatch, $\Delta k \neq 0$. For instance the wavelength acceptance tolerance, $\Delta\lambda_{\text{FWHM}}/2$ is defined as the deviation from the proper phase matching wavelength ($\Delta k = 0$) leading to a halving of output power.

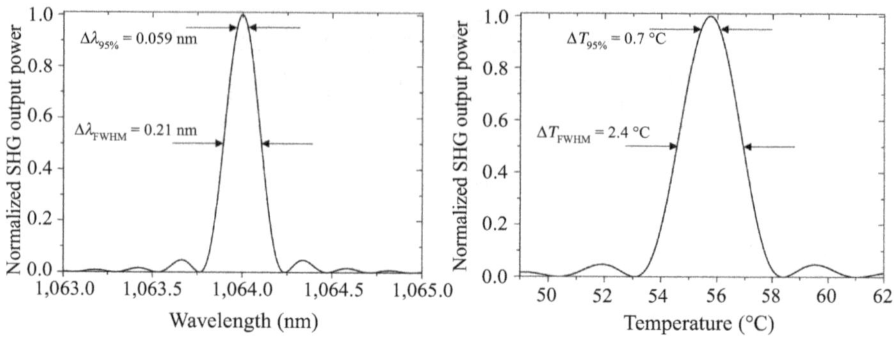

Figure 7.5 Simulated wavelength and temperature tolerances for QPM with a 10 mm long periodically poled MgO:LiNbO$_3$ crystal. The calculations are based on temperature and wavelength dependent extraordinary refractive indices according to the Sellmeier equations, and a thermal expansion coefficient of $\alpha = 1.59 \times 10^{-5}$/K [12]

The phase matching tolerances in terms of wavelength and temperature for SHG in periodically poled crystals are given by the following equations [11]:

$$\Delta\lambda_{FWHM} = \frac{0.4429\lambda_1}{L}\left|\frac{n_2 - n_1}{L} + \frac{\partial n_1}{\partial \lambda} - \frac{1}{2}\frac{\partial n_1}{\partial \lambda}\right|^{-1} \tag{7.20}$$

$$\Delta T_{FWHM} = \frac{0.4429\lambda_1}{L}\left|\frac{\partial(n_2 - n_1)}{\partial T} - \alpha(n_2 - n_1)\right|^{-1} \tag{7.21}$$

As an example, Figure 7.5 shows the wavelength and temperature tuning curves for a 10 mm long, periodically poled 5% MgO:LiNbO$_3$ (PPMgOLN) crystal with a poling period of 6.92 μm. The simulations show a spectral acceptance bandwidth of $\Delta\lambda_{\mathrm{FWHM}} = 210$ pm and a temperature acceptance of $\Delta T_{\mathrm{FWHM}} = 2.4$ °C. At 95% of maximum intensity the tolerances are significantly reduced to $\Delta\lambda_{95\%} = 59$ pm and $\Delta T_{95\%} = 0.7$ °C, respectively. Nonlinear interactions consequently require precise control of crystal temperatures and lasers with narrow spectral emission, such as DBR-tapered diode lasers. From (7.20) and (7.21), it is seen that the acceptance tolerances are approximately inversely proportional to the crystal length. Similar expression can be found for birefringent phase matching.

In (7.15), we have given the efficiency of SHG under the plane wave approximation. In reality, laser emission is finite in extent, exhibits a nonuniform intensity and phase distribution across the beam and is typically focused into the nonlinear crystal to increase the intensity and thus the efficiency. For the optimal case of focused Gaussian beams, the parametric interaction was studied by Boyd

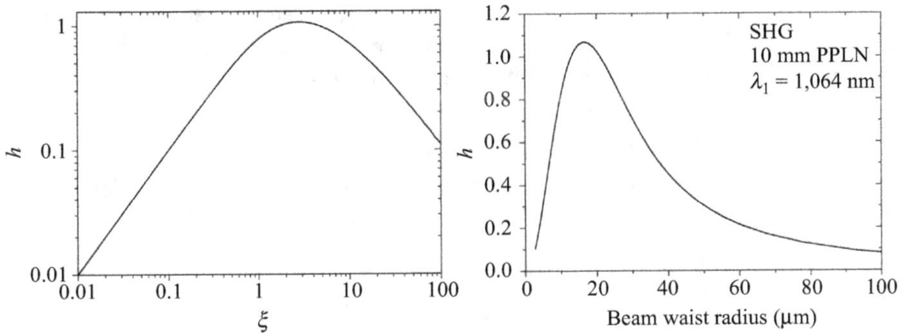

Figure 7.6 Boyd–Kleinman focusing factor h for zero walk-off as a function of the focusing parameter ξ and beam waist radius for SHG of 1,064 nm light in a 10 mm long PPLN crystal

and Kleinman as early as 1968 [13]. They found that in the un-depleted pump approximation, the second harmonic output power can be written as

$$P_2 = \eta_{\text{SHG}} P_1^2 \tag{7.22}$$

$$\eta_{\text{SHG}} = \frac{16\pi^2 d_{\mathit{eff}}^2}{\varepsilon_0 c \lambda_1^3 n_1 n_2} L h(\sigma, \beta, \kappa, \xi, \mu) \tag{7.23}$$

In (7.23), h is a function depending on a phase mismatch parameter σ, a walk-off parameter β, a crystal absorption parameter κ, a focusing parameter ξ and a focus displacement parameter μ. In most practical cases, we can assume that optimum phase matching can be obtained, the absorption is negligible and the focal position is in the middle of the crystal. In that case h depends on the walk-off angle in the nonlinear crystal and on the focusing. In the absence of walk-off, such as in the cases of temperature phase matching or quasi-phase-matching, the efficiency is maximized by focusing the incident light to a beam waist radius w_0 corresponding to $\xi = \lambda L / 2\pi n {w_0}^2 = 2.84$ as seen in Figure 7.6. The corresponding optimum beam waist radius can be found as

$$w_0 = \sqrt{\frac{\lambda L}{2 \cdot 2.84 \cdot \pi n}} \tag{7.24}$$

From Figure 7.6, it is observed that a slightly smaller ξ, indicating larger beam waist only results in a small reduction in efficiency. In practice, a larger than optimal beam waist is often preferred in order to prevent optical damage to the nonlinear crystal and reduce intensity-induced adverse effects. In case walk-off is present, the optimal beam waist radius increases and the maximal efficiency is reduced as described in [13].

In case of strong conversion, we can no longer assume that the fundamental beam is not depleted. By solving (7.14), we can find the following expression for the power generated in the second harmonic beam [14].

$$P_2 = P_1 \tanh^2 \sqrt{\eta_{SHG} P_1} \tag{7.25}$$

In a similar fashion, the expressions for sum and DFG in the un-depleted pump approximation can be found as [14,15]

$$P_3 = \frac{32\pi^2 d_{\mathrm{eff}}^2}{\varepsilon_0 c \lambda_1 \lambda_2 \lambda_3 n_3^2} P_1 P_2 L h_{GF}(\sigma, \beta, \kappa, \xi, \mu) \tag{7.26}$$

$$P_3 = \frac{32\pi^2 d_{\mathrm{eff}}^2}{\varepsilon_0 c \lambda_3^2 n_3 (n_1 \lambda_2 + n_2 \lambda_1)} P_1 P_2 L h_{CB}(\sigma, \beta, \kappa, \xi, \mu) \tag{7.27}$$

In these expressions, the focusing function h has been modified according to the interactions [16,17].

7.2 Methods for nonlinear frequency conversion

Different methods exist for nonlinear frequency generation that all have their advantages and drawbacks. Probably the simplest approach is to have the light from the laser pass the nonlinear crystal in a single pass and in this way generate new wavelengths by, for example, SHG. The inherent advantages of this approach are its simplicity and robustness while the lower conversion efficiency especially for continuous-wave (CW) lasers is the main drawback. Increase of the conversion efficiency can be obtained by either passing the beam through multiple crystals or having multiple passes in the same crystal. These methods provide higher efficiency and are still robust but at the expense of a more complex setup and possible multiple nonlinear crystals. The ultimate example of multiple passes in the nonlinear crystal is by use of external cavity approaches, where the incoming laser beam is coupled to a resonant cavity increasing the fundamental power in the nonlinear crystal and thus increasing the efficiency significantly. External cavity approaches have the drawback of requiring delicate alignment and locking of the laser frequency and cavity resonance frequency. Another possible way of taking advantage of the high circulating power in a cavity is intracavity frequency conversion where the nonlinear material is positioned inside the cavity of a laser. In this way the efficiency is high and no electronic locking is necessary. However, these lasers tend to have power stability problems due to competition and coupling of different longitudinal modes in the laser cavity. In the following, we will describe the different methods in more detail.

7.2.1 Single-pass frequency conversion

In single-pass frequency conversion, the light from a laser is passed through a nonlinear crystal and a part of it is converted to light at a new wavelength following

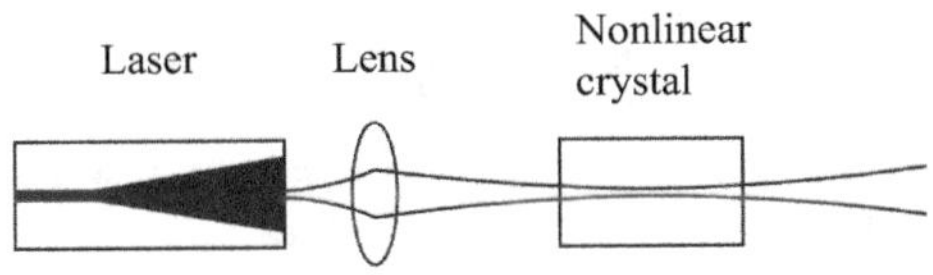

Figure 7.7 Principle sketch of single-pass frequency conversion

the relations in (7.25–7.27) for SHG, SFG or DFG, respectively. A principle sketch is shown in Figure 7.7. In order to obtain high conversion efficiency some relations between the nonlinear material and the incoming laser beams must be optimized. The nonlinear material should have a high figure of merit ($d_{eff}^2/n_1n_2n_3$), the losses should be low and the length of the nonlinear material must be as long as possible [18]. Furthermore, the input power and the spatial and spectral quality from the lasers must be high. Finally, focusing of the input lasers has a large influence on the conversion efficiency as described by Boyd and Kleinman [13] for SHG, Guha and Falk for SFG [16] and Chu and Broyer for DFG [17], respectively. In the Boyd–Kleinman analysis it is assumed that the interacting beams are diffraction limited Gaussian beams without ellipticity and astigmatism. For real laser beams and especially for diode laser beams, this assumption is not fulfilled. The emitting aperture of a diode laser is asymmetric leading to elliptical beams and for high power diode lasers like tapered diode lasers, the beam will exhibit a certain degree of astigmatism. These nonideal beam properties all lead to reduced conversion efficiencies as described by several authors [19–22].

In bulk materials, the theory of Boyd and Kleinman propose a trade-off between the high intensity of a small spot size and the requirement for a long interaction length. A strongly focused beam will be highly diverging making the interaction length short. In an optical waveguide, the intensity can be kept high over an extended region and as such the nonlinear conversion efficiency is increased. The generated power in a nonlinear waveguide was calculated by Regener and Sohler [23] for SHG and the generated power was found to be

$$P_3 = P_1 \tanh^2\left(\sqrt{\frac{8\pi^2 d_{eff}^2 \Im_{ov}^2 L}{\lambda_1^2 \epsilon_0 c N_{eff}^3} P_1}\right) \tag{7.28}$$

with $\Im_{ov}$ as the overlap integral between the fundamental and second harmonic beam and N_{eff} as the effective refractive index. The high intensity inside the optical waveguide is also the main limitation of them as optical damage limits the generated power level to a few hundred mW.

In single-pass SHG, the conversion efficiency is often limited by the length of the nonlinear material. Some types of nonlinear crystals are available in long lengths while others are limited to lengths of a few mm or cm. Especially periodically poled crystals are often limited to a few cm by the poling process used. A specific drawback of using very long crystals is that the acceptance bandwidths

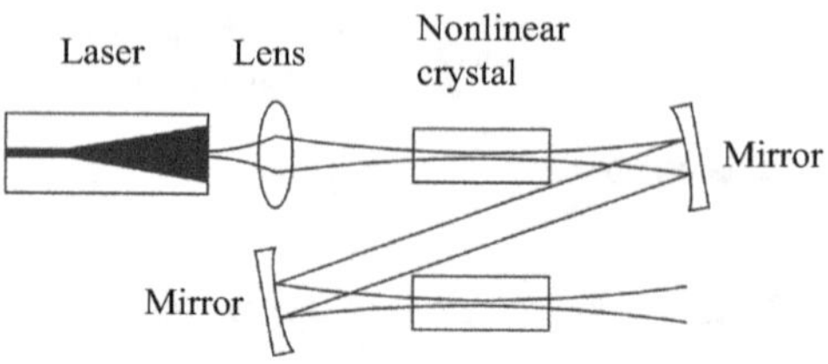

Figure 7.8 Cascaded frequency conversion in a series of crystals. Phase compensation between the different waves is necessary

regarding wavelength, temperature and angle are reduced proportional to the crystal length making phase matching more difficult as seen from (7.20) and (7.21).

A method for increasing the conversion efficiency is to make multiple passes in the same crystal or to use several crystals in series [24], as shown schematically in Figure 7.8. In order to obtain high conversion efficiency in multi-pass or multi-crystal schemes, phase compensation is often required in order to maintain constructive interference between the contributions from each subsequent pass or crystal. This phase compensation can be achieved by using the natural dispersion in air or by use of a phase plate [24].

For a setup where the fundamental beam passes once through N identical crystals of length L or N passes through one such crystal, the general expression for output power at optimal phase mismatch in the individual crystals and optimal phase compensation between each pair of crystals or passes is

$$P_{2\omega} = \bar{\eta} L N^2 P_{\omega}^2 \tag{7.29}$$

where $\bar{\eta}$ denotes the nonlinear efficiency per unit length and the effects of depletion are assumed to be negligible. The expression assumes that the beam is focused into each individual crystal or for each pass through the crystal, and that the phase matching is noncritical. The acceptance bandwidths scale with $(\mathrm{NL})^{-1}$ [25]. By adjusting the phase compensation, it is possible to alternate between constructive and destructive interference between the contributions in the individual crystals or passes. Additionally, even if the phase compensation is nonoptimal, one can compensate by shifting the phase mismatches such that even in the case where the phase compensation in a two-crystal cascade is set to the worst possible value, it is nevertheless possible to obtain a total output power higher than twice that of a single crystal [26]. The principle can also be applied to cascaded sum frequency generation, in which the same scaling law applies [27].

7.2.2 External cavity frequency conversion

The ability to enhance the intensity of a CW laser beam by coupling it into an external resonant cavity opens the door for high nonlinear conversion efficiencies at low (mW–W) input powers due to the linear dependence of conversion efficiency on the intensity of the fundamental fields.

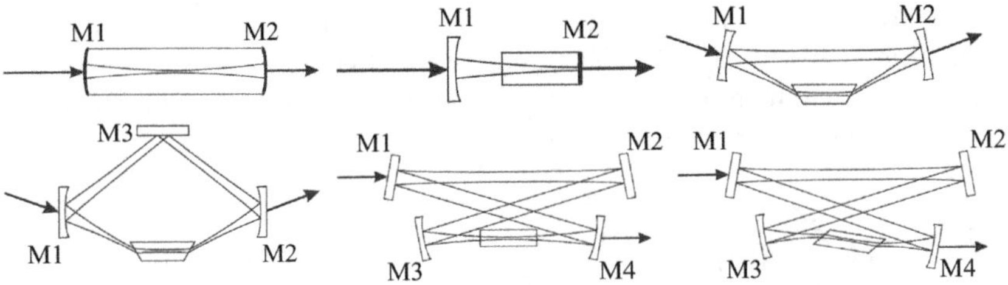

Figure 7.9 *Figure with monolithic, semi-monolithic, two-, three-, and four-mirror designs, with one four-mirror design being normal incidence and one being Brewster's angle incidence (tangential plane)*

While crystal choice was also very important in the previous section due to the great variation in η_{SHG}, use in external cavities place additional demands on transparency for the fundamental light and the ability to handle the large intracavity intensities without adverse effects in the crystal such as thermal lensing, intensity-induced absorption, grey-tracking, photorefractive damage, coating damage etc. The intensity thresholds for onset of most of these effects are not well known, not least because they depend strongly on wavelength and crystal purity and doping, so the expected severity of adverse effects must be assessed on a case-by-case basis.

A variety of designs have been successfully demonstrated for external cavity enhanced nonlinear frequency conversion, including monolithic [28], semi-monolithic [29,30], two-mirror [31–33], three-mirror [34] and four-mirror [35–52] designs as exemplified in Figure 7.9. Some implementations use crystals with the interfaces cut at normal incidence to the beam and some use crystals cut at Brewster's angle. In the former case, anti-reflective coatings are important to reduce reflection losses at the interfaces. Nearly all cavity designs use curved mirrors as the focusing elements rather than intracavity lenses. One or more mirrors have a dichroic coating to transmit the output radiation or an additional non-circulating input beam while reflecting the circulating beam.

Compared to SFG and DFG, SHG is the nonlinear process most straightforwardly enhanced in an external cavity and since many of the principles and considerations remain the same for design of a cavity for the other processes, we will here provide a walkthrough of the considerations relevant for choice and design of a SHG cavity only. SHG cavity theory can also be found in references [52] and [53]. Theory on SFG cavities can be found in references [54] and [55]. Examples of all types of cavity-enhanced processes will be provided in the next section.

The single-pass SHG power conversion factor η_{SHG} is defined in the absence of fundamental depletion by (7.23). The power enhancement factor of a cavity is defined as the ratio P_c/P_{in} of circulating intracavity power P_c and the power incident on the cavity P_{in}. We can write the SHG power output of a cavity in terms of the circulating power as $P_{SHG} = \gamma\eta_{SHG}P_c^2$, where γ is the fraction of second harmonic power that escapes the cavity. γ includes the reflection loss at the crystal exit interface and at the output coupler mirror.

The power enhancement factor is connected to the reflectivity of the input coupler, which we will call M1. Denoting the input coupler reflectivity R_1 and the fractional power remaining after one round trip of all *other* components in the cavity ρ, the power enhancement factor is given by

$$\frac{P_c}{P_{\text{in}}} = \delta \frac{1 - R_1}{\left(1 - \sqrt{R_1 \rho}\right)^2} \tag{7.30}$$

where ρ contains both linear and nonlinear losses and the coupling constant for mode matching δ is a factor between 0 and 1 representing the nonperfect overlap between the incident fundamental beam and the cavity mode. For a good mode matching with an approximately Gaussian input beam, $\delta \geq 90\%$. For incident beams with non-Gaussian profile or profiles with different ellipticity to the cavity mode, the highest achievable δ will be lower. This is especially important to keep in mind when using a diode laser as the source, which often start out elliptic and may be non-Gaussian.

The optimal input coupler reflectivity maximizing the expression above is the one that satisfies the condition of impedance matching, the condition that the component of incident fundamental radiation that is reflected off M1 must cancel out by destructive interference with the component of the circulating power that is transmitted out through M1. When this condition is fulfilled, the radiation incident on the cavity effectively experiences full transmission on M1 while all the intra-cavity light effectively experiences full reflection on M1. The optimal, impedance matched input coupler reflectivity is $R_{1,\text{opt}} = \rho$. By insertion, we see that the power enhancement factor at perfect impedance matching is

$$\frac{P_c}{P_{\text{in}}} = \frac{\delta}{1 - R_{1,\text{opt}}} \tag{7.31}$$

We separate the total round trip linear loss ε and nonlinear loss $\eta_{\text{SHG}} P_c$ by writing

$$\rho = (1 - \epsilon)(1 - \eta_{\text{SHG}} P_c) \tag{7.32}$$

Solving (7.30) and (7.31) for $R_{1,\text{opt}}$ we get

$$R_{1,\text{opt}} = 1 - \left(\epsilon/2 + \sqrt{(\epsilon/2)^2 + (1 - \epsilon)\eta_{\text{SHG}} \delta P_{\text{in}}}\right) \tag{7.33}$$

Since this is a function of the incident power, the input coupler reflectivity should be chosen to correspond either to the expected typical incident power or, alternatively, the maximum achievable incident power. If possible, it is best to base this calculation on actual single pass measurements of the mirror reflectivity's and the (nonphase-matched) transmission of the crystal as well as the experimentally achievable (phase-matched) η_{SHG} rather than trusting manufacturer specifications. It is also not uncommon to test the cavity performance with input couplers of two or three different reflectivity's around the expected optimum.

Under the assumption of impedance matching at any particular input power, the internal conversion efficiency $\psi = \frac{P_{\mathrm{SHG}}/\gamma}{\delta P_{\mathrm{in}}}$ can be written as [37,54]

$$\psi = 1 + (\chi\delta P_{\mathrm{in}})^{-1} - \sqrt{(\chi\delta P_{\mathrm{in}})^{-2} + 2(\chi\delta P_{\mathrm{in}})^{-1}}\,,\ \chi = 2\eta_{\mathrm{SHG}}/\epsilon^2 \tag{7.34}$$

χ is an important figure of merit describing the general performance of the cavity when impedance matched, together with δ and γ which describe a simple linear downscaling of the input and output powers, respectively. Adverse effects will cause deviations from this formula at higher powers.

Let us now take a look at some of the specific cavity designs there are to choose from. Monolithic and semi-monolithic designs have the benefit of reducing the number of separate mirrors per round trip in the cavity to zero or one by having the beam reflect off one or more of the coatings on the crystal. While this helps reduce the linear losses, the reduction in degrees of freedom complicate and compromise alignment. Additionally, many monolithic and semi-monolithic designs are linear and thus form a standing wave for the fundamental field, which may induce unwanted effects in the crystal and means that the nonlinear field is also emitted in both directions.

When two-or three-mirror designs are chosen over four mirror designs, it is usually in the interest of minimizing the total optical path length L of the cavity, thereby increasing the free spectral range $\frac{c}{L}$ and the bandwidth acceptance of the cavity $\Delta\nu = \frac{c}{L}\frac{1}{\mathcal{F}}$, where the cavity finesse for round trip loss 1 - ρ is $\mathcal{F} \approx \frac{\pi}{1-\sqrt{\rho}}$. With a cavity length of 400 mm and round trip loss 4% we get free spectral range $\sim$ 750 MHz, finesse $\mathcal{F} \sim 150$, power enhancement factor $P_c/P_{\mathrm{in}} \sim 25$ and bandwidth $\Delta\nu \sim 5$ MHz. This bandwidth must be larger than the laser linewidth, which for diode lasers is often in the MHz range. For high-linewidth lasers, the small two- or three-mirror designs or (semi-)monolithic designs may provide a high enough cavity bandwidth. Highly efficient crystals will also increase the acceptance bandwidth by lowering the finesse.

By far the most commonly used cavity design, however, is the four mirror bow-tie design using two flat and two spherical mirrors. Generally, if low-loss anti-reflective coatings with high enough damage thresholds are available for the crystal at the fundamental and second harmonic wavelengths, a cavity with a normal-incidence crystal is preferred. If anti-reflective coatings are problematic at the powers and wavelengths involved, for example, for doubling the ultraviolet, a crystal cut at Brewster's angle may be preferable. While the loss at the crystal interfaces in the latter case can be made less than the loss at a normal-incidence anti-reflective coated interface, the trade-off is that the resulting foci in the cavity are elliptic and, unless quasi-phase-matching is employed, the vertically polarized second harmonic will suffer a large loss due to Fresnel reflection at the crystal exit interface (typically >17%) before it can exit the cavity. If periodically poled crystals are used, the quasi-phase-matching results in a horizontally polarized second harmonic which will experience a much lower reflectivity.

The following design considerations are limited to the case of a four-mirror bow-tie cavity, such as the one shown in Figure 7.10, with an anti-reflective coated crystal

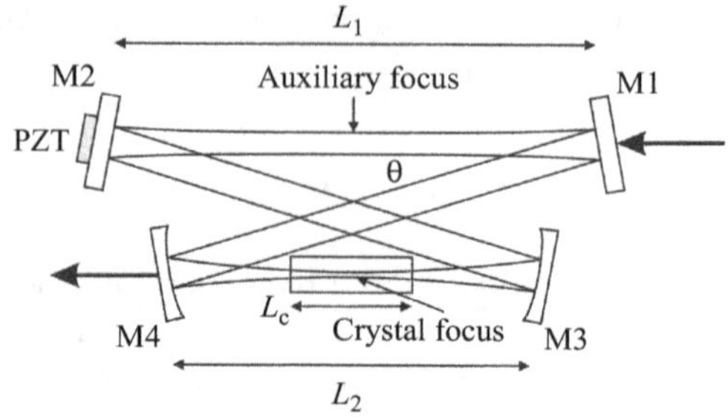

Figure 7.10 Figure of normal-incidence bow-tie cavity with symbols

cut at normal incidence. The reader is encouraged to consult the references for design considerations for crystals cut at Brewster's angle [53] or the other cavity designs.

For the layout in Figure 7.10, the fraction of power remaining after one round trip is $\rho = R_2R_3R_4T_{AR}^2T_c(1 - \eta_{SHG}P_c)$ where R_2, R_3, and R_4 are the reflectivity's of the mirrors M2, M3, and M4, respectively; T_{AR} is the transmission of the anti-reflective coating on one crystal interface; T_c is the transmission through the crystal including linear absorption and scattering and $\eta_{SHG}P_c$ is the relative loss due to nonlinear conversion in the crystal. The linear loss in this design is then $\epsilon = 1 - R_2R_3R_4T_{AR}^2T_c$.

Once the crystal material, phase matching scheme and length has been decided on, the optimal waist can be calculated from the theory of Boyd and Kleinman [13] as in the previous sections. It is in practically all cases prudent to use a significantly larger waist, for example, 50%–100% larger. This larger waist is often experimentally found to be closer to the actual power optimum and furthermore reduces the severity of adverse effects mentioned previously as well as the impact of local structural inhomogeneities such as periodic poling defects.

Once the target crystal waist is decided on, cavity stability must be considered by numerically calculating the round trip ABCD matrix. For ABCD matrix theory, see reference [56] or an optics textbook. For a symmetric four-mirror design such as Figure 7.10, the so-called crystal focus or primary focus is formed halfway between the curved mirrors and the so-called auxiliary focus or secondary focus is formed halfway between the flat mirrors. Assuming a folding angle θ sufficiently small to ignore astigmatism in the focusing of the curved mirrors with radius of curvature R, the cavity mode is circular and the ABCD matrix for propagation from the crystal focus to the auxiliary focus is given by

$$M_1 = \overbrace{\begin{bmatrix} 1 & \frac{L_1}{2} \\ 0 & 1 \end{bmatrix}}^{\text{M1}\to\text{aux}} \overbrace{\begin{bmatrix} 1 & \frac{L_1+L_2}{2\cos(\theta)} \\ 0 & 1 \end{bmatrix}}^{\text{M4}\to\text{M1}} \overbrace{\begin{bmatrix} 1 & 0 \\ -\frac{2}{R} & 1 \end{bmatrix}}^{\text{M4}} \overbrace{\begin{bmatrix} 1 & \frac{L_2-L_c}{2} \\ 0 & 1 \end{bmatrix}}^{\text{surf}\to\text{M4}} \overbrace{\begin{bmatrix} 1 & 0 \\ 0 & \frac{n}{1} \end{bmatrix}}^{\text{refr.}} \overbrace{\begin{bmatrix} 1 & \frac{L_c}{2} \\ 0 & 1 \end{bmatrix}}^{\text{cry}\to\text{surf}} \tag{7.35}$$

and the matrix for propagation from the auxiliary focus back to the crystal focus is

$$M_2 = \overbrace{\begin{bmatrix} 1 & \frac{L_c}{2} \\ 0 & 1 \end{bmatrix}}^{\text{surf}\to\text{cry}} \overbrace{\begin{bmatrix} 1 & 0 \\ 0 & \frac{1}{n} \end{bmatrix}}^{\text{refr.}} \overbrace{\begin{bmatrix} 1 & \frac{L_2 - L_c}{2} \\ 0 & 1 \end{bmatrix}}^{\text{M3}\to\text{surf}} \overbrace{\begin{bmatrix} 1 & 0 \\ -\frac{2}{R} & 1 \end{bmatrix}}^{\text{M3}} \overbrace{\begin{bmatrix} 1 & \frac{L_1 + L_2}{2\cos(\theta)} \\ 0 & 1 \end{bmatrix}}^{\text{M2}\to\text{M3}} \overbrace{\begin{bmatrix} 1 & \frac{L_1}{2} \\ 0 & 1 \end{bmatrix}}^{\text{aux}\to\text{M2}} \tag{7.36}$$

The round trip matrix $M = \begin{bmatrix} A & B \\ C & D \end{bmatrix}$ is then $M = M_2 M_1$ and is a function of L_1 and L_2.

The crystal waist is then $w_c = \sqrt[4]{-\frac{\lambda^2}{\pi^2 n^2} \frac{B}{C}}$ where λ is the vacuum wavelength of the fundamental. The stability region is comprised of the points where $\frac{B}{C}$ is negative. With the crystal waist known, the auxiliary waist can be found through the ABCD matrix $M_1 = \begin{bmatrix} A_1 & B_1 \\ C_1 & D_1 \end{bmatrix}$ as

$$w_a = \sqrt{\frac{\sqrt{B}A_1 + \sqrt{C}B_1}{\sqrt{B}C_1 + \sqrt{C}D_1} \frac{\lambda}{i\pi}} \tag{7.37}$$

When plotting the crystal waist as a function of L_1 and L_2, it is generally recommended to make the choice of lengths such that the crystal waist assumes the desired value near the middle of the region of stability, providing relative insensitivity of the actually achieved crystal waist to variations in L_1 and L_2 and, indirectly, to uncertainties in other parameters such as radii of curvature of the curved mirrors.

Lastly, a means to scan or lock the cavity length is needed. This is typically done by mounting one mirror on a piezoelectric transducer (PZT). To increase the speed of cavity length adjustment, the mirror mounted on the PZT is typically small, reducing its weight. Linearly scanning the mirror back and forth and monitoring the intracavity fundamental power is essential to optimize the alignment and mode-matching of the cavity, since the transverse mode structure of the cavity is then visible as separated peaks in the scan. By proper alignment of the cavity and of the incident beam as well as proper focusing of the incident beam to match the cavity mode auxiliary focus, the power in the Gaussian TEM_{00} mode can be maximized.

When in operation, the cavity length needs to be locked to an integer multiple of the laser wavelength. For this purpose, a variety of locking schemes exist. The two most frequently used are the Hänsch–Couillaud (HC) locking scheme [57] and the Pound–Drever–Hall (PDH) locking scheme [58], which requires frequency modulation of the incident light. Generally, the HC scheme requires fewer and simpler optical and electronic elements while the PDH scheme is generally considered more robust in operation because of a greater error signal gradient around the resonance. While any laser beam can be frequency modulated by sending it through an electro-optic modulator (EOM), the emission from most diode lasers have the benefit of being easily frequency modulated by current modulation of the diode itself, thus eliminating the need for a separate EOM and an EOM driver. PDH locking is thus cheaper and simpler with diode lasers than with most other lasers. Examples of both locking schemes are provided in the next section.

In the HC scheme, the error signal used to lock the cavity is generated by measuring whether the generally elliptically polarized light exiting the cavity through M1 is left-or right-handed, which in turn depends on whether the cavity is too short or too long. This is done by sending the beam through a quarter-wave plate and a polarizing beam splitter and then measuring the two polarizations with differentially coupled photodiodes.

In the PDH scheme, frequency modulation with a frequency typically in the MHz range creates two small spectral sidebands around the fundamental frequency. If the cavity is off resonance, a slight beat at the modulation frequency appears in the light leaked from the cavity with a phase shift depending on whether the cavity is too short or too long. The amplitude and sign of this beat component is extracted with an electronic mixer and comprises the error signal for the PZT.

7.2.3 Intracavity frequency conversion

In the previous section, we saw that an increase of the fundamental power by use of an enhancement cavity has a large effect on the conversion efficiency. In many laser sources, only a small portion of the light is transmitted through the output coupler and hence a much higher power is resonating within the laser cavity. The basic concept of intracavity frequency conversion is to place the nonlinear material inside the laser cavity in order to take advantage of this large circulating field to achieve high conversion efficiency as shown schematically in Figure 7.11. This has the inherent advantage over external cavity approaches that the laser wavelength will automatically adapt to the conditions without any electronic locking and need for single-frequency operation. Most commercially available green lasers are based on intracavity frequency doubling of solid-state lasers.

In the example of intracavity SHG, it is theoretically possible to extract the same amount of power in the second harmonic beam as is possible from the fundamental laser beam. This can be achieved by use of a nonlinear crystal with the

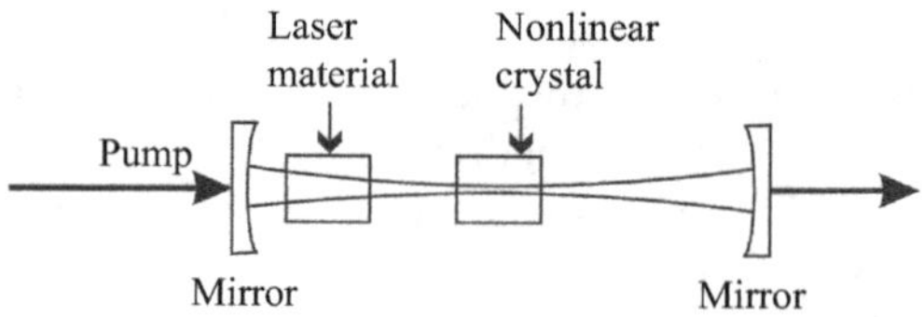

Figure 7.11 Principle sketch of intracavity frequency conversion. Energy is pumped into the laser material and laser oscillation takes place in the laser cavity. A nonlinear crystal is positioned inside the laser cavity

same conversion efficiency as the optimum mirror transmission for fundamental operation. In practice, however, this is not the case as the insertion loss of the nonlinear crystal is significant in the low-gain lasers where intracavity frequency doubling is often employed in high finesse cavities. A further drawback of intracavity SHG is the increased noise under multi-longitudinal mode operation, named the "green problem" [59]. Coupling of the different longitudinal modes in the nonlinear crystal through SFG will cause the intensity of the different longitudinal modes to fluctuate strongly giving excess noise in the output. Some different methods exist to circumvent the "green problem" including single-frequency operation, operation in a high number of longitudinal modes or operation in a few modes that do not perform SFG.

Diode lasers are rarely used in intracavity frequency conversion for two main reasons. Firstly, the intracavity field inside a diode laser is limited by optical damage to the diode laser facets. Second, the gain in diode lasers is so high that very high output coupling ($T > 90\%$) is optimum meaning that very high conversion efficiency in the nonlinear material is required to obtain efficient operation. Such high conversion efficiencies are difficult to obtain and if possible, it may prove as efficient to perform the frequency conversion in a single pass through the crystal.

A special example of intracavity frequency conversion is single-pass SFG or DFG through a nonlinear crystal placed inside the high finesse cavity of a solid-state laser. This has the advantage that high conversion efficiency can be achieved for the single-pass SFG. Furthermore, there is no coupling between the longitudinal modes of the solid-state laser through SFG in the nonlinear crystal as this is not phase matched. In this way, low-noise operation is ensured. A further advantage is that the mixing process will have a positive effect on the beam quality of the generated light as light is only generated in the overlapping parts of the beams [60]. The large diversity and tunability of diode lasers enables generation of light in many spectral regions that are otherwise difficult to access.

7.3 Nonlinear frequency converted laser systems

In all the different methods briefly explained in Section 7.3.1, all have different advantages and disadvantages. Nevertheless, all methods are used to produce laser systems suitable for the vast amount of applications within biophotonics. These applications have different requirements in terms of, for example, wavelength, power, spectral linewidth and spatial properties. Here, we will present different examples of diode laser based systems developed using nonlinear frequency conversion.

7.3.1 Single-pass frequency conversion

7.3.1.1 Second harmonic generation

Diode lasers have been used within frequency doubling since 1970, where the first experiments were performed [61,62] with conversion efficiencies of about 10^{-5}.

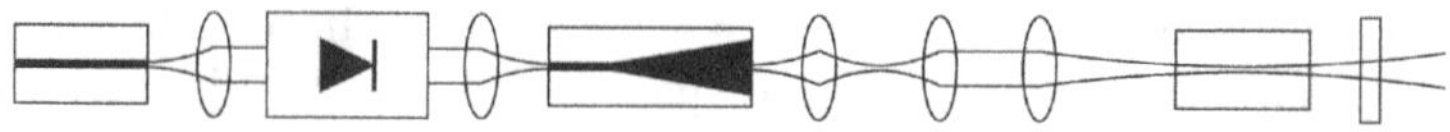

Figure 7.12 Sketch of the experimental setup used in Ref.[68]

Since then, both diode lasers and nonlinear materials have evolved tremendously. The output power of CW single-mode diode lasers has passed 1 W and with the introduction of tapered diode lasers, more than 10 W in a near diffraction limited beam is possible. Frequency doubling of single-mode diode lasers is rarely performed in a single pass through a bulk nonlinear crystal owing to the low conversion efficiency achievable. With single-mode diode lasers, SHG in waveguides are often employed to achieve tighter confinement and thus higher conversion efficiency. Alternatively, cavity enhanced SHG is used to increase the conversion efficiency. Here, we will mainly focus on SHG of tapered diode lasers as this has proven to be a promising route to high conversion efficiency and output power [63,64].

SHG of tapered diode lasers was first demonstrated in 1993 by Waarts *et al.* [65]. An output power of 3.7 mW was obtained at 491 nm with 1 W of input laser power. The conversion efficiency was limited mainly by the laser power and degradation of the beam quality at high power levels. The introduction of quasi-phase-matched nonlinear crystals provided access to significantly larger nonlinearities and the output power levels increased dramatically. External cavity tapered diode lasers and MOPA configurations have been introduced to increase the available narrow linewidth output power. Batchko *et al.* [66] used a MOPA configuration to generate 60 mW of 465 nm blue light in a PPLN crystal and at 404 nm, 24 mW was demonstrated by Chi *et al.* using a tapered diode laser operated in a tunable Littrow external cavity configuration [67]. Maiwald *et al.* [68] increased the available frequency-doubled output power dramatically to 600 mW at 488 nm by SHG of a MOPA setup emitting 4 W output power as sketched in Figure 7.12. A master oscillator power amplifier system consisting of a low power, narrow linewidth DFB laser coupled to a tapered amplifier for amplification while conserving the spatial and spectral properties. The output from the tapered amplifier was collimated and corrected for astigmatism using an aspheric lens and a cylindrical lens and focused into the 30 mm long PPMgOLN crystal for efficient frequency doubling. 15% power conversion and 4% wall-plug efficiency was achieved. High power DBR tapered diode lasers have further increased the available fundamental output power with narrow spectral linewidth and good beam quality suitable for SHG. Second harmonic output powers above 1.5 W in the 488 and 532 nm wavelength ranges have been demonstrated using these lasers [69,70] and compact modules using this technology has proven some of the advantages of diode laser technology [71]. Such laser sources have been introduced as pump sources for mode-locked Ti:sapphire lasers giving access to sub-20 fs pulses suitable for biophotonics applications like, for example, optical coherence tomography [72,73].

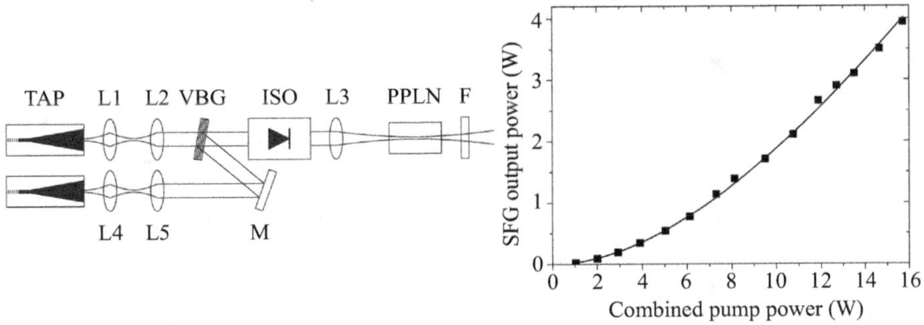

Figure 7.13 *(Left) Sketch of the experimental setup for spectral beam combining of DBR tapered diode lasers and subsequent SFG. (Right) Output power generated by SFG in the experiments [78]*

7.3.1.2 Sum frequency generation

Some regions in the visible spectral range are still not easily accessible by direct SHG of diode lasers. This is especially true for the yellow spectral range as no suitable high power diode lasers exist that can be frequency-doubled. SFG between two diode lasers or a diode laser and a solid-state laser may be used to efficiently generate light in these wavelength ranges. Single-pass SFG between two single-mode diode lasers has been used extensively to generate light for spectroscopy, see, for example, [74–76]. Higher power is possible using power amplifiers to boost the power of the interacting beams. The ultraviolet spectral region can also be reached by SHG of diode lasers. An example of this is generation of 222.5 nm light by frequency doubling of an external cavity blue diode laser [77].

A new approach was shown by Müller *et al.* [78] for SFG between two closely spaced wavelengths as shown in Figure 7.13. They used spectral beam combination in a volume Bragg grating to efficiently combine the two high power DBR tapered diode lasers. Subsequent single-pass SFG in a PPMgOLN crystal generated 3.9 W of near diffraction-limited green light. This method can easily be extended to other spectral regions as long as high power diode lasers are available.

7.3.1.3 Difference frequency generation

For many spectroscopic methods, tunable light in the mid-infrared spectral region is required. DFG between a fixed wavelength laser and a tunable laser is often employed to generate the desired wavelengths. Many examples exist using both gas lasers, solid-state lasers and diode lasers. Here we will give a few examples of laser systems using DFG between two diode lasers and refer the interested reader to the vast literature on the topic including [79,80] and references therein.

Sumpf *et al.* [81] mixed a single-mode diode laser at 1,290 nm and an external cavity tunable diode laser at 1,550 nm in an $AgGaSe_2$ crystal. With 8 mW and 6 mW from the two lasers, respectively, they reported generation of 10 nW radiation in the 7,138–7,373 nm wavelength range. This radiation was used for spectroscopic measurements on SO_2. The relatively low power is due to the low input power.

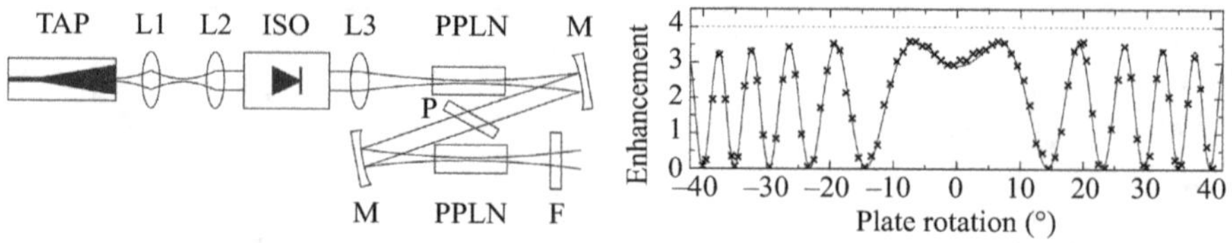

Figure 7.14 (Left) Sketch of the experimental setup for cascaded SHG of a DBR tapered diode laser used in [26]. (Right) Phase-matched cascade enhancement as a function of the phase plate rotation angle

Generation of widely tunable light using two tunable diode lasers was demonstrated by Sanders *et al.* [82] By mixing 510 mW from a 778 nm diode laser with 540 mW from a 972 nm diode laser, up to 7.1 μW was generated at 3,890 nm in a 7.8 mm long PPLN crystal. By simultaneous tuning of the two diode lasers, a tuning range of 3,600 nm to 4,300 nm was realized at constant crystal temperature in a single PPLN period.

7.3.1.4 Cascaded frequency conversion

In order to increase the conversion efficiency over single-pass configurations, it is possible to have multiple passes in the same nonlinear crystal or to have the beams pass several nonlinear crystals in series. A thorough investigation of multipass frequency doubling was performed in periodically poled KTP using both single-and multifrequency lasers [25]. They achieved a very high conversion efficiency of 27.3%/W by using 4 passes through a 3 cm long PPKTP crystal. A small wedge on the crystal cut angle was used to obtain phase compensation between the fundamental light and the second harmonic light.

Up to 56% conversion efficiency was achieved using three 30 mm long periodically poled MgO:$LiTaO_3$ crystals for SHG of a 1,064 nm fiber laser [83]. 5.6 W of green light was obtained from 10 W of input light and up to 13 W was reached at 25.1 W input power using a cascade of only two crystals. Here, the phase shift in air was used to obtain phase compensation. However, fiber lasers were used.

Tapered diode lasers have proven efficient in cascaded frequency conversion. Using the setup shown in Figure 7.14 Frequency doubling of a DBR tapered diode laser at 1,064 nm in a cascade of two nonlinear crystals resulted in 3.7 W of diffraction limited green light at 39% conversion efficiency [26]. In this paper, the influence of the phase offset between the fundamental and second harmonic beams were investigated and 361% conversion enhancement was demonstrated. The wavelength versatility of diode lasers has enabled cascaded SHG at 515 nm and 562 nm with high efficiency and excellent performance. 1.9 W has been demonstrated at 562 nm and 3.5 W was demonstrated at 515 nm [84,85].

Cascaded sum frequency generation has also been demonstrated as an efficient way of enhancing the conversion efficiency despite the need for phase compensation of three beams simultaneously. 5.5 W of diffraction limited light at 532 nm was demonstrated [27]. Here an experimental verification of the efficiency enhancement of the cascade principle in both SHG and SFG was presented.

7.3.1.5 Waveguide frequency conversion

In principle, frequency conversion in waveguides can be performed in the same manner as in bulk materials. However, waveguides are not used as frequently because of more difficult coupling into the waveguides and the needs of the application. Here, we will give a few examples of results obtained using nonlinear waveguides for SHG of diode lasers and DFG to generate mid-infrared wavelengths.

Frequency doubling of the light from diode lasers in nonlinear waveguides has been demonstrated using many different types of lasers and nonlinear materials. In recent years, most focus has been on the use of periodically poled waveguides as they exhibit very high conversion efficiencies. First-order periodically poled waveguides for the generation of visible light were first produced in 1992 [86] and used for the generation of 20.7 mW at 426 nm. Since then, diode lasers have become the lasers of choice for compact and efficient laser systems based on frequency doubling in nonlinear waveguides. Different types of diode lasers have been used including DFB lasers [87,88], DBR lasers [89], external cavity diode lasers [90–92], broad area lasers [93], master-oscillator power amplifier lasers [94] and DBR tapered lasers [95]. In single-mode waveguides, output power reaching 100–200 mW have been demonstrated in the blue-green spectral range and the nonlinear conversion efficiency can exceed $200\%/\mathrm{W}\cdot\mathrm{cm}^2$ [87,88,93]. In the yellow spectral range the power generated has been increased to 860 mW by use of a DBR tapered laser emitting at 1,177 nm [96]. A route to increased output power from optical waveguides is by using planar waveguides where the light is only confined in one dimension and allowed to diffract freely in the other dimension. By using either a MOPA laser system or a DBR tapered laser coupled to a PPMgOLN planar waveguide, up to 1.07 W of green light has been generated [94,95].

Nonlinear waveguides can also be used for efficient SFG between two incoming lasers. Laurell *et al.* [97] used a segmented KTP waveguide to generate up to 2.7 mW output power at 454 nm from fundamental input powers of 149 mW at 942 nm and 106 mW at 875 nm corresponding to a nonlinear conversion efficiency of 17%/W. By using different periods in the KTP and different interacting wavelengths from the tunable Ti:sapphire lasers, light in the 393–478 nm wavelength range was generated.

Waveguides have been used extensively to generate mid-infrared light through DFG. These light sources have an important application in spectroscopy of many different molecules having strong absorption lines in this spectral region. The easy tunability of diode lasers enables widely tunable mid-infrared light sources in almost any wavelength region. 34 μW of 3.5 μm radiation was generated by DFG of a DBR laser at 1,063 nm and a DFB laser at 1,525 nm [98]. This laser source was used for spectroscopy of methane and formaldehyde. In a similar approach for the detection of ethene, a DFG based light source emitting 200 μW around 3,240 nm was constructed by mixing a 1,064 nm DBR laser and a 1,583 nm DFB laser in a

PPLN waveguide [99]. O-H absorption measurements were demonstrated using a tunable DFG based light source emitting 450 μW around 2,600 nm where the two lasers were a fiber Bragg grating stabilized diode laser at 976 nm and an external cavity diode laser at 1,550 nm [100]. Tuning was demonstrated in the 2,600–2,950 nm range and the influence of water absorption was investigated.

Although these light sources have not been implemented in biophotonics applications, they show a clear potential within for instance breath analysis.

7.3.2 External cavity frequency conversion

Various examples of external cavities for nonlinear frequency conversion will be presented now. Solving equation (7.33) for χ yields $\chi = 2\frac{P_{SHG}}{\gamma} \Big/ \left(\delta P_{\text{in}} - \frac{P_{SHG}}{\gamma}\right)^2$, which we will calculate for each SHG cavity for comparison.

Described in [32] was a SHG two-mirror ring cavity with a 20 mm long PPKTP crystal converting up to 840 mW of 972 nm diode laser light to 700 mW of 486 nm light. This corresponded to an efficiency of 83%. The input coupler transmission of 9.5% and coupling efficiency $\delta = 96\%$ indicated a power enhancement factor of about 10 assuming perfect impedance matching. Because the crystal was periodically poled, the second harmonic was polarized horizontally and thus experienced only a 0.3% Fresnel reflection loss on the exit interface of the crystal. Since the second harmonic exited the cavity without passing through an output coupler, γ was determined purely by the Fresnel reflection loss, so $\gamma = 99.7\%$. From P_{SHG}, P_{in}, δ and γ, we calculate an impressive cavity performance figure of merit $\chi = 1.3 \cdot 10^2\ \text{W}^{-1}$. This value is high mostly because of the high single pass conversion coefficient, which was measured at $\eta_{\text{SHG}} = 1.06\%/\text{W}$. The ellipticity in the crystal focus caused by the Brewster's angle cut of the crystal was only slight, in this case just 44 μm horizontal waist by 51.2 μm vertical waist, which had a negligible impact on conversion efficiency. 5% of the generated second harmonic was estimated to be absorbed inside the crystal, which poses one limiting factor potentially making further power scaling in such a crystal difficult due to local thermal heating.

In reference [47], frequency quadrupling of a fiber laser was performed with two external resonant SHG cavities. Both were four-mirror bowtie cavities, but the first step employed an antireflection coated 18 mm long LBO crystal while the second contained a 10 mm long Brewster's angle cut BBO crystal. Consequently, the folding angle in the first cavity was kept as small as possible to minimize astigmatism while the folding angle in the second cavity was chosen to compensate for the crystal refraction astigmatism. The first step converted 1.8 W of 1,118 nm light to 950 mW of 559 nm light, which was then further converted to 275 mW of output at 280 nm. The respective efficiencies were 52.7% and 28.9%. From the respective input coupler reflectivity's of 97.5% and 98.4%, we can estimate power enhancement factors of about 36 and 56, respectively, assuming $\delta = 90\%$ and optimal impedance matching. For the first cavity, the SHG exit loss was determined by the output coupler transmission for SHG, $\gamma = 95\%$. For the second cavity, theoretical Fresnel reflection loss at the Brewster's angle cut crystal was about 27%

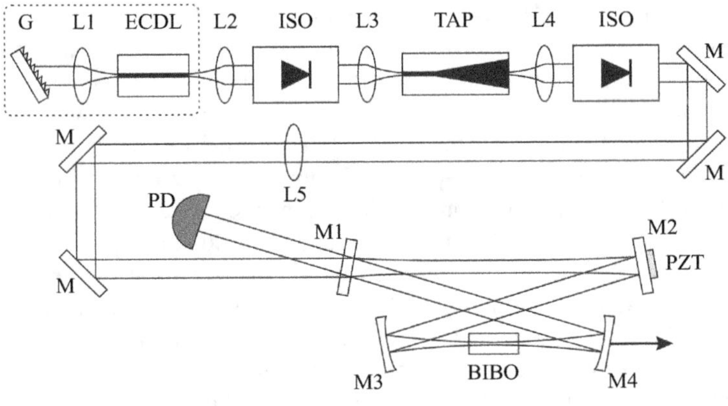

Figure 7.15 Experimental setup for SHG of a master oscillator power amplifier laser diode reported in [50]

and the output coupler transmission was 94%. For the second cavity we therefore estimate $\gamma = 69\%$. From this we calculate $\chi = 5.2\ \mathrm{W}^{-1}$ for the LBO cavity and $\chi = 3.8\ \mathrm{W}^{-1}$ for the BBO cavity. Keep in mind that χ does not include the SHG output coupling loss. Both cavities were locked using the Hänsch–Couillaud scheme. The reason that these χ's are lower than the previous example is primarily that the crystals were not periodically poled and thus utilized nonlinear coefficients lower than that of PPKTP by a factor of roughly 10 for LBO and 5 for BBO. Recall that the nonlinear coefficient is squared in the expression for η_{SHG}. Furthermore, the critical phasematching necessary for the BBO crystal resulted in a walk-off which further reduced the efficiency through reduction of the Boyd and Kleinman h parameter by more than an order of magnitude. This was compensated for by a higher power enhancement factor, which was made possible by the high transmissivity of LBO and BBO and good mirror coatings.

The last SHG example is of a cavity locked with the PDH scheme, described in reference [50] and shown schematically in Figure 7.15. They used a normal-cut critically phase matched BIBO crystal in a bow-tie cavity. An output of 400 mW of 390 nm light was achieved with 900 mW of 780 nm incident radiation from an amplified diode laser, corresponding to a maximum efficiency of 44%. They specified $\delta = 74\%$ and input coupler transmission 2.4%, so we can estimate an optimal power enhancement factor of roughly 31. The SHG exit loss was determined by the output coupler SHG transmission, which was not specified. If we assume $\gamma = 95\%$ we get the cavity performance figure of merit $\chi = 14\ \mathrm{W}^{-1}$. They estimated a single pass efficiency of roughly 0.02%/W, which is not unexpected considering that the nonlinear coefficient is roughly a factor 4 lower than PPKTP's and the Boyd and Kleinman h factor is, as in the previous example, up to an order of magnitude lower than in noncritically phase matched crystals such as periodically poled crystals. The walk-off angle in BIBO is, however, somewhat smaller than in BBO [101].

Now an SFG example: As detailed in reference [102], an external cavity was constructed for doubly-resonant SFG between radiation at 1,319 nm and radiation at 1,064 nm. They used a bow-tie cavity with normal-cut LBO and used the PDH scheme for locking the cavity to the 1,319 nm laser and the 1,064 nm laser to the cavity. Noncritical phase matching is achieved by heating the crystal to 40 °C. The generated sum frequency radiation is coupled out of the cavity with a so-called scraper mirror inserted just after the crystal, transmitting the circulating p-polarized infrared light and reflecting the s-polarized sum frequency light. While the article does not specify the exact maximum incident powers of each source on the cavity, the 1,319 nm source delivered up to 15.7 W and the 1,064 nm source up to 24.0 W, and the cavity's maximum output was 20 W of light at 589 nm. Their stated efficiency is 55%. Aside from a relatively harmless thermal shift of the resonance frequency of the cavity when the lasers were introduced, no limiting adverse effects were seen, suggesting the setup is suitable for further power scaling. It is, however, interesting to compare to the SHG cavity in the second example (reference [47]), in which they accomplished an efficiency of 52.7% at a tenth of the input power, despite SFG being theoretically a factor four more efficient than SHG. Intracavity losses due to the presence of the scraper mirror and the difficulty of achieving low loss coatings for the two significantly different circulating wavelengths may very well have posed the greatest limitation to the final efficiency of the SFG cavity.

Finally we take a look at a DFG example. In reference [49], a singly resonant DFG system is described in which up to 105 μW of light at 5.1 μm wavelength is generated from a 1.06 μm Nd:YAG laser with 5.7 W power and a quantum dot external cavity diode laser delivering 60 mW of light at 1.34 μm. The 1.06 μm light is used only in single pass while the light from the diode laser is resonantly enhanced in an external cavity with a specified power enhancement factor of up to 60. The crystal used was a 7.4 mm long MgO doped PPLN crystal with 6 channels of different poling periods from 21 to 25.4 μm. A strongly limiting factor on the efficiency is the significant absorption of the generated mid-infrared within the crystal, necessitating a shift of the foci towards the far side of the crystal to generate the mid infrared light as close as possible to the exit interface.

7.3.3 Intracavity frequency conversion

7.3.3.1 Second harmonic generation

Intracavity frequency doubling of diode lasers is intriguing because of the high gain and the large tunability available in diode lasers. The main limitation of this method is the high intracavity power that is required for efficient SHG, which will damage the diode laser material or facets. Some early demonstrations of the principles were performed using waveguides as the nonlinear material and grating feedback for wavelength selection [103,104]. An elegant demonstration of a wide tuning range using a fan-structured periodically poled $LiTaO_3$ (PPLT) crystal was performed by Zimmermann *et al.* [105] achieving 10 nm tuning in the blue spectral region. Single parameter tuning was performed by translation of a slit located in a Littmann external cavity and the fan angle of the PPLT crystal was chosen to fit this tuning. Kozlovsky *et al.* introduced an external enhancement resonator in a linear laser cavity to increase the circulating power in the nonlinear crystal and to

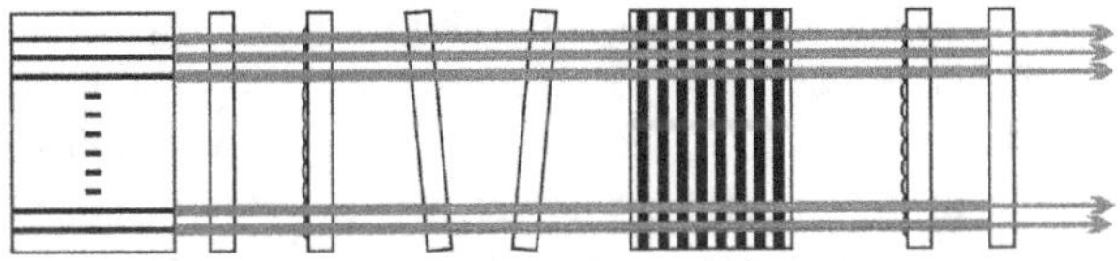

Figure 7.16 Sketch of the experimental setup for intracavity SHG of a 49-emitter diode laser bar [110]

ensure single-frequency operation [106]. The latter approach was pursued recently using tapered amplifiers as gain media in a unidirectional ring cavity and has the inherent advantage of achieving single-frequency operation with the possibility of tuning of the wavelength [107,108]. Furthermore, the wavelength of the laser automatically adapts to the resonance frequency of the external resonator and no electronic locking is necessary as opposed to external cavity frequency conversion. More than 500 mW of tunable visible light has been generated using single emitters.

Higher output power has been demonstrated using diode laser bars. With 49 single-mode emitters up to 1.3 W of second harmonic light has been extracted in the blue and green spectral ranges [109,110]. A PPMgOLN crystal contained in a compact external cavity performed the frequency doubling of the light from the 49 individual emitters. An intracavity filter ensured narrow line width operation required for efficient SHG. A principle sketch is shown in Figure 7.16.

7.3.3.2 SFG and DFG

The advantages of intracavity frequency conversion for SFG were first demonstrated by Risk *et al.* [111]. They used noncritical phase matching in KTP for SFG between a 1,064 nm Nd:YAG laser and the 809 nm diode laser pump light to generate about 1 mW of blue light at 460 nm. They also showed that the blue output power was stable demonstrating the absence of longitudinal mode coupling in the solid-state laser. In this particular example, the noncritical phase matching occurred exactly at the pump and lasing wavelengths of Nd:YAG. A larger flexibility in the laser wavelengths can be achieved by using a separate laser for single-pass SFG through the nonlinear crystal. This approach was demonstrated early to generate 20 mW of blue light from 100 mW of power in the single-pass diode laser [112]. A similar approach was pursued by several authors to generate visible light at many different wavelengths generating up to 300 mW of output power [113–115]. The use of tapered diode lasers as the single-pass laser has increased the available input power and consequently also the generated power.

SFG internal to a diode laser cavity was demonstrated recently [116]. Here, a tapered amplifier was operated in a unidirectional ring cavity containing an enhancement cavity with a nonlinear crystal and an additional tapered laser was single-passed through the nonlinear crystal. In this way more than 300 mW of tunable blue light at 460 nm was generated.

A method similar to the one used for singly resonant SFG presented above can also be used for DFG. This was pursued by Høgstedt *et al.* [117] using a tunable external cavity tapered diode laser to single-pass through a PPMgOLN crystal enclosed in a 1,064 nm Nd:YVO_4 laser. Up to 3 mW of single-frequency mid-infrared light tunable in the 2.9–3.4 μm range was generated.

A more exotic example is intracavity DFG inside the semiconductor material of the diode lasers. This was first demonstrated by Zvonkov *et al.* [118] with two diode lasers operating at 1,130 nm and 990 nm, respectively. Besides generation of 0.1 μW of radiation at 8,000 nm, they also showed SHG and SFG signals from this laser system. A similar approach utilizing quantum cascade laser has recently been employed to generate up to 120 μW THz radiation tunable in the 1.7–5.25 THz range [119].

7.4 Outlook

Diode lasers offer a vast selection of wavelengths and power levels in large parts of the optical spectrum. The development of diode lasers will continue to increase the available output power levels and also reach other spectral regions. For certain applications requiring high-quality light in specific spectral regions, however, nonlinear frequency conversion will continue to be the method of choice. With the increasing power levels from the diode lasers, also the frequency converted power levels will increase. This will enable frequency converted diode lasers to compete directly with solid-state lasers in many applications and due to the favorable cost structure of diode laser based systems, it is expected that these will replace solid-state lasers for many applications. The use of diode lasers will also enable the users to select the optimum wavelength for the application. As of today, the users are often limited to specific wavelengths given by the atomic transitions in solid-state laser materials like the 532 nm from frequency-doubled Nd:YAG lasers.

Laser systems will become significantly more compact with the use of micro-optics and single-and multi-pass frequency conversion techniques. In contrast to solid-state lasers, there are less strict requirements on the positioning of components for high finesse laser cavities. This also enables automated production, significantly lowering the cost of these laser systems. This may enable the laser systems to come out of the clinics and hospitals and be introduced as diagnostics tools for home use. Within ophthalmology, green and yellow lasers are used for photocoagulation [120]. Frequency-doubled diode lasers in this spectral region have reached power levels making them interesting for this application [26,69,121] and the expected lower cost of such lasers may enable use of such laser systems in the developing countries.

Frequency-converted diode lasers have a further advantage over solid-state lasers. The noise properties of solid-state lasers are governed by the long excited state lifetime giving relaxation oscillation noise typically located in the kHz frequency range. Many biophotonics imaging techniques have a detection bandwidth located exactly in the kHz frequency range and thus the relaxation oscillation noise induces a limitation on the achievable signal to noise ratio. Diode lasers on the other hand have a very short excited state lifetime and the relaxation oscillation noise is located in the GHz frequency range. The detection bandwidth of imaging systems will not detect these frequencies and a better signal to noise ratio can be achieved improving the image quality and thus the ability to diagnose diseases. The use of frequency converted diode lasers for pumping of ultrafast Ti:sapphire lasers expand the application range. Ti:sapphire lasers pumped by frequency-doubled tapered diode lasers have been demonstrated for optical coherence tomography with good performance [122] and a similar laser system has been demonstrated for multiphoton microscopy [123,124].

References

[1] Sutherland, R.L.: *'Handbook of Nonlinear Optics'* (New York, NY: Marcel Dekker, 2003, Second edition).

[2] Miller, R.C.: 'Optical second harmonic generation in piezoelectric crystals' *Appl. Phys. Lett.*, 1964, 5, (1), pp. 17–19.

[3] Byer, R.L.: 'Nonlinear optical phenomena and materials' *Annu. Rev. Mater. Sci.*, 1974, 4, (August), pp. 147–190.

[4] Franken, P.A., Hill, A.E., Peters, C.W., Weinreich, G.: 'Generation of optical harmonics' *Phys. Rev. Lett.*, 1961, 7, (4), pp. 118–120.

[5] Armstrong, J.A., Bloembergen, N., Ducuing, J., Pershan, P.S.: 'Interactions between light waves in a nonlinear dielectric' *Phys. Rev.*, 1962, 127, (6), pp. 1918–1939.

[6] Myers, L.E., Eckardt, R.C., Fejer, M.M., Byer, R.L., Bosenberg, W.R., Pierce, J.W.: 'Quasi-phase-matched optical parametric oscillators in bulk periodically poled $LiNbO_3$' *J. Opt. Soc. Am. B*, 1995, 12, (11), pp. 2102–2116.

[7] Eckardt, R.C., Fan, Y.X., Byer, R.L., Marquardt, C.L., Storm, M.E., Esterowitz, L.: 'Broadly tunable infrared parametric oscillator using $AgGaSe_2$' *Appl. Phys. Lett.*, 1986, 49, (11), pp. 608–610.

[8] Jundt, D.H.: 'Temperature-dependent Sellmeier equation for the index of refraction, n(e), in congruent lithium niobate' *Opt. Lett.*, 1997, 22, (20), pp. 1553–1555.

[9] Jackel, J.L., Rice, C.E., Veselka, J.J.: 'Proton exchange for high-index waveguides in $LiNbO_3$' *Appl. Phys. Lett.*, 1982, 41, (7), pp. 607–608.

[10] Ito, H., Takyu, C., Inaba, H.: 'Fabrication of periodic domain grating in $LiNbO_3$ by electron beam writing for application of nonlinear optical processes' *Electron. Lett.*, 1991, 27, (14), pp. 1221–1222.

[11] Fejer, M.M., Magel, G.A., Jundt, D.H., Byer, R.L.: 'Quasi-phase-matched second harmonic generation: tuning and tolerances' *IEEE J. Quantum Electron.*, 1992, 28, (11), pp. 2631–2654.

[12] Kim, Y.S., Smith, R.T.: 'Thermal expansion of lithium tantalate and lithium niobate single crystals' *J. Appl. Phys.*, 1969, 40, (11), pp. 4637–4641.

[13] Boyd, G.D., Kleinman, D.A.: 'Parametric interaction of focused Gaussian light beams' *J. Appl. Phys.*, 1968, 39, (8), pp. 3597–3639.

[14] Risk, W.P., Gosnell, T.R., Nurmikko, A.V.: *'Compact Blue-Green Lasers'* (Cambridge: Cambridge University Press, 2003).

[15] Borri, S., Cancio, P., De Natale, P., Giusfredi, G., Mazzotti, D., Tamassia, F.: 'Power-boosted difference-frequency source for high-resolution infrared spectroscopy' *Appl. Phys. B Lasers Opt.*, 2003, 76, (4), pp. 473–477.

[16] Guha, S., Falk, J.: 'The effects of focusing in the three-frequency parametric upconverter' *J. Appl. Phys.*, 1980, 51, (1), pp. 50–60.

[17] Tran-Ba-Chu, Broyer, M.: 'Intracavity CW difference frequency generation by mixing three photons and using Gaussian laser beams' *J. Phys.*, 1985, 46, pp. 523–533.

[18] Lin, J.T.: 'Non-linear crystals for tunable coherent sources' *Opt. Quantum Electron.*, 1990, 22, pp. 283–313.

[19] Agrawal, G.P.: 'Second-harmonic generation with arbitrary pump-beam profiles' *Phys. Rev. A*, 1981, 23, (4), pp. 1863–1868.

[20] Cai, Y., Peschel, U.: 'Second-harmonic generation by an astigmatic partially coherent beam' *Opt. Express*, 2007, 15, (23), pp. 15480–15492.

[21] Uebernickel, M., Güther, R., Blume, G., Fiebig, C., Paschke, K., Erbert, G.: 'Study of the properties of the SHG with diode lasers' *Appl. Phys. B*, 2010, 99, (3), pp. 457–464.

[22] Freegarde, T., Coutts, J., Walz, J., Leibfried, D., Hänsch, T.W.: 'General analysis of type I second-harmonic generation with elliptical Gaussian beams' *J. Opt. Soc. Am. B*, 1997, 14, (8), pp. 2010–2016.

[23] Regener, R., Sohler, W.: 'Efficient second-harmonic generation in Ti: $LiNbO_3$ channel waveguide resonators' *J. Opt. Soc. Am. B*, 1988, 5, (2), pp. 267–277.

[24] Fluck, D., Günter, P.: 'Efficient second-harmonic generation by lens waveguiding in $KNbO_3$ crystals' *Opt. Commun.*, 1998, 147, (4–6), pp. 305–308.

[25] Spiekermann, S., Laurell, F., Pasiskevicius, V., Karlsson, H., Freitag, I.: 'Optimizing non-resonant frequency conversion in periodically poled media' *Appl. Phys. B*, 2004, 79, (2), pp. 211–219.

[26] Hansen, A.K., Tawfieq, M., Jensen, O.B., *et al.*: 'Concept for power scaling second harmonic generation using a cascade of nonlinear crystals' *Opt. Express*, 2015, 23, (12), pp. 15921–15934.

[27] Hansen, A.K., Andersen, P.E., Jensen, O.B., Sumpf, B., Erbert, G., Petersen, P.M.: 'Highly efficient single-pass sum frequency generation by cascaded nonlinear crystals' *Opt. Lett.*, 2015, 40, (23), pp. 5526–5529.

[28] Kozlovsky, W.J., Nabors, C.D., Byer, R.L.: 'Efficient second harmonic generation of a diode-laser-pumped CW Nd:YAG laser using monolithic MgO:$LiNbO_3$ external resonant cavities' *IEEE J. Quantum Electron.*, 1988, 24, (6), pp. 913–919.

[29] Klappauf, B.G., Bidel, Y., Wilkowski, D., Chanelière, T., Kaiser, R.: 'Detailed study of an efficient blue laser source by second-harmonic

generation in a semimonolithic cavity for the cooling of strontium atoms' *Appl. Opt.*, 2004, 43, (12), pp. 2510–2527.

[30] Ast, S., Nia, R.M., Schönbeck, A., *et al.*: 'High-efficiency frequency doubling of continuous-wave laser light' *Opt. Lett.*, 2011, 36, (17), pp. 3467–3469.

[31] Feng, J., Li, Y., Tian, X., Liu, J., Zhang, K.: 'Noise suppression, linewidth narrowing of a master oscillator power amplifier at 1.56 μm and the second harmonic generation output at 780 nm' *Opt. Express*, 2008, 16, (16), pp. 11871–11877.

[32] Danekar, K.: '83% net efficiency in generating 700 mW of CW blue by IR diode laser doubling with PPKTP', in 'CLEO 2010 Conference on Lasers and Electro-Optics, 2010' (2010), p. Paper JTuD69.

[33] Lundeman, J.H., Jensen, O.B., Andersen, P.E., *et al.*: 'High power 404 nm source based on second harmonic generation in PPKTP of a tapered external feedback diode laser' *Opt. Express*, 2008, 16, (4), pp. 2486–2493.

[34] Bhawalkar, J.D., Mao, Y., Po, H., *et al.*: 'High-power 390-nm laser source based on efficient frequency doubling of a tapered diode laser in an external resonant cavity' *Opt. Lett.*, 1999, 24, (12), pp. 823–825.

[35] Scholz, M., Opalevs, D., Leisching, P., *et al.*: '1.3-mW tunable and narrow-band continuous-wave light source at 191 nm' *Opt. Express*, 2012, 20, (17), pp. 18659–18664.

[36] Wilson, A.C., Ospelkaus, C., VanDevender, A.P., *et al.*: 'A 750-mW, continuous-wave, solid-state laser source at 313 nm for cooling and manipulating trapped $^{9}Be^{+}$ ions' *Appl. Phys. B*, 2011, 105, (4), pp. 741–748.

[37] Ruseva, V., Hald, J.: 'Generation of UV light by frequency doubling in BIBO' *Opt. Commun.*, 2004, 236, (1–3), pp. 219–223.

[38] Jurdik, E., Hohlfeld, J., van Etteger, A.F., *et al.*: 'Performance optimization of an external enhancement resonator for optical second-harmonic generation' *J. Opt. Soc. Am. B*, 2002, 19, (7), pp. 1660–1667.

[39] Kim, J.-I., Meschede, D.: 'Continuous-wave coherent ultraviolet source at 326 nm based on frequency trippling of fiber amplifiers' *Opt. Express*, 2008, 16, (14), pp. 10803–10808.

[40] Ruseva, V., Hald, J.: 'High-power 457-nm light source by frequency doubling of an amplified diode laser' *Appl. Opt.*, 2003, 42, (27), pp. 5500–5507.

[41] Ricciardi, I., Rosa, M. De, Rocco A., Ferraro, P., Natale, P. De: 'Cavity-enhanced generation of 6 W cw second-harmonic power at 532 nm in periodically-poled MgO: $LiTaO_3$' *Opt. Express*, 2010, 18, (11), pp. 10985–10994.

[42] Raab, C., Bolle, J., Oberst, H., Eschner, J., Schmidt-Kaler, F., Blatt, R.: 'Diode laser spectrometer at 493 nm for single trapped Ba^{+} ions' *Appl. Phys. B Lasers Opt.*, 1998, 67, (6), pp. 683–688.

[43] Torabi-Goudarzi, F., Riis, E.: 'Efficient CW high-power frequency doubling in periodically poled KTP' *Opt. Commun.*, 2003, 227, (4–6), pp. 389–403.

[44] Song, X., Li, Z., Zhang, P., *et al.*: 'Frequency doubling with periodically poled $KTiOPO_4$ at the fundamental wave of cesium D transition' *Chinese Opt. Lett.*, 2007, 5, (10), pp. 596–598.

[45] Goyal, A.K., Bhawalkar, J.D., Conturie, Y., *et al.*: 'High beam quality of ultraviolet radiation generated through resonant enhanced frequency doubling of a diode laser' *J. Opt. Soc. Am. B*, 1999, 16, (12), pp. 2207–2216.

[46] Lodahl, P., Sørensen, J.L., Polzik, E.S.: 'High efficiency second harmonic generation with a low power diode laser' *Appl. Phys. B*, 1997, 64, pp. 383–386.

[47] Friedenauer, A., Markert, F., Schmitz, H., *et al.*: 'High power all solid state laser system near 280 nm' *Appl. Phys. B*, 2006, 84, (3), pp. 371–373.

[48] Kumagai, H.: 'Fine frequency tuning in sum-frequency generation of continuous-wave single-frequency coherent light at 252 nm with dual-wavelength enhancement' *Opt. Lett.*, 2007, 32, (1), pp. 62–64.

[49] Bressel, U., Ernsting, I., Schiller, S.: '5 μm laser source for frequency metrology based on difference frequency generation' *Opt. Lett.*, 2012, 37, (5), pp. 918–920.

[50] Schmitt, T., Deninger, A., Lison, F., Kaenders, W.: 'Recent advances in nonlinear frequency conversion of high-power, single-mode diode lasers' *Proc. SPIE*, 2005, 5707, pp. 16–22.

[51] Manoel, D. de A., Figueira, D.S.L., Pereira, D., Cruz, F.C.: 'Single-frequency blue light source based on optically injected diode lasers' *Opt. Eng.*, 2004, 43, (6), pp. 1311–1313.

[52] Freegarde, T., Zimmermann, C.: 'On the design of enhancement cavities for second harmonic generation' *Opt. Commun.*, 2001, 199, pp. 435–446.

[53] Hald, J.: 'Second harmonic generation in an external ring cavity with a Brewster-cut nonlinear crystal: theoretical considerations' *Opt. Commun.*, 2001, 197, pp. 169–173.

[54] Moore, G.T.: 'Resonant sum-frequency generation' *IEEE J. Quantum Electron.*, 2002, 38, (1), pp. 12–18.

[55] Kaneda, Y., Kubota, S.: 'Theoretical treatment, simulation, and experiments of doubly resonant sum-frequency mixing in an external resonator' *Appl. Opt.*, 1997, 36, (30), pp. 7766–7775.

[56] Kogelnik, H., Li, T.: 'Laser beams and resonators' *Appl. Opt.*, 1966, 5, (10), pp. 1550–1567.

[57] Hansch, T.W., Couillaud, B.: 'Laser frequency stabilization by polarization spectroscopy of a reflecting reference cavity' *Opt. Commun.*, 1980, 35, (3), pp. 441–444.

[58] Drever, R.W.P., Hall, J.L., Kowalski, F. V., *et al.*: 'Laser phase and frequency stabilization using an optical resonator' *Appl. Phys. B Photophysics Laser Chem.*, 1983, 31, (2), pp. 97–105.

[59] Baer, T.: 'Large-amplitude fluctuations due to longitudinal mode coupling in diode-pumped intracavity-doubled Nd:YAG lasers' *J. Opt. Soc. Am. B*, 1986, 3, (9), pp. 1175–1180.

[60] Karamehmedović, E., Pedersen, C., Jensen, O.B., Tidemand-Lichtenberg, P.: 'Nonlinear beam clean-up using resonantly enhanced sum-frequency mixing' *Appl. Phys. B*, 2009, 96, (2–3), pp. 409–413.

[61] Chi, C.: 'Second-harmonic generation of the GaAs laser' *J. Appl. Phys.*, 1970, 41, (7), p. 3184.

[62] Edmonds, H.D., Smith, A.W.: 'Second-harmonic generation with the GaAs laser' *IEEE J. Quantum Electron.*, 1970, 6, (6), pp. 356–360.

[63] Jechow, A., Menzel, R., Paschke, K., Erbert, G.: 'Blue-green light generation using high brilliance edge emitting diode lasers' *Laser Photon. Rev.*, 2010, 4, (5), pp. 633–655.

[64] Jensen, O.B., Hansen, A.K., Müller, A., *et al.*: 'Power scaling of nonlinear frequency converted tapered diode lasers for biophotonics' *IEEE J. Sel. Top. Quantum Electron.*, 2014, 20, (2), p. 7100515.

[65] Waarts, R., Sanders, S., Parke, R., *et al.*: 'Frequency-doubled monolithic master oscillator power amplifier laser diode' *IEEE Photonics Technol. Lett.*, 1993, 5, (10), pp. 1122–1125.

[66] Batchko, R.G., Fejer, M.M., Byer, R.L., *et al.*: 'Continuous-wave quasi-phase-matched generation of 60 mW at 465 nm by single-pass frequency doubling of a laser diode in backswitch-poled lithium niobate' *Opt. Lett.*, 1999, 24, (18), pp. 1293–1295.

[67] Chi, M., Jensen, O.B., Holm, J., *et al.*: 'Tunable high-power narrow-linewidth semiconductor laser based on an external-cavity tapered amplifier' *Opt. Express*, 2005, 13, (26), pp. 10589–10596.

[68] Maiwald, M., Schwertfeger, S., Güther, R., *et al.*: '600 mW optical output power at 488 nm by use of a high-power hybrid laser diode system and a periodically poled MgO: LiNbO3 bulk crystal' *Opt. Lett.*, 2006, 31, (6), pp. 802–804.

[69] Jensen, O.B., Andersen, P.E., Sumpf, B., Hasler, K.-H., Erbert, G., Petersen, P.M.: '1.5 W green light generation by single-pass second harmonic generation of a single-frequency tapered diode laser' *Opt. Express*, 2009, 17, (8), pp. 6532–6539.

[70] Sahm, A., Uebernickel, M., Paschke, K., Erbert, G., Tränkle, G.: 'Thermal optimization of second harmonic generation at high pump powers' *Opt. Express*, 2011, 19, (23), pp. 23029–23035.

[71] Fiebig, C., Fricke, J., Uebernickel, M., Jedrzejczyk, D., Sahm, A., Paschke, K.: 'Watt-class green-emitting laser modules using direct second harmonic generation of diode laser radiation' *Opt. Rev.*, 2012, 19, (6), pp. 405–408.

[72] Müller, A., Jensen, O.B., Unterhuber, A., *et al.*: 'Frequency-doubled DBR-tapered diode laser for direct pumping of Ti: sapphire lasers generating sub-20 fs pulses' *Opt. Express*, 2011, 19, (13), pp. 12156–12163.

[73] Unterhuber, A., Považay, B., Müller, A., *et al.*: 'Simultaneous dual wavelength eye-tracked ultrahigh resolution retinal and choroidal optical coherence tomography' *Opt. Lett.*, 2013, 38, (21), pp. 4312–4315.

[74] Sugiyama, K., Yoda, J., Sakurai, T.: 'Generation of continuous-wave ultraviolet light by sum-frequency mixing of diode-laser and argon-ion-laser radiation in beta-BaB(2)O(4)' *Opt. Lett.*, 1991, 16, (7), pp. 449–451.

[75] Alnis, J., Gustafsson, U., Somesfalean, G., Svanberg, S.: 'Sum-frequency generation with a blue diode laser for mercury spectroscopy at 254 nm' *Appl. Phys. Lett.*, 2000, 76, (10), p. 1234.

[76] Sakata, H., Furukawa, Y., Yagi, T.: 'Type II quasi-phase-matched sum-frequency mixing of diode lasers in $KTiOPO_4$ with broad spectral and temperature acceptance bandwidths' *Appl. Opt.*, 2004, 43, (25), pp. 4922–4928.

[77] Ruhnke, N., Müller, A., Eppich, B., *et al.*: 'Single-pass UV generation at 222.5 nm based on high-power GaN external cavity diode laser' *Opt. Lett.*, 2015, 40, (9), pp. 2127–2129.

[78] Müller, A., Jensen, O.B., Hasler, K.-H., *et al.*: 'Efficient concept for generation of diffraction-limited green light by sum-frequency generation of spectrally combined tapered diode lasers' *Opt. Lett.*, 2012, 37, (18), pp. 3753–3755.

[79] Curl, R.F., Tittel, F.K.: 'Tunable infrared laser spectroscopy' *Annu. Reports Sect. 'C' (Physical Chem.)*, 2002, 98, pp. 219–272.

[80] Richter, D., Fried, A., Weibring, P.: 'Difference frequency generation laser based spectrometers' *Laser Photonics Rev.*, 2009, 3, (4), pp. 343–354.

[81] Sumpf, B., Rehle, D., Kelz, T., Kronfeldt, H.-D.: 'A tunable diode-laser spectrometer for the MIR region near 7.2 μm applying difference-frequency generation in $AgGaSe_2$' *Appl. Phys. B Lasers Opt.*, 1998, 67, (3), pp. 369–373.

[82] Sanders, S., Lang, R.J., Myers, L.E., Fejer, M.M., Byer, R.L.: 'Broadly tunable mid-IR radiation source based on difference frequency mixing of high power wavelength tunable laser diodes in bulk periodically poled $LiNbO_3$' *Electron. Lett.*, 1996, 32, (3), pp. 218–219.

[83] Samanta, G.K., Kumar, S.C., Devi, K., Ebrahim-Zadeh, M.: 'Multicrystal, continuous-wave, single-pass second-harmonic generation with 56% efficiency' *Opt. Lett.*, 2010, 35, (20), pp. 3513–3515.

[84] Hansen, A.K., Christensen, M., Noordegraaf, D., *et al.*: 'Efficient generation of 1.9 W yellow light by cascaded frequency doubling of a DBR tapered diode laser' *Appl. Opt.*, 2016, 55, (32), pp. 9270–9274.

[85] Jensen, O.B., Hansen, A.K., Müller, A., Sumpf, B., Petersen, P.M., Andersen, P.E.: 'Efficient generation of 3.5 W laser light at 515 nm by frequency doubling a single-frequency high power DBR tapered diode laser' *Opt. Commun.*, 2017, 392, pp. 167–170.

[86] Yamada, M., Nada, N., Saitoh, M., Watanabe, K.: 'First-order quasi-phase matched $LiNbO_3$ waveguide periodically poled by applying an external field for efficient blue second-harmonic generation' *Appl. Phys. Lett.*, 1993, 62, (5), pp. 435–436.

[87] Nguyen, H.K., Hu, M.H., Nishiyama, N., *et al.*: '107-mW low-noise green-light emission by frequency doubling of a reliable 1060-nm DFB semiconductor laser diode' *IEEE Photonics Technol. Lett.*, 2006, 18, (5), pp. 682–684.

[88] Jechow, A., Schedel, M., Stry, S., Sacher, J., Menzel, R.: 'Highly efficient single-pass frequency doubling of a continuous-wave distributed feedback laser diode using a PPLN waveguide crystal at 488 nm' *Opt. Lett.*, 2007, 32, (20), pp. 3035–3037.

[89] Bhatia, V., Bauco, A.S., Oubei, H.M., Loeber, D.A.S.: 'Efficient green lasers for high-resolution scanning micro-projector displays', in Powers, P.E. (Ed.): *'Proc. SPIE'* (2010), pp. 758205–758205–12.

[90] Fedorova, K.A., Cataluna, M.A., Battle, P.R., *et al.*: 'Orange light generation from a PPKTP waveguide end pumped by a CW quantum-dot tunable laser diode' *Appl. Phys. B*, 2010, 103, (1), pp. 41–43.

[91] Kim, E.B., Lee, W.-K., Park, C.Y., Yu, D.-H., Park, S.E.: 'Narrow linewidth 578 nm light generation using frequency-doubling with a waveguide PPLN pumped by an optical injection-locked diode laser' *Opt. Express*, 2010, 18, (10), pp. 10308–10314.

[92] Fedorova, K.A., Sokolovskii, G.S., Khomylev, M., Livshits, D.A., Rafailov, E.U.: 'Efficient yellow-green light generation at 561 nm by frequency-doubling of a QD-FBG laser diode in a PPLN waveguide' *Opt. Lett.*, 2014, 39, (23), pp. 6672–6674.

[93] Jechow, A., Skoczowsky, D., Menzel, R.: '100 mW high efficient single pass SHG at 488 nm of a single broad area laser diode with external cavity using a PPLN waveguide crystal' *Opt. Express*, 2007, 15, (11), pp. 6976–6981.

[94] Sakai, K., Koyata, Y., Shimada, N., *et al.*: 'Master-oscillator power-amplifier scheme for efficient green-light generation in a planar MgO: PPLN waveguide' *Opt. Lett.*, 2008, 33, (5), pp. 431–433.

[95] Jedrzejczyk, D., Güther, R., Paschke, K., Jeong, W.-J., Lee, H.-Y., Erbert, G.: 'Efficient high-power frequency doubling of distributed Bragg reflector tapered laser radiation in a periodically poled MgO-doped lithium niobate planar waveguide' *Opt. Lett.*, 2011, 36, (3), pp. 367–369.

[96] Bege, R., Jedrzejczyk, D., Blume, G., *et al.*: 'Watt-level second-harmonic generation at 589 nm with a PPMgO:LN ridge waveguide crystal pumped by a DBR tapered diode laser' *Opt. Lett.*, 2016, 41, (7), pp. 1530–1533.

[97] Laurell, F., Brown, J.B., Bierlein, J.D.: 'Sum-frequency generation in segmented KTP waveguides' *Appl. Phys. Lett.*, 1992, 60, (9), pp. 1064–1066.

[98] Ciaffoni, L., Grilli, R., Hancock, G., Orr-Ewing, A.J., Peverall, R., Ritchie, G.A.D.: '3.5-μm high-resolution gas sensing employing a $LiNbO_3$ QPM-DFG waveguide module' *Appl. Phys. B*, 2008, 94, (3), pp. 517–525.

[99] Grilli, R., Ciaffoni, L., Hancock, G., Peverall, R., Ritchie, G.A.D., Orr-Ewing, A.J.: 'Mid-infrared ethene detection using difference frequency generation in a quasi-phase-matched $LiNbO_3$ waveguide' *Appl. Opt.*, 2009, 48, (30), pp. 5696–5703.

[100] Tadanaga, O., Nishida, Y., Yanagawa, T., *et al.*: 'Efficient 2.7-μm difference frequency generation using direct-bonded quasi-phase-matched LiNbO3 ridge waveguide and investigation of O–H absorption influence' *Jpn. J. Appl. Phys.*, 2007, 46, (10A), pp. 6643–6646.

[101] Akbari, R., Major, A.: 'Optical, spectral and phase-matching properties of BIBO, BBO and LBO crystals for optical parametric oscillation in the visible and near-infrared wavelength ranges' *Laser Phys.*, 2013, 23, (3), p. 035401.

[102] Bienfang, J.C., Denman, C.A., Grime, B.W., Hillman, P.D., Moore, G.T., Telle, J.M.: '20 W of continuous-wave sodium D2 resonance radiation from sum-frequency generation with injection-locked lasers' *Opt. Lett.*, 2003, 28, (22), pp. 2219–2221.

[103] Laurell, F.: 'Stable blue second harmonic generation in a KTP waveguide with a diode laser in an external cavity' *Electron. Lett.*, 1993, 29, (18), pp. 1629–1630.

[104] Risk, W.P., Kozlovsky, W.J., Lau, S.D., Bona, G.L., Jaeckel, H., Webb, D.J.: 'Generation of 425-nm light by waveguide frequency doubling of a GaAlAs laser diode in an extended-cavity configuration' *Appl. Phys. Lett.*, 1993, 63, (23), pp. 3134–3136.

[105] Zimmermann, J., Struckmeier, J., Hofmann, M.R., Meyn, J.-P.: 'Tunable blue laser based on intracavity frequency doubling with a fan-structured periodically poled $LiTaO_3$ crystal' *Opt. Lett.*, 2002, 27, (8), pp. 604–606.

[106] Kozlovsky, W.J., Risk, W.P., Lenth, W., *et al.*: 'Blue light generation by resonator-enhanced frequency doubling of an extended-cavity diode laser' *Appl. Phys. Lett.*, 1994, 65, (5), pp. 525–527.

[107] Skoczowsky, D., Jechow, A., Menzel, R., Paschke, K., Erbert, G.: 'Efficient second-harmonic generation using a semiconductor tapered amplifier in a coupled ring-resonator geometry' *Opt. Lett.*, 2010, 35, (2), pp. 232–234.

[108] Jensen, O.B., Petersen, P.M.: 'Generation of single-frequency tunable green light in a coupled ring tapered diode laser cavity' *Opt. Express*, 2013, 21, (5), pp. 6076–6081.

[109] Li, K., Wang, H., Copner, N.J., *et al.*: '465 nm Laser Sources By Intra-cavity Frequency Doubling Using a 49-Edge-Emitters Laser Bar' *Opt. Lett.*, 2011, 36, (3), pp. 361–363.

[110] Li, K., Yao, A., Copner, N.J., *et al.*: 'Compact 1.3 W green laser by intra-cavity frequency doubling of a multi-edge-emitter laser bar using a MgO: PPLN crystal' *Opt. Lett.*, 2009, 34, (22), pp. 3472–3474.

[111] Risk, W.P., Baumert, J.-C., Bjorklund, G.C., Schellenberg, F.M., Lenth, W.: 'Generation of blue light by intracavity frequency mixing of the laser and pump radiation of a miniature neodymium: yttrium aluminum garnet laser' *Appl. Phys. Lett.*, 1988, 52, (2), pp. 85–87.

[112] Kean, P.N., Standley, R.W., Dixon, G.J.: 'Generation of 20 mW of blue laser radiation from a diode-pumped sum-frequency laser' *Appl. Phys. Lett.*, 1993, 63, (3), pp. 302–304.

[113] Karamehmedovic, E., Pedersen, C., Andersen, M.T., Tidemand-Lichtenberg, P.: 'Efficient visible light generation by mixing of a solid-state laser and a tapered diode laser' *Opt. Express*, 2007, 15, (19), pp. 12240–12245.

[114] Johansson, S., Wang, S., Pasiskevicius, V., Laurell, F.: 'Compact 492-nm light source based on sum-frequency mixing' *Opt. Express*, 2005, 13, (7), pp. 2590–2595.

[115] Sørensen, K.P., Tidemand-Lichtenberg, P., Pedersen, C.: 'Efficient near diffraction limited blue light source by sum-frequency mixing of a BAL and a solid-state laser' *Laser Phys. Lett.*, 2011, 8, (3), pp. 209–213.

[116] Jensen, O.B., Petersen, P.M.: 'Single-frequency blue light generation by single-pass sum-frequency generation in a coupled ring cavity tapered laser' *Appl. Phys. Lett.*, 2013, 103, p. 141107.

[117] Høgstedt, L., Jensen, O.B., Dam, J.S., Pedersen, C., Tidemand-Lichtenberg, P.: '500 nm continuous wave tunable single-frequency mid-IR light source for C-H spectroscopy' *Laser Phys.*, 2012, 22, (11), pp. 1676–1681.

[118] Zvonkov, B.N., Biryukov, A. A., Ershov, A. V., *et al.*: 'Room-temperature intracavity difference-frequency generation in butt-joint diode lasers' *Appl. Phys. Lett.*, 2008, 92, (2), p. 21122.

[119] Vijayraghavan, K., Jiang, Y., Jang, M., *et al.*: 'Broadly tunable terahertz generation in mid-infrared quantum cascade lasers' *Nat. Commun.*, 2013, 4, (May), p. 2021.

[120] Inagaki, K., Ohkoshi, K., Ohde, S., Deshpande, G.A., Ebihara, N., Murakami, A.: 'Comparative efficacy of pure yellow (577-nm) and 810-nm subthreshold micropulse laser photocoagulation combined with yellow (561–577 nm) direct photocoagulation for diabetic macular edema' *Jpn. J. Ophthalmol.*, 2015, 59, (1), pp. 21–28.

[121] Müller, A., Jensen, O.B., Hasler, K.-H., *et al.*: 'Efficient concept generating 3.9 W of diffraction-limited green light with spectrally combined tapered diode lasers' 2013, 8604, pp. 860404–860404–9.

[122] Unterhuber, A., Považay, B., Müller, A., *et al.*: 'Simultaneous dual wavelengths eye tracked ultrahigh resolution retinal and choroidal OCT' *Opt. Lett.*, 2013, 38, (21), pp. 4312–4315.

[123] König, K., Andersen, P., Le, T., Breunig, H.G.: 'Multiphoton imaging with a novel compact diode-pumped Ti:Sapphire oscillator' *Microsc. Res. Tech.*, 2015, 78, (12), pp. 1154–1158.

[124] Marti, D., Djurhuus, M., Jensen, O.B., Andersen, P.E.: 'Multi-photon microscope driven by novel green laser pump', in *'Proc. SPIE'* (2016), p. 97121M.

Chapter 8
Semiconductor disk lasers

Mircea Guina[1], Emmi Kantola[1], Tomi Leinonen[1], and Antti Rantamäki[1]

Semiconductor disk lasers (SDLs), also known as optically pumped vertical external-cavity surface-emitting lasers (VECSELs), offer unparalleled features combining high beam quality, large wavelength coverage, multi-watt output power, narrow linewidth, and broad tuning range. Moreover, relative to the features they provide, SDLs are compact and cost-effective systems. Their unique properties have already been recognized in biophotonics, for example, in photocoagulation treatment of macular degeneration using yellow radiation and multiphoton imaging, yet their full spread to applications is still to come. This chapter provides an overview of key technology concepts related to SDLs and a summary of the most important developments in terms of wavelength coverage and output powers. Specific developments are presented for SDLs emitting continuous wave (CW) radiation at the yellow–orange–red optical spectrum and for SDLs emitting picosecond pulses at red wavelengths. The chapter is concluded with a review of several biophotonics applications and their needs for laser sources, which are then put in a perspective of current SDL achievements.

8.1 Introduction

Continuous progress in developing less expensive and more versatile lasers has fostered important advances in life science. On the other hand, rapidly growing biophotonics applications, such as flow cytometry and advanced imaging techniques, have generated demand for high beam quality lasers emitting at new wavelengths that are not readily available. Generally speaking, the processes that laser light induces in tissues are wavelength sensitive [1] and depend largely on the intensity of the beam and the mode of operation, i.e., continuous mode or pulsed mode. Therefore, different biophotonics applications have largely different requirements for the laser source, depending on the targeted tissue components and

[1]Optoelectronics Research Centre, Tampere University of Technology, Finland

the desired effect. For example, laser treatment of cutaneous vascular disorders, such as port wine stains (PWS), has triggered the development of novel laser sources emitting in the yellow–orange band of the optical spectrum. In PWS treatment, the absorbed light energy heats the target tissue coagulating the blood inside the micro-vessels; the most efficient way to coagulate blood is to target deoxygenated and oxygenated hemoglobin that have high absorption peaks in the yellow spectral range, around 577 nm [2]. For the same reason, yellow lasers would be ideal for imaging blood flow [3] and lasers at slightly longer wavelengths (585–595 nm) can be used to penetrate deeper into skin [4]. However, the yellow–orange spectral range cannot be reached via direct emission from semiconductor lasers due to material restrictions. Alternative yellow laser solutions make use of amplification and frequency conversion in solid-state systems [5,6] but their complexity, price, and often their limitations in power and wavelength coverage, render them unsuitable for a wide exploitation.

Optically pumped semiconductor disk lasers (OP-SDLs), also known as OP-VECSELs (optically pumped vertical external-cavity surface-emitting lasers) and OPSLs (optically pumped semiconductor lasers), have been recently developed as a compromise between the simple directly emitting semiconductor lasers and the more complicated solid-state lasers [7,8]. The OP-SDLs draw important advantages from both semiconductor and solid-state lasers: the simplicity to engineer the emission properties of semiconductors is combined with the functionality of solid-state lasers which is enabled by the use of external cavity elements. Ultimately, this concept renders possible emission in a broad wavelength range with a high output power and a good beam quality, narrow linewidth operation, and efficient intra-cavity frequency conversion. Concerning intracavity frequency conversion, we should note that semiconductor gain media exhibit a short lifetime of the excited state, which leads to superior noise properties compared to intracavity frequency doubled solid-state lasers. Another benefit of the external cavity is that it enables the incorporation of nonlinear components to generate ultrashort pulses [9]. Moreover, the advances in the availability, cost, and performance of near infrared laser diodes used to pump the SDLs have made the optical pumping concept practical and cost effective.

In this chapter, we first focus on introducing the main SDL concepts, including power scaling, thermal management, common cavity architectures, and wavelength coverage for different material systems. These aspects are discussed in Section 8.2. In Section 8.3, we present more detailed examples concerning the development of SDLs emitting yellow–orange–red radiation: a spectral range that seems to be of particular importance for biophotonics. The technology overview covers mainly CW operation and optical pumping. A more detailed description of underlying technologies, for example, concerning wafer bonding technique for wavelength extension and characterization methods used for SDL gain chip developments, can be found in a recent review article [10]. Finally, in Section 8.4 we provide an overview of several biophotonics applications and indicate the main benefits the SDL technology could bring.

8.2 Technology overview

8.2.1 SDL operation and thermal management

The laser resonator of an SDL is formed between a gain mirror and one or more external-cavity mirrors (see Figure 8.1, additional details on cavity architectures are provided in Section 8.2.2). The concept of an OP-SDLs dates back to 1966 when Basov *et al.* brought forth the idea for lasers with radiating mirrors [11]. However, it was not until the 1990s before the proposed concept was fully acknowledged and the first working devices were demonstrated [12,13]. Several more years were then needed before the benefits of the technology were fully proven and a surge of new developments were triggered [7,8].

Structurally the SDL gain chip comprises a high reflectivity mirror and a semiconductor gain region. The gain region usually includes several quantum-well (QW) or quantum-dot (QD) layers separated by spacer/barrier layers. A typical mirror structure consists of a stack of quarter-wavelength semiconductor layers, forming a distributed Bragg reflector (DBR), although metallic, dielectric, or hybrid [14] mirror structures have been used in some cases. Typically, the total thickness of the gain mirror structure is less than 10 μm. The single transverse mode of operation is achieved by controlling the fundamental mode size via cavity design and matching it with the pumped area. It should be noted, however, that SDLs can also be electrically pumped to bring more compactness; however, in this case the semiconductor technology is more demanding and the power scaling capability is significantly lower compared to the optically pumped scheme [15].

The inadequate heat dissipation techniques in the early SDL experiments lead to the requirements of very low operating temperatures for high-power operation, which made these devices unpractical for use. The gain mirror heating arises from the portion of the pump energy that is not converted to useful photons, either because of the quantum defect (i.e., the photon energy difference between the pump

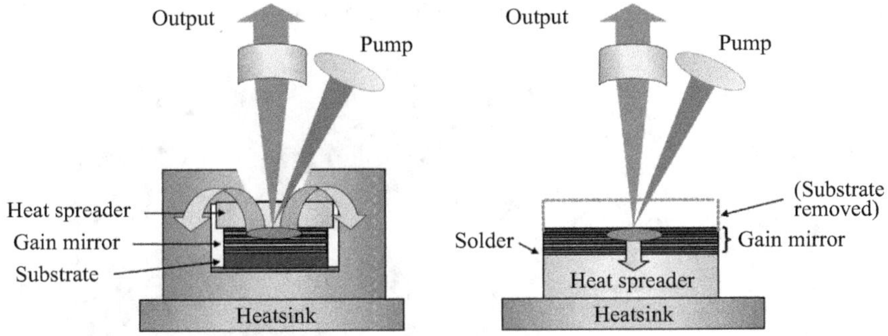

Figure 8.1 Generic description of the OP-SDL concept emphasizing the common strategies used for heat extraction. Left: Gain chip with intracavity heat spreader. Right: Flip-chip device with substrate removed

photon and the laser photon) or because of non-radiative recombination. Typically, the pump photon has markedly higher photon energy than the emitted laser photon, for example, for a 1,060 nm laser pumped with 808 nm radiation the quantum defect is about 24% of the pump photon energy. In other words, the optical-to-optical conversion efficiency of such a laser cannot exceed 76% even under theoretically perfect conditions. Usually, the laser is designed to absorb the pump radiation in the spacer/barrier layers separating the QWs. This is because most of the pump energy can be absorbed in a single pass interaction owing to the efficient absorption in semiconductors. On the other hand, there must be a notable bandgap contrast between the spacer layers and the QWs in order to ensure a good carrier confinement and, hence, efficient operation at elevated temperatures. From this perspective a high quantum defect can also be beneficial. In most spacer-pumped SDLs, the quantum defect is between 15% and 50% of the pump photon energy.

Excess heating reduces the emission efficiency via increased nonradiative recombination and carrier leakage, but also red-shifts the emission wavelength, which leads to a mismatch between the emission wavelength and the resonant wavelength in a resonant periodic gain structure, further reducing the gain [16]. Effectively, such heat-induced processes create a positive feedback loop with very negative impact on the laser performance. Consequently, the output power of the laser exhibits a roll-over characteristic when the pump power is increased beyond a critical point. As a general strategy for achieving high-power operation, one should try to maximize the heat transfer from the gain and at the same time minimize the heat generation by reducing the quantum defect and non-radiative processes.

In order to extract the heat, the gain chip is mounted on a heatsink, which is cooled externally. Figure 8.2 shows a cross section of a water-cooled mount on which the gain chip is clamped. The heat load generated by pumping the gain mirror leads to a slightly higher temperature of the mount compared to the coolant temperature. Moreover, the mount temperature increases usually by a few degrees Celsius with the pump power due to its finite thermal conductivity at the available cooling power. Another detailed observable in Figure 8.2 is that the mount has a recess region to enable angle pumping of the gain at an angle of about 30–40°.

Figure 8.2 Left: Schematic of the water-cooled mount. Right: Picture of an SDL cavity with two gain chips

For high-power operation, it is also necessary to ensure a minimal thermal resistance between the heatsink and the active region. One should therefore aim at minimizing the distance between the heatsink and the gain region, and at the same time use materials that have high thermal conductance. Using a planar gain mirror geometry, the pumping is concentrated on an area that has typically a diameter of some tens or hundreds of micrometers, while the overall thickness of the semiconductor layer structure is only a few microns. In other words, the heated area is very large compared to the thickness of the layers. Thermal simulations show that in such a structure the heat flow is essentially one-dimensional and is directed normal to the sample surface [17].

The most common heat dissipation method used in research, the so called *intracavity heat spreader* method, is conceptually simple and involves contacting a transparent heat spreader element onto the gain mirror [18]. A schematic representation of an SDL mounted in this way is shown in Figure 8.1 (left). In this method, the heat spreader is located right next to the gain region. The practical limitations arise from the fact that the heat spreader is located inside the laser cavity and that the number of transparent materials with high thermal conductance is limited; their cost is also significant compared to the cost of the gain chip. These material requirements are best fulfilled by diamond that has an extremely high thermal conductivity (up to 2,000 W/m·K) and a wide transmission window. The other material options include sapphire, i.e., crystalline Al_2O_3 [19], and silicon carbide [20]. The intracavity heat spreader is usually bonded onto the SDL gain element using capillary bonding with de-ionized water [21]. In this approach, two smooth and flat surfaces (here the gain mirror and the heat spreader) are pulled together by surface tension of a liquid, and as the liquid evaporates the two surfaces are brought to close optical contact and held together by surface forces. Simple mechanical clamping can also be used for optical contacting as long as the surfaces are sufficiently smooth, flat and free from particles or other contaminants. However, capillary bonding is a good way to make sure that the surfaces meet these requirements and can be brought to close optical and thermal contact. The success of the bonding process can be simply monitored by observing the disappearance of the Newton interference rings as the surfaces are brought closer together.

Another option for efficient heat dissipation, described, for example, in the seminal paper of Kuznetsov [7], is schematically shown in Figure 8.1 (right). In this approach, the gain mirror is grown in reverse order with the active region first and the DBR section second. The structure is then bonded "upside-down" onto a heatsink and the substrate is removed using wet etching. Effectively this method transfers the epitaxial layers from a semiconductor substrate onto a substrate with higher thermal conductance. The process leaves only the DBR layers between the heated active region and the heatsink, which greatly reduces the thermal resistance in comparison to the situation where the semiconductor substrate would be located between the gain mirror and the heatsink. This process is often referred to as the "flip-chip" process or the "thin-device" process. From processing point of view, the flip-chip process requires longer overall time but can be done in batches of many devices. The major challenge of the flip-chip approach relates to the fact that the epitaxial layers are mechanically very fragile without the support of the original substrate. The bonding process requires usually the use of temperatures exceeding 150 °C. Therefore, any differences in the coefficients of thermal expansion between the epitaxial layers and the heatsink may translate to mechanical stress as the sample cools down and the solder hardens. This is particularly critical issue with large samples bonded with hard solders, such as AuSn, that have high melting points. To alleviate the mechanical stress one can resort to soft solders such as indium. However, it is a well-known fact that in high-power diode lasers indium solder tends to fail due to the thermal diffusion and other effects. Such effects should be considered in connection with the lifetime of high-power SDLs as the gain region is operated at relatively high temperatures. There are also many alternative bonding methods, such as InAu bonding [14,22], that can be used instead. An important aspect related to soldering concerns the presence of voids within the solder; any voids in the solder will likely result in physical damage to the gain mirror under pumping. The voids can be monitored prior to substrate removal using a scanning acoustic microscope, for example. To further improve the heat dissipation, one can use a heat spreader, such as a diamond, between the sample and the metallic heatsink; the thermal energy is thus rapidly spread from a point source (i.e. the pumped spot) to a larger area over which it is conducted to the actual heatsink. The cost of the heat spreader is also markedly reduced when there is no need for optical-grade quality as in the case of the intracavity heat spreader approach. For high volume SDL chip production (flip-chip or intracavity heat spreader) one can use direct optical contacting of the gain mirror and the heat spreader on wafer scale level [23].

Substrate removal is also an important step in the flip-chip process. To some extent, the substrate can be thinned by lapping prior to bonding, but in any case tens or hundreds of microns of semiconductor substrate must be removed by etching. This is usually done by wet etching employing an etch stop layer. It is important, for the success of the process, that the selectiveness of the etching is sufficiently high and that the process can be carried out in a reasonable time. For GaAs removal one can use, for example, InGaP or Al(Ga)As etch stop layers and $NH_4OH:H_2O_2$-based etchants. InP etchants are often based on HCl [24], which may limit or hinder the use of indium as a solder for sample bonding. For GaSb-based compounds good etchant-etch stop combinations are less developed; yet successful flip-chip SDLs based on this material system has been reported [22].

In practice, the intracavity heat spreader approach is very convenient to implement in laboratory conditions due to its processing quickness and simplicity. Moreover, the high cost of intracavity diamond heat spreaders is overcome by the possibility of recycling them, at least in research use. The biggest drawback of intracavity heat spreader is that it inevitably introduces optical losses and acts as an intracavity etalon within the laser cavity. The optical spectrum is therefore modulated, which can hinder mode-locking and continuous wavelength tuning properties. The etalon effect can be removed by using wedged heat spreader with an antireflection coating [25], but the wedge angle usually increases the reflection losses despite the AR layer and hence decreases the output power. Nevertheless, both heat management strategies have been successfully used to achieve over 10 W of output power from standard InGaAs/GaAs gain structures, though, the highest output powers have been achieved with the flip-chip components [26–28]. However, if the thickness of the DBR stack increases (due to longer operation wavelength or poor index contrast of the materials), or if the DBR layers have very poor thermal conductance, it may be more advantageous to use the intracavity heat spreader technique. This is particularly true for InP and GaSb-based SDLs. Finally, a more advanced technique to ensure efficient cooling involves double-side cooling [29] of the SDL chip with a complete removal of the substrate, or even also of the DBR [30]; such gain membrane technology provides important means for future wavelength extension and power scaling of SDLs.

In order to minimize the generation of heat, and hence allow the use of less critical mounting processes, one can use direct "in-well" pumping [31,32] where the spacer layers between the quantum wells are transparent to the pump radiation and the pump wavelength closely matches the QW emission wavelength. This approach minimizes the quantum defect, but another technical difficulty arises from a short light-matter interaction length. The thickness of one QW is typically some nanometers and the total absorptive path length is rather small, since a gain mirror would typically include 5–15 QWs. To some extent, the pump absorption can be improved by adding more QWs to the structure, but usually either a resonant pumping scheme or external pump recirculation optics is required for efficient pump absorption. We should note that an in-well pumped gain mirror provides by default a high reflection for the unabsorbed pump light, thus avoiding pump absorption in the DBR and providing double-pass of pump radiation through the gain region. The selection of the pump laser for in-well pumping is more critical (and possibly more expensive) than in a spacer pumped lasers, where low-cost 808 nm diodes can be used for pumping 920 nm SDLs as well as 2,000 nm SDLs. Nevertheless, in-well pumping offers an interesting option for reducing the quantum defect and the heating related to it. One should also notice that quantum defect optimization makes sense only if the quantum efficiency of the laser is already high. If a significant part of the pump photons is anyway lost to non-radiative processes, the benefits of the quantum defect optimization become marginal to the overall performance of the system. Therefore, high quality gain materials and proper structural designs are prerequisites for efficient operation of SDLs.

To summarize, efficient heat removal is highly important for high-power operation of SDLs. The use of high thermal conductance heat spreader materials, such as diamond, greatly improves the heat extraction from the SDL gain element. The heat spreader should be placed as close as possible to the gain region either by optically contacting the heat spreader to the light-emitting surface or by flip-chip bonding the bottom of the SDL gain element onto the heat spreader. The SDL materials and the applications determine which process is more suitable in each case. Furthermore, the heat generation can be minimized by reducing the quantum defect between the pump and the SDL emission wavelength.

8.2.2 Cavity architectures

The most common SDL cavity architectures are summarized in Figure 8.3. The simplest cavity has an I-shape that is formed between the gain mirror and a single external output coupler (OC) mirror. However, in research environment it is often easier to use a V-shaped cavity that is formed between the gain mirror, a curved folding mirror, and a planar OC. The advantage of the V-cavity is that planar OCs with various coupling ratios are often cheaper and more widely available in stock than equivalent curved couplers. Another practical advantage of the V-shaped cavity is related to the alignment of the laser; if the final alignment is done by monitoring the output of a photodiode placed behind the OC, the folding mirror in a V-shaped cavity collects the light efficiently to the photodiode enhancing the available signal, while in I-shaped cavity the spontaneous emission from the gain is rapidly dispersed to all directions. More complex cavity configurations are often used for frequency conversion and mode-locking. For efficient frequency conversion, the nonlinear crystal is often placed at a location near or at the mode waist. This is usually easier to do in a V-shaped or Z-shaped cavity than in an I-shaped cavity. More complicated Z-shaped cavities are typically used in mode-locking SDLs to accommodate also a semiconductor saturable absorber mirror (SESAM); the challenge there is to produce a sufficiently small mode diameter on the absorber mirror, while at the same time maintaining a reasonably large mode diameter on the gain [9].

To scale up the power of an SDL, it is possible to deploy multiple gain elements in a single cavity [7,33]. SDLs can also employ ring cavities [34], but probably due to the added complexity such lasers have not gained much popularity. In addition to different external cavity configurations, one can also produce an SDL with a semi-monolithic cavity that may include a plane-plane design stabilized by a thermal lens [35]. Such a laser may be more limited in power and brightness but does possess an extremely rugged design. Furthermore, the semi-monolithic cavity can be processed to have curved surfaces with mirror structures, thus avoiding the need for cavity stabilization by a thermal lens [36]. The integrated cavity concept has been refined for mode-locking operation in the form of a mode-locked integrated external-cavity surface-emitting laser (MIXSEL). In MIXSEL concept, the

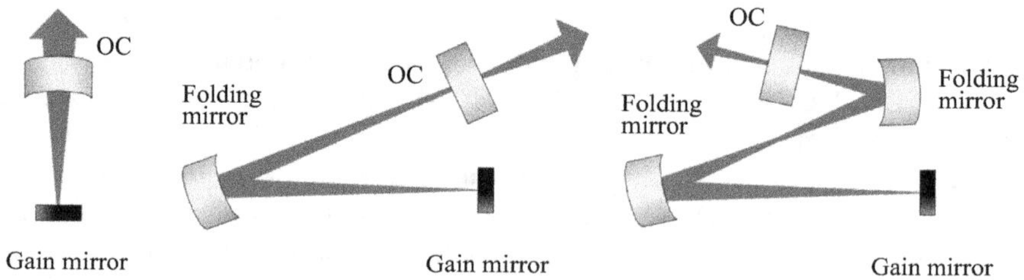

Figure 8.3 Typical SDL cavities. Left: I-shaped cavity. Center: V-shaped cavity. Right: Z-shaped cavity. OC = output coupler (partially reflecting mirror)

gain and the saturable absorber mirrors are integrated monolithically, with the ultimate goal of avoiding the use of an external cavity mirror and achieving ultra-high repetition rates in a monolithic optically pumped laser chip [37,38]. We should also note that the gain element can be used in transmission, as demonstrated, for example, in [30].

8.2.3 Wavelength coverage

During the last decade, the SDL research has progressed along three major directions: (i) power scaling, (ii) extending the wavelength coverage, and (iii) generation of ultrashort pulses. In terms of available output power, the 10-W level has been reached in many experiments using both single and multiple gain elements. The record output power for an SDL at the time of writing is 106 W and was set at 1 μm wavelength range [28]. Other more relevant parameters for biophotonics applications, such as wavelength tunability and narrow linewidth operation have been continuously improved. Owing to the ability to incorporate intracavity filters, SDLs can be easily tuned by several tens of nanometers [39,40]. Assuming that the DBR stop-band is broad enough, the QW region can be engineered to achieve a tuning range as broad as 150 nm [41]. Narrow linewidth operation can be easily achieved by inserting suitable etalons and filters within the cavity and by employing cavity stabilization techniques [42,43]. Combined continuous tuning and narrow linewidth has also been demonstrated, even when using intracavity heat spreaders [44]. Excellent results have also been obtained for ultrashort pulse generation [45], as well as in generation of pulses with high average output power and high repetition rate [38]. Moreover, the spectral coverage of mode-locked SDLs has recently been extended outside the common InGaAs wavelengths near 1 μm to cover both visible wavelengths [46] and wavelengths in the 2 μm range [47,48].

With regard to the spectral coverage, the CW SDLs (fundamental and frequency converted emission) spans today from 244 to 5,000 nm [49,50], although not without gaps. Table 8.1 gives an overview of output powers achieved for different wavelength bands for CW SDLs. It should be noted that these results are not readily comparable neither do they cover all technology approaches (the summary is mainly aimed at revealing maximum power at a certain spectral range). First of all, the efforts to develop various material systems and perfect the high-power capabilities are not equal, having most of the developments focused on InGaAs/GaAs material system and fundamental operation at 1 μm wavelength range operation. Moreover, the mount temperature as well as the mode quality and the spectral properties, which would impact the output power, are not comparable. Another fundamental difference relates to the fact that not all of the SDLs have direct emission. In particular, for wavelengths below 670 nm they are usually covered by frequency doubled infrared systems, which are the most mature and efficient way of generating fundamental light using well established semiconductor materials. There are also some fundamental differences concerning the technology as not all SDLs included in the summary have a monolithic DBR. Highest powers for wavelengths in the range of 1.3–1.5 μm have been achieved using a wafer-bonding technology that combines InP gain regions with GaAs DBRs.

8.3 SDLs emitting yellow–orange–red radiation

In this chapter, we present details concerning the development of SDLs with emission at the yellow–orange–red wavelengths, which are of particular interest for medical applications. To date, yellow SDLs have been of high interest for several biophotonics applications, including flow cytometry [93,94], ophthalmology [95], fluorescence imaging using continuous-wave stimulated emission depletion (STED) [96], and medical therapy [97]. The development of SDLs emitting at the red wavelengths of 600–800 nm could further expand the scope of their applications. For instance, efficient SDLs at 650 nm could be used for photodynamic therapy (PDT) [98,99] and fluorescence imaging [100], whereas 700–800 nm SDLs would be particularly suitable for applications that require large penetration depths into tissue [101–103] and possibly for isotope separation which is a very important application area in pharmacology [104]. Regardless of the application, the advantage of SDLs arises from their compact footprint and wavelength flexibility that enable targeting specific applications.

In general, the wavelength versatility is related to the possibility to tailor the emission wavelength simply by selecting the appropriate material combinations and compositions for fabricating the SDL structure and inserting intracavity filters for wavelength conditioning. However, the yellow–orange SDLs are only available as frequency doubled systems, because currently there are no semiconductor gain materials to ensure direct emission in this wavelength range. Here, we will review advances concerning frequency doubled high-power yellow–orange SDLs working in CW and pulsed mode. On the other hand, the red spectral range can be achieved using both directly emitting and frequency doubled SDLs. In this context, we will summarize recent results concerning the development of frequency doubled red SDLs as well as progress in directly red-emitting pulsed SDLs, including the shortest emission wavelength obtained with a mode-locked system so far.

8.3.1 Frequency-doubled yellow SDLs at 560-590 nm

To reach emission in the wavelength range of 560 nm to 590 nm (i.e., yellow–orange) by second harmonic generation (SHG), the starting point is to develop gain mirrors with fundamental emission at 1,120–1,180 nm. As indicated in Table 8.1, the highest powers at these wavelengths can be reached by using GaInAs/GaAs and GaAs/AlAs DBRs. However, there are certain challenges that need to be considered when manufacturing reliable gain mirrors with emission in this spectral range. First, for conventional InGaAs/GaAs QW materials a relatively large content of indium (>30%), required to achieve wavelength above 1,100 nm, increases the lattice constant of the material causing a build-up of strain. Excessive material strain, arising from the difference between the lattice constants of the semiconductor layers, can lead to formation of crystalline defects and ultimately to the relaxation of the layered structure and catastrophic failure of the SDL. To tackle this, one would need to employ more advanced growth methods such as growing InGaAs at low temperatures and strain compensation techniques.

Table 8.1 An overview of spectral coverage and output power for SDLs. The SDLs are also categorized in terms of emission mode: direct for emission at fundamental wavelengths, SHG for the second harmonic generation, THG for the third harmonic generation, and FHG for the fourth harmonic generation

Wavelength (nm)	**Gain region QW/barrier**	**Generation mode**	**CW power (W)**	**Mount temp (°C)**	**M^2 or linewidth**	**Ref.**
229	InGaAS/GaAs	FHG	0.56			[51]
244	InGaAs/GaAs	FHG	0.21			[52]
254	InGaAS/GaAs	serial SHG	0.125			[53]
295	InGaAs/GaAs	FHG	0.13			[54]
330	GaInP/GaAs	SHG	0.26	−25		[55]
338	AlGaInP/GaAs	SHG	0.12	−10		[49]
355	InGaAs/GaAs	THG	0.45			[56]
460	GaInAs/GaAs	SHG	5.2	0		[57]
488	GaInAs/GaAs	SHG	15			[58]
505	InGaAs/GaAs	SHG	8			[26]
532	InGaAs/GaAs	SHG	30			[59]
532	GaInAs/GaAs 3 gain chips	SHG	64			[26]
559	InGaAs/GaAs	SHG	1.1		300 kHz	[60]
575	GaInAs /GaAs	SHG	8			[61]
588	GaInAs /GaAs	SHG	20	8	<1.5@10 W	[62]
615	InGaNAs/GaAs	SHG	10.5	8		[63]
625	InGaNAs/GaAs	SHG	6.5	20		
650	AlGaInAs/InP wafer-fused DBR	SHG	3	7		[64]
665	GaInP/GAs	Direct	2.5	10		[65]
739	QDs:InP/AlGaInP	Direct	0.05	7		[66]
740-790 nm	AlGaAs QWs	Direct	4.24	14		[67]
750	AlGaInAs/InP wafer-fused	SHG	1.5	15		[68]
785	AlGaInAs/InP wafer-fused DBR	SHG	1	15	1.45	[69]
850	GaAs/AlGaAs	Direct	0.5	10		[70]
853	GaAs/AlGaAs	Direct	1.6	12	<1.1@1W	[32]
920	InGaAs/GaAs	Direct	20			[26]
960	GaInAs/GaAs	Direct	20.2	−20	<1.1	[71]
980	InGaAs/GaAs	Direct	40			[26]
1,028	GaInAs/GaAs	Direct	106	3		[28]
1,040	GaInAs/GaAs	Direct	40.7	0	<1.5@23 W	[27]
1,064	InGaAs/GaAS	Direct	60			[59]
1,119	InGaAs/GaAs	Direct	4		900 kHz	[44]
1,140	InGaAs/GaAs	Direct	12			[72]
1,170	GaInAs/GaAs	Direct	7	5	<1.46	[73]
1,180	GaInAs/GaAs	Direct	50	−15		[74]

(Continues)

Table 8.1 (*Continued*)

Wavelength (nm)	Gain region QW/barrier	Generation mode	CW power (W)	Mount temp (°C)	M^2 or linewidth	Ref.
1,200	GaAs/GaInAs/ GaAsSb	Direct	1.6	15		[75]
1,220	GaInNAs/GaAs	Direct	5	15		[76]
1,228	InGaAs/GaAs Raman conversion	Direct	4.4	7		[77]
1,255	QDs: InAs/GaAs	Direct	4.6	10		[78]
1,270	AlGaInAs/InP wafer-fused	Direct	33	−15		[79]
1,305	AlGaInAs/InP wafer-fused DBR	Direct	6.6	7		[64]
1,315	InGaAs/GaAs Raman conversion	Direct	2.5	10		[80]
1,541	GaInAs/InP	Direct	0.78	−30	<1.2	[19]
1,480	AlGaInAs/InP wafer-fuser	Direct	4.8	8		[81]
1,580	AlGaInAs/InP wafer-fused DBR	Direct	4.6	8		[69]
1,980	GaInSb/GaSb	Direct	4.2	10		[41]
2,000	GaInSb/GaSb	Direct	5	−15		[82]
2,020	GaInAsSb/GaSb pumped at 1,470 nm	Direct	17	20		[83]
2,020	GaInAsSb/GaSb pumped at 980 nm	Direct	20.7	−3		[84]
2,360	GaInAsSb/GaSb	Direct	3.2	−15		[85]
2,500	GaInAsSb/ AlGaAsSb	Direct	0.6	5	<1.6	[86]
2,800	GaInAsSb/ AlGaAsSb	Direct	0.12	20		[20]
3,300	PbTe/Si pulsed	Direct	0.02	−20		[87]
3,500−4,300	QDs: PbTe pulsed	Direct	0.01−0.05	<−93		[88]
4,500	PbSe/BaF_2 pulsed	Direct	0.006	27		[89]
5,000	PbSe/BaF_2	Direct	0.017	−173		[90]
5,260	PbSe/Si	Direct	0.026	−173		[91]
5,400	InGaAs/GaAs	DFG	0.005			[92]

If the strain energy accumulated in the crystal lattice exceeds that needed for a dislocation formation, the strain energy cannot be accommodated elastically and a misfit dislocation will form in the SDL crystal. The dislocations introduce non-radiative recombination centers to the QWs that severely impair their luminescence efficiency. Furthermore, dislocations formed during the crystal growth can extend during the laser operation, and will detrimentally influence the laser lifetime. The strain energy accumulated in the crystal lattice depends on the strain magnitude and the thickness of the strained layer. Therefore, for a given amount of strain, the

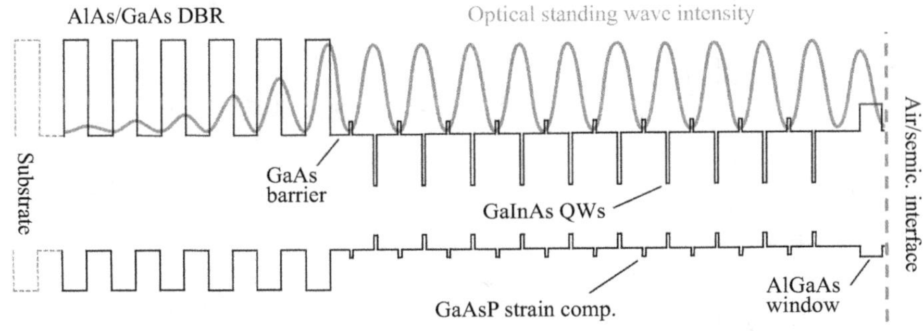

Figure 8.4 Layer structure of a strain-compensated gain chip developed for 1,180 nm emission. Reproduced with permission from [62]

dislocation formation begins after the critical thickness is exceeded. The critical thickness can be increased by lowering the growth temperature of the GaInAs layer. Unfortunately, the lower growth temperature will also introduce non-radiative recombination centers to the material. With careful optimization of the growth parameters, however, it is possible to minimize the concentration of the non-radiative recombination centers and dislocations in the material.

Even if one can grow GaInAs QWs of high quality by using low temperature, the overall strain accumulated in an SDL active region, incorporating a relatively high number of strained QWs, would lead to the dislocation formation and associated problems with device performance and lifetime. This problem can be solved by resorting to the use of the so-called *strain compensation layers of opposite strain* as compared to the quantum wells to the gain chip active region [105]. By proper design, this keeps the net strain close to zero and removes the limit to the number of QWs that can be incorporated to a resonant periodic gain. The strain compensation of a GaInAs SDL fabricated by Molecular Beam Epitaxy (MBE) has been reported, for example, in Ref. [106], where the effect of the strain compensation was assessed by measuring the curvature of the wafer and by mapping the photoluminescence to identify the non-emissive dark areas. The active regions in the study comprised a total of six 7-nm thick $Ga_{0.69}In_{0.31}As$ QWs embedded within GaAs barriers. The *strain-thickness* product of a QW was balanced with the opposite sign strain-thickness product of a GaAsP layer. The amount of compensation as well as the composition and thickness of the GaAsP layers were chosen so that the tensile strain of individual GaAsP layers would be below the relaxation limit and the net strain of the structure would be near zero. Therefore, 27-nm thick $GaAs_{0.85}P_{0.15}$ layers, exhibiting ~0.54% misfit tensile strain were used to compensate for about ~91% of the compressive strain associated with the QWs.

This strain-compensation strategy was used in [107] to obtain high-power 1,180 nm radiation. In this case, the SDL gain chip was designed to be used with an intracavity heat spreader. The active region comprised 10 QWs that were embedded between GaAs barrier layers as depicted in Figure 8.4. The strain compensation layers placed between the QW groups consisted of 32-nm tensile-strained GaAsP and the

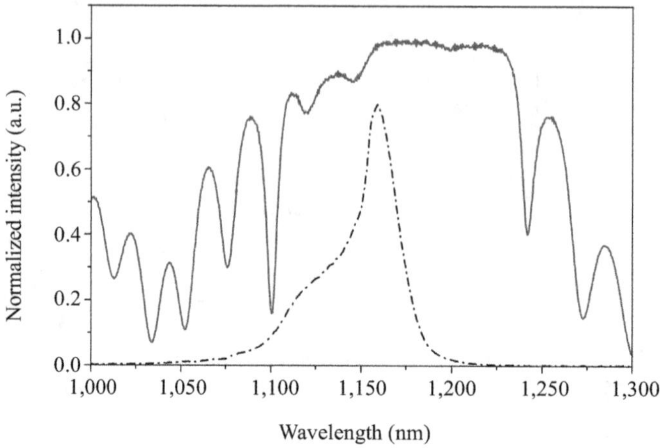

Figure 8.5 Reflectance (solid line) and edge photoluminescence signal (dashed line) of the 1,180-nm gain chip

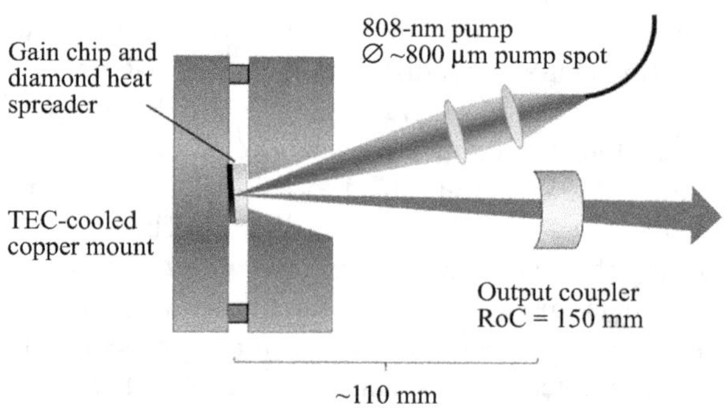

Figure 8.6 Cavity illustration of the 50-W SDL system. Reproduced with permission from [74]

7-nm GaInAs QWs had an indium mole fraction of ~35%. The structure was terminated by a 20-nm AlGaAs window layer and a 10-nm GaAs capping layer. The gain structure was grown by all-solid-source molecular beam epitaxy on a GaAs substrate. The photoluminescence and reflectance graphs are shown in Figure 8.5.

After the growth, the gain structure was diced into 3 mm × 3 mm chips and bonded onto an intracavity diamond heat spreader. The diamond was attached to a large TEC-cooled copper mount to further extract the heat away from the chip and to cool it to low temperatures. The chip was incorporated into an I-shaped cavity formed by the gain chip and a partially reflecting mirror ($R \sim 97\%$) with a radius of curvature of 150 mm, shown in Figure 8.6. The optical pumping was achieved using a fiber coupled 808-nm diode laser and lenses were used to focus the pump light onto the surface of the chip in a diameter of ~810 μm. The laser operation was assessed at different mount temperatures: −15, −10, −5, 0, 10, and 20 °C. The highest output

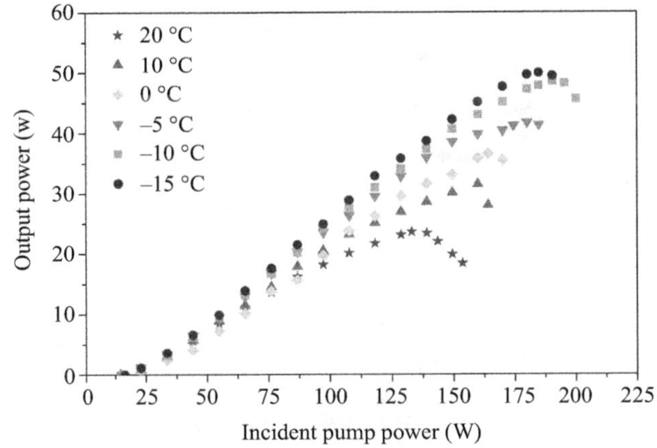

Figure 8.7 Output power curves for the 1,180-nm SDL measured at various mount temperatures. Reproduced with permission from [74]

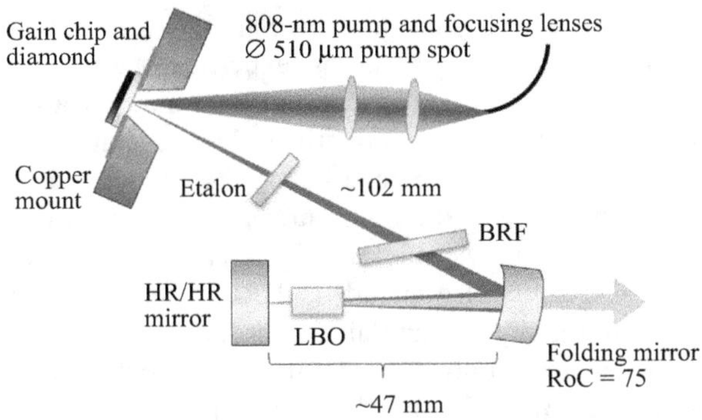

Figure 8.8 Cavity illustration of the frequency-doubled yellow SDL. Adapted and reproduced with permission from [62]

power of 50 W was measured for a mount temperature of −15 °C. In this case, the optical-to-optical conversion efficiency was 27%. The lasing spectrum of the SDL system was centered at around 1,180 nm throughout the measurements. Figure 8.7 shows the output power curves measured at different mount temperatures.

The tuning range of the gain material was tested in a setup with a more modest cooling assembly, i.e., a smaller water-cooled copper mount with a thinner diamond heat spreader. The cavity configuration was similar to the 50-W cavity, but a 1.5-mm birefringent filter (BRF) was added inside the laser cavity for wavelength control. The filter was rotated to tune the wavelength of the laser and a maximum tuning range of 67 nm was achieved with the cavity assembly [107].

In order to reach the yellow spectral range, the same gain structure was used to build a V-shaped cavity (shown in Figure 8.8) that incorporated a nonlinear crystal

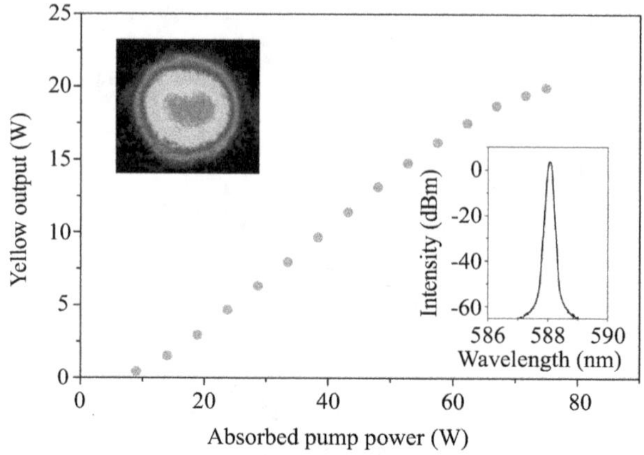

Figure 8.9 Output power curve for the yellow-orange SDL emitting at 588 nm. The insets show the beam profile and the spectrum at the maximum output power. Adapted and reproduced with permission from [62]

for frequency doubling the 1,180-nm fundamental emission [62]. The same water-cooled mount as in the wavelength tuning experiment was employed for heat extraction. The nonlinear crystal was chosen to be a noncritically phase matched lithium triborate (LBO), because of its relatively high damage threshold and non-linear coefficient. The crystal was placed inside the cavity, near the mode waist, where it could take advantage of the high intracavity power of the 1,180-nm gain chip. Efficient phase matching was achieved by temperature controlling the crystal with a copper oven. The cavity also included a BRF, which also set the polarization for the frequency doubling crystal, and an etalon for wavelength control and line-width narrowing. A maximum output power of 20 W at 588 nm was reached in this experiment. The output power vs. absorbed pump power is plotted in Figure 8.9, revealing also high optical slope and conversion efficiencies of 36% and 28%, respectively. The BRF was also used to tune the wavelength, which yielded a tuning range of ~26 nm for the frequency-doubled yellow emission.

In addition to CW mode, the yellow–orange laser was also tested in pulsed mode. Pulsed operation was achieved by directly modulating the pump laser and for this purpose a critically phase matched LBO was used to convert the output to the yellow–orange spectral range (the non-critically phase matched LBO had suffered damage in use). In this setup, it was possible to produce pulse widths in the microsecond range and down to 570 ns with ~14 W of peak power. The repetition rate was kept constant at 10 kHz. The maximum optical-to-optical conversion efficiency was calculated to be ~21%. The output power of the laser was also measured in CW mode using the same critically phase matched crystal and coolant temperature (20 °C) as in pulsed operation, which yielded a maximum of 8.5 W. This shows a 65% improvement in power in pulsed operation. The power characteristics for pulsed and CW operation in the same laser setup are shown in Figure 8.10, as well as the output spectrum for 570 ns pulse width.

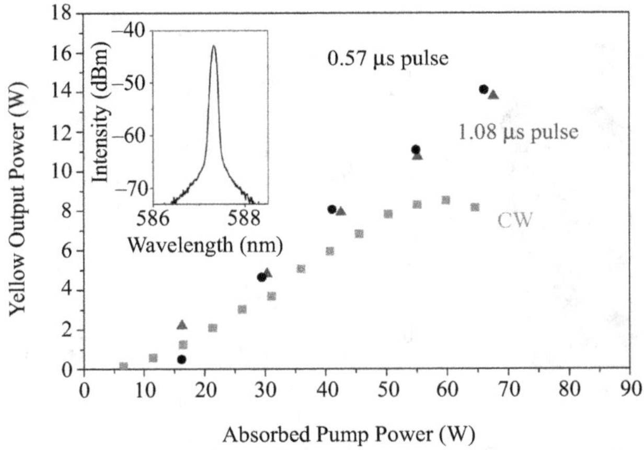

Figure 8.10 Output characteristics for 20 °C (coolant temperature) CW (average power) and pulsed operation (peak power). Inset: Output spectrum measured for pulsed operation for 0.57 μs pulse width. Adapted and reproduced with permission from [62]

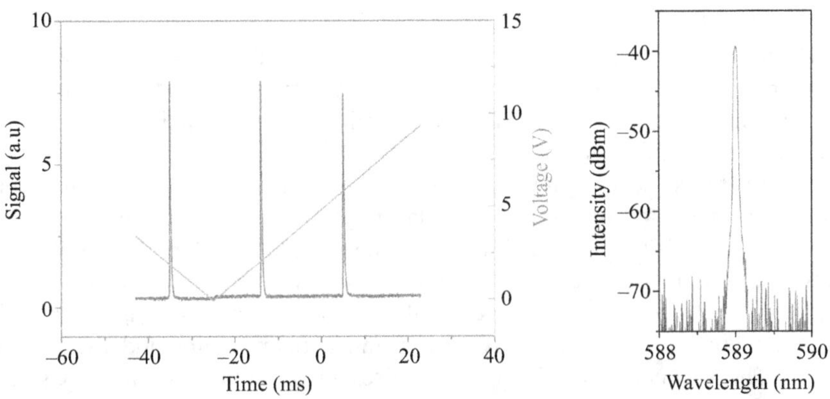

Figure 8.11 Left: Oscilloscope trace of scanning Fabry - Perot interferometer signal and its scanning voltage. Right: Output spectrum of the single-frequency SDL with its emission tuned to 589 nm with the BRF. Reproduced with permission from [109]

Single-wavelength operation at 590 nm wavelength range has also been demonstrated [108]. In this case, the gain chip was a dilute nitride structure, which will be presented more in depth in the next section. The maximum output power of 2.7 W was reached with a linewidth <30 MHz. The mode and the emission spectra of the laser are shown in Figure 8.11. The emission wavelength was controlled by using a 1-mm thick YAG etalon and a 3-mm thick Quartz etalon, and the second harmonic radiation was generated in 10-mm long LBO crystal. In a similar

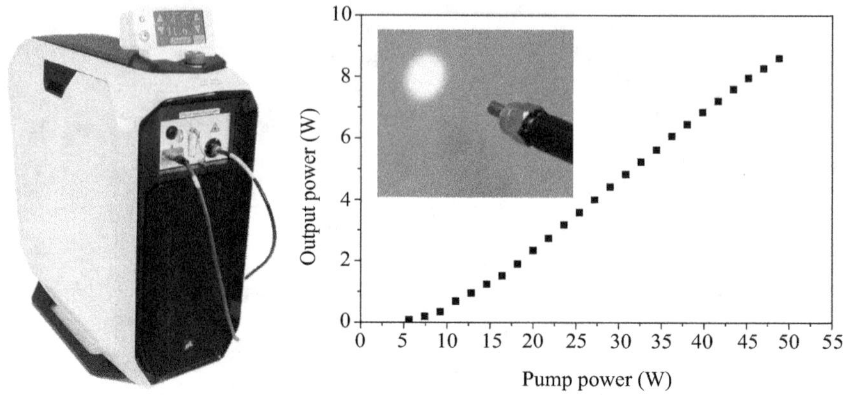

Figure 8.12 Left: The yellow–orange SDL system (638 mm × 262 mm × 601 mm) for dermatology applications developed in collaboration with INSERM, France. Right: The output power of the laser system as a function of the 808-nm pump diode power (enclosed within the system). The system is not yet CE-marked, but clinical trials are planned for 2018

experiment [44], continuously tunable operation was achieved over a range of 10 GHz around 589 nm. The maximum output power of this laser was 3 W and the short-term (in order of seconds) emission linewidth was 10 MHz. With further optimization of the gain chip and cavity architectures, even higher powers for single-frequency operation should be soon available.

Finally, the yellow–orange SDL was developed into a compact and mobile module and integrated into a prototype laser system targeting dermatology applications, shown in Figure 8.12. The system delivered up to 8.6 W of CW output power at a wavelength of 585 nm through a multimode fiber, which was attached onto a handheld laser scanner. For clinical use, the system was designed to deliver light pulses in the order of 10–100 ms using a fiber shutter and the handheld scanner. The scanner can deliver light pulses in different patterns and sizes to accommodate for different size targets.

8.3.2 Frequency-doubled orange–red SDLs at 590–635 nm

In order to reach the orange–red wavelength range at > 590 nm via frequency doubling, the fundamental SDL emission needs to be extended beyond 1.18 μm. The most promising approach for generating multi-watt emission in this spectral range is frequency doubling dilute nitride structures emitting at a fundamental wavelength of 1,180–1,270 nm. In these heterostructures, a small fraction of N (typically below 1.5%) is alloyed within the GaInAs QWs, so that the bandgap and the lattice constant of GaIn(N)As/GaAs QWs are simultaneously decreased. Thus, the use of such dilute nitride structures offers a higher flexibility in designing the gain mirrors and reduces the high compressive lattice strain present in GaInAs/GaAs QWs emitting at wavelengths above 1,100 nm [110]. We should note that this wavelength range can also be covered with quantum-dot (QD) gain

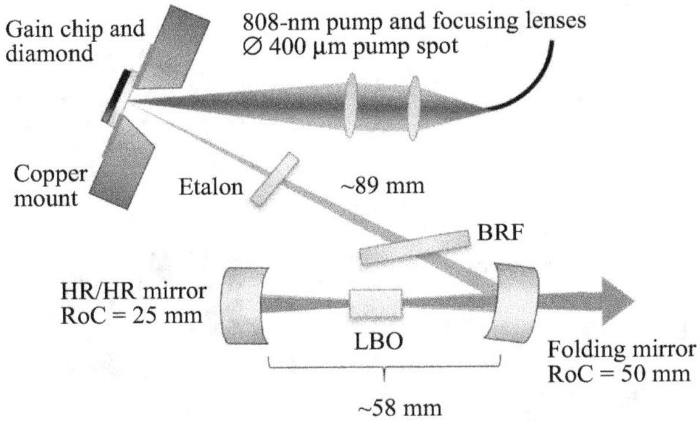

Figure 8.13 Cavity schematic for > 10 W SDL at 615 nm. Adapted and reproduced with permission from [63]

heterostructures [111,112]. An important aspect for SDL operation is that GaIn-NAs/GaAs QWs exhibit a large conduction band offset and higher electron effective mass providing improved carrier confinement and, thus, temperature behavior [113]. This positive feature is present also for QD SDLs, although the modal gain is, indeed, lower for the QD heterostructures than for the QWs [114]. On the other hand, the incorporation of N is usually accompanied by the introduction of non-radiative defects, which ultimately decreases the performance of dilute nitride lasers. Nevertheless, dilute nitride SDLs with emission wavelengths in the range of 1,180–1,240 nm require a N-content of less than 1% and can exhibit multi-watt output powers for both fundamental [115] and frequency doubled operation [116].

The most impressive performance from a dilute nitride SDL was demonstrated in [63] with > 10 W of output power at the red wavelength of 615 nm. The structure was grown by MBE on a GaAs substrate and was designed to be resonant at 1,230 nm. It comprised 10 GaInNAs QWs with 0.5% N composition that were placed at the antinodes of the optical field. The structure was grown at a temperature of 375 °C, after which the wafer was annealed at 700 °C for improving the material quality and diced again into 3 mm × 3 mm-sized chips. The mounting was similar as for the yellow–orange SDL, and the V-shaped cavity configuration is schematically illustrated in Figure 8.13. The lasing wavelength was further controlled with a BRF and an etalon. The fundamental radiation was then converted to the red spectral range by placing a non-critically phase matched LBO crystal near the mode waist of the cavity. The frequency doubled light exited the cavity through the folding mirror, which had an antireflection coating for the red spectral range.

The red SDL was measured in CW and in pulsed mode. In CW mode, the maximum output power of 10.5 W was measured at a chip mount temperature of 5–8 °C. In this case, the optical-to-optical conversion efficiency was 17.5%. The power and spectral characteristics are shown in Figure 8.14. The full width at half maximum of the spectrum was 0.4 nm, and the shape of the beam was elliptical most likely due to multi-transversal mode operation. In pulsed operation, the mount temperature was raised and kept near 20 °C. The maximum measured peak output

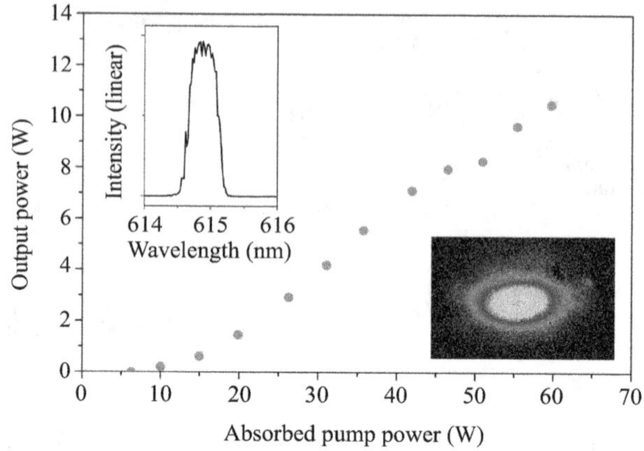

Figure 8.14 Output characteristics in CW operation. Insets: Output spectrum and beam profile measured at 7 W of output power. Adapted and reproduced with permission from [63]

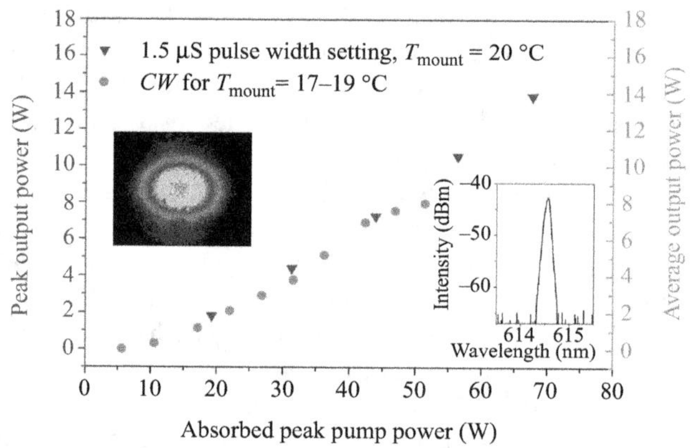

Figure 8.15 Output power characteristics for pulsed operation at 20 °C. Insets: Output spectrum and beam profile measured at the maximum pulsed pump power. Adapted and reproduced with permission from [63]

power was 13.8 W for a pulse width of 1.16 μs (see Figure 8.15) and it was pump power limited (the pump laser was only capable of producing a maximum of 70.6 W of peak power). The maximum optical-to-optical conversion efficiency was measured to be 20.4%. The higher conversion efficiency can be attributed to the reduced thermal load on the chip due to the pulsing scheme.

The same SDL structure was also used to demonstrate 6.6 W of output power from a compact, application-ready prototype [117]. The prototype emitted at a slightly longer wavelength of 625 nm and is shown in Figure 8.16. The temperature of the gain chip was controlled with a TEC-cooled copper heatsink and a diamond heat spreader. Similar cavity configuration and components were used to generate the red emission as in the laboratory setup. The optical-to-optical efficiency was approximately 10%.

Figure 8.16 Photo of the 6.5 W red SDL prototype. The laser cavity was built inside a compact air-cooled aluminum casing (220 mm × 160 mm × 77 mm)

When compared to the frequency doubled yellow–orange SDLs, the lower output power at this red spectral range can be attributed to the less mature dilute nitride technology and inherently higher amount of defects associated to higher lattice strain and/or higher nitride content. However, it is expected that with additional optimization of the semiconductor structure, achieving output powers in excess of 10 W is possible.

8.3.3 Directly emitting red SDLs at 630–750 nm

In addition to frequency doubling, red wavelengths above 600 nm can also be reached with directly emitting AlGaInP/GaAs heterostructures [118]. However, with this material system the shortest achievable wavelength is limited by the lack of sufficiently high QW barriers which would suppress carrier leakage from the QWs. The red edge-emitting lasers, quite well established today, are limited to slightly longer wavelengths than 600 nm, which implies that this is the short-wavelength limit attainable with SDLs too. The long wavelength limit for the AlGaInP material system is set by the increasing strain in the QWs at wavelengths longer than 700 nm; instead, the wavelength range of 700–850 nm could be reached directly with AlGaAs material system.

To compromise between the two categories of constrains, the first high-power red SDLs emitted at around 670 nm [119]. The design of the gain structure used for this demonstration is shown in Figure 8.17. It incorporated 10 pairs of QWs with In composition of 54%. Besides the poor carrier confinement, typical for emission at red wavelengths, the fabrication of such a structure has to tackle one more problem, which is the lack of transparent DBR materials with a high refractive index contrast at the lasing wavelength. In fact, the poor carrier confinement and low refractive index contrast are manifestations of the same physical phenomena, the bandgap structure. A possible alternative for the DBR materials would be to use AlGaInP layers, but in this case the refractive index contrast would be only about 0.12 resulting in insufficient reflection and a narrow stop-band [120]. Instead, AlAs/AlGaAs layers were utilized, ensuring a reasonable refractive index contrast of

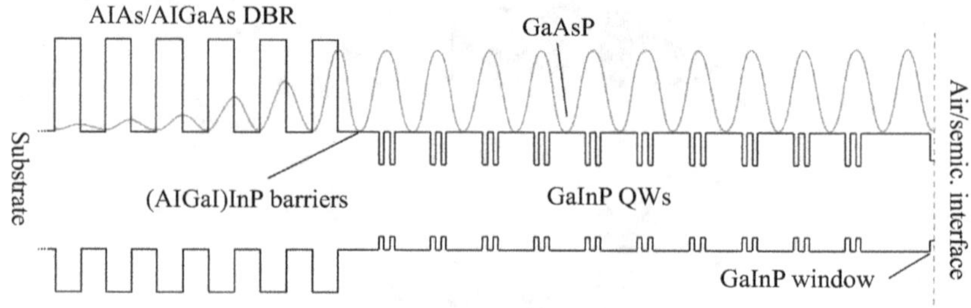

Figure 8.17 Layer structure of an SDL emitting in the spectral region of 670 nm

about 0.49 [121]. Yet, the number of mirror periods had to be relatively high (40) in order to reach high reflectance.

We should note one more challenge, this time related to the fabrication and linked to the As/P interchange. This interface is located near the active region. Any disruption in crystal quality at this sensitive interface causes dramatic reduction in device performance. The difficulty arises from the high volatility of the group-V species. Volatility makes any rapid evacuation of P or As from the growth chamber difficult, causing small but significant amounts of residual As or P to incorporate into the growing lattice even if the As or P source is switched off. To alleviate this problem, the As/P interchange was performed during a short growth interruption, which allowed a little more time for evacuation of residual background gases. During the growth interruption the group-III beams had to be switched-off too, which stopped the growth as the group-V species alone cannot incorporate into the lattice. In general, controlling the growth is further complicated by the fact that too long a growth interruption leads to piling up of impurity atoms at the interface. This is especially detrimental in the proximity of the QWs. A proper choice of the duration of the interruption depends on the reactor cleanliness.

The structure was characterized by bonding a chip to a diamond heat spreader attached to a water-cooled copper mount. A V-shaped cavity was built around it, shown in Figure 8.18. Green 532-nm laser was used to pump the structure instead of the 808-nm diode laser, because of the shorter lasing wavelength (670 nm) compared to the frequency doubled infrared lasers.

The output wavelength peaked at 674 nm and the maximum output power was ~400 mW in CW operation. The output power characteristics and spectrum at high power are shown in Figure 8.19. For a more practical pumping alternative, one could use barrier pumping with blue InGaN-based laser diodes [122] or in-well pumping with 635–660 nm AlGaInP-based diodes. The blue InGaN-based laser diodes allow the pump light to be absorbed by the AlGaInP barriers, but as compared to the 532-nm pump laser, will reduce the absorption length and increase the quantum defect due to the shorter pump wavelength. On the other hand, powerful short-wavelength AlGaInP laser diodes would be required for in-well pumping, because the pump wavelength has to be 20–40 nm shorter than the emission wavelength of the SDL.

Mode-locking of a red direct-emitting SDL using an identical gain chip was also demonstrated in [46]. To initiate mode-locking, a SESAM with a structure similar to the gain chip but having only one QW was utilized. The cavity configuration and the output

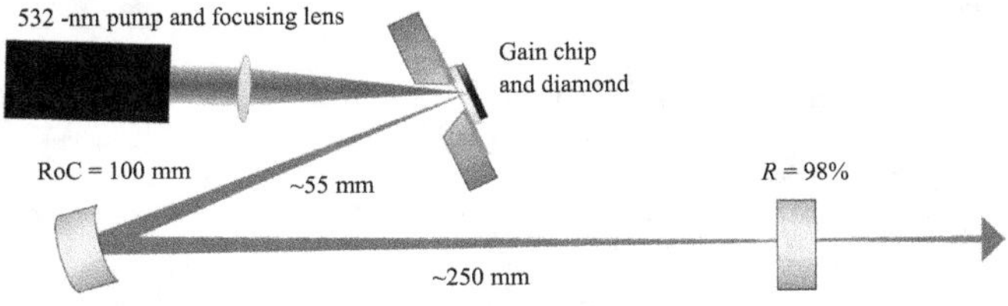

Figure 8.18 Cavity illustration of the 670-nm SDL. Adapted and reproduced with permission from [119]

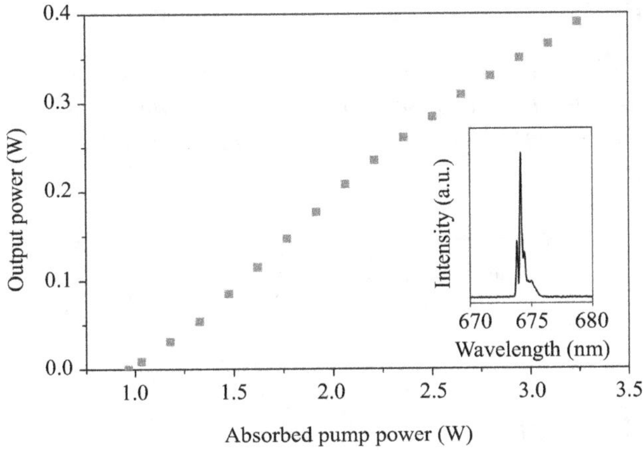

Figure 8.19 Output power curve of the 670-nm red SDL. Inset: Corresponding output spectrum. Adapted and reproduced with permission from [119]

power characteristics in mode-locked and CW operation (in this case the SESAM was replaced with a highly reflective mirror) are shown in Figure 8.20. The slope efficiency was rather modest owing to the poor carrier pumping and high carrier escape in/from the active region. The optical spectrum corresponding to the mode-locked operation is shown in Figure 8.21 (left). The spectrum was centered at 675.8 nm and the spectral width was 0.14 nm (FWHM). The pulse shape was characterized with an SHG-based intensity autocorrelator. The autocorrelation trace is shown in Figure 8.21 (right) and reveals a width of $\sim$10 ps with extended wings that are incompatible with symmetric Gaussian or hyperbolic secant pulses. The PICASO method [123] was employed to obtain an estimate for the underlying pulse structure and duration. The retrieved FWHM width was 5.1 ps, which is 3.75 times above the transform limit.

Although the average power was quite modest and the repetition rate was close to 1GHz, the peak power was estimated to be 15 W, which could be sufficient for fluorescent lifetime microscopy [124] and imaging techniques exploiting nonlinear optical absorption [125].

When it comes to extending the wavelengths beyond 750-nm range, we should note recent demonstration of emission in the 740–90 nm wavelength region [67].

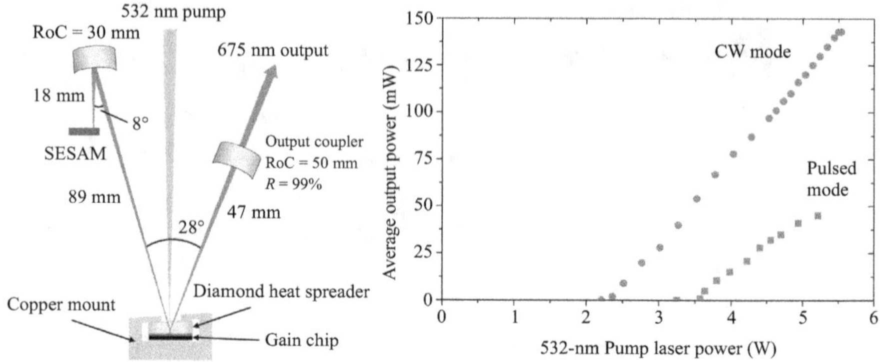

Figure 8.20 Left: Cavity illustration of the red mode-locked SDL. Right: Output power versus incident pump power (532 nm) for CW and mode-locked operation. Reproduced with permission from [46]

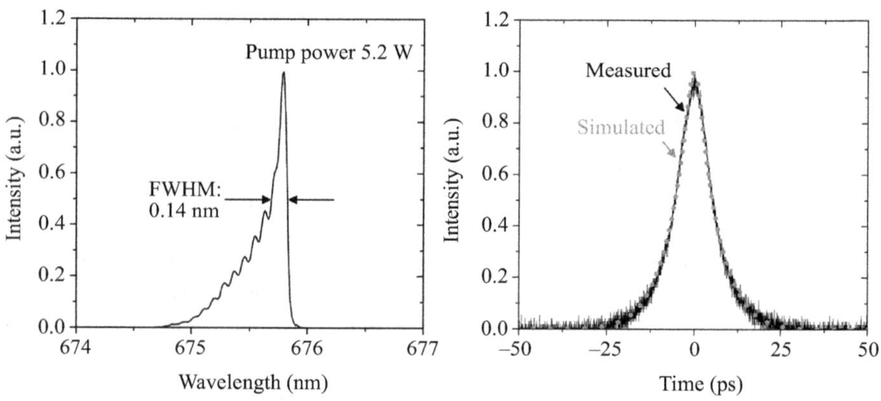

Figure 8.21 Left: Output spectrum of the laser. Right: Measured intensity autocorrelation trace (solid line) and the simulated autocorrelation of the optical pulses (dotted line). Reproduced with permission from [46]

In this case, the gain structure was based on 12 AlGaAs quantum wells enabling tuning between 747 nm to 788 nm and free running operation with a maximum power of 4.24 W (pump power limited) for a heat sink temperature of 14 °C. This laser system fills a gap that so far has been eluded for direct emitting SDLs, yet one which will potentially be of interest for medical applications.

8.3.4 Frequency-doubled red SDLs at 630–800 nm

In addition to the directly emitting AlGaInP QWs and AlGaAs QWs, the 650–800 nm wavelength range can also be reached via frequency doubling SDLs emitting at 1,300–1,600 nm. SDLs in this wavelength range are preferably not fabricated using GaAs-based materials, such as InAs QDs [114] or dilute nitride GaInNAs QWs

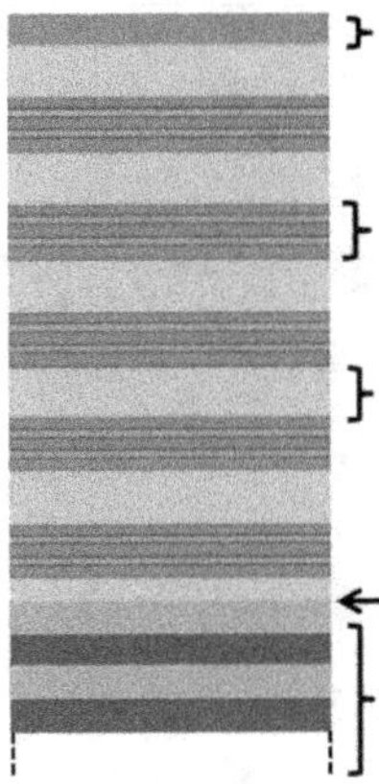

Figure 8.22 A generic schematic of a wafer-fused VECSEL structure. In this case, there are 10 QWs that are placed at the antinodes of the optical field with a distribution of 2-2-2-2-2

[126,127], because the gain of these materials decreases with increasing wavelength and therefore hinders their use in practical applications. Higher gain is provided by InP-based QWs, but this material system suffers from poor performance DBRs; the performance of these DBRs is hindered by the lack of materials that are lattice-matched to InP with high refractive index contrast. This shortcoming impairs the reflective properties by limiting the reflectance bandwidth, but also leads to thick mirror structures since a relatively high number of layer pairs is needed to reach high reflectivity. These thick mirror structures entail increased absorption losses [128] and reduced thermal conductance, especially since the materials lattice-matched to InP possess low thermal conductivity.

The issues with InP-based DBRs can be circumvented via wafer bonding, also called wafer fusion, that allows the integration of high-performance GaAs-based DBRs onto InP-based active regions as described in [129]. A generic schematic of a wafer-fused SDL with an InP-based active region and a GaAs-based DBR is shown in Figure 8.22. The illustrated structure contains 10 QWs that are placed at the antinodes of the optical field with a distribution of 2-2-2-2-2 [130,131], but other configurations have also been demonstrated with QW distributions of 3-3-2-2 [132,133] and 4-2-2 [81]. The DBR has usually comprised of 35 pairs of AlGaAs/GaAs layers.

In most cases, the thermal management of these structures has been performed using 300 μm thick intracavity diamond heat spreaders [130,131,81], though flip-chip structures have also been demonstrated with high output powers [132,133]. In contrast to GaAs-based SDLs that are pumped at 808 nm, these InP-based SDLs are usually pumped using fiber-coupled diode lasers emitting at 980 nm. The InP-based active regions typically contain compressively strained AlGaInAs QWs and GaAs-based DBRs with AlGaAs/GaAs layer pairs. Such active regions and the DBR sections have been fabricated using both low pressure metalorganic vapor phase epitaxy (LP MOVPE) and solid-source MBE. The integration of these structures via wafer fusion technology has been the pivotal step in advancing the output

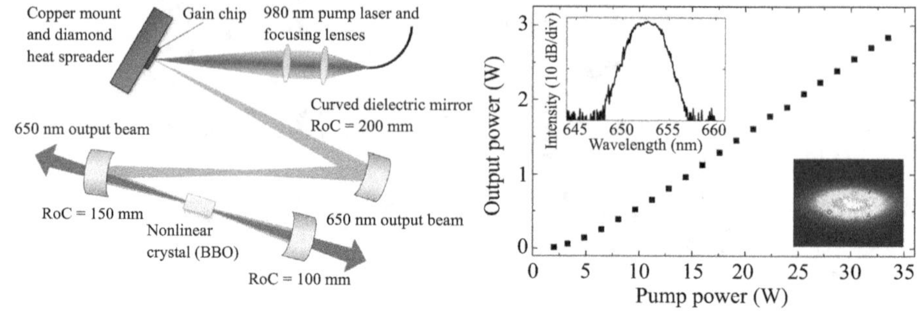

Figure 8.23 Left: A cavity schematic of the 650-nm frequency-doubled SDL. Right: The output power characteristics of the 650-nm SDL. The optical spectrum and the output beam profile at the highest output power are shown as insets. Adapted and reproduced with permission from [64]

power of 1.3–1.6 μm SDLs into the multiwatt levels with output powers reaching 33 W at 1.3 μm [133], 5 W at 1.5 μm [81], and 4.6 W at 1.6 μm [69].

These developments in InP-based SDLs have also led to the demonstration of watt-level emission in the wavelength range of 650–800 nm via intracavity frequency-doubling. However, the demonstrated output powers via frequency doubling at 650–800 nm are still below the levels obtained at the fundamental wavelengths, but further work in this area is expected to yield power levels exceeding 5 W throughout the red wavelength range. The following will cover the results that have been obtained from intracavity frequency doubling of these SDL structures. The frequency doubling has been performed using Z-cavities in each case, though in slightly different configurations and optical components. Nevertheless, the cavity mirrors were always highly reflective for the fundamental wavelengths, and in most cases low reflective (>5%) for the second harmonic wavelengths.

8.3.4.1 Wafer-fused SDLs emitting at 650 nm

The cavity schematic for a frequency-doubled red wafer-fused SDL is shown in Figure 8.23 (left) [64]. The gain element was pumped using a pump spot diameter of about 300 μm. It had a photoluminescence peak at 1,263 nm at room temperature and was targeted to operate at 1,310 nm under optical pumping. The frequency-doubled output was obtained using a 4-mm long Type I critically phase matched BBO crystal that was antireflection coated for the fundamental and second harmonic wavelength. The combined output power of the two output beams is shown in Figure 8.23 (right), with the optical spectrum and the transverse beam profile at the highest output power shown as insets. The maximum output power reaches 3 W at a pump power of about 35 W. The output beam is slightly asymmetric due to spatial walk-off in the BBO crystal.

8.3.4.2 Wafer-fused SDLs emitting at 750 nm

A frequency-doubled SDL emitting at a wavelength of 750 nm was fabricated using a wafer-fused structure with fundamental emission at 1,500 nm [68].

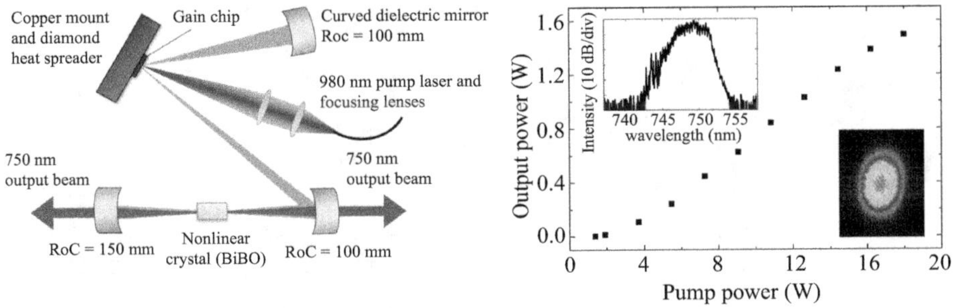

Figure 8.24 Left: A cavity schematic of the 750-nm frequency-doubled SDL. Right: The output power characteristics of the 750-nm DSL. The optical spectrum and the output beam profile at 1.5 W of output power are shown as insets. Adapted and reproduced with permission from [68]

In this case, the gain element's PL peak was located at 1,445 nm at room temperature and it was pumped using a 200 μm pump spot diameter. The frequency-doubled output was obtained using a 5-mm long Type I critically phase matched BiBO crystal, as shown in the schematic in Figure 8.24. The maximum output power of the combined output beams reached 1.5 W at a pump power of about 18 W. As before, the inset of Figure 8.24 also shows the output spectrum and the transverse beam profile at the highest output power. The output beam is still slightly asymmetric, but not as much as with the 650 nm SDL due to the smaller walk-off in BiBO than in BBO.

8.3.4.3 Wafer-fused SDLs emitting at 785 nm

Finally, as with the two previous wafer-fused SDLs, the highest output power at 785 nm has been obtained using a frequency-doubled SDL with a fundamental emission at 1,580 nm [69]. Here, the PL peak of the gain element was located at 1,520 nm at room temperature and it was pumped with a 300 μm pump spot diameter. The second harmonic output was obtained using a 6-mm long Type I critically phase matched LBO crystal. However, in this case, the end mirror of the laser cavity was highly reflective for both the fundamental and the second harmonic wavelength, so that the single output beam was obtained through the folding mirror with 150 mm curvature. The output power characteristics are shown in Figure 8.25 (right) with the maximum output power of 1 W at 785 nm. The inset of Figure 8.25 (right) also shows the transverse profile of the output beam, which is symmetric due to the absence of spatial walk-off in the LBO crystal.

8.4 Biophotonics applications of SDLs

So far, SDLs have not been much deployed in practical biophotonics applications. This is not surprising, as most SDLs listed in Table 8.1 are only laboratory demonstrations, and with only a few exceptions (mainly SDLs at blue, green, and 577 nm

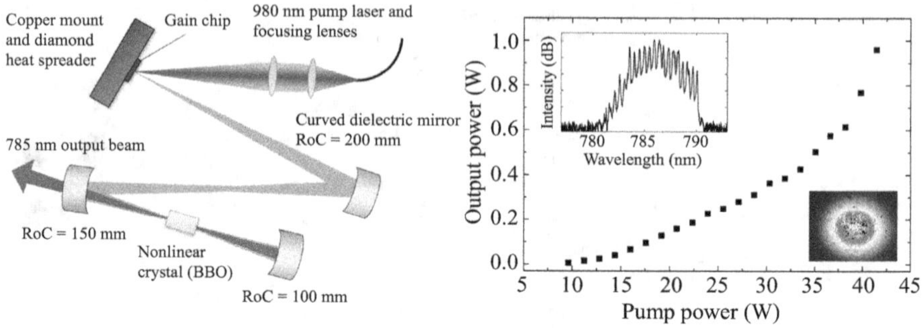

Figure 8.25 Left: A cavity schematic of the frequency-doubled 785-nm SDL. Right: The output characteristics of the 785-nm SDL. The optical spectrum and the output beam profile at 1.5 W of output power are shown as insets. Adapted and reproduced with permission from [69]

wavelengths), these lasers are not yet available as commercial, application-ready sources. Moreover, SDLs are relatively new type of lasers and their properties have not yet been recognized by the application sector, though this status is expected to rapidly change. For example, the application in ophthalmology, exploiting the high absorption peaks of hemoglobin in the yellow–orange spectral range, has already been recognized by Coherent Inc., which markets an OP-SDL (or OPSL) with emission at 577 nm [97]. There are also other applications that could benefit from SDL technology. In order to successfully develop practical SDLs for these applications, it is important to understand the requirements the applications set on the laser light source. In Table 8.2 we make a modest attempt to review several biophotonics applications of lasers and discuss their requirements from the perspective of the current achievements of SDL technology. In Figure 8.26 we attempt to illustrate the wavelength and power requirements for the light source of each biophotonics application.

In conclusion, the main features that SDL technology has to offer for biophotonics are: multi-watt output power with possibility to scale up the power to 100 W level; high beam quality, capability to fill spectral gaps that cannot be reached by traditional solid state disk lasers (current wavelength coverage is from 240 nm to 5,000 nm); broad wavelength tuning (easily more than 10 nm but could be as high as 150 nm) with controlled spectral features (i.e. narrow linewidth); a small footprint; and all coming for a relatively low cost compared to what it delivers. With reference to diagram in Figure 8.26, it seems that SDLs have unique positioning for ophthalmic and dermatology applications, while other applications can be served by lower laser diode technologies or mature pulse laser technologies, such as supercontinuum sources. This chapter provided an overview of the key technology concepts related to SDLs, as well as of the most important features that could be relevant for wider use in biophotonics. Owing to their unique features and the intense development efforts in the last decade, SDLs have reached a

Table 8.2 A prospective table of biophotonics applications where SDLs could be used

Application	Short description and laser specifications	What SDLs can offer
Eye surgery and retinal laser therapy	Laser light is used to coagulate blood vessels, for example, to seal tears, which can lower the risk of vision loss in various eye diseases. The laser wavelength is determined by hemoglobin absorption (577 nm). Ideal pulse duration is close to a microsecond, but also few tens of milliseconds have been used. The risk of ruptures increases as pulse length decreases. Peak powers of ~2 W have been utilized [134].	SDLs have a compact footprint and are more affordable than, for example, pulsed dye lasers. Emission at 577 nm with a good beam quality and several watts of output power can be achieved via intracavity SHG. Microsecond pulses have been obtained by directly modulating the pump laser [62]. It should be noted that yellow diode lasers producing microsecond pulses are also available and have proved to be effective in retinal therapy [135].
Photodynamic therapy (PDT)	In addition to laser light, a photosensitizer is used to cause a photochemical process that creates free radicals to destroy targeted cells [136]. Routinely used in PWS treatment or cancer therapy. The ideal laser wavelength is dependent on the photosensitizer's absorption spectrum. Hematoporphyrin derivatives and temoporfin are common photosensitizers used in the treatment of PWSs [137] and cancer therapy [138]. Typically photosensitizers have absorption peaks within 630–740 nm range. Both CW and pulsed operation are utilized.	SDLs can produce red light via SHG, and recently a pulsed direct emitting red SDL was reported [46]. The wavelength versatility and broad tuning range of SDLs could provide important benefits for dealing with a wider range of photosensitizers while employing only one laser system. On the other hand, the needs in terms of power level and brightness are met by simpler laser diode technology, whenever available.
Selective Photothermolysis (PT)	Laser light is directed to a certain target, for example, a blood vessel. The target absorbs the light producing enough heat to cause controlled damage. Often used to treat cutaneous vascular diseases such as PWSs [2], and PT is also the method behind tattoo removal. Hemoglobin in blood has an absorption peak at 577 nm. However, longer wavelengths are often used for deeper penetration (585–595 nm). Pulse durations in the order of few milliseconds are used to avoid unwanted damage [2,139].	The preferred 577-nm emission is already available as a commercial SDL product (Genesis Taipan 577 commercialized by Coherent delivers multi-watt output power [140]). Longer wavelengths (e.g., 585–595 nm) are also available, at least for research demonstrators. SDL's compact footprint enables an easy integration into medical equipment. For instance, medical laser company Asclepion has designed a table-top yellow laser for dermatology based on SDL technology [141].

(Continues)

Table 8.2 (Continued)

Application	Short description and laser specifications	What SDLs can offer
Photoacoustics (PA)	Photoacoustics is an imaging method in which laser light is absorbed by tissue causing it to heat up. This temperature change induces sound waves which can be detected to form an image [142]. Pulsed NIR lasers (1,064 nm) are often used for deeper penetration, but visible wavelengths can be better for more targeted absorption. Pulse duration is in the nanosecond range, repetition rate in tens of Hz and typical pulse energy is a few tens of millijoules [143].	The development of SDL technology continuously aims for higher power. However, we should note that unlike the solid-state system the fast carrier dynamics in SDL mirrors does not allow for energy accumulation between pulses. Cavity damping schemes have led to generation of ns pulses with peak powers of 57 W [144].
Stimulated Emission Depletion (STED) Microscopy	Aims at microscopy imaging on a smaller scale than the diffraction limit. Two laser beams are overlapped; the first laser beam excites fluorophores in a sample and the second depletes fluorescence around the excited spot through stimulated emission. The absorption spectrum of the fluorophores determines the laser wavelength. Common green and yellow fluorescent proteins require excitation with blue laser and depletion with yellow/orange laser (~560–600 nm). Both lasers can be pulsed (with pulse duration in the picosecond range) or the depleting laser can be CW for simplicity and lower cost. A few hundred milliwatts of CW power is needed on the sample which translates to a few watts of laser power [96].	SDLs are power scalable and hence they can operate at a chosen power level from tens of milliwatts to several watts. They can emit continuous yellow–orange radiation and be wavelength tuned to better match a fluorophore's absorption peak. Therefore, they would be suitable to operate as the depletion laser in STED microscopy. Other advantages are low cost for a high-power source, small footprint and reliability. Preliminary experiments with an SDL as the depleting laser have been reported [145].

Flow Cytometry	Cells (or other particles) are passed through laser light one at a time and the resulting forward scatter, side scatter and fluorescence light are collected by detectors and analyzed to determine multiple characteristics of cells [146,147]. Tens of milliwatts of CW output power is sufficient and several wavelengths are required, for example, blue, green, yellow, and red.	SDLs can reach the visible spectrum via SHG and a power scale of tens of milliwatts is easily achievable. A possible advantageous feature is the relative broad tuning range (i.e. 10-20 nm) for the SHG operation. In addition, for direct emitting SDLs at red wavelengths, tuning of more than 10 nm can be achieved [148].
NIR Laser Surgery	Laser light is absorbed to an extent that it vaporizes tissue leading to tissue ablation. The advantage over a conventional scalpel is that, since it is a thermal effect, the laser cut tissue is burnt and hence less bleeding occurs. Can be used to, for example, remove cancerous [149] tissue or resurface skin. Water absorbs light increasingly from about 800 nm until 11 μm and there is a high peak around 2 μm [150]. Hence, NIR lasers are optimal for tissue ablation. High-power, usually pulsed, lasers are needed.	SDLs emitting at 2 μm are mature. Power-wise SDLs cannot yet match the currently used lasers in NIR laser surgery (such as Er:YAG laser), but average power in the range of 10 W could already be used in nano-surgery. Main advantages over solid-state systems are compactness and overall simple solution.
Multiphoton spectroscopy	Based on nonlinear excitation in which light effect is confined to a volume smaller than in the case of one-photon fluorescence. This provides benefits in terms of photo-toxicity and allows optical sectioning of the sample with 3D reconstruction [151].	High repletion rate (1 GHz range) providing better factor of merit with lower average power, and their wavelength can be tailored to match a wide range of targets [152,153]. In addition, they exhibit a high amplitude stability. Potentially lower cost systems compared to established ultrafast solid-state lasers.

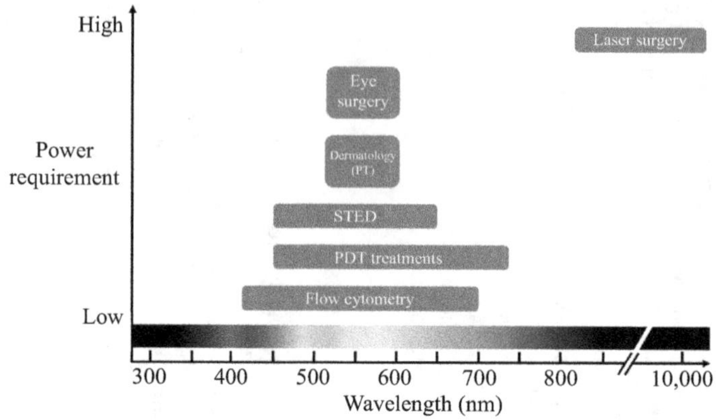

Figure 8.26 Illustration of the laser wavelength and output power requirements for biophotonics applications

development stage that makes them very attractive for application deployments. We have a firm belief that in the years to come, SDLs will render possible important development opportunities for many life science applications.

Acknowledgments

This work was financially supported by the Finnish Funding Agency for Technology and Innovation TEKES (BRIGHTLASE project #40048, and Photolase project #40152/14), FP7 project APACOS (GA 315711), and Academy of Finland (project grants #251018 and #128844). The authors would like to thank numerous colleagues and external collaborators for their continuous contribution to the development of SDLs reviewed here. In particular, we thank Jussi-Pekka Penttinen, Antti Härkönen, Sanna Ranta, and former colleagues Miki Tavast and Ville-Markus Korpijärvi. Professor Serge Mordon (INSERM) is acknowledged for insights on using yellow lasers in dermatology and assistance in building the prototype shown in Figure 8.12.

References

[1] M. Mainster, "Continuous-wave and Micropulse 577 nm Yellow–orange Laser Photocoagulation: A Laser for All Reasons," *Retina Today*, 2010.

[2] K. Kelly, B. Choi, S. McFarlane, *et al.*, "Description and analysis of treatments for port-wine stain birthmarks," *Archives of Facial Plastic Surgery*, vol. 7, no. 5, pp. 287–294, 2005.

[3] L. Song, K. Maslov and L. Wang, "Multi-focal optical-resolution photoacoustic microscopy in vivo," *Optics Letters*, vol. 36, no. 7, pp. 1236–1238, 2011.

[4] O. Tan, S. Murray and A. Kurban, "Action spectrum of vascular specific injury using pulsed irradiation," *Journal of Investigative Dermatology*, vol. 92, pp. 868–871, 1989.

[5] J. Janousek, S. Johansson, P. Tidemand-Lichtenberg, *et al.*, "Efficient all solid-state continuous-wave yellow-orange light source," *Optics Express*, vol. 13, no. 4, pp. 1188–1192, 2005.

[6] Y. Feng, L. Taylor and D. Calia, "25 W Raman-fiber-amplifier-based 589 nm laser for laser guide star," *Optics Express*, vol. 17, no. 21, pp. 19021–19026, 2009.

[7] M. Kuznetsov, F. Hakimi, R. Sprague and A. Mooradian, "High-power (>0.5-W CW) diode-pumped vertical-external-cavity surface-emitting semiconductor lasers with circular TEM00 beams," *IEEE Photonics Technology Letters*, vol. 9, no. 8, pp. 1063–1065, 1997.

[8] M. Kuznetsov, F. Hakimi, R. Sprague and A. Mooradian, "Design and characteristics of high-power (>0.5-W CW) diode-pumped vertical-external-cavity surface-emitting semiconductor lasers with circular TEM00 beams," *IEEE Journal of Selected Topics in Quantum Electronics*, vol. 5, no. 3, pp. 561–573, 1999.

[9] U. Keller and A. Tropper, "Passively modelocked surface-emitting semiconductor lasers," *Physics Reports*, vol. 429, no. 2, p. 67–120, 2006.

[10] M. Guina, A. Rantamäki and A. Härkönen, "Optically pumped VECSELs: review of technology and progress," *Journal of Physics D: Applied Physics*, vol. 50, no. 38, p. 383001, 2017.

[11] N. Basov, O. Bogdankevich and A. Grasyuk, "9B4–Semiconductor lasers with radiating mirrors," *Quantum Electronics*, vol. 2, no. 9, pp. 594–597, 1966.

[12] W. Jiang, S. Friberg, H. Iwamura and Y. Yamamoto, "High powers and subpicosecond pulses from an external-cavity surface-emitting InGaAs/InP multiple quantum well laser," *Applied Physics Letters*, vol. 58, no. 8, pp. 807–809, 1991.

[13] H. Le, S. Cecca and A. Mooradian, "Scalable high-power optically pumped GaAs laser," *Applied Physics Letters*, vol. 58, no. 18, pp. 1967–1969, 1991.

[14] C. Symonds, J. Dion, I. Sagnes, M. S. M. Dainese, L. Leroy and J. Oudar, "High performance 1.55 μm vertical external cavity surface emitting laser with broadband integrated dielectric-metal mirror," *Electronics Letters*, vol. 40, no. 12, pp. 734–735, 2004.

[15] J. Orchard, D. T. Childs, L. Lin, B. Stevens, D. Williams and R. Hogg, "Tradeoffs in the realization of electrically pumped vertical external cavity surface emitting lasers," *IEEE Journal of Selected Topics in Quantum Electronics*, vol. 17, no. 6, pp. 1745–1752, 2011.

[16] A. Tropper and S. Hoogland, "Extended cavity surface-emitting semiconductor lasers," *Progress in Quantum Electronics*, vol. 30, no. 1, pp. 1–43, 2006.

[17] A. Kemp, G. Valentine, J.-M. Hopkins, *et al.*, "Thermal management in vertical-external-cavity surface-emitting lasers: Finite-element analysis of a

heatspreader approach," *IEEE Journal of Quantum Electronics*, vol. 41, no. 2, pp. 148–155, 2005.

[18] W. Alford, T. Raymond and A. Allerman, "High power and good beam quality at 980 nm from a vertical external-cavity surface-emitting laser," *Journal of the Optical Society of America B*, vol. 19, no. 4, pp. 663–666, 2002.

[19] H. Lindberg, M. Strassner, E. Gerster and A. Larsson, "0.8 W optically pumped vertical external cavity surface emitting laser operating CW at 1550 nm," *Electronics Letters*, vol. 40, no. 10, pp. 601–602, 2004.

[20] B. Rösener, M. Rattunde, R. Moser, S. Kaspar, T. Töpper, C. Manz, K. Köhler and J. Wagner, "Continuous-wave room-temperature operation of a 2.8 μm GaSb-based semiconductor disk laser," *Optics Letters*, vol. 36, no. 3, pp. 319–321, 2011.

[21] Z. Liau, "Semiconductor wafer bonding via liquid capillarity," *Applied Physics Letters*, vol. 77, no. 5, pp. 651–653, 2000.

[22] J. Perez, A. Laurain, L. Cerutti, I. Sagnes and A. Garnache, "Technologies for thermal management of mid-IR Sb-based surface emitting lasers," *Semiconductor Science and Technology*, vol. 25, no. 4, p. 045021, 2010.

[23] S. Butterworth, M. Reed and D. Fischer, "Contact-bonded optically pumped semiconductor laser structure". Coherent Inc. Patent WIPO Patent Application WO/2005/093915, 6 October 2005.

[24] A. Clawson, "Guide to references on III-V semiconductor chemical etching.," *Materials Science and Engineering: R: Reports*, vol. 31, no. 1, pp. 1–438, 2001.

[25] A. Maclean, A. Kemp, S. Calvez, *et al.*, "Continuous tuning and efficient intracavity second-harmonic generation in a semiconductor disk laser with an intracavity diamond heatspreader," *IEEE Journal of Quantum Electronics*, vol. 44, no. 3, pp. 216–225, 2008.

[26] J. Chilla, Q.-Z. Shu, H. Zhou, E. Weiss, M. Reed and L. Spinelli, "Recent advances in optically pumped semiconductor lasers," in *Proc. of SPIE 6451*, San Jose, 2007.

[27] T.-L. Wang, Y. Kaneda, J. Yarborough, *et al.*, "High-power optically pumped semiconductor laser at 1040 nm," *IEEE Photonics Technology Letters*, vol. 22, no. 9, pp. 661–663, 2010.

[28] B. Heinen, T.-L. Wang, M. Sparenberg, *et al.*, "106 W continuous-wave output power from vertical-external-cavity surface-emitting laser," *Electronics Letters*, vol. 48, no. 9, pp. 516–517, 2012.

[29] V. Iakovlev, J. Walczak, M. Gębski, *et al.*, "Double-diamond high-contrast-gratings vertical external cavity surface emitting laser," *Journal of Physics D: Applied Physics*, vol. 47, no. 6, p. 065104, 2014.

[30] H. Kahle, C. Mateo, U. Brauch, *et al.*, "Semiconductor membrane external-cavity surface-emitting laser (MECSEL)," *Optica*, vol. 3, no. 12, pp. 1506–1512, 2016.

[31] M. Schmid, S. Benchabane, F. Torabi-Goudarzi, R. Abram, A. Ferguson and E. Riis, "Optical in-well pumping of a vertical-external-cavity surface-emitting laser," *Applied Physics Letters*, vol. 84, no. 24, p. 4860, 2004.

[32] S. Beyertt, U. Brauch, F. Demaria, *et al.*, "Efficient gallium-arsenide disk laser," *IEEE Journal of Quantum Electronics*, vol. 43, no. 10, pp. 869–875, 2007.

[33] L. Fan, M. Fallahi, J. Hader, *et al.*, "Multichip vertical-external-cavity surface-emitting lasers: a coherent power scaling scheme," *Optics Letters*, vol. 31, no. 24, pp. 3612–3614, 2006.

[34] T. Ochalski, A. Burca, G. Huyet, *et al.*, "Passively modelocked bi-directional vertical external ring cavity surface emitting laser," *Conference on Lasers and Electro-Optics (LEOS)*, vol. CWL6, 2008.

[35] J. Hastie, J.-M. Hopkins, C. Jeon, *et al.*, "Microchip vertical external cavity surface emitting lasers," *Electronics Letters*, vol. 39, no. 18, pp. 1324–1326, 2003.

[36] S. Smith, J.-M. Hopkins, J. Hastie, *et al.*, "Diamond-microchip GaInNAs vertical external-cavity surface-emitting laser operating CW at 1315 nm," *Electronics Letters*, vol. 40, no. 15, pp. 935–936, 2004.

[37] D. Maas, A.-R. Bellancourt, B. Rudin, *et al.*, "Vertical integration of ultrafast semiconductor lasers," *Applied Physics B*, vol. 88, no. 4, p. 493–497, 2007.

[38] B. Rudin, V. Wittwer, D. Maas, *et al.*, "High-power MIXSEL: an integrated ultrafast semiconductor laser with 6.4 W average power," *Optics Express*, vol. 18, no. 26, pp. 27582–27588, 2010.

[39] L. Fan, M. Fallahi, J. Murray, *et al.*, "Tunable high-power high-brightness linearly polarized vertical-external-cavity surface-emitting lasers," *Applied Physics Letters*, vol. 88, no. 2, p. 021105, 2006.

[40] C. Borgentun, J. Bengtsson, A. Larsson, F. Demaria, A. Hein and P. Unger, "Optimization of a broadband gain element for a widely tunable high-power semiconductor disk laser," *IEEE Photonics Technology Letters*, vol. 22, no. 13, pp. 978–980, 2010.

[41] J. Paajaste, S. Suomalainen, R. Koskinen, A. Härkönen, M. Guina and M. Pessa, "High-power and broadly tunable GaSb-based optically pumped VECSELs emitting near 2 μm," *Journal of Crystal Growth*, vol. 311, no. 7, pp. 1917–1919, 2011.

[42] M. Holm, D. Burns, A. Ferguson and M. Dawson, "Actively stabilized single-frequency vertical-external-cavity AlGaAs laser," *IEEE Photonics Technology Letters*, vol. 11, no. 12, pp. 1551–1553, 1999.

[43] A. Laurain, M. Myara, G. Beaudoin, I. Sagnes and A. Garnache, "Multiwatt–power highly– coherent compact single–frequency tunable vertical–external–cavity–surface–emitting– semiconductor–laser," *Optics Express*, vol. 18, no. 14, pp. 14627–14636, 2010.

[44] W. Alford, G. Fetzer, R. Epstein, *et al.*, "Optically pumped semiconductor lasers for precision spectroscopic applications," *IEEE Journal of Quantum Electronics*, vol. 49, no. 8, p. 719–727, 2013.

[45] P. Klopp, U. Griebner, M. Zorn and M. Weyers, "Pulse repetition rate up to 92 GHz or pulse duration shorter than 110 fs from a mode-locked semiconductor disk laser," *Applied Physics Letters*, vol. 98, no. 7, p. 071103, 2011.

[46] S. Ranta, A. Härkönen, T. Leinonen, *et al.*, "Mode-locked VECSEL emitting 5 ps pulses at 675 nm," *Optics Letters*, vol. 38, no. 13, pp. 2289–2291, 2013.

[47] A. Härkönen, J. Paajaste, S. Suomalainen, *et al.*, "Picosecond passively mode-locked GaSb-based semiconductor disk laser operating at 2 μm," *Optics Letters*, vol. 35, no. 24, pp. 4090–4092, 2010.

[48] A. Härkönen, C. Grebing, J. Paajaste, *et al.*, "Modelocked GaSb disk laser producing 384 fs pulses at 2 μm wavelength," *Electronics Letters*, vol. 47, no. 7, pp. 454–456, 2011.

[49] J. Hastie, L. Morton, A. Kemp, M. Dawson, A. Krysa and J. Roberts, "Tunable ultraviolet output from an intracavity frequency-doubled red vertical-external-cavity surface-emitting laser," *Applied Physics Letters*, vol. 89, no. 6, p. 061114, 2006.

[50] M. Rahim, F. Felder, M. Fill and H. Zogg, "Optically pumped 5 μm IV-VI VECSEL with Al-heat spreader," *Optics Letters*, vol. 33, no. 24, pp. 3010–3012, 2008.

[51] Y. Kaneda, J. Yarborough, Y. Merzlyak, *et al.*, "Continuous-wave, single-frequency 229 nm laser source for laser cooling of cadmium atoms," *Optics Letters*, vol. 41, no. 4, pp. 705–708, 2016.

[52] Y. Kaneda, J. Yarborough, L. Li, *et al.*, "Continuous-wave all-solid-state 244 nm deep-ultraviolet laser source by fourth-harmonic generation of an optically pumped semiconductor laser using CsLiB6O0 in an external resonator," *Optics Letters*, vol. 33, no. 15, pp. 1705–1707, 2008.

[53] J. Paul, Y. Kaneda, T.-L. Wang, C. Lytle, J. Moloney and R. Jones, "Doppler-free spectroscopy of mercury at 253.7 nm using a high-power, frequency-quadrupled, optically pumped external-cavity semiconductor laser," *Optics Letters*, vol. 36, no. 1, pp. 61–63, 2011.

[54] Y. Kaneda, M. Fallahi, J. Hader, *et al.*, "Continuous-wave single-frequency 295 nm laser source by a frequency-quadrupled optically pumped semiconductor laser," *Optics Letters*, vol. 34, no. 22, pp. 3511–3513, 2009.

[55] H. Kahle, R. Bek, M. Heldmaier, T. Schwarzbäck, M. Jetter and P. Michler, "High optical output power in the UVA range of a frequency-doubled, strain-compensated AlGaInP-VECSEL," *Applied Physics Express*, vol. 7, no. 9, p. 092705, 2014.

[56] Q.-Z. Shu, A. Caprara, J. Berger, D. Anthon, H. Jerman and L. Spinelli, "Intracavity-tripled optically-pumped semiconductor laser at 355 nm," in *Proc. of SPIE 7193*, San Jose, 2009.

[57] A. Hein, F. Demaria, A. Kern, *et al.*, "Efficient 460-nm second-harmonic generation with optically pumped semiconductor disk laser," *IEEE Photonics Technology Letters*, vol. 23, no. 3, pp. 179–181, 2011.

[58] J. Chilla, S. Butterworth, A. Zeitschel, *et al.*, "High-power optically pumped semiconductor lasers," in *Proc. of SPIE 5332*, San Jose, 2004.

[59] J. Berger, D. Anthon, A. Caprara, *et al.*, "20 Watt CW TEM00 intracavity doubled optically pumped semiconductor laser at 532 nm," in *Proc. of SPIE 8242*, San Franciso, 2012.

[60] S. Ranta, M. Tavast, T. Leinonen, R. Epstein and M. Guina, "Narrow linewidth 1118/559 nm VECSEL based on strain compensated GaInAs/GaAs quantum-wells for laser cooling of Mg-ions," *Optical Materials Express*, vol. 2, no. 8, pp. 1011–1019, 2012.

[61] S. Hilbich, W. Seelert, V. Ostroumov, *et al.*, "New wavelengths in the yellow-orange range between 545 nm and 580 nm generated by an intracavity frequency doubled optically pumped semiconductor laser," in *Proc. of SPIE 6451*, San Jose, 2007.

[62] E. Kantola, T. Leinonen, S. Ranta, M. Tavast and M. Guina, "High-efficiency 20 W yellow VECSEL," *Optics Express*, vol. 22, no. 6, pp. 6372–6380, 2014.

[63] E. Kantola, T. Leinonen, J.-P. Penttinen, V.-M. Korpijärvi and M. Guina, "615 nm GaInNAs VECSEL with output power above 10 W," *Optics Express*, vol. 23, no. 16, pp. 20280–20287, 2015.

[64] A. Rantamäki, A. Sirbu, A. Mereuta, E. Kapon and O. Okhotnikov, "3 W of 650 nm red emission by frequency doubling of wafer-fused semiconductor disk laser," *Optics Express*, vol. 18, no. 21, p. 21 645–21 650, 2010.

[65] T. Schwarzbäck, H. Kahle, M. Eichfelder, R. Roßbach, M. Jetter and P. Michler, "Wavelength tunable ultraviolet laser emission via intra-cavity frequency doubling of an AlGaInP vertical external-cavity surface-emitting laser down to 328 nm," *Applied Physics Letters*, vol. 99, no. 26, p. 261101, 2011.

[66] P. Schlosser, J. Hastie, S. Calvez, A. Krysa and M. Dawson, "InP/AlGaInP quantum dot semiconductor disk lasers for CW TEM00 emission at 716 – 755 nm," *Optics Express*, vol. 17, no. 24, p. 21782–21787, 2009.

[67] H. Kahle, K. Nechay, J.-P. Penttinen, A. Tukiainen, S. Ranta and M. Guina, "AlGaAs-based vertical-external-cavity surface-emitting laser exceeding 4 W of direct emission power in the 740–790 nm spectral range," *Optics Letters*, vol. 43, no. 7, p. 1578–1581, 2018.

[68] E. Saarinen, J. Lyytikäinen, S. Ranta, *et al.*, "750 nm 1.5 W frequency-doubled semiconductor disk laser with a 44 nm tuning range," *Optics Letters*, vol. 40, no. 19, p. 4380–4383, 2015.

[69] A. Rantamäki, J. Rautiainen, J. Lyytikäinen, *et al.*, "1 W at 785 nm from a frequency-doubled wafer-fused semiconductor disk laser," *Optics Express*, vol. 20, no. 8, pp. 9046–9051, 2012.

[70] J. Hastie, J.-M. Hopkins, S. Calvez, *et al.*, "0.5-W single transverse-mode operation of an 850-nm diode-pumped surface-emitting semiconductor laser," *IEEE Photonics Technology Letters*, vol. 15, no. 7, pp. 894–896, 2003.

[71] B. Rudin, A. Rutz, M. Hoffmann, *et al.*, "Highly efficient optically pumped vertical-emitting semiconductor laser with more than 20 W average output power in a fundamental transverse mode," *Optics Letters*, vol. 33, no. 22, pp. 2719–2721, 2008.

[72] S. Hackett, A. R. Albrecht, Z. Yang, *et al.*, "Development of vertical external cavity surface emitting lasers (VECSELs) for use as monochromatic

and polychromatic sodium guidestars," in *Proc. of SPIE 9909*, San Francisco, 2016.

[73] L. Fan, C. Hessenius, M. Fallahi, *et al.*, "Highly strained InGaAs/GaAs multiwatt vertical-external-cavity surface-emitting laser emitting around 1170 nm," *Applied Physics Letters*, vol. 91, no. 13, p. 1114, 2007.

[74] E. Kantola, T. Leinonen, S. Ranta, M. Tavast, J.-P. Penttinen and M. Guina, "1180 nm VECSEL with 50 W output power," in *Proc. of SPIE 9349*, San Francisco, 2015.

[75] C. Möller, C. Fuchs, C. Berger, *et al.*, "Type-II vertical-external-cavity surface-emitting laser with Watt level output powers at 1.2 μm," *Applied Physics Letters*, vol. 108, no. 7, p. 071102, 2016.

[76] V.-M. Korpijärvi, M. Guina, J. Puustinen, *et al.*, "MBE grown GaInNAs-based multi-Watt disk lasers," *Journal of Crystal Growth*, vol. 311, no. 7, pp. 1868–1871, 2009.

[77] D. Parrotta, A. Kemp, M. Dawson and J. Hastie, "Multiwatt, continuous-wave, tunable diamond Raman laser with intracavity frequency-doubling to the visible region," *IEEE Journal of Selected Topics in Quantum Electronics*, vol. 19, no. 4, p. 1400108, 2013.

[78] A. Albrecht, T. Rotter, C. Hains, *et al.*, "High-power 1.25 μm InAs QD VECSEL based on resonant periodic gain structure," in *Proc. of SPIE 7919*, San Francisco, 2011.

[79] V. Iakovlev, T. Leinonen, A. Sirbu, E. Kapon and M. Guina, "33 W continuous output power of 1275 nm semiconductor disk laser," *Optics Express*, vol. 25, no. 6, pp. 7008–7013, 2017.

[80] D. Parrotta, R. Casula, J.-P. Penttinen, *et al.*, "InGaAs-QW VECSEL emitting >1.300-nm via intracavity Raman conversion," in *Proc. of SPIE 9734*, San Francisco, 2016.

[81] J. Lyytikäinen, J. Rautiainen, A. Sirbu, *et al.*, "High-power 1.48-μm wafer-fused optically pumped semiconductor disk laser," *IEEE Photonics Technology Letters*, vol. 23, no. 13, p. 917–919, 2011.

[82] J.-M. Hopkins, N. Hempler, B. Rösener, *et al.*, "High-power, (AlGaIn)(AsSb) semiconductor disk laser at 2.0 μm," *Optics Letters*, vol. 33, no. 2, pp. 201–203, 2008.

[83] P. Holl, M. Rattunde, S. Adler, *et al.*, "GaSb-based 2.0 μm SDL with 17 W output power at 20°C," *Electronics Letters*, vol. 52, no. 21, pp. 1794–1795, 2016.

[84] P. Holl, M. Rattunde, S. Adler, *et al.*, "GaSb-based 2.0 μm SDL with 17 W output power at 20 °C," *Electronics Letters*, vol. 52, no. 21, pp. 1794–1795, 2016.

[85] N. Schulz, M. Rattunde, C. Ritzenthaler, *et al.*, "Resonant optical in-well pumping of an (AlGaIn)(AsSb)-based vertical-external-cavity surface-emitting laser emitting at 2.35μm," *Applied Physics Letters*, vol. 91, no. 9, p. 091113, 2007.

[86] J. Nikkinen, J. Paajaste, R. Koskinen, S. Suomalainen and O. Okhotnikov, "GaSb-based semiconductor disk laser with 130-nm tuning range at 2.5 μm," *IEEE Photonics Technology Letters*, vol. 23, no. 12, pp. 777–779, 2011.

[87] M. Rahim, A. Khiar, M. Fill, F. Felder and H. Zogg, "Continuously tunable singlemode VECSEL at 3.3 μm wavelength for spectroscopy," *Electronics Letters*, vol. 47, no. 18, pp. 1037–1039, 2011.

[88] A. Khiar, M. Eibelhuber, V. Volobuev, *et al.*, "Vertical extrenal cavity surface emitting PbTe/CdTe quantum dot lasers for the mid-infrared spectral region," *Opt. Lett.*, vol. 39, no. 23, pp. 6577–6580, 2014.

[89] M. Rahim, A. Khiar, F. Felder, M. Fill and H. Zogg, "4.5 μm wavelength vertical external cavity surface emitting laser operating above room temperature," *Applied Physics Letters*, vol. 94, no. 20, p. 201112, 2009.

[90] M. Rahim, M. Arnold, F. Felder, K. Behfar and H. Zogg, "Midinfrared lead-chalcogenide vertical external cavity surface emitting laser laser with 5 μm wavelength," *Applied Physics Letters*, vol. 91, no. 15, p. 151102, 2007.

[91] M. Rahim, F. Felder, M. Fill, D. Boye and H. Zogg, "Lead chalcogenide VECSEL on Si emitting at 5 μm," *Electronics Letters*, vol. 44, no. 25, pp. 1467–1469, 2008.

[92] M. Lukowski, C. Hessenius, R. Bedford and M. Fallahi, "Tunable type II intracavity difference frequency generation at 5.4 μm in a two chip vertical external cavity surface emitting laser," *Optics letters*, vol. 40, no. 17, pp. 4174–4177, 2015.

[93] M. Schulze and A. Lepert, "Flow cytometry: OPSL technology meets critical needs," *Optik & Photonik*, vol. 4, no. 4, pp. 29–32, 2009.

[94] M. Schulze, "OPSLs for quick flow cytometry," *Optik & Photonik*, vol. 5, no. 4, pp. 41–43, 2010.

[95] S. Morioka, "High-power optically pumped semiconductor laser applications," *Proceedings of SPIE*, vol. 7919, p. 791913, 2011.

[96] A. Honigmann, C. Eggeling, M. Schulze and A. Lepert, "Biophotonics: Super-resolution STED microscopy advances with yellow CW OPSL," *Laser Focus World*, 1 January 2012.

[97] C. Kannengiesser, V. Ostroumov, V. Pfeufer, *et al.*, "Ten years optically pumped semiconductor lasers: review, state-of-the-art, and future developments," in *Proc. of SPIE 7578*, San Francisco, 2010.

[98] S. Brown, E. Brown and I. Walker, "The present and future role of photodynamic therapy in cancer treatment," *The Lancet Oncology*, vol. 5, no. 8, pp. 497–508, 2004.

[99] M. Alexiades-Armenakas, "Laser-mediated photodynamic therapy," *Clinics in Dermatology*, vol. 24, no. 1, pp. 16–25, 2006.

[100] M. Heilemann, S. Van De Linde, M. Schüttpelz, *et al.*, "Subdiffraction-resolution fluorescence imaging with conventional fluorescent probes," *Angewandte Chemie International Edition*, vol. 47, no. 33, pp. 6172–6176, 2008.

[101] S. Jacques, "Corrigendum: optical properties of biological tissues: a review," *Physics in Medicine & Biology*, vol. 58, no. 14, pp. 5007–5008, 2013.

[102] S. Matcher, M. Cope and D. Delpy, "Use of the water absorption spectrum to quantify tissue chromophore concentration changes in near-infrared spectroscopy," *Physics in Medicine and Biology*, vol. 38, no. 1, p. 177, 1994.

[103] W. Zijlstra, A. Buursma and O. van Assendelft, Visible and near infrared absorption spectra of human and animal haemoglobin, VSP, 2000.

[104] T. R. Mazur, B. Klappauf and M. G. Raizen, "Demonstration of magnetically activated and guided isotope separation," *Nature Physics*, vol. 10, pp. 601–605, 2014.

[105] P. Thiis, L. Tiemeijer, J. Binsma and T. Van Dongen, "Progress in long-wavelength strained-layer InGaAs (P) quantum-well semiconductor lasers and amplifiers," *IEEE Journal of Quantum Electronics*, vol. 30, no. 2, pp. 477–499, 1994.

[106] S. Ranta, T. Hakkarainen, M. Tavast, J. Lindfors, T. Leinonen and M. Guina, "Strain compensated 1120 nm GaInAs/GaAs vertical external-cavity surface-emitting laser grown by molecular beam epitaxy," *Journal of Crystal Growth*, vol. 335, no. 1, p. 4–9, 2011.

[107] S. Ranta, M. Tavast, T. Leinonen, N. Van Lieu, G. Fetzer and M. Guina, "1180 nm VECSEL with output power beyond 20 W," *Electronics Letters*, vol. 49, no. 1, pp. 59–60, 2013.

[108] T. Leinonen, V.-M. Korpijärvi, J. Puustinen, R. Epstein and M. Guina, "11 W single gain element dilute nitride disk laser emitting at 1180 nm," in *Proc. of SPIE 7919*, San Francisco, 2011.

[109] T. Leinonen, V.-M. Korpijärvi, A. Härkönen and M. Guina, "Recent advances in the development of yellow-orange GaInNAs-based semiconductor disk lasers," in *Proc. of SPIE 8242*, San Francisco, 2012.

[110] M. Guina, T. Leinonen, A. Härkönen and M. Pessa, "High-power disk lasers based on dilute nitride heterostructures," *New Journal of Physics*, vol. 11, no. 12, p. 125019, 2009.

[111] T. Germann, A. Strittmatter, U. Pohl, *et al.*, "Quantum-dot semiconductor disk lasers," *Journal of Crystal Growth*, vol. 310, no. 23, p. 5182–5186, 2008.

[112] J. Rautiainen, I. Krestnikov, J. Nikkinen and O. Okhotnikov, "2.5 W orange power by frequency conversion from a dual-gain quantum-dot disk laser," *Optics Letters*, vol. 35, no. 12, pp. 1935–1937, 2010.

[113] M. Kondow, T. Kitatani, S. Nakatsuka, *et al.*, "GaInNAs: a novel material for long-wavelength semiconductor lasers," *IEEE Journal of Selected Topics in Quantum Electronics*, vol. 3, no. 3, pp. 719–730, 1997.

[114] A. Rantamäki, G. Sokolovskii, S. Blokhin, *et al.*, "Quantum dot semiconductor disk laser at 1.3 μm," *Optics Letters*, vol. 40, no. 14, pp. 3400–3403, 2015.

[115] V.-M. Korpijärvi, T. Leinonen, J. Puustinen, A. Härkönen and M. Guina, "11 W single gain-chip dilute nitride disk laser emitting around 1180 nm," *Optics Express*, vol. 18, no. 25, pp. 25633–25641, 2010.

[116] T. Leinonen, A. Härkönen, V.-M. Korpijärvi, *et al.*, "High-power narrow-linewidth optically pumped dilute nitride disk laser with emission at 589 nm," in *Proc. of SPIE 7720*, Brussels, 2010.

[117] J.-P. Penttinen, T. Leinonen, A. Rantamäki, *et al.*, "High-power 625 nm VECSEL prototype," in *Proc. of SPIE 10087*, San Francisco, 2017.

[118] P. Blood, "Material issues in AlGaInP red-emitting laser diodes," *Materials Science and Engineering: B*, vol. 66, no. 1, pp. 174–180, 1999.

[119] J. Hastie, S. Calvez, M. Dawson, *et al.*, "High power CW red VECSEL with linearly polarized TEM00 output beam," *Optics Express*, vol. 13, no. 1, pp. 77–81, 2005.

[120] H. Tanaka, Y. Kawamura and H. Asahi, "Refractive indices of In0.49Ga0.51-xAlxP lattice matched to GaAs," *Journal of Applied Physics*, vol. 59, no. 3, pp. 985–986, 1986.

[121] S. Adachi, "GaAs, AlAs, and AlxGa1-xAs: Material parameters for use in research and device applications," *Journal of Applied Physics*, vol. 58, no. 3, pp. R1–R29, 1985.

[122] A. Smith, J. Hastie, H. Foreman, T. Leinonen, M. Guina and M. Dawson, "GaN diode-pumping of red semiconductor disk laser," *Electronics Letters*, vol. 44, no. 20, pp. 1195–1196, 2008.

[123] J. Nicholson and W. Rudolph, "Noise sensitivity and accuracy of femto-second pulse retrieval by phase and intensity from correlation and spectrum only (PICASO)," *Journal of the Optical Society of America B*, vol. 19, no. 2, pp. 330–339, 2002.

[124] A. Periasamy, P. Wodnicki, X. Wang, S. Kwon, G. Gordon and B. Herman, "Time-resolved fluorescence lifetime imaging microscopy using a picosecond pulsed tunable dye laser system," *Review of Scientific Instruments*, vol. 67, no. 10, pp. 3722–3731, 1996.

[125] T. Liu, J. Wang, G. Petrov, V. Yakovlev and H. Zhang, "Photoacoustic generation by multiple picosecond pulse excitation," *Medical Physics*, vol. 37, no. 4, pp. 1518–1521, 2010.

[126] J.-M. Hopkins, S. Smith, C.-W. Jeon, *et al.*, "0.6 W CW GaInNAs vertical external-cavity surface emitting laser operating at 1.32 μm," *Electronics Letters*, vol. 40, no. 1, pp. 30–31, 2004.

[127] V.-M. Korpijarvi, E. Kantola, T. Leinonen, R. Isoaho and M. Guina, "Monolithic GaInNAsSb/GaAs VECSEL operating at 1550 nm," *IEEE Journal of Selected Topics in Quantum Electronics*, vol. 21, no. 6, pp. 480–484, 2015.

[128] A. Karim, S. Björlin, J. Piprek and J. Bowers, "Long-wavelength vertical-cavity lasers and amplifiers," *IEEE Journal of Selected Topics in Quantum Electronics*, vol. 6, no. 6, p. 1244–1253, 2000.

[129] A. Sirbu, A. Mereuta, A. Caliman, *et al.*, "High power optically pumped VECSELs emitting in the 1310 nm and 1550 nm wavebands," in *Proc. of SPIE 7355*, San Francisco, 2011.

[130] J. Rautiainen, J. Lyytikäinen, A. Sirbu, *et al.*, "2.6 W optically-pumped semiconductor disk laser operating at 1.57-μm using wafer fusion," *Optics Express*, vol. 16, no. 26, pp. 21881–21886, 2008.

[131] J. Lyytikäinen, J. Rautiainen, L. Toikkanen, *et al.*, "1.3-μm optically-pumped semiconductor disk by wafer fusion," *Optics Express*, vol. 17, no. 11, pp. 9047–9052, 2009.

[132] A. Rantamäki, A. Sirbu, E. Saarinen, *et al.*, "High-power flip-chip semiconductor disk laser in the 1.3 μm wavelength band," *Optics Letters*, vol. 39, no. 16, pp. 4855–4858, 2014.

[133] T. Leinonen, V. Iakovlev, A. Sirbu, E. Kapon and M. Guina, "33 W continuous output power semiconductor disk laser emitting at 1275 nm," *Optics Express*, vol. 25, no. 6, pp. 7008–7013, 2017.

[134] C. Sramek, L. Leung, Y. Paulus and D. Palanker, "Therapeutic window of retinal photocoagulation with green (532-nm) and yellow (577-nm) lasers," *Ophthalmic Surgery, Lasers and Imaging Retina*, vol. 43, no. 4, pp. 341–347, 2012.

[135] J. D. Walker, "Initial experience with a yellow diode laser," *Retina: Advanced Ocular Care*, pp. 34–36, 2011.

[136] S. Brown, E. Brown and I. Walker, "The present and future role of photodynamic therapy in cancer treatment," *The Lancet Oncology*, vol. 5, no. 8, pp. 497–508, 2004.

[137] K.-H. Yuan, Q. Li, W.-L. Yu and Z. Huang, "Photodynamic therapy in treatment of port wine stain birthmarks — recent progress," *Photodiagnosis and Photodynamic Therapy*, vol. 6, no. 3, pp. 189–194, 2009.

[138] T. Dougherty, C. Gomer, B. Henderson, *et al.*, "Photodynamic therapy," *Journal of the National Cancer Institute*, vol. 90, no. 12, pp. 889–905, 1998.

[139] C. Dierickx, J. Casparian, V. Venugopalan, W. Farinelli and R. Anderson, "Thermal relaxation of port-wine stain vessels probed in vivo: the need for 1–10-millisecond laser pulse treatment," *Journal of Investigative Dermatology*, vol. 105, no. 5, pp. 709–714, 1995.

[140] Coherent Inc., "Genesis Taipan HD-Series; High-Definition, High-Power Optically Semiconductor Lasers (OPSL)," [Online]. Available: https://cohrcdn.azureedge.net/assets/pdf/Genesis_Taipan_HD_Series_Datasheet_2016.pdf. [Accessed September 2017].

[141] "QuadroStarPRO YELLOW," Asclepion, [Online]. Available: http://www.asclepion.com/asclepion_product/quadrostar-pro-yellow/. [Accessed January 2018].

[142] L. Wang and S. Hu, "Photoacoustic tomography: in vivo imaging from organelles to organs," *Science*, vol. 335, no. 6075, pp. 1458–1462, 2012.

[143] M. Xu and L. Wang, "Photoacoustic imaging in biomedicine," *Review of Scientific Instruments*, vol. 77, no. 4, p. 041101, 2006.

[144] V. Savitski, J. Hastie, S. Calvez and M. Dawson, "Cavity-dumping of a semiconductor disk laser for the generation of wavelength-tunable micro-Joule nanosecond pulses," *Optics Express*, vol. 18, no. 11, pp. 11933–11941, 2010.

[145] E. Bailey, R. Marsh, S. Culley, E. Kantola, M. Guina and A. Bain, "Time and polarization resolved CW STED photodeselection in molecular probes," in *Proc. of SPIE 8950*, San Francisco, 2014.

[146] M. Brown and C. Wittwer, "Flow cytometry: principles and clinical applications in hematology," *Clinical Chemistry*, vol. 46, no. 8, pp. 1221–1229, 2000.

[147] S. Cho, J. Godin, C.-H. Chen, W. Qiao, H. Lee and Y.-H. Lo, "Review article: Recent advancements in optofluidic flow cytometer," *Biomicrofluidics*, vol. 4, no. 4, p. 043001, 2010.

[148] S. Calvez, J. Hastie, M. Guina, O. Okhotnikov and M. Dawson, "Semiconductor disk lasers for the generation of visible and ultraviolet radiation," *Laser & Photonics Reviews*, vol. 3, no. 5, p. 407–434, 2009.

[149] N. Fried and K. Murray, "High-power thulium fiber laser ablation of urinary tissues at 1.94 μm," *Journal of Endourology*, vol. 19, no. 1, pp. 25–31, 2005.

[150] A. Welch and M. Van Gemert, Eds., Optical-thermal response of laser-irradiated tissue, vol. 2, Springer, Netherlands, 2001, p. 45.

[151] C. Xu, W. Zipfel, J. B. Shear, R. M. Williams and W. W. Webb, "Multiphoton fluorescence excitation: new spectral windows for biological nonlinear microscopy," *Proceedings of the National Academy of Sciences of the USA*, vol. 90, no. 20, pp. 10763–10768, 1996.

[152] R. Aviles-Espinosa, G. Filippidis, C. Hamilton, *et al.*, "Compact ultrafast semiconductor disk laser: targeting GFP based nonlinear applications in living organisms," *Biomedical Optics Express*, vol. 2, no. 4, pp. 739–747, 2011.

[153] F. F. Voigt, F. Emaury, P. Bethge, *et al.*, "Multiphoton in vivo imaging with a femtosecond semiconductor disk laser," *Biomedical Optics Express*, vol. 8, no. 7, pp. 3213–3231, 2017.

Chapter 9

Potential biomedical use of diode-laser-induced luminescence from upconverting nanoparticles

Monirehalsadat Mousavi[1] *and Stefan Andersson-Engels*[2,3]

9.1 Introduction

The intention with this chapter is to briefly review the capabilities made possible in the field of biomedical diagnostics and treatment by employing upconverting nanoparticles (UCNPs), as well as the importance of appropriate light sources for such applications. The field of UCNPs in biomedicine has grown rapidly the last decade, since they first were made sufficiently small and bright to be of real interest for biomedical applications. This research has been reviewed previously (see, e.g., [1–7], while this chapter will focus on recent advances in the use of UCNPs for preclinical applications and diagnostic assays using laser diode excitation.

9.1.1 Present challenges in biomedical imaging

The development of novel imaging modalities is of high importance in the field of medicine, for the purposes of clinical diagnostics and therapy [8]. It is of equally importance for preclinical studies in biomedical research and for screening of pharmaceutical drugs. The ability to visualise morphological structures of various sizes, from subcellular structures to organs, can provide essential diagnostic and prognostic information for a diseased patient, such as the cause, extent and prognosis of a disease. In clinical care, and also to an increased extent in preclinical studies, diagnostic information is today provided by amongst other techniques various diagnostic imaging modalities (*in vivo* or *ex vivo*). These modalities include optical X-ray, magnetic resonance imaging (MRI), ultrasound and histopathological examinations. In particular, the quickly growing incidence of detected malignancies will require further extensive development and refinement of various imaging techniques, both *in vivo* and for histopathological samples, to facilitate adequate care [9]. It would for these patients be of particular interest to benefit from

[1]Department of Physics, Lund University, Sweden
[2]IPIC, Tyndall National Institute, Cork, Ireland
[3]Department of Physics, University College Cork, Ireland

improved possibilities for visualisation of the confinement and spread of the malignant cells in the treatment planning and prognostic assessments. In 2008, malignant deceases became the leading cause of death in the world and lead to 13% of deaths (equal to 7.6 million deaths annually) [9,10]. Furthermore, the trend is that the death toll of malignant deceases will continue to grow and is expected to become 13.1 million deaths annually by the year 2030.

A key prognostic factor for malignant deceases is the time of initiating adequate treatment. Early treatments require early detection – yielding that improved diagnostic potential provides a potential to significantly reduce mortality and morbidity of malignancies. Thus, malignant deceases alone motivates the development of improved imaging and other diagnostic tools, both in terms of sensitivity and specificity of malignant decease at a very early stage [11]. Today the most used anatomical imaging modalities for deep tissue imaging includes primarily X-ray computed tomography (CT), MRI, ultrasonography (US), positron emission tomography (PET) and single-photon emission computed tomography (SPECT) [12,13]. So far, optical imaging (OI) has mainly been used for endoscopic and microscopic imaging. Subsurface imaging is much more challenging due to the high light scattering and absorption properties of tissue, limiting the imaging depths [14,15]. However, with the development of much improved clinically adapted sources and detectors, imaging beyond millimetres and even centimetres has today become possible. As shown in Figure 9.1, OI techniques have proven to be significantly faster, more compact and cost effective, and possess excellent molecular

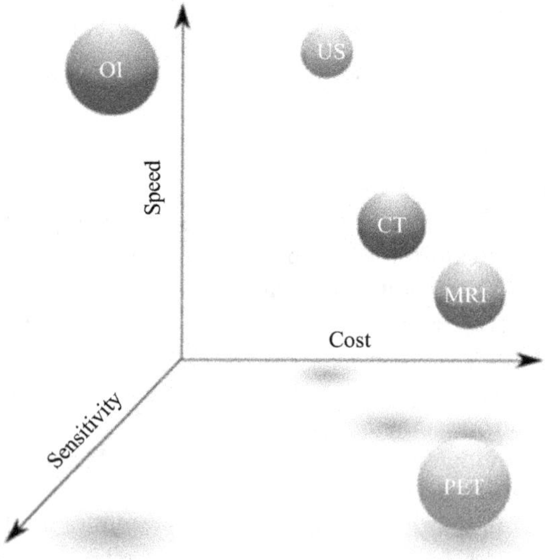

Figure 9.1 This image illustrates how optical imaging (OI) is compared to conventional biomedical imaging techniques in terms of sensitivity, acquisition speed and cost. The illustration is reprinted with permission from the PhD thesis of Can Xu [23]

sensitivity through the use of fluorescent contrast agents, as compared to conventional non-optical techniques [12,15,16]. Today, diffuse optical imaging (DOI), diffuse optical tomography and diffuse optical spectroscopy (DOS) are used to, for example, monitor and visualise tumour protease activity [17], apoptosis [18], brain activity [19,20], breast cancer [21] and to control photodynamic therapy treatments of cancer [22].

In a similar way, preclinical, often whole-body, imaging has rapidly grown in use and importance. This is primarily in an effort to make preclinical studies less expensive and time-consuming. The vision and trend is to adhere to the 3R principle (replace, reduce and refine) animals studies by designing an increasingly fraction of longitudinal studies, where each individual animal is followed throughout the study. This strategy would facilitate studies of any disease development or treatment response based on few subjects only.

9.1.2 Motivation of interest for optical bioimaging

Despite the many merits of OI techniques, several challenges and limitations exist, in particular related to the spatial resolution and limited penetration depths in certain tissues illustrated in Figure 9.1. There exists a general ambition to find ways to improve the OI and spectroscopy techniques by employing novel contrast agents known as UCNPs, as well as models for light-tissue interaction. This may potentially enhance several aspects of current optical techniques, including the sensitivity, depth penetration, spatial resolution, quantitative accuracy and system complexity. The OI techniques can provide excellent spatial maps of quantities of interest, however, this is often at the cost of little to no spectral information. DOS can be used to obtain spectral information and has already been used heavily in numerous tissue diagnostics applications [24–26].

When employing OI, one can rely on a number of different contrast mechanisms. OI employs typically molecular contrast rather than morphological structures for imaging. The molecular contrast can either be through absorption (such as in DOI), luminescence or scattering (such as in various Raman-based techniques). This chapter will concern luminescence-based techniques. The luminescence emission is generally originating in molecules present in the tissue, either tissue endogenous molecules present naturally in the tissue, or exogenous agents used as contrast agents for improved imaging qualities. The exogenous drugs used for imaging are often based on fluorescence dyes that are preferentially located in the tissue structures to be imaged. Important characteristics for the contrast agent is that it should provide a characteristic signal easily discriminated from tissue autofluorescence, it should not bleach easily when illuminated and obviously be selectively bound to the tissue structures to image. The excitation and emission wavelengths are also important as they define the probe volume and spatial resolution that will be obtained in the image. Also genetically encoded markers are used, such as fluorescence proteins. These contrast agents are produced by genetically modified cells, and will thus label these cells very uniquely. The drawback with this technique is that the cells labelled are genetically modified, and thus only applicable in specially engineered preclinical studies.

9.1.3 Genetically modified and exogenous molecular labelling

Controlled activation of biomolecules is very crucial in medicine, for example, controllable gene delivery/knockdown in gene therapy and anti-cancer drug delivery in cancer chemotherapy. Photoactivation provides an elegant approach for on-demand site-specific molecule release with high spatio-temporal precision. Unfortunately, photoactivable molecule vectors mostly respond to ultraviolet (UV) and blue radiation, which has very limited penetration depths and light toxicity to biological tissues. This makes direct-UV-irradiation-based photoactivation method unrealistic in deep tissues in *in vivo* systems. A very unique and recently suggested photoactivation approach, employing lanthanide doped UCNPs as energy transducers, opens the possibility of implementing light-controlled biomolecule activation in deep tissues [27,28]. UCNPs, upon near-infrared (NIR) excitation, can generate UV emission, which acts as a switch to activate biomolecules in vicinity. The NIR excitation guarantees maximised penetration depth and minimised adverse effects of excitation light in tissues, while the localisation of UV or blue emission vanishes its toxic effect on non-targeted tissues.

9.1.4 Motivation of interest for UCNPs for bioimaging

With photon upconversion we mean a process in a medium where energy of the photons emitted are higher than for those absorbed. Typically two or more photons are absorbed and their combined energy is higher than the emitted photon they generate through the upconverting process. This reduction in number of photons makes the process possible as it guarantees conservation of energy. The process is schematically shown in Figure 9.2(b). The generated, the so-called upconverted, emission is also known as anti-Stokes shifted emission. The upconversion phenomenon was first observed by Bloembergen in 1959. He proposed that the concept could be used for infrared quantum counters [29]. Today, upconversion is, for example, used in infrared markers to visualise near-infrared light, otherwise invisible for the human eyes. While the upconversion mechanism proposed by Bloembergen relied on meta-stable real states in crystal hosts, with the availability of lasers, upconversion is today routinely realised in multiphoton microscopy [30], triplet–triplet annihilation [31–33] and other nonlinear effects in crystals and gases [34]. The processes of upconversion were thoroughly investigated by Auzel [35–37]. The interest of these materials came with the development of these materials in form of nanoparticles. Faris *et al.* pioneered this early biomedical research of UCNPs [38]. The field gained increased interest once UCNPs were produced with considerable improved quantum yield (QY) [39–48]. The UCNPs used today for biomedical applications include mainly rare-earth metal doped nanocrystals. Frequently $NaYF_4$ is used as a host material in these nanoparticles, due to the low losses of the emission in these materials. The UCNPs have been demonstrated to provide unique properties, clearly providing potentials for improved optical bioimaging, as reviewed in this chapter. These advantages will be discussed in more detail under the *Photophysical properties* and *Applications* subchapters below.

9.1.5 Use of diode lasers for upconversion nanoparticle-based bioimaging

In order to gain full interest of UCNP-based imaging for the biomedical community, it has to be safe, provide better results than other markers and also be cost-effective. As far as the community so far has been observing, the particles are biocompatible, in the sense that they can be administered systemically without complication. No toxicity at concentrations of biomedical interest has been observed [49], while more studies in this topic obviously are necessary before any use in human studies. Such studies are underway.

Cost-effectiveness is ensured by availability of commercial particles, either coated with specified functional material to specifically bind to a number of pre-defined and identified tissue targets, or as bare particles, that can be coated by the users with their own specific targeting coating. Instrumentation-wise the technique can also be kept at low cost through the use of inexpensive diode lasers for excitation and conventional laboratory cameras for detection. An intention with this chapter is also to discuss the specific properties of the excitation sources to enable optimised detection possibilities. In particular, wavelength, power and pulse duration properties are essential. For many applications, the possibility to couple the light into large core-size multi-mode optical fibres is crucial. The most important identified properties of the excitation source are discussed in the subchapter *Sources*. In this respect, both the spectral and temporal properties of the light source is of importance, as will be further discussed below.

9.1.6 Organisation of chapter

The organisation of this chapter is as follows. After the introduction, a subchapter entitled *Photophysical properties* is provided. Here the main properties of the particles are discussed, together with a brief description of the unique possibilities these properties provide for biomedical imaging. This subchapter is relatively more extensive than the others, as most of our own research has been within characterising UCNPs for biomedical applications, and identify the unique properties, so they can be utilised as well as possible.

Next follows a subchapter briefly discussing some of many potential biomedical applications, where UCNPs can become an enabler for successful studies or biomedical procedures. This subchapter will review selected applications, to illustrate the vast possibilities with UCNPs for biomedical use.

Before the final summary of the chapter, being a part of *Handbook of Semiconductor Lasers and Diode-Based Light Sources in Biophotonics*, we will also discuss the essential properties for the sources to generate as useful information as possible. These properties are primarily wavelength, pulse-length, power and beam properties. In addition, the importance of short- and long-term stability will be addressed.

9.2 Optical properties of UCNPs

The interaction between light and tissue is strong, making the depth of light penetration limited to a rather shallow region. The interaction is due to both absorption and scattering. The light-tissue interaction and thus light penetration is dependent on its wavelength. The near-infrared wavelength region is beneficial for deep light penetration, due to low absorption and scattering for light in this wavelength range. Selecting the best contrast agent and light source thereby ensures that the light penetrates the furthest through tissue, enabling that deeply located target cells can be investigated. Furthermore, in luminescence imaging employing UCNPs as a contrast agent, the signals also clearly depends on the concentration of UCNPs, their absorption cross section and QY. Techniques for characterisation and synthesis of UCNPs and various factors for optimisation of their luminescence efficiency are addressed in this subchapter.

9.2.1 Upconversion process

The photophysical pathway associated with upconversion (UC) photoluminescence is relatively well understood. UC refers to non-linear optical processes in which sequential absorption of two or more low-energy photons results in higher-energy emission. It is thereby described as an anti-Stokes mechanism. There are three main processes causing upconversion (shown in Figure 9.2): Excited state absorption (ESA), also known as sequential two photon absorption; energy transfer upconversion (ETU); and photon avalanche effect (PA). The principles of them are all schematically presented and discussed in several review articles [37,50–52]. In ESA (illustrated in Figure 9.2(a)) is a pumping mechanism suitable for single-doped materials for which the emitting ion first absorbs one photon of proper energy to be prompted from the ground to the first excited state (E1). Sequentially it is then excited again now from E1 to emitting level (E2) by absorption of another pump photon. The two pump photons can be generated from the same source if the

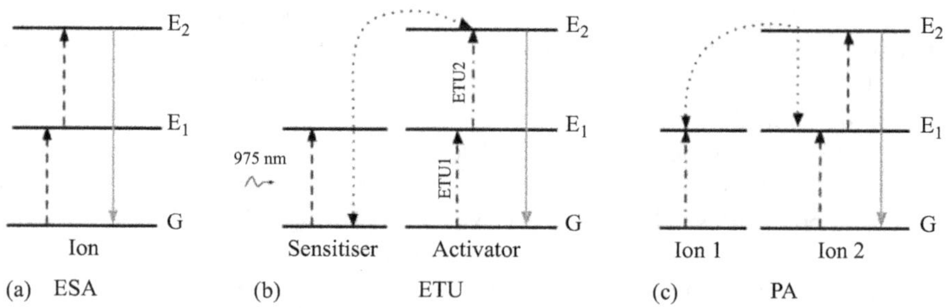

Figure 9.2 This image shows simplified energy diagram of the upconversion photoluminescence mechanism: (a) excited state absorption (ESA); (b) energy transfer upconversion (ETU); and (c) photon avalanche (PA). Dashed, solid and dotted lines represent photon absorption, photon emission and energy transfer processes, respectively

two energy gaps are equal. UC emission achieved via ETU follows a similar principle as those of ESA, whilst the UC process involves a non-radiative energy transfer between neighbouring ions in the crystal (sensitiser and activator), as depicted in Figure 9.2(b). In the ETU process, the sensitiser absorbs a photon and is excited to an excited state. The activator is through the energy transfer in a first step excited to an intermediate level (E1), and then, following a second absorbed photon by the sensitiser, sequentially transferred to a higher excited level (E2). The photon avalanche effect, is also based on sequential energy transfers between ions as result of cross relaxation process, simplified in Figure 9.2(c). All of these mechanisms take advantage of employing real intermediate energy levels. These intermediate levels also possess a relatively long lifetime (in the order of milliseconds). Therefore, the simultaneous absorption of two and more photons is not required, and the excitation intensity for observing upconversion can be achieved by using low-cost lasers suitable for biological applications. This is in sharp contrast to the mechanisms employed in multiphoton photoluminescence, requiring very intense excitation (typically femtosecond lasers) as all excitation photons need to be present simultaneously due to the virtual intermediate energy level unable to store the absorbed photon energy. Competition for the depopulation of the intermediate state of the involved activator governs the power density dependence of the upconversion luminescence intensity and consequently, the QY.

9.2.2 Lanthanide-doped UCNPs

The Lanthanide (Ln) ions, also known as rare-earth metals, with a ladder-like energy level structure are the most commonly used luminescent centres for upconversion luminescence, owning their special electron configuration with a filled outer 4f electron shell and stable excitation state. Ln-doped ions generally comprise a nanoscale inorganic host, doped with sensitiser and activator ions [53]. The most popular UCNP represents $NaYF_4$ as inorganic crystal matrix, co-doped with lanthanide ions in the form of the sensitiser ytterbium Yb^{3+} and activator Er^{3+} or thulium Tm^{3+}. The widely accepted energy level diagram of $NaYF_4$:Yb^{3+}, Er^{3+} is shown in Figure 9.3. The sensitisers absorb energies from NIR photons and then transfer their excitation energy to the activators from where radiation can be emitted.

Apart from the upconversion phenomena, UCNPs benefit from their unique properties including, sharp emission bandwidth, long lifetime, high photostability and low cytoxicity, all being very desirable properties in biotechnology. Since the optical properties of lanthanide ions are greatly determined by the f-f electron transitions, their forbidden nature (Laporte rule) [54] leads to a long lifetime – in the order of milliseconds. Distinguished from traditional fluorescent labelling agents, RE-doped UCNPs can be excited by sequential absorption of two or more NIR photons and emit light at shorter wavelengths (range of NIR-UV to NIR). This upconversion mechanism possesses several advantages which include: (i) high signal-to-background ratio and thereby high detection sensitivity due to the absence of autofluorescence; (ii) deep penetration of NIR light into biological tissue and less photo damage to biological samples. In contrast to direct two photon absorption, upconversion processes are often related to energy stepwise transfer between a

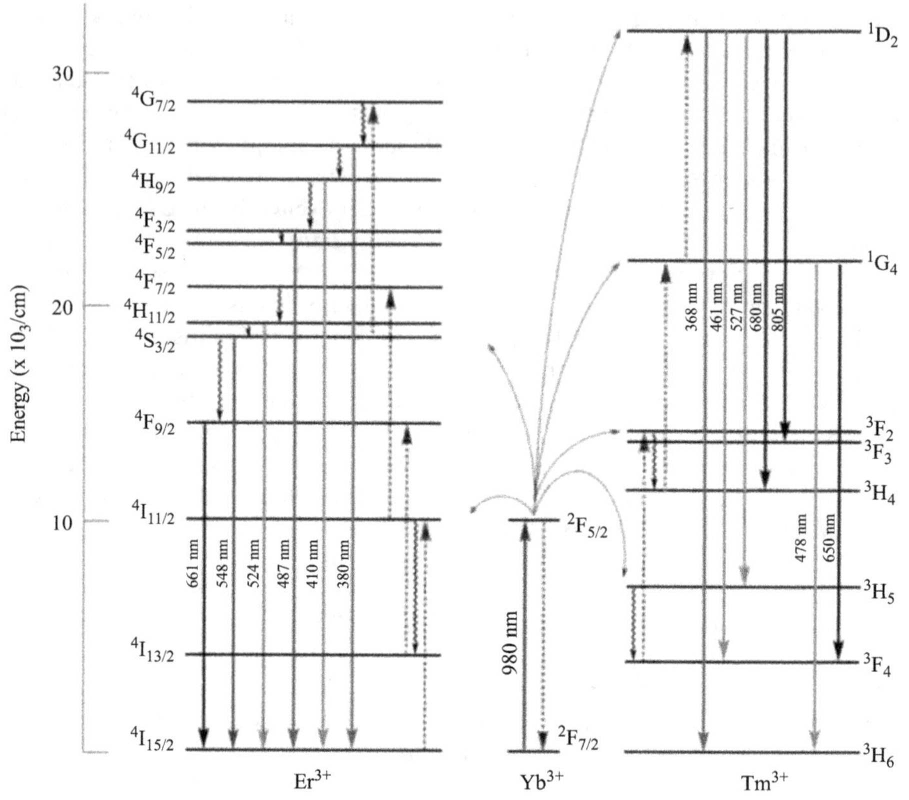

Figure 9.3 Energy-level diagram and upconversion scheme for the Yb^3+-sensitized Er^3+ and Tm^3+ systems is shown. The figure is taken from [7] with permission

sensitiser and activator. This results in much higher efficiency of upconversion, excitation can be achieved at substantially lower power continues wave diode laser. Individual UCNPs have been spectroscopically investigated and no photobleaching or photodamage was reported after 1 h of continues laser illumination [55]. Moreover, UCNPs showed non-blinking behaviour since the upconversion process accruing among numerous randomly positioned ions in a single UCNPs which produce steady-state emission after continues exposure to a high excitation energy level.

The absorption probability of UCNPs is determined by the doped lanthanide ions. The majority of lanthanide ions exhibit low absorption cross sections leading to poor excitation efficiency. To enhance upconversion efficiency, co-doping with a second ion as a sensitiser with a high absorption cross section in NIR region is generally utilised. Among the lanthanides, Yb^{3+} (1.2×10^{-20} cm^2 at 980 nm) and Nd^{3+} (1.2×10^{-19} cm^2) have relatively large absorption cross section. The absorption band of Yb^{3+} is due to $^2F_{\frac{7}{2}}$–$^2F_{\frac{5}{2}}$ transition and well resonant with f-f transition of many activator ions, such as Er^{3+}, Tm^{3+} and Ho^{3+} [56,57]. The selection of different emitters in terms of combinations of lanthanide dopants is a novel approach to generate tunable multicolour conversion emission [56,58]. It is notable that the emission spectrum of an UCNP is associated with the host composition, particle size and particle surface properties. Contrary to the Yb which has a very narrow band absorption and only one excited state $({}^2F_{\frac{5}{2}})$, Nd has multiple NIR excitation bands at wavelength 730, 808 and 865 nm corresponding to transitions from ${}^4I_{\frac{9}{2}}$ to ${}^4F_{\frac{7}{2}}$, ${}^4F_{\frac{5}{2}}$ and ${}^4F_{\frac{3}{2}}$, respectively. The absorption by water around these excitation wavelengths is much lower than that at 980 nm, allowing not only light to penetrate deeper into biological tissue, but also minimising tissue damage due to heating effect. Additionally, the large absorption cross section and efficient energy transfer from excited Nd^{3+} to Yb^{3+} ions suggest Nd as a promising candidate for bioimaging applications [59–62]. Another interesting novel approach to address overheating by shifting the excitation to shorter wavelengths is to utilise an organic broadband dye-sensitised antenna molecules to increase the absorptivity and broadens the absorption spectrum of upconverter as well as enhance the overall upconversion efficiency by a factor of 3.3 [63].

While many Lanthanide-doped host materials have been observed to exhibit upconversion luminescence, a significant effort has been focused on the optimal selection of host material, dopant ions and dopant concentration to achieve most efficient UC luminescence with controllable emission profiles. The choice of host lattice determines the distance between ions and consequently, concentration of ions. In the case of sensitised UC luminescence the average distance between dopant ions significantly influence the upconversion efficiency. The sensitiser concentration must be high enough to optimise energy transfer between sensitiser and activators, but not so high to be influenced by deleterious cross-relaxation effects. The three-dimensional lattice structure of host material brings the luminescence centres into the optimal position with respect to the ion-to-ion distance and spatial distribution of dopant ions. An ideal host material should be optically transparent, chemically and thermally stable and possess low phonon cut-off energies [64]. The local structure of lattice is known to have an impact on its optical properties. It was found that the intensity of the upconversion luminescence in hexagonal-phase $NaYF_4{:}Yb^{3+}, Er^{3+}$ is approximately 10 times stronger than that in cubic one for otherwise identical particles [65]. The emission intensity of an upconverting material depends also on its dimension. Although the small nanoparticles are advantageous to the biological applications, the reduced particles size leads to an increase of surface-to-volume ratio of materials and thus more defects on the nanoparticles surface is introduced [66]. There are now many methodologies to synthesise nanoparticles with desired size and chemical compositions.

9.2.3 Synthetic methods of UCNPs

In the use of UCNPs as luminescent labels in biological applications, suitable size and uniform shape of the nanocrystals and high aqueous solubility are the primary prerequisites for imaging. The size of the particles plays a key role in the uptake, biodistribution and clearance of nanoparticles in living organisms. Sizes comparable to the biomolecules are required so as not to interfere with cellular systems. Great efforts have been dedicated to exploring the efficiency of various methods for the synthesis of UCNPs [57,67,68]. The thermal decomposition and hydro(solvo)thermal synthesis have become the most popular methods as they provide excellent control over shape, particle size and phase of fluoride UCNPs.

The thermal composition method. Thermal composition method was first employed by Yan and colleagues for fluoride-based nanocrystals via the co-thermolysis of the trifluoroacetate precursors in a solution of surface ligands, octadecene (ODE), and capping ligands, oleic acid (OA), at high temperature ($\approx$300 °C) [69]. In this reaction, the octadecene was used as the primary high boiling surface solvent to provide high temperature environment. Oleic acid acted as a capping ligand to control the particles growth and stabilise the growing particles in solution. Yi replaced the ODE and OA by a single olelyamine (OM) and produced uniform and small size $NaYF_4$:Yb^{3+}, Er^{3+} and $NaYF_4$:Yb^{3+}, Tm^{3+} [70]. In addition, Mai *et al.* synthesised monodisperse high-quality cubic-(α) and hexagonal-(β) phase particles with controllable size in OA/OM/ODE mixture [71]. The thermal decomposition synthesis procedure has been refined by several groups and now is a general route to synthesis monodisperse and well-defined size nanocrystals in different phases. So far, a series of UCNPs based on $NaYF_4$ [67,70–72], $NaGdF_4$ [71,73], $NaYbF_4$ [74], $LiYF_4$ [75,76] and $BaGdF_5$ [77] hosts have been synthesised through this particular method. The thermal decomposition method requires expensive and toxic air-sensitive metal precursors and control of high reaction temperature. The need for non-polar capping ligands on the surface of UCNPs entails further surface modification to make them water-soluble.

Hydro(solvo)thermal synthesis. The solvothermal techniques operate at much lower temperatures and produce fewer toxic by-products than the thermal decomposition methods. This approach utilises a solvent in a sealed environment under pressure and temperature above their critical point to facilitate the interaction of precursors during synthesis. Typically, the reaction takes place in specialised vessels known as autoclave involving mixing of lanthanide and fluoride precursors in an aqueous solution. Li and co-workers reported hydrothermal synthesis of β$-NaYF_4$:Yb^{3+}, Er^{3+} UCNPs of controlled size and morphology in different solvents aided by cetyltrimethylammonium bromide (CTAB) and ethylenediaminetetraacetic acid (EDTA) [78]. It was observed that the NPs size and their morphology is not easy to control when water is used as a solvent (hydrothermal). However, spherical nanoparticles on the order of 20–30 nm are produced in acetic acid and uniform size of 50 nm were generated in ethanol (solvothermal). A "one pot process" with heat-resistant organic polymer (e.g., PVP, PEI) added to the solvent and polyacrylic acid (PAA) as capping ligands was reported by Wang *et al.* [79]. The pure hexagonal phase UCNPs with a uniform size of about 40 nm β-$NaYF_4$, with the capability of readily dispersing in aqueous media, was fabricated. The feasibility of this method has been demonstrated by several groups.

Coprecipitation is another friendly and convenient approach for synthesising lanthanide doped $NaYF_4$ UCNPs [80]. The benefits of this method are operational simplicity, easy control of particle size and composition, and the lack of toxic by-products.

In addition to discussed methods, also other less well known synthetic routes such as sol-gel [81], microemulsion [82,83] and flame synthesis [84] have also been employed for the generation of luminescence non-crystalline nanoparticles of varying host materials.

9.2.4 Surface modification

The UCNPs fabricated by many of these available methods are prepared using an organic environment, such as oleic acid, and are surrounded by hydrophobic capping ligands. A preparation with hydrophilic nanoparticle surfaces is required for biomedical applications (or they will aggregate immediately after in vivo administration to the water rich tissue [85–88]. Therefore, surface modification is necessary to convert the hydrophobic UCNPs into hydrophilic ones. In addition to hydrophilicity, surface functionality is required to facilitate the conjugation between UCNPs and biomolecules. Surface modification of UCNPs can be classified into two broad categories: (1) coating with an inorganic shell layer and (2) modification with organic capping ligands. The most frequently employed method in inorganic surface functionalisation is surface silanisation of silicon. Silica is an attractive coating with a negligible influence on the luminescence properties of UCNPs. The surface silanisation technique can be used to modify the surface of both hydrophobic and hydrophilic UC nanocrystals. The Stöber method has been widely used to apply a uniform coating of silica on hydrophilic NPs while UCNPs with hydrophobic ligands surface are coated with well-established reverse microemulsion method. Li and Zhang reported a uniform size (10 nm) silica-coated polyvinylpyrrolidone (PVP) stabilised $NaYF_4$:Yb, Er/Tm nanoparticles in a mixture of ethanol-water solution via Stöber method [89]. The thickness of silica shell can be adjusted within 1–3 nm by varying the initial amount of tetraethoxysilane (TEOS), see Figure 9.4. Silica-coated UCNPs are highly stable, biocompatible and well dispersible in aqueous buffers [90,91]. Silanisation method provides a core shell structure which creates an easily accessible platform for introducing the various functional group (e.g., $-NH_2$, $-COOH$ or $-SH$, etc.) for the further conjugation

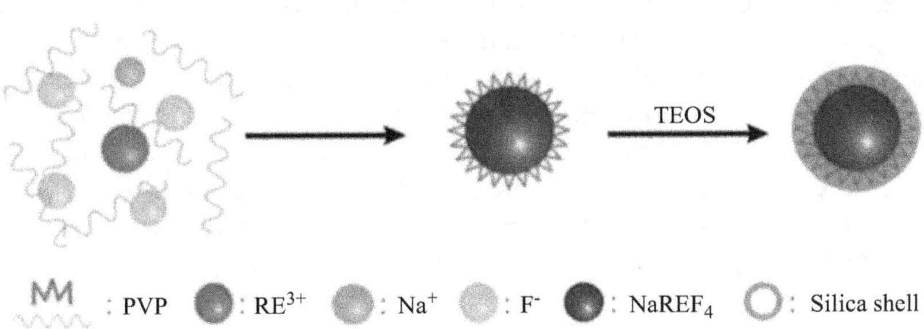

Figure 9.4 Schematic illustration of the $NaYF_4$:Yb,Er@PVP silica-coated. The concentration of TEOS controls the size of silica shell thickness [92]

with biological molecules. Another inorganic surface modification scheme is core-shell concept discussed in Section 9.2.5.

The most common surface modification strategies with organic capping ligands investigated are ligand-exchange, oxidation, absorption and free modification, as well as what is called the layer-by-layer (LbL) method. In the ligand exchange process, bifunctional organic molecules are utilised to replace the original hydrophobic OA ligands. To date, hexanedioic acid [93], 1,10-Decanedicarboxylic acid [94], 11-mercaptoundecanoic acid [94], citric acid [55,73,95], poly(ethyleneglycol)-phosphate [96], polyacrilic acid (PAA) [72,97,98] and polyvinylpyrrolidone (PVP) [99] are the most common acids and polymers used in a ligand exchange process. The ligand oxidation is an alternative method based on the chemical reaction of surface ligands [100]. The capped OA ligands of UCNPs are oxidised into azelaic acid and nonanoic acid with the Lemieux–von Rudloff reagent. This results in the generation of free carboxylic acid group on the surface due to the cleavage of unsaturated carbon-carbon bonds. In the OA ligands, the oxidised UCNPs modified with carboxylic acids exhibit the same size, shape and luminescence properties as the original UCNP. The transformation of hydrophobic UCNPs to hydrophilic materials can be achieved also through the absorption of an amphiphilic polymer onto the surface of the nanoparticles as a result of hydrophobic-hydrophobic van-der-Waals interactions. Chow *et al.* [101] reported modifying the coating of $NaYF_4$:Yb,Er@$NaYF_4$ to become an amphiphilic layer consisting of 25% octylamine and 40% isopropylamine modified PAA. Speghini *et al.* reported that the formation of an oleate double layer around colloidal CaF_2:Yb,Er/Ho/Tm nanoparticles rendered stable water suspension [102]. Wang *et al.* modified this strategy and developed a method of depositing LbL to successfully generate water-soluble UC nanoparticles. This technique involved the electrostatic attraction between oppositely charged species [103]. The LbL assembly technique permits the preparation of coated colloids of different size and shape with high stability and biocapability.

9.2.5 The core-shell concept

The dopant ions distributed at the surface of UCNPs are readily quenched through non-radiative decay pathway due to the presence of surface ligands with high vibrational oscillators, weakly bound surface impurities and crystal imperfections, as well as solvent molecules. The luminescence quenching can be reduced by forming a uniform core without sensitiser ions with small lattice mismatch around the core, to protect the host lattice and greatly minimise the surface quenching-induced emission losses. The most straightforward choice for the shell materials is the same pure host materials of the core particles with no optical active ions, called as active core/inert shell. Chow *et al.* [101] reported a significant luminescence enhancement of nearly 30 times by 1.5 nm tick shell of undoped shell materials on the surface of $NaYF_4$ Yb,Tm core particles and 7.4 times for $NaYF_4$:Yb,Er core particles. These small core/shell nanoparticles (11 nm) were rendered hydrophilic with an amphiphilic layer of polymer. A heterogeneous core-shell nanostructure can also be synthesised with different core and shell materials. In this case, the quality of core-shell structure can be characterised in transfer electron microscopy (TEM) as the core and shell will differ. To further enhance upconversion luminescence and also tunability of the upconversion luminescence, optically active centres have been introduced in the shell layers, fabricating active core/active shell structure. For example, Capobianco and co-workers have demonstrated a significant increase in the intensity of upconversion in $NaGdYF_4$:Yb^{3+}, Er^{3+} active core $NaGdYF_4$:Yb^{3+} active shell compared to $NaGdYF_4$:Yb^{3+}, Er^{3+} active core $NaGdYF_4$ inert shell and $NaGdYF_4$:Yb^{3+} active core only nanoparticles [104].

9.2.6 Quantum yield

The luminescence QY determines the efficiency of the conversion of absorbed photons into emitted fluorescence photons. Since the intensity of an upconversion process has a nonlinear dependency on absorbed pump power, the QY will be power dependence. As will be seen below, this will be a linear dependence for a 2-photon process below saturation excitation levels, gradually going towards a constant value for saturating power densities. Thus, it is critical to measure the exact excitation power density to evaluate the QY precisely. Determination of QY of conventional fluorophore has been carried out via two widely used experimental setups: a fluorometer and integrating sphere. However, there have not been any report on measuring the absolute QY of UCNPs until recently where Boyer *et al.* have successfully designed a setup based on integrating sphere using a basic spectrometer and an inexpensive diode laser [105]. In their study, they have measured the QY of $NaYF_4$ UCNPs in various sizes ranging from 10 to 100 nm. QY was measured to be in the range of 0.005% to 0.3% for particles and 3% for bulk material. A fluorometer can also be utilised to measure relative QY of UCNPs through careful optical characterisation by comparison to a reference fluorophore with known QY. Xu *et al.* reported a QY of the 800 nm emission of hexagonal-phase core-shell $NaYF_4$:Yb^{3+}, Tm^{3+} $NaYF_4$ using spectrofluorometer-based setup. A QY of 3.2% was measured under an excitation intensity of 78 W/cm^2. They theoretically compared the QY of the UCNPs with the most efficient two-photons dye under the same experimental conditions. It was observed that the two-photon dye is 8 orders of magnitude less bright for realistic power densities a few centimetres deep in tissue.

In almost all reports so far, QY values are published without consideration of their complex dependency to the power density. This ignorance leads to incomparable results from different studies. Thus, a standard protocol for QY measurement of UCNPs to provide complete information about UC is desirable. Liu *et al.* proposed a simple approach to fully describe the QY of UCNPs for arbitrary excitation power densities. In this method, a simplified steady-state condition for the population of excited energy state of Yb^{3+} is considered from a rate-equation analysis. As depicted in Figure 9.2(b)) the activator is described by a quasi-three-level model including ground state G, the intermediate state E1 and the emitting state E2. The incoming photons excite the sensitiser ground state which is well populated. The laser power (photon flux) (ρ), dopant concentration and absorption cross sections (σ)

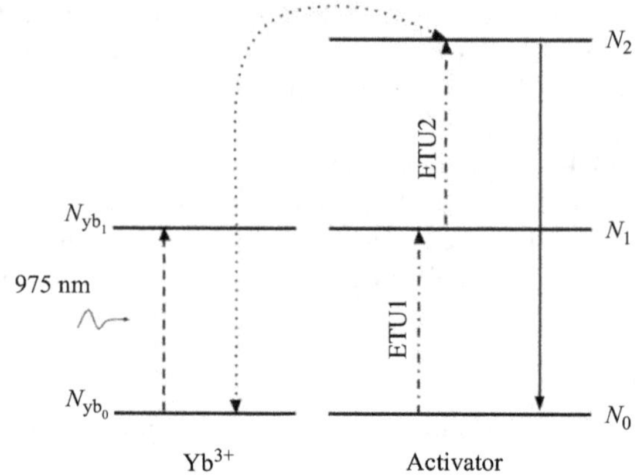

Figure 9.5 Simplified energy levels model used for the rate equation modeling

determine the ground state excitation rate. As shown in Figure 9.5, these parameters can be indicated in the following steady-state rate equations:

$$\frac{dN_{\mathrm{Yb}_1}}{dt} = \sigma \rho N_{\mathrm{Yb}_0} - \frac{N_{\mathrm{Yb}_1}}{\tau_{\mathrm{Yb}_1}} = 0 \tag{9.1}$$

$$\frac{dN_1}{dt} = C_0 N_0 N_{\mathrm{Yb}_1} - C_1 N_1 N_{\mathrm{Yb}_1} - \frac{N_1}{\tau_1} = 0 \tag{9.2}$$

$$\frac{dN_2}{dt} = C_1 N_1 N_{\mathrm{Yb}_1} - \frac{N_2}{\tau_2} = 0 \tag{9.3}$$

Where τ_{Yb_1}, τ_1 and τ_2 are the lifetimes of sensitiser excited level, activator intermediate level and activator emission level, respectively. In this equation, a term representing the depletion of the population of sensitiser excited state due to ETU is neglected since the linear decay is the dominant depletion mechanism [106]. This is also the case for the population depletion of activator emission state E2 due to ETU to even higher energy levels. After rearranging and solving the above system of equations, the QY of two-photon upconversion emission can be obtained as follows:

$$\eta \equiv \frac{P_{\mathrm{lum}}}{\sigma N_{\mathrm{Yb}_0} \rho h\nu} = \eta_s \cdot \frac{\frac{\rho}{\rho_b}}{1 + \frac{\rho}{\rho_b}} \tag{9.4}$$

with

$$\rho_{\mathrm{b}} = \frac{h\nu}{\sigma \tau_1 \tau_{\mathrm{Yb}_1} N_{\mathrm{Yb}_0} C_1} \tag{9.5}$$

and

$$\eta_s = C_0 N_0 \tau_2 \tau_{\mathrm{Yb}_1} / \tau_2^{\mathrm{rad}} \tag{9.6}$$

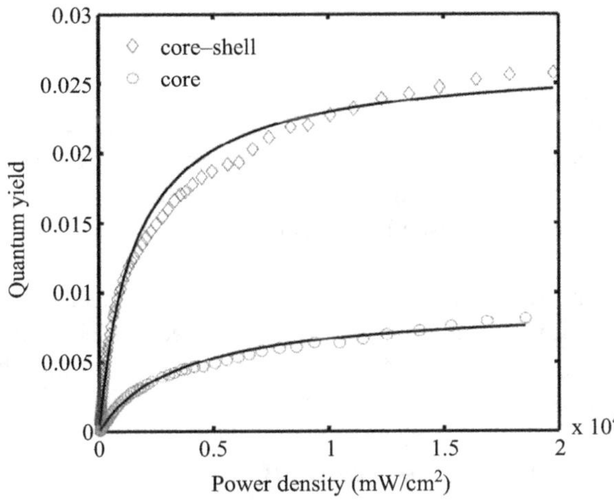

Figure 9.6 Experimentally determined QYs for the 800 nm upconversion emission band of core $NaYF_4$:Yb^{3+}/ Tm^3+ and core-shell $NaYF_4$: Yb^{3+}/Tm^{3+}@$NaYF_4$ UCNPs. The solid lines are the fitted curves using (9.4). The fitting parameters used for the core particles are $\rho_b = 3{,}800$ mW/cm² and $\eta_s = 0.91\%$, and for the core-shell particles they are $\rho_b = 1{,}300$ mW/cm² and $\eta_s = 2.6\%$. Reproduced from ref [107] with permission from The Royal Society of Chemistry

where ρ_b is referred as balancing point when the ETU rate and linear decay rate equally contribute to the depopulation of the intermediate state E1, i.e., $\frac{N_1}{\tau_1} = C_1 N_1 N_{Yb_1}$. The η_s is the maximum attainable QY of the system when the pump power density is at the saturation level and τ_2^{rad} is the radiative lifetime of excited state E2. When the excitation intensity is at the balancing power density, $\rho = \rho_b$, the QY, η, is the half of the maximum QY, η_s. As can be seen in Figure 9.6, experimental QY data acquired from measurement on core $NaYF_4$:Yb^{3+},Tm^{3+} and core-shell $NaYF_4$:Yb^{3+}, Tm^{3+} $NaYF_4$ is well fitted by the expression in (9.4).

The power density dependence of the QY can be extracted by the derivative of luminescence power, P_{lum}, over excitation photon flux, ρ, in a double log scale.

$$k \equiv \frac{d\log P_{lum}}{d\log \rho} = 1 + \frac{1}{1 + \tau_1 \cdot C_1 \tau_{Yb_1} \sigma N_{Yb_0} \rho} \tag{9.7}$$

The slope efficiency, k, has a value of 1.5 at the balancing power density point. Under low excitation intensities, where the linear decay from E1 to G state is dominant over the ETU rate, the slope efficiency will be close to 2, indicating a quadratic dependence on the excitation intensity. An extra consideration needs to be applied in UCNPs QY measurement where non-uniform excitation spatial radiation, typically with a Gaussian profile, is illuminating the sample. A CCD-based detector was used to record the intensity distribution of a laser beam cross-section in real time to compensate for change on excitation power density to provide accurate QY information [108,109].

9.3 Applications

The advantages of UCNPs as imaging probes are derived from their unique upconverting luminescence (UCL) properties and also from the NIR radiation used for exciting UCNPs. Compared to visible light excitation, NIR excitation for *in vivo* imaging provides several advantages such as remarkable light penetration depth, weak autofluorescence, low tissue heating and weak non-specific photo-damage to biological samples. Therefore, the Ln-doped UCNPs are an attractive choice for biomedical applications, primarily for *in vivo* imaging and assays. To date, few aspects within small animal imaging have been studied, which include whole small animal body imaging, *in vivo* tumour targeting imaging, multimodal animal imaging, immunoassay, PDT and photo-activate ions.

9.3.1 Whole body small animal imaging

An early study on imaging of small animals using Ln-doped nanoparticles is reported by Zhang *et al.* injecting 100 μl of polyethyleneimine (PEI)-coated $NaYF_4$:Yb,Er nanoparticles subcutaneously in anaesthetised Wistar rats [39]. Luminescence signals were observed up to 10 mm beneath the rat's skin. The *ex vivo* biodistribution of PEI coated nanoparticles was investigated 0.5 h, 24 h and 7 days after injection. A rapid accumulation in the lung was observed 30 min after injection while the concentration of nanoparticles in all tissues was decreased 24 h after injection with highest concentration in the spleen. The nanoparticles were undetectable by 7 days post-intravenous injection. A long-term *in vivo* distribution study of PAA-UCNPs ($NaYF_4$:Yb,Tm) with NIR-to-NIR UCNP emission was conducted by Li's group by an UCNP imaging method [97]. The distribution analysis shows a long-term retention of nanoparticles in the lungs and spleen, similar to other reports. To date, a high contrast UCL imaging of whole body small animals with a penetration depth of 20 mm has been achieved [73]. Moreover, ultra-high sensitive stem cell labelling for long-term *in vivo* tracking was generated and used to detect as few as 10 cells in a mouse [110]. Biodistribution studies are useful indicators for studying the toxicity effects of the UCNPs. The absence of short term toxicity has been proven so far in various studies, encouraging further exploration of UCNPs for *in vivo* biomedical applications.

9.3.2 Tumour targeting imaging

Development of tumour targeting molecular imaging using upconversion materials is still a great challenge due to insufficient tumour accumulation which limits their further clinical application. However, their ability to easily bind targeting molecules is one important aspects of the use of UCNPs as bioprobes for bioimaging. The utilisation of Ln-UCNPs have been documented for tumour targeting imaging in few cases based on ligand-acceptor and antigen-antibody interactions. For instance, folic acid modified UCNPs and arginine-glycine-aspartic acid (RGD) peptide have been employed as a targeting agent for *in vivo* tumour targeting. On the basis of high affinity between folic acid (FA) and folate receptors (FR), the first successful example of *in vivo* targeting imaging with mice bearing HeLa tumours was reported [87]. Strong upconversion emission signals were collected from the location of the tumour. Afterwards, several FA-modified UCNP have been employed in various studies for targeted bioimagining of tumour cells [39,111–113]. Li and co-workers also used the RGD (RGD = arginine-glycine-aspartic acid) peptide conjugated UCNPs for targeted imaging of mice bearing U87MG tumours [114]. These probes show a significant increase in the avidity of α_v/β_3 integrin receptors, being over expressed in human glioblastoma U87MG tumours. It has been well known that most antibodies have high affinity for their antigens. The antigen-antibody reaction can be widely used as a useful strategy for targeted recognition for luminescence bioimaging. For example, anti-Her2 antibody was conjugated with nanoparticles to image cancer cells over-expressed with a protein called Her2 [115]. An obvious fluoresce pattern under a modified epifluorescence microscope was observed in the tumour as a direct result of Her2 recognition, whereas no significant fluorescence could be detected in a control group. In addition, the use of UCNPs as a feasible cell tracker tool was demonstrated by Zhang *et al.* The long-term stability of UCNPs has allowed to obtain information on cell dynamics at high resolution inside living animals [116]. Multicolour *in vivo* UCNP imaging is demonstrated in multiplexed lymph node mapping and multicolour cancer cell tracking by using a series of UCNPs with three different emission colours [117]. Moreover, vascular imaging of the mouse ear was performed. The probe was clearly visible more than 2 h post injection [118].

9.3.3 Multimodal imaging

Within the vision of future medical imaging to detect life-threatening diseases long before clinical manifest, even at the time of the earliest cellular changes that define the disease, improved diagnostics are needed. Two main lines towards meeting this vision are the use of molecular and multimodal imaging. Molecular imaging is the labelling of contrast agents with tumour/tissue specific targeting agents, for example, antibodies, affibodies, peptides and aptamers, by which it is possible to detect changes that occur on a cellular or molecular level. Multi-modal imaging is the combination of several imaging technologies to extract more information by combining the advantages of each technology, such as high spatial resolution, quantification ability, and portability. Key for this would be to accomplish additional value in using a combination of techniques, not to just get the same information in a multitude of ways. This development is expected to have a major impact on the detection and individualised treatment guidance of many diseases, the development of new drugs and our understanding of how disease arises.

UCNPs can be very interesting in the context of molecular sensitivity and multimodal approach, as they offer a contrast agent that could be used for both purposes. The molecular sensitivity can be achieved by coating the particles with specific targeting agents. Appropriate modification of the surface can provide selective binding to specific molecules. For this, the size of the particles is also of interest, as this is a critically important parameter for the biodistribution and pharmacokinetics of the compound following administration. Bigger particles will not be able to reach all parts of the body, and thereby not the molecules they are aimed to target. As a consequence most molecules targeted should be localised outside or on the cell membranes, i.e., for instance, being expressed on the cell membrane of the cells to be labelled. To obtain the capabilities of multimodal approach, the nanoparticles should be active in several respects. Upconversion nanoparticles provide an ideal platform for achieving multifunction, for example, multimodal imaging, sensing and therapy. This can be accomplished primarily by combining various nanomaterials to construct complicated heterogeneous nanostructures. This can be possible in multiple ways, depending on the techniques to combine in the imaging. By exchanging the commonly used matrix from $NaYF_4$ to one with a crystal lattice with Gd ions, like for instance NaGdF4, the nanoparticles would be detectable also with MRI [119–122]. The nanoparticles can also be optimised for simultaneous upconverting and downconverting imaging [123], providing also an opportunity to dual optical imaging.

Another approach that could be considered for multifunctional particles, would be to consider developing superparagmagnetic iron oxide (SPIO) nanoparticles coated with a UCNP shell [124,125]. Such particles would be possible to detect with magnetomotive ultrasound imaging [126], and with PET/SPECT if also attached to a radionucleid [127].

9.3.4 Bioassays

Bioassays are experiments that measure the effects of the substance on a living reporter system and are essential to provide information about biochemical substance such as a drug or bioactive substance. To achieve a high sensitive detection and accurate analysis of single biomolecules, efficient bioassay methods based on Ln-doped UCNPs probes have been developed. UCNPs are photostable and exhibit long-lived emissions even in an excitation field of relatively high power. In addition, the narrow emission line width of lanthanide ions in hosts (with full width at half maximum of 10–20 nm) increases the potential of changing the lanthanide dopants concentration to achieve distinguishable emission bands, enabling a large number of multiplexed detections [128]. The currently established biological assays are generally categorised into heterogeneous and homogeneous bioassays. The first use of upconverting phosphors in heterogeneous bioassay technique was reported by Zijlman and co-workers in the year of 1999 [129]. Two UCL reporters, including a green emitting ytterbium/erbium Y_2O_2S:Yb,Er and blue-emitting ytterbium/thulium Y_2O_2S:Yb,Tm particles conjugated to NeutrAvidin were used in a model system consisting of human prostate specific antigen. Later, Tanke *et al.* used the green emitting phosphor particles (Y_2O_2S:Yb,Er) to identify human papillomavirus type 16 infection via detection of specific nucleic acid molecules sequences [130]. Cooper *et al.* used phosphor particles with a composition of Y_2O_2S:Yb,Er and Y_2O_2S:Yb,Tm in lateral flow (LF) assays on human chorionic gonadotropin (hCG) to benefit from the advantages of this assay format with those of the UC phosphors reports system [131]. A detection limit of 10pg hCG in a

100 μl sample was achieved, which represents at least a 10-fold improvement over conventional reporter systems. The possibility of detecting two different targets simultaneously on the same membrane strip colour coded with two different UC phosphors compositions was proven as a concept of assay multiplexing. This was furthermore demonstrated by Niedbala *et al.*, where they developed a lateral flow strip assay format for simultaneous detection of amphetamine, methamphetamine, phencyclidine and opiates in saliva [132]. The 400-nm diameter spherical UCP particles were covalently conjugated to antibodies and were placed into the test by premixing with the sample or dried into a sample pad. Smaller magnetic and upconversion luminescent nanoparticles were produced through magnetic-field associated biochemical separation and concentration technology [133]. The magnetic nanoparticles were approximately 150 nm and were conjugated with nucleic acid for sensitive detection of trace amounts of DNA. The possibility of printing multiple antibodies was demonstrated by Hanna Päkkila *et al.* [134]. Three model analytes, including prostate specific antigen, thyroid-stimulating hormone and luteinising hormone, were selected to be used in evaluation of the performance of array-based immunoassay using a mix of secondary antibody-coated UCNP. Very recently, Kale has demonstrated spectrally and spatially multiplexed array-in-wall immunoassay for simultaneous classification and detection of serological samples utilising green-emitting IgG-coated UCNPs and blue emitting IgM coated to detect human IgG and IgM, respectively [135]. In this method, the differentiation between antigen specificity and antibody class was achieved based on the position and the colour of the signal, respectively. UCNP-based bioassays can achieve much more sensitive detection than conventional ones due to their desirable photophysical properties. Moreover, they are inexpensive and allow for multiplexing and quantitative detection. The development of new optical encoding strategies for multiple analyte detection is well established endowing the unique properties of UCNPs. Tunable luminescence lifetime of UCNPs in the microseconds region is an additional attractive parameter that can be used for optical encoding. By manipulating the luminescence decay lifetime, one can create a temporal coding dimension from the micro to milliseconds range in a single colour band [136]. Each code can be well-separated at this time scale to avoid overlapping of the signals.

A homogeneous assay platform generally utilises a sandwich structure to measure the analyte concentration and biological activities in biological fluid based on sensitive and reliable fluorescence resonance energy transfer (FRET) technique. In a conventional FRET process, the labelled donor and acceptor molecules are capable to interact in the presence of analyte through a luminescence energy transfer. UC luminescence resonance energy transfer (LRET) biodetection utilises UC nanoparticles as ideal energy donor insomuch as the noise of the autofluorescence background can be thoroughly removed as a result of NIR excitation. Moreover, unwanted detection signal due to direct excitation of acceptor can be completely ruled out, thus offering high detection sensitivity and signal-to-noise ratio. In 2005, Soukka's group has successfully demonstrated the sensitive detection of biotin with a limit of detection (LOD) of 0.7–9 nM by utilising Y_2O_2S:Yb,Er phosphors as a donor and a fluorescent phycobiliprotein (bio-PBE) as an acceptor [137]. Subsequently, they applied this technology for simultaneous detection of multiple analyte without interference from autofluorescence by using sub-micrometre-sized particles which produce multiple narrow emission band at visible wavelength under IR excitation [138]. However, due to the strong distance dependency of the energy transfer efficiency the use of large size particles limits the detection sensitivity. A highly sensitive bio-sensor to quantitatively detect target DNA with LOD of 1.3 nM has been exploited [139]. In addition to organic dyes to be used as acceptor, metallic NPs such as gold NPs and graphene oxide have also been utilised as superior quenchers of UC donor. For instance, Amino-functionalised $NaYF_4$:Yb, Er UCNPs and gold NPs (Au NPs) were prepared and conjugated with the human immunoglobulin G (IgG) and rabbit anti-goat IgG, respectively. This sandwich-type LRET-based approach can reach a low detection limit (0.88 μg/mL) of goat antihuman IgG [140]. Recently, Chu and co-workers designed a bio-sensor comprising the peptide-functionalised $NaYF_4$:Yb, Er UCNPs as energy donor and graphene oxide as energy quencher for the sensitive and selective determination of human immunodeficiency virus (HIV) antibody [141]. A detection limit of 2 nM was reported. For the detection of cholesterol level in human body, Ding and co-workers developed a upconversion nanocomposite composed of β-cyclodextrin (β-CD) derivative-modified $NaYF_4$:Yb, Er UCNPs and Rhodamine B [142]. It was observed that the detection accuracy reaches down to approximately 3.0 μM in the concentration range of 10–110 μM. Latterly, these efforts clearly demonstrate that Ln-doped UCNPs are excellent donor labels in homogeneous LRET and are capable to extend the application of FRET techniques to be engaged in clinical, food and environmental monitoring.

9.3.5 Photo-activation

There is a great potential in activating molecular reactions with light as it enables another means to remotely control molecular events in tissue. A slight complication is that many reactions only can be initiated with short, energy-rich wavelengths. An issue with this is that such short wavelengths do not penetrate tissue well, limiting the activation capability to superficial locations. For deeper tissue activations, UCNPs can be employed as mediators, facilitating local conversion of NIR-to-violet light. As NIR light penetrates much deeper into tissue, UCNP can provide the means to conduct non-invasive photo-activation below the very first superficial cell layers without artificially guiding the light into the tissue. The idea of using UCNPs for this purpose was generated already about 10 years ago, when UCNPs started to become efficient for biomedical use. Zhou *et al.* demonstrated then an NIR-controllable luminescence switch based on UCNPs. They were utilising intermolecular energy transfer within a photochromic diarylethene derivative and UCNPs loaded in a poly(methyl methacrylate) (PMMA) film [143]. The use of UCNPs for *in vivo* photo-activation has since gained a considerable interest. Many studies have been conducted in the development and evaluation of the potential of UCNP-based photo-activated chemical reactions. Examples of such research projects include activation of carboxylic acid, therapeutic nitric oxide, chlorambucil (an anti-cancer drug), doxorubicin (another anti-cancer drug) [144–147].

Besides small molecules, complex biomolecules like proteins and nucleic acids have also been photo-caged and photo-activated using UCNPs. The ability to use of UCNPs for uncaging biomolecules has been shown both *in vitro* and *in vivo* [148]. In that study thiolated silane-modified $NaYF_4$:Yb,Tm@$NaYF_4$ core-shell UCNPs were caged with D-luciferin. When incubated with the enzyme luciferase, uncaging of D-luciferin occurred and produced bioluminescence. This type of research was advanced by the group of Zhang, successfully demonstrating the use of NIR-to-UV UCNPs for light-controlled gene expression and knockdown [149]. In that study, plasmid DNA and siRNA were first converted to a non-active form by chemically photo-caging. The nucleic acids were bound to $NaYF_4$:Yb,Tm UCNPs coated with mesoporous silica. By exciting the UCNPs with NIR light, they demonstrated it is possible to locally generate sufficient amount of UV light to photo-reactivate the nucleic acids and expressed green fluorescent protein in B16F0 cells (a melanoma cell line). This was demonstrated both on a cell culture and when the cells were located in tissue at depths down to 4 mm. In a next step, this group successfully showed the use of UCNPs as mediators for gene therapy [150]. A similar methodology as above was employed. They managed to accomplish an efficient tumour reduction in a mouse model following NIR excitation. The idea behind the project was to employ NIR light to knock down the expression of the STAT3 (signal transducer and activator of transcription 3) gene. To accomplish such an activation, NIR-to-UV converting UCNPs were co-loaded with STAT3 photo-morpholino (a UV-triggered short nucleic acid analogue) and delivered into B16F0 cells. These cells were then used to induce tumours in the C57BL/6 mice model. The results illustrate the potential of utilising UCNPs for efficient photoactivation deep in tissue for gene therapy purposes. Further another study illustrated the possibility for employing UCNPs in optogenetics, to optically activate the channel Rhodopsin 2 modified mechanosensory neurons in *Caenorhabditis elegans* [151]. The activation was conducted using NIR excitation of UCNPs. The locally emitted blue light following the NIR absorption in the UCNPs stimulated the channel Rhodopsin in the neurons, triggering the *C. elegans* to change direction in its movement, see Figure 9.7. These studies provide hopes that NIR diode-laser excited UCNPs can be used as a tool to photo-activate vital chemical reactions, useful in fundamental biological research as well as in the future potentially also in medical applications.

9.3.6 Photodynamic therapy

Photodynamic therapy constitutes a special branch of photo-activation, and is given a special subchapter here because of its potential and special academic challenges in fully understanding the processes involved, as well as in optimising the treatment parameters. Photodynamic therapy is a non-thermal treatment modality involving a catalysing photosensitiser in combination with light to initiate a photochemical reaction involving oxygen. The aim of this photochemical reaction is to cause selective tissue ablation or cell death. Photodynamic therapy has mainly been utilised as a treatment option for malignant tumour destruction, but is increasingly frequented also in the management of infections. The latter capability is of particular

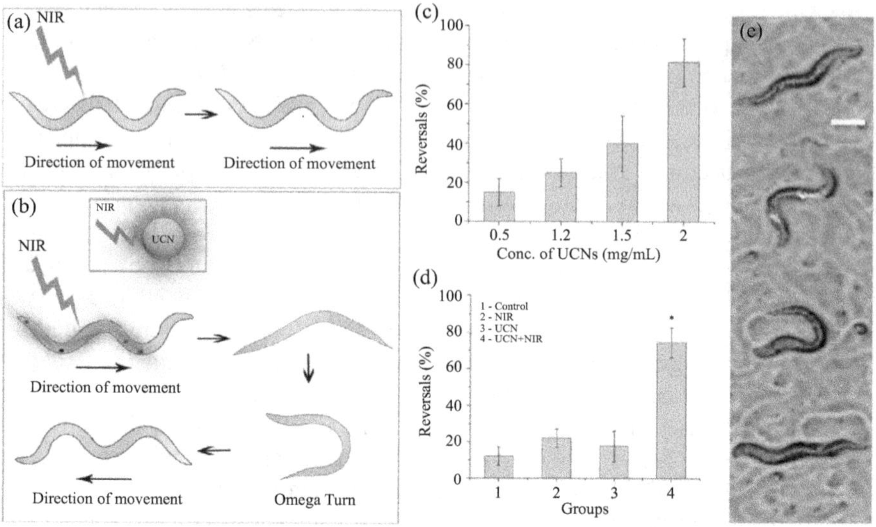

Figure 9.7 This image illustrates how it is possible to trigger a C. elegans to change direction of movement with light. This figure is reproduced with permission from Bansal et al. [151]. Here follows a slightly modified figure caption from the paper: Schematic showing differences in movement behaviour in worms expressing ChR2 in their mechanosensory neurons (a) when exposed to NIR light alone (no UCNPs) and (b) when exposed to NIR light in the presence of UCNPs. The worms move fluidly without drastic changes in direction in panel (a). However in panel (b), the worms show a drastic change in direction akin to being touched due to the activation of the mechanosensory neurons caused by the activation of ChR2 via the blue light produced by UCNPs upon NIR irradiation. (c) Effect of varying UCNP concentration on percentage of worms showing reversal phenotype when irradiated with NIR light at 50% duty cycle and peak power 4 W. n = 20 per group. (d) Percentage of worms showing the reversal phenotype upon incubation with UCNPs and subsequent NIR irradiation, along with the controls (untreated worms, worms exposed to NIR light alone, or worms exposed to UCNPs alone). p-value = 0.0004 between the control and UCNP + NIR groups. n = 90 per group. (e) Representative images (video stills) showing the reversal process in the UCNP+NIR group

interest in view of the increase in methicillin-resistant *Staphylococcus aureus* infections. Photodynamic therapy provides a new tool with the potential to be efficient also for resistant antibacterial and anti-fungal infections. A reason for that is that the PDT mechanisms are not primarily targeting the cell nucleus, resulting in poor ability for microbes to modify themselves to develop resistance against PDT.

The potential for a unique treatment selectivity is obtained through the light delivery, photosensitiser localisation and presence of oxygen in the tissues. This multiparameter sensitivity function is valuable as it provides great means to generate a very selective treatment, while it is also making the modality complex and the treatment planning challenging. Well planned and correctly conducted, PDT can provide a treatment with high precision for optimised outcomes in each individual treatment. The modality promises a minimally invasive treatment (repeated if necessary) as no *per se* harmful substances are employed. It is thereby offering selective and efficient ablation of malignant tissue volumes or infections, while sparing surrounding tissues and yielding minimal side effects.

Typically PDT employs NIR activatable photosensitisers, as the NIR light penetrates tissue more readily, facilitating treatment of more than very superficial and thin lesions. Interstitial light delivery is another route increasingly frequently employed to be enable treatment of deeply located treatment volumes or thick lesions. A remaining challenge is that the photosensitisers utilised today still have rather limited absorption cross sections and QYs, requiring high fluence to achieve the desired treatment. This is in particular challenging away from the location of light delivery, when light has been attenuated in the tissue in passing to the target volume. Also, the photosensitisers are typically activated at a still non-ideal wavelength in light penetration aspects, as in the selection of appropriate wavelength to employ, one has to balance the light penetration with the absorption properties of the photosensitiser used. An unmet need is consequently to develop a photosensitiser with properties as close as possible to what is desired in terms of efficiency and wavelength for the activation with optimal light penetration properties. Employing UCNPs as local wavelength converters could be a means to accomplish these needs, as it provides a tool to somewhat uncouple the absorption wavelength (influencing the light penetration) and the quantum efficiency in the subsequent photochemical reaction generating the cell depth. The efficiency can also be addressed by modifying the mechanistic reactions involved. One such very interesting development of PDT is what is called photochemical internalisation (PCI) [152,153].

The intention with using UCNPs as mediators in PDT is simple, they will convert far NIR to violet light locally inside the tissue to treat. In this way NIR light at a wavelength that penetrates tissue well can be utilised to convert the absorbed light to a wavelength that is optimal for the photosensitiser. Gain can be achieved as the treatment absorption cross section multiplied with treatment QY for the conventionally used photosensitisers are generally much higher for violet light than for the most used red or NIR wavelengths. This gain has to overcome the losses we have in the absorption cross section times the QY for the violet emission by the UCNPs employed. This might be non-trivial to accomplish as the QY for the UCNPs is generally low (in the per cent range) and also dropping for low light fluence rates. For that reason one needs to critically evaluate the potential gain before proposing this as a general tool to improve PDT treatment at deep locations in tissue. Figure 9.8 illustrates this issue, presenting a back-of-an-envelope calculation for which the condition that the QY is unaltered for the photosensitiser for the two treatment scenarios, meaning it is independent whether it is excited at red

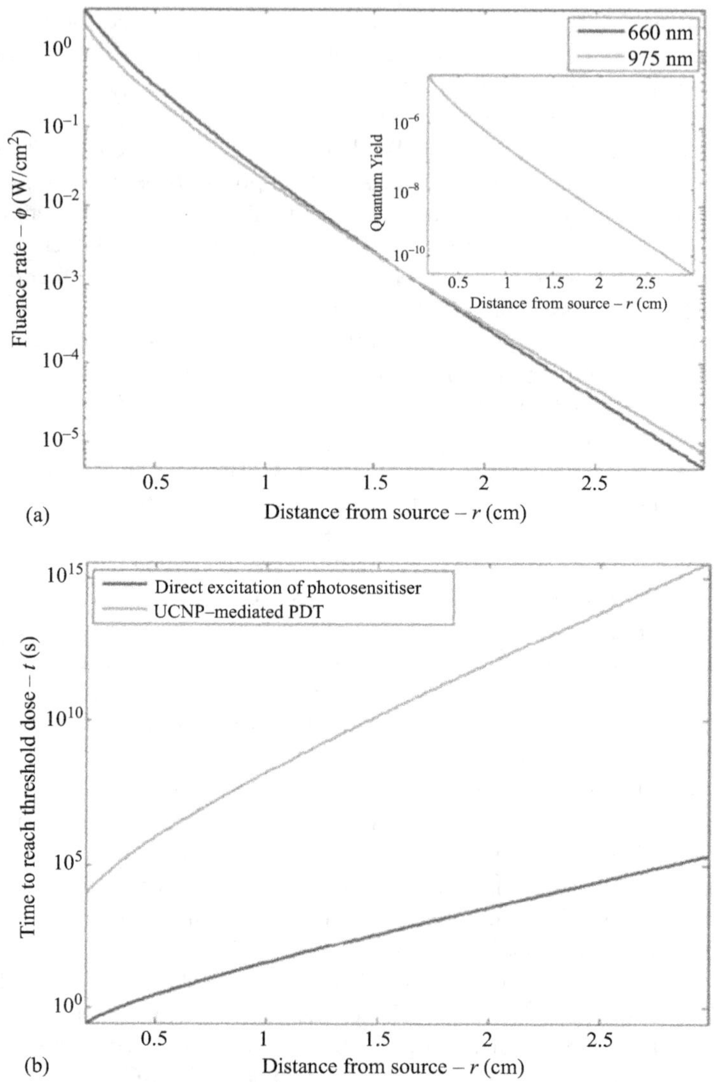

Figure 9.8 The panels are used with permission from Xu et al. [4], and illustrates plots from simple estimates of light fluence rates and treatment times versus depth in simulated prostate tissue. (a) Calculated fluence rates within human prostate tissue following 660 nm and 975 nm excitation, respectively. The modelling is based on light diffusion theory using the optical properties of the tissue detailed in the reference. The inset shows the corresponding QY for the UCNPs (the QY for the photosensitiser at 660 nm is constant) as a function of depth. (b) The treatment time needed to reach the light threshold dose of 1 J/cm^2

or violet wavelengths. In this case the necessary light energy density for reaching the treatment threshold is much higher for the UCNPs-mediated treatment than for the conventional photosensitiser.

The reality is obviously more complex than the simple model used to present Figure 9.8. This is clearly illustrated by the results presented by the group of Zhang [154–157] and others, see, for example [158]. They demonstrate enhanced treatment response by employing UCNPs mediated PDT as compared to conventional photosensitisers. To obtain these results they used UCNPs coated with a thin layer of TiO_2 as a photosensitiser. This method avoids the previously identified limitation with leaking and uneven loading of the photosensitising molecules, as well as the possibility for the generated radicals can reach the critical cellular target structures.

As for the other photoactivation ideas, there are still many remaining challenges and questions, before this modality of PDT will gain clinical interest. Such questions include:

- How efficient is the singlet oxygen production?
- How far away from a structure can the particles be to still be able to destroy critical cellular structures? Does this constitute a challenge? Where do you need to have the particles located in order to obtain optimal damage?
- Treatment at depth:
  - How is the treatment depth normally limited in conventional PDT?
  - What are the properties to compare for the photosensitiser and UCNP construct to be able to predict their efficiency for treating deep tissue locations?
  - What are the mechanisms that potentially will make the use of UCNP-mediated PDT at deep tissue locations favourable?

That said, it is remarkable promising results, given the doubts of the potential given in Figure 9.8. We are looking forward to follow the future development of UCNP-mediated photoactivation and PDT.

9.4 Sources

In order to make the UCNPs tools attractive and user-friendly, the optical arrangements employed must be acceptable both in terms of size and costs. Semiconductor sources would thereby be ideal to use as they are both small and cost efficient. What are then the additional desired properties of the light source to efficiently excite the UCNPs, sometimes located deeply into tissues?

- Wavelength – the light needs to penetrate tissue well and cause minimal tissue heating. It should also be efficiently absorbed by the UCNPs.
- Temporal profile – due to the multiphoton stepwise excitation, the pulse length should match the lifetime of the intermediate state.
- Optical integration – the light source should allow that the light can be efficiently delivered to the target tissue
- Power stability – the source should be sufficiently stable to provide a robust signal. This includes both short and long-term stability.

Firstly, for the wavelength considerations the near infrared region yields the best light tissue penetration between 700 and 900 nm. With low absorption comes also minimal tissue heating, as the heating is generated by the light absorbed in the tissue. There are a number of strategies for obtaining good absorption of the UCNPs as described above. Most frequent is to use Yb as sensitiser ion. This ion absorbs best at 975 nm. Alternatively Nd ions can be employed, absorbing around 808 nm. Some studies have also proposed to use molecular dyes in the UCNP coating to increase the absorption cross section. This scheme relies on transferring the absorbed energy from the dye to the rare-earth ions, yielding that also these should absorb in the NIR wavelength range. The benefit would be a higher absorption cross section of the dye as compared to the rare-earth ion, and also more flexibility in the absorption wavelength as compared to the discrete number of lines available for rare-earth ions.

Secondly, the light pulse length has to match the intrinsic lifetime of the UCNPs. This is because following the first excitation event in an ETU process, the second should provide an energy transfer to an already excited ion. This timing has been understood by using rate-equation analysis [107,159]. The first energy transfer must thus have been completed and the intermediate state been populated, and at the same time it should not have been depopulated through other spontaneous relaxations. This process is thus quite different from how one can employ chromophores for multi-photon excitations, as such excitation does not rely on an intermediate state, but the various photons yielding the multi-photon excitation must be present simultaneously, within the temporal uncertainty determined by the line width of the light. Typically, the intermediate state is best populated within the timeframe of milliseconds. Thus, the optimal light pulse is of that pulse length. Longer pulses will dilute the photon density within the active excitation process, while shorter pulses will provide a considerable fraction of the photons before the intermediate state is populated. Following this reasoning, longer pulses should be employed the more energy transfer process used in the excitation – meaning three or more photon processes. The pulse energy and repetition rate is then limited by the permitted tissue exposure limits. For skin at 975 nm, this is 200 $mWcm^{-2}$ average power. This also provides a relation between pule energy and repetition frequency. Here it is an advantage to maximise the peak power, on the expense of low repetition rate, as the upconversion process is characterised by a non-linear behaviour up to a saturation limit.

Thirdly, for most applications, the light will be guided through an optical fibre. This means the beam quality from the source should be efficiently coupled into the fibre. The diameter of the fibre is usually of less importance. Frequently 400–600 μm core diameter fibres are used. One part of the reason for this is to allow simple and efficient coupling of the light source into the fibre. Another reason is that one typically will illuminate a larger area of the tissue with some known, usually top-hat illumination profile to just balance the illumination to what is permitted. Therefore, the tip of the fibre is frequently imaged with a lens on the tissue surface, with a magnification that provides the desired illumination beam size. For practical reasons a larger core fibre diameter makes such magnification optics simpler and, most of all, more compact.

Lastly, the light source needs to be sufficiently stable to secure reliability and repeatability in measurements or photoactivation. Short-term stability means the power profile within a millisecond pulse. The source needs to provide a controllable power profile within each pulse. The first intention would naturally be to provide a constant power throughout each light pulse. This is what the relatively course rate-equation analysis suggests. With improved rate-equation analysis we do not exclude that it may be better to design the temporal pulse profile differently to further improve the possibility to generate as high populations in the desired energy level in the activator ion as possible, within the limits presented by the light economy determined by the permitted light exposure, as well as by the properties of the light source employed. The long term stability is obvious to guarantee repeatability. When discussing repeatability one need to understand that this is a multi-photon process, sometimes with many photons involved. The results can thus be very sensitive to the excitation power.

In order for UCNP-based technique to become translated to become well-adopted scientific tools, these requirements must be fulfilled and especially affordable instruments need to be developed. With the great promise in the techniques UCNPs can mediate, our strong believe is that instrumentation will be developed to fully facilitate the potential of the techniques described above as well as techniques yet to be invented. The instruments obviously include appropriate excitation sources. Here cost effective diode lasers seem to be very appropriate.

9.5 Summary

We have in this chapter tried to provide a brief overview of the field of UCNPs for biomedical applications and given a flavour of its promising potential and future. Our interpretation is that UCNPs will become a valuable agent for several applications, some of them still to be invented/studied. For imaging it has great promises mainly due to the possibility to avoid any background from tissue autofluorescence, as well as the ability to simultaneously label several targets by UCNPs with different dopants. In this chapter we have not discussed applications within immunohistochemistry and digital pathology, as these fields have not been studied in any details. We believe these fields could become important application areas of UCNP labelling. Furthermore, we predict a very rapidly rising and fascinating field in general UCNP-photoactivation, with the demonstrated possibility to employ UCNPs as mediators to convert NIR light to violet light *in vivo,* and the rapidly growing interest in photo-activation in basic research in situations where one easily can access the tissue location of interest with NIR light. Here is, in our opinion, a promise for novel applications to emerge, see, for example [27,160].

By describing not only the biomedical applications possible benefiting from the properties of the UCNPs, but also providing their photophysical properties, we hope this chapter can form a valuable source for introducing the field. We have aimed to stress properties of an optimal light source, and thereby established a clear link to the use of diode lasers and LEDs. In the field of UCNPs the excitation is particularly relevant, as the upconverting excitation scheme is a non-linear to its character, with multiple sequential excitation steps. Due to the long lived intermediate state, or states, if more than a two-photon process is employed, saturation of the excitation can easily be observed when the excitation power density increases. Also, the lifetime(s) of the intermediate state(s) determines how short pulses should be used, typically in the range of milliseconds. This means that modulated CW lasers are preferred, suggesting the use of diode lasers for small illumination areas when focussing is important. For broad field illumination LEDs might be suggested, if any emission in the detection band can be avoided.

Acknowledgement

We would like to thank several previous students in the group and other collaborators in Lund who have been central in much of our research directed towards UCNPs as novel contrast agents for bioimaging. Especially we would like to express our warmest gratitude to Haichun Liu, Pontus Svenmarker and Can X. Xu for their key contributions to our research at Lund University, forming the background for this book chapter. In addition, we very much appreciate the collaboration with Peter E Andersen and Ole B. Jensen at DTU, as well as with the group of Yong Zhang at National University of Singapore, in the area of UCNPs. The work was financially supported by the Science Foundation Ireland (SFI/15/RP/2828) and the Swedish Research Council (VR/621-2013-6038).

References

[1] Heike S Mader, Peter Kele, Sayed M Saleh, and Otto S Wolfbeis. Upconverting luminescent nanoparticles for use in bioconjugation and bioimaging. *Current Opinion in Chemical Biology*, 14(5):582–596, 2010.

[2] Daniela E Achatz, Reham Ali, and Otto S Wolfbeis. Luminescent chemical sensing, biosensing, and screening using upconverting nanoparticles. In *Luminescence Applied in Sensor Science*, pp. 29–50. Springer, 2010.

[3] Min Lin, Ying Zhao, ShuQi Wang, *et al.* Recent advances in synthesis and surface modification of lanthanide-doped upconversion nanoparticles for biomedical applications. *Biotechnology Advances*, 30(6):1551–1561, 2012.

[4] Can T Xu, Qiuqiang Zhan, Haichun Liu, *et al.* Upconverting nanoparticles for pre-clinical diffuse optical imaging, microscopy and sensing: Current trends and future challenges. *Laser & Photonics Reviews*, 7(5):663–697, 2013.

[5] Anna Gnach and Artur Bednarkiewicz. Lanthanide-doped up-converting nanoparticles: Merits and challenges. *Nano Today*, 7(6):532–563, 2012.

[6] Hans H Gorris and Otto S Wolfbeis. Photon-upconverting nanoparticles for optical encoding and multiplexing of cells, biomolecules, and microspheres. *Angewandte Chemie International Edition*, 52(13):3584–3600, 2013.

[7] Matthew V DaCosta, Samer Doughan, Yi Han, and Ulrich J Krull. Lanthanide upconversion nanoparticles and applications in bioassays and bioimaging: A review. *Analytica Chimica Acta*, 832:1–33, 2014.

[8] Frederik Maes, Andre Collignon, Dirk Vandermeulen, Guy Marchal, and Paul Suetens. Multimodality image registration by maximization of mutual information. *IEEE Transactions on Medical Imaging*, 16(2):187–198, 1997.

[9] Bernard W Stewart and Christopher P Wild. *World Cancer Report 2014*. International Agency for Research on Cancer, WHO, 2014.

[10] Ahmedin Jemal, Freddie Bray, Melissa M Center, Jacques Ferlay, Elizabeth Ward, and David Forman. Global cancer statistics. *CA: A Cancer Journal for Clinicians*, 61(2):69–90, 2011.

[11] Ruth Etzioni, Nicole Urban, Scott Ramsey, *et al*. The case for early detection. *Nature Reviews Cancer*, 3(4):243–252, 2003.

[12] Vasilis Ntziachristos, Jorge Ripoll, Lihong V Wang, and Ralph Weissleder. Looking and listening to light: The evolution of whole-body photonic imaging. *Nature Biotechnology*, 23(3):313, 2005.

[13] Ralph Weissleder and Mikael J. Pittet. Imaging in the era of molecular oncology. *Nature*, 452(7187):580–589, 2008.

[14] Simon R Arridge and John C Schotland. Optical tomography: Forward and inverse problems. *Inverse Problems*, 25(12):123010, 2009.

[15] John V Frangioni. In vivo near-infrared fluorescence imaging. *Current Opinion in Chemical Biology*, 7(5):626–634, 2003.

[16] Rebecca Richards-Kortum and Eva Sevick-Muraca. Quantitative optical spectroscopy for tissue diagnosis. *Annual Review of Physical Chemistry*, 47(1):555–606, 1996.

[17] Vasilis Ntziachristos, Ching-Hsuan Tung, Christoph Bremer, and Ralph Weissleder. Fluorescence molecular tomography resolves protease activity in vivo. *Nature Medicine*, 8(7):757–760, 2002.

[18] Vasilis Ntziachristos, Eyk A Schellenberger, Jorge Ripoll, *et al*. Visualization of antitumor treatment by means of fluorescence molecular tomography with an annexin v–cy5. 5 conjugate. *Proceedings of the National Academy of Sciences of the United States of America*, 101(33):12294–12299, 2004.

[19] Arno Villringer and Britton Chance. Non-invasive optical spectroscopy and imaging of human brain function. *Trends in Neurosciences*, 20(10):435–442, 1997.

[20] David A Boas, Anders M Dale, and Maria Angela Franceschini. Diffuse optical imaging of brain activation: Approaches to optimizing image sensitivity, resolution, and accuracy. *Neuroimage*, 23:S275–S288, 2004.

[21] Stephanie van de Ven, Sjoerd Elias, Andrea Wiethoff, *et al*. Diffuse optical tomography of the breast: Initial validation in benign cysts. *Molecular Imaging and Biology*, 11(2):64–70, 2009.

[22] Dennis EJGJ Dolmans, Dai Fukumura, and Rakesh K Jain. Photodynamic therapy for cancer. *Nature Reviews Cancer*, 3(5):380–387, 2003.

[23] Can T. Xu. Diffuse optical spectroscopy and imaging of turbid media, PhD thesis, Department of Physics, Lund University, Sweden, 2013.

[24] Bruce J Tromberg, Brian W Pogue, Keith D Paulsen, Arjun G Yodh, David A Boas, and Albert E Cerussi. Assessing the future of diffuse optical imaging technologies for breast cancer management. *Medical physics*, 35(6):2443–2451, 2008.

[25] George Zonios, Lev T Perelman, Vadim Backman, *et al.* Diffuse reflectance spectroscopy of human adenomatous colon polyps in vivo. *Applied Optics*, 38(31):6628–6637, 1999.

[26] Konstantin Sokolov, Jesse Aaron, Betsy Hsu, *et al.* Optical systems for in vivo molecular imaging of cancer. *Technology in Cancer Research & Treatment*, 2(6):491–504, 2003.

[27] Niagara Muhammad Idris, Muthu Kumara Gnanasammandhan Jayakumar, Akshaya Bansal, and Yong Zhang. Upconversion nanoparticles as versatile light nanotransducers for photoactivation applications. *Chemical Society Reviews*, 44(6):1449–1478, 2015.

[28] Emmanuel Ruggiero, Silvia Alonso-de Castro, Abraha Habtemariam, and Luca Salassa. Upconverting nanoparticles for the near infrared photoactivation of transition metal complexes: New opportunities and challenges in medicinal inorganic photochemistry. *Dalton Transactions*, 45(33):13012–13020, 2016.

[29] Nicolaas Bloembergen. Solid state infrared quantum counters. *Physical Review Letters*, 2(3):84, 1959.

[30] Warren R Zipfel, Rebecca M Williams, and Watt W Webb. Nonlinear magic: Multiphoton microscopy in the biosciences. *Nature Biotechnology*, 21(11):1369–1377, 2003.

[31] Tanya N Singh-Rachford and Felix N Castellano. Photon upconversion based on sensitized triplet–triplet annihilation. *Coordination Chemistry Reviews*, 254(21):2560–2573, 2010.

[32] Marc A Baldo, Chihaya Adachi, and Stephen R Forrest. Transient analysis of organic electrophosphorescence. ii. transient analysis of triplet-triplet annihilation. *Physical Review B*, 62(16):10967, 2000.

[33] Jianzhang Zhao, Shaomin Ji, and Huimin Guo. Triplet–triplet annihilation based upconversion: From triplet sensitizers and triplet acceptors to upconversion quantum yields. *RSC Advances*, 1(6):937–950, 2011.

[34] Thomas Brabec and Ferenc Krausz. Intense few-cycle laser fields: Frontiers of nonlinear optics. *Reviews of Modern Physics*, 72(2):545, 2000.

[35] Francois F Auzel and Dario C Pecile. Comparison and efficiency of materials for summation of photons assisted by energy transfer. *Journal of Luminescence*, 8(1):32–43, 1973.

[36] Francois F Auzel and Dario C Pecile. Absolute efficiency for ir to blue conversion materials and theoretical prediction for optimized matrices. *Journal of Luminescence*, 11(5–6):321–330, 1976.

[37] Francois F Auzel. Upconversion and anti-Stokes processes with f and d ions in solids. *Chemical Reviews*, 104(1):139–173, 2004.

[38] Xudong Xiao, Jeanne P. Haushalter, and Gregory W. Faris. Upconversion from aqueous phase lanthanide chelates. *Optical Letters*, 30(13):1674–1676, 2005.

[39] Dev K Chatterjee, Abdul J Rufaihah, and Yong Zhang. Upconversion fluorescence imaging of cells and small animals using lanthanide doped nanocrystals. *Biomaterials*, 29(7):937–943, 2008.

[40] Dev K Chatterjee and Zhang Yong. Upconverting nanoparticles as nanotransducers for photodynamic therapy in cancer cells. *Nanomedicine*, 3(1): 73–82, 2008.

[41] Christopher Salthouse, Scott Hilderbrand, Ralph Weissleder, and Umar Mahmood. Design and demonstration of a small-animal up-conversion imager. *Optics Express*, 16(26):21731–21737, 2008.

[42] Can T Xu, Niclas Svensson, Johan Axelsson, *et al.* Autofluorescence insensitive imaging using upconverting nanocrystals in scattering media. *Applied Physics Letters*, 93(17):171103, 2008.

[43] Can T. Xu, Johan Axelsson, and Stefan Andersson-Engels. Fluorescence diffuse optical tomography using upconverting nanoparticles. *Applied Physics Letters*, 94(25):251107–251107–3, 2009.

[44] Hisataka Kobayashi, Mikako Ogawa, Raphael Alford, Peter L. Choyke, and Yasuteru Urano. New strategies for fluorescent probe design in medical diagnostic imaging. *Chemical Reviews*, 110(5):2620–2640, 2010. PMID: 20000749.

[45] Hisataka Kobayashi, Mikako Ogawa, Raphael Alford, Peter L Choyke, and Yasuteru Urano. New strategies for fluorescent probe design in medical diagnostic imaging. *Chemical Reviews*, 110(5):2620–2640, 2010.

[46] Haichun Liu, Can T Xu, and Stefan Andersson-Engels. Multibeam fluorescence diffuse optical tomography using upconverting nanoparticles. *Optics Letters*, 35(5):718–720, 2010.

[47] Pontus Svenmarker, Can T Xu, and Stefan Andersson-Engels. Use of nonlinear upconverting nanoparticles provides increased spatial resolution in fluorescence diffuse imaging. *Optics Letters*, 35(16):2789–2791, 2010.

[48] Qiuqiang Zhan, Jun Qian, Huijuan Liang, *et al.* Using 915 nm laser excited $Tm^{3+}/Er^{3+}/Ho^{3+}$-doped $NaYbF_4$ upconversion nanoparticles for *in vitro* and deeper *in vivo* bioimaging without overheating irradiation. *ACS Nano*, 5(5):3744–3757, 2011.

[49] Anna Gnach, Tomasz Lipinski, Artur Bednarkiewicz, Jacek Rybka, and John A Capobianco. Upconverting nanoparticles: Assessing the toxicity. *Chemical Society Reviews*, 44(6):1561–1584, 2015.

[50] Marie-France Joubert. Photon avalanche upconversion in rare earth laser materials. *Optical Materials*, 11(2):181–203, 1999.

[51] Daniel R Gamelin and Hans U Güdel. Design of luminescent inorganic materials: New photophysical processes studied by optical spectroscopy. *Accounts of Chemical Research*, 33(4):235–242, 2000.

[52] Daniel R Gamelin and Hans U Güdel. Topics in current chemistry, vol. 214, 2000.

[53] Feng Wang and Xiaogang Liu. Recent advances in the chemistry of lanthanide-doped upconversion nanocrystals. *Chemical Society Reviews*, 38(4):976–989, 2009.

[54] Brian R Judd. Optical absorption intensities of rare-earth ions. *Physical Review*, 127(3):750, 1962.

[55] Shiwei Wu, Gang Han, Delia J. Milliron, *et al.* Non-blinking and photostable upconverted luminescence from single lanthanide-doped nanocrystals. *Proceedings of the National Academy of Sciences*, 106(27):10917–10921, 2009.

[56] Xue Teng, Yihan Zhu, Wei Wei, *et al.* Lanthanide-doped Na_xScF_{3+x} nanocrystals: Crystal structure evolution and multicolor tuning. *Journal of the American Chemical Society*, 134(20):8340–8343, 2012. PMID: 22506543.

[57] Xingchen Ye, Joshua E. Collins, Yijin Kang, *et al.* Morphologically controlled synthesis of colloidal upconversion nanophosphors and their shape-directed self-assembly. *Proceedings of the National Academy of Sciences*, 107(52):22430–22435, 2010.

[58] Feng Wang and Xiaogang Liu. Upconversion multicolor fine-tuning: Visible to near-infrared emission from lanthanide-doped $NaYF_4$ nanoparticles. *Journal of the American Chemical Society*, 130(17):5642–5643, 2008. PMID: 18393419.

[59] Xiaomin Li, Rui Wang, Fan Zhang, *et al.* Nd^{3+} sensitized up/down converting dual-mode nanomaterials for efficient in-vitro and in-vivo bioimaging excited at 800 nm. *Scientific Reports*, 3:3536, 2013.

[60] Ye-Fu Wang, Gao-Yuan Liu, Ling-Dong Sun, Jia-Wen Xiao, Jia-Cai Zhou, and Chun-Hua Yan. Nd^{3+}-sensitized upconversion nanophosphors: Efficient in vivo bioimaging probes with minimized heating effect. *ACS Nano*, 7(8):7200–7206, 2013. PMID: 23869772.

[61] Hugo Söderlund, Monirehalsadat Mousavi, Haichun Liu, and Stefan Andersson-Engels. Increasing depth penetration in biological tissue imaging using 808-nm excited $Nd^{3+}/Yb^{3+}/Er^{3+}$-doped upconverting nanoparticles. *Journal of Biomedical Optics*, 20(8):086008, 2015.

[62] Qiuqiang Zhan, Baoju Wang, Xuanyuan Wen, and Sailing He. Controlling the excitation of upconverting luminescence for biomedical theranostics: Neodymium sensitizing. *Optical Materials Express*, 6(4):1011–1023, 2016.

[63] Wenqiang Zou, Cindy Visser, Jeremio A Maduro, Maxim S Pshenichnikov, and Jan C Hummelen. Broadband dye-sensitized upconversion of near-infrared light. *Nature Photonics*, 6(8):560–564, 2012.

[64] Guanying Chen, Hailong Qiu, Paras N Prasad, and Xiaoyuan Chen. Upconversion nanoparticles: Design, nanochemistry, and applications in theranostics. *Chemical Reviews*, 114(10):5161–5214, 2014.

[65] Karl W. Kramer, Daniel Biner, Gabriela Frei, Hans U. Gudel, Markus P. Hehlen, and Stefan R. Luthi. Hexagonal sodium yttrium fluoride based green and blue emitting upconversion phosphors. *Chemistry of Materials*, 16(7):1244–1251, 2004.

[66] Xiaojie Xue, Shinya Uechi, Rajanish N Tiwari, *et al.* Size-dependent upconversion luminescence and quenching mechanism of $LiYF_4$:Er^{3+}/Yb^{3+} nanocrystals with oleate ligand adsorbed. *Optical Materials Express*, 3 (7):989–999, 2013.

[67] John-Christopher Boyer, Fiorenzo Vetrone, Louis A. Cuccia, and John A. Capobianco. Synthesis of colloidal upconverting $NaYF_4$ nanocrystals doped with Er^{3+}, Yb^{3+} and Tm^{3+}, Yb^{3+} via thermal decomposition of lanthanide trifluoroacetate precursors. *Journal of the American Chemical Society*, 128 (23):7444–7445, 2006. PMID: 16756290.

[68] Jingning Shan and Yiguang Ju. A single-step synthesis and the kinetic mechanism for monodisperse and hexagonal-phase $NaYF_4$:Yb, Er upconversion nanophosphors. *Nanotechnology*, 20(27):275603, 2009.

[69] Ya-Wen Zhang, Xiao Sun, Rui Si, Li-Ping You, and Chun-Hua Yan. Single-crystalline and monodisperse LaF_3 triangular nanoplates from a single-source precursor. *Journal of the American Chemical Society*, 127(10):3260–3261, 2005.

[70] Guang Shun Yi and Gan Moog Chow. Synthesis of hexagonal-phase $NaYF_4$: Yb, er and $NaYF_4$: Yb, Tm nanocrystals with efficient up-conversion fluorescence. *Advanced Functional Materials*, 16(18):2324–2329, 2006.

[71] Hao-Xin Mai, Ya-Wen Zhang, Rui Si, *et al.* High-quality sodium rare-earth fluoride nanocrystals: Controlled synthesis and optical properties. *Journal of the American Chemical Society*, 128(19):6426–6436, 2006. PMID: 16683808.

[72] Rafik Naccache, Fiorenzo Vetrone, Venkataramanan Mahalingam, Louis A Cuccia, and John A Capobianco. Controlled synthesis and water dispersibility of hexagonal phase $NaGdF_4$: Ho^{3+}/Yb^{3+} nanoparticles. *Chemistry of Materials*, 21(4):717–723, 2009.

[73] Qian Liu, Yun Sun, Tianshe Yang, Wei Feng, Chenguang Li, and Fuyou Li. Sub-10 nm hexagonal lanthanide-doped $NaLuF_4$ upconversion nanocrystals for sensitive bioimaging in vivo. *Journal of the American Chemical Society*, 133(43):17122–17125, 2011.

[74] Feng Wang, Dev K Chatterjee, Zhengquan Li, Yong Zhang, Xianping Fan, and Minquan Wang. Synthesis of polyethylenimine/$NaYF_4$ nanoparticles with upconversion fluorescence. *Nanotechnology*, 17(23):5786, 2006.

[75] Venkataramanan Mahalingam, Rafik Naccache, Fiorenzo Vetrone, and John A Capobianco. Sensitized Ce^{3+} and Gd^{3+} ultraviolet emissions by Tm^{3+} in colloidal $LiYF_4$ nanocrystals. *Chemistry–A European Journal*, 15(38):9660–9663, 2009.

[76] Bastian Henke, Bernd Ahrens, Paul-T Miclea, Christian Eisenschmidt, Jacqueline Johnson, and Stefan Schweizer. Erbium-and chlorine-doped fluorozirconate-based glasses for up-converted fluorescence. *Journal of Non-Crystalline Solids*, 355(37):1916–1918, 2009.

[77] Dongmei Yang, Chunxia Li, Guogang Li, Mengmeng Shang, Xiaojiao Kang, and Jun Lin. Colloidal synthesis and remarkable enhancement of the

upconversion luminescence of $BaGdF_5$:Yb^{3+}/Er^{3+} nanoparticles by active-shell modification. *Journal of Materials Chemistry*, 21(16):5923–5927, 2011.

[78] Jing-H Zeng, Ji Su, ZH Li, Ruoxue Yan, and Yadong Li. Synthesis and upconversion luminescence of hexagonal-phase $NaYF_4$:Yb, Er^{3+} phosphors of controlled size and morphology. *Advanced Materials*, 17(17):2119–2123, 2005.

[79] Meng Wang, Cong-Cong Mi, Jin-Ling Liu, *et al.* One-step synthesis and characterization of water-soluble $NaYF_4$: Yb, er/polymer nanoparticles with efficient up-conversion fluorescence. *Journal of Alloys and Compounds*, 485(1):L24–L27, 2009.

[80] Wilson Sacchi Peternele, Victoria Monge Fuentes, Maria Luiza Fascineli, *et al.* Experimental investigation of the coprecipitation method: An approach to obtain magnetite and maghemite nanoparticles with improved properties. *Journal of Nanomaterials*, 2014:1–10, Article ID 682985, 2014.

[81] Rakhi Gupta and Nabo K Chaudhury. Entrapment of biomolecules in sol–gel matrix for applications in biosensors: Problems and future prospects. *Biosensors and Bioelectronics*, 22(11):2387–2399, 2007.

[82] Maqsood Ahmad Malik, Mohammad Younus Wani, and Mohd Ali Hashim. Microemulsion method: A novel route to synthesize organic and inorganic nanomaterials: 1st nano update. *Arabian Journal of Chemistry*, 5(4):397–417, 2012.

[83] Anna Zielińska-Jurek, Joanna Reszczyńska, Ewelina Grabowska, and Adriana Zaleska. Nanoparticles preparation using microemulsion systems. *Microemulsions—an introduction to properties and applications.* InTech Publisher, Rijeka, Croatia, 2012.

[84] Xiao Qin, Takeshi Yokomori, and Yiguang Ju. Flame synthesis and characterization of rare-earth (Er^{3+}, Ho^{3+}, and Tm^{3+}) doped upconversion nanophosphors. *Applied Physics Letters*, 90(7):073104, 2007.

[85] Jing Hui Zeng, Zhi Hua Li, Ji Su, Leyu Wang, Ruoxue Yan, and Yadong Li. Synthesis of complex rare earth fluoride nanocrystal phosphors. *Nanotechnology*, 17(14):3549, 2006.

[86] Feng Wang, Dev K Chatterjee, Zhengquan Li, Yong Zhang, Xianping Fan, and Minquan Wang. Synthesis of polyethylenimine/$NaYF_4$ nanoparticles with upconversion fluorescence. *Nanotechnology*, 17(23):5786, 2006.

[87] Li-Qin Xiong, Zhi-Gang Chen, Meng-Xiao Yu, Fu-You Li, Chun Liu, and Chun-Hui Huang. Synthesis, characterization, and in vivo targeted imaging of amine-functionalized rare-earth up-converting nanophosphors. *Biomaterials*, 30(29):5592–5600, 2009.

[88] Cheng Chen, Ling-Dong Sun, Zhen-Xing Li, *et al.* Ionic liquid-based route to spherical $NaYF_4$ nanoclusters with the assistance of microwave radiation and their multicolor upconversion luminescence. *Langmuir*, 26(11):8797–8803, 2010. PMID: 20085339.

[89] Zhengquan Li and Yong Zhang. Monodisperse silica-coated polyvinylpyrrolidone/$NaYF_4$ nanocrystals with multicolor upconversion fluorescence emission. *Angewandte Chemie*, 118(46):7896–7899, 2006.

[90] Zhiyao Hou, Chunxia Li, Pingan Ma, *et al.* Electrospinning preparation and drug-delivery properties of an up-conversion luminescent porous $NaYF_4$: Yb^{3+}, Er^{3+}@ silica fiber nanocomposite. *Advanced Functional Materials*, 21(12):2356–2365, 2011.

[91] Meng Wang, Congcong Mi, Yixin Zhang, *et al.*. NIR-responsive silica-coated $NaYBF_4$: Er/Tm/Ho upconversion fluorescent nanoparticles with tunable emission colors and their applications in immunolabeling and fluorescent imaging of cancer cells. *The Journal of Physical Chemistry C*, 113(44):19021–19027, 2009.

[92] Meng Wang, Gopal Abbineni, April Clevenger, Chuanbin Mao, and Shukun Xu. Upconversion nanoparticles: Synthesis, surface modification and biological applications. *Nanomedicine: Nanotechnology, Biology and Medicine*, 7(6):710–729, 2011.

[93] Qingbin Zhang, Kai Song, Junwei Zhao, *et al.* Hexanedioic acid mediated surface–ligand-exchange process for transferring $NaYF_4$:Yb/Er (or Yb/Tm) up-converting nanoparticles from hydrophobic to hydrophilic. *Journal of Colloid and Interface Science*, 336(1):171–175, 2009.

[94] Jie Shen, Ling-Dong Sun, Ya-Wen Zhang, and Chun-Hua Yan. Superparamagnetic and upconversion emitting Fe_3O_4/$NaYF_4$: Yb, Er heteronanoparticles via a crosslinker anchoring strategy. *Chemical Communications*, 46(31):5731–5733, 2010.

[95] Tianye Cao, Tianshe Yang, Yuan Gao, Yang Yang, He Hu, and Fuyou Li. Water-soluble $NaYF_4$:Yb/Er upconversion nanophosphors: Synthesis, characteristics and application in bioimaging. *Inorganic Chemistry Communications*, 13(3):392–394, 2010.

[96] John-Christopher Boyer, Marie-Pascale Manseau, Jill I Murray, and Frank CJM van Veggel. Surface modification of upconverting $NaYF_4$ nanoparticles with PEG-phosphate ligands for NIR (800 nm) biolabeling within the biological window. *Langmuir*, 26(2):1157–1164, 2009.

[97] Liqin Xiong, Tianshe Yang, Yang Yang, Congjian Xu, and Fuyou Li. Long-term in vivo biodistribution imaging and toxicity of polyacrylic acid-coated upconversion nanophosphors. *Biomaterials*, 31(27):7078–7085, 2010.

[98] Stephanie J Budijono, Jingning Shan, Nan Yao, *et al.* Synthesis of stable block-copolymer-protected $NaYF_4$: Yb^{3+}, Er^{3+} up-converting phosphor nanoparticles. *Chemistry of Materials*, 22(2):311–318, 2009.

[99] Noah JJ Johnson, Neralagatta M Sangeetha, John-Christopher Boyer, and Frank CJM van Veggel. Facile ligand-exchange with polyvinylpyrrolidone and subsequent silica coating of hydrophobic upconverting β-$NaYF_4$: Yb^{3+}/Er^{3+} nanoparticles. *Nanoscale*, 2(5):771–777, 2010.

[100] Zhigang Chen, Huili Chen, He Hu, *et al.* Versatile synthesis strategy for carboxylic acid-functionalized upconverting nanophosphors as biological labels. *Journal of the American Chemical Society*, 130(10):3023–3029, 2008.

[101] Guang-Shun Yi and Gan-Moog Chow. Water-soluble $NaYF_4$: Yb, Er(Tm)/ $NaYF_4$/polymer core/shell/shell nanoparticles with significant enhancement of upconversion fluorescence. *Chemistry of Materials*, 19(3):341–343, 2007.

[102] Marco Pedroni, Fabio Piccinelli, Tiziana Passuello, *et al.* Lanthanide doped upconverting colloidal CaF_2 nanoparticles prepared by a single-step hydrothermal method: Toward efficient materials with near infrared-to-near infrared upconversion emission. *Nanoscale*, 3(4):1456–1460, 2011.

[103] Leyu Wang, Ruoxue Yan, Ziyang Huo, *et al.* Fluorescence resonant energy transfer biosensor based on upconversion-luminescent nanoparticles. *Angewandte Chemie International Edition*, 44(37):6054–6057, 2005.

[104] Fiorenzo Vetrone, Rafik Naccache, Venkataramanan Mahalingam, Christopher G Morgan, and John A Capobianco. The active-core/active-shell approach: A strategy to enhance the upconversion luminescence in lanthanide-doped nanoparticles. *Advanced Functional Materials*, 19(18): 2924–2929, 2009.

[105] John-Christopher Boyer and Frank CJM Van Veggel. Absolute quantum yield measurements of colloidal $NaYF_4$: Er^{3+}, Yb^{3+} upconverting nanoparticles. *Nanoscale*, 2(8):1417–1419, 2010.

[106] Markus Pollnau, Daniel R Gamelin, Stefan R Luthi, Hans U Gudel, and Markus P Hehlen. Power dependence of upconversion luminescence in lanthanide and transition-metal-ion systems. *Physical Review B*, 61(5):3337, 2000.

[107] Haichun Liu, Can T Xu, David Lindgren, *et al.* Balancing power density based quantum yield characterization of upconverting nanoparticles for arbitrary excitation intensities. *Nanoscale*, 5(11):4770–4775, 2013.

[108] Björn Thomasson. High accuracy relative luminescence quantum yield measurements of upconverting nanoparticles, 2016. Student Paper.

[109] Monirehalsadat Mousavi, Björn Thomasson, Meng Li, *et al.* Beam-profile-compensated quantum yield measurements of upconverting nanoparticles. *Physical Chemistry Chemical Physics*, 19(33):22016–22022, 2017.

[110] Chao Wang, Liang Cheng, Huan Xu, and Zhuang Liu. Towards whole-body imaging at the single cell level using ultra-sensitive stem cell labeling with oligo-arginine modified upconversion nanoparticles. *Biomaterials*, 33(19):4872–4881, 2012.

[111] Tianye Cao, Yang Yang, Yuan Gao, Jing Zhou, Zhengquan Li, and Fuyou Li. High-quality water-soluble and surface-functionalized upconversion nanocrystals as luminescent probes for bioimaging. *Biomaterials*, 32(11):2959–2968, 2011.

[112] Liang Cheng, Kai Yang, Yonggang Li, *et al.* Facile preparation of multifunctional upconversion nanoprobes for multimodal imaging and

dual-targeted photothermal therapy. *Angewandte Chemie*, 123(32):7523–7528, 2011.

[113] Qingtao Chen, Xin Wang, Fenghua Chen, *et al.* Functionalization of upconverted luminescent $NaYF_4$:Yb/Er nanocrystals by folic acid–chitosan conjugates for targeted lung cancer cell imaging. *Journal of Materials Chemistry*, 21(21):7661–7667, 2011.

[114] Young-Hwa Kim, Jongho Jeon, Su Hyun Hong, *et al.* Tumor targeting and imaging using cyclic RGD-PEGylated gold nanoparticle probes with directly conjugated iodine-125. *Small*, 7(14):2052–2060, 2011.

[115] Guangshun Yi, Yanfen Peng, and Zhiqiang Gao. Strong red-emitting near-infrared-to-visible upconversion fluorescent nanoparticles. *Chemistry of Materials*, 23(11):2729–2734, 2011.

[116] Niagara Muhammad Idris, Zhengquan Li, Lei Ye, *et al.* Tracking transplanted cells in live animal using upconversion fluorescent nanoparticles. *Biomaterials*, 30(28):5104–5113, 2009.

[117] Liang Cheng, Kai Yang, Shuai Zhang, Mingwang Shao, Shuittong Lee, and Zhuang Liu. Highly-sensitive multiplexed in vivo imaging using pegylated upconversion nanoparticles. *Nano Research*, 3(10):722–732, 2010.

[118] Scott A Hilderbrand, Fangwei Shao, Christopher Salthouse, Umar Mahmood, and Ralph Weissleder. Upconverting luminescent nanomaterials: Application to in vivo bioimaging. *Chemical Communications*, 2009(28): 4188–4190, 2009.

[119] Jing Zhou, Mengxiao Yu, Yun Sun, *et al.* Fluorine-18-labeled $Gd^{3+}/Yb^{3+}/Er^{3+}$ co-doped $NaYF_4$ nanophosphors for multimodality PET/MR/UCL imaging. *Biomaterials*, 32(4):1148–1156, 2011.

[120] Feng Wang, Renren Deng, Juan Wang, *et al.* Tuning upconversion through energy migration in core-shell nanoparticles. *Nature Materials*, 10(12):968–973, 2011.

[121] Qiyue Shao, Lilai Ouyang, Lifei Jin, and Jianqing Jiang. Multifunctional nanoheater based on $NaGdF_4$:Yb^{3+}, Er^{3+} upconversion nanoparticles. *Optics Express*, 23(23):30057–30066, 2015.

[122] Zengxia Zhao, Yuning Han, Chenghong Lin, *et al.* Multifunctional core–shell upconverting nanoparticles for imaging and photodynamic therapy of liver cancer cells. *Chemistry – An Asian Journal,* 7(4):830–837, 2012.

[123] Dandan Ma, Xiang Xu, Min Hu, *et al.* Rare-earth-based nanoparticles with simultaneously enhanced near-infrared (NIR)-visible (vis) and NIR-NIR dual-conversion luminescence for multimodal imaging. *Chemistry – An Asian Journal*, 11(7):1050–1058, 2016.

[124] Renata Madru, Pontus Svenmarker, Christian Ingvar, *et al.* Development of a hybrid nanoprobe for triple-modality MR/spect/optical fluorescence imaging. *Diagnostics*, 4(1):13–26, 2014.

[125] Tomas Jansson, Stefan Andersson-Engels, Sarah Fredriksson, Freddy Ståhlberg, and Sven-Erik Strand. Superparamagnetic iron oxide nanoparticles as a multimodal contrast agent for up to five imaging modalities. *Clinical and Translational Imaging*, 3(3):247–249, 2015.

[126] Maria Evertsson, Pontus Kjellman, Magnus Cinthio, *et al.* Multimodal detection of iron oxide nanoparticles in rat lymph nodes using magnetomotive ultrasound imaging and magnetic resonance imaging. *IEEE Transactions on Ultrasonics, Ferroelectrics, and Frequency Control*, 61(8):1276–1283, 2014.

[127] Renata Madru, Pontus Kjellman, Fredrik Olsson, *et al.* 99m Tc-labeled superparamagnetic iron oxide nanoparticles for multimodality spect/mri of sentinel lymph nodes. *Journal of Nuclear Medicine*, 53(3):459–463, 2012.

[128] Lei Zhou, Rui Wang, Chi Yao, *et al.* Single-band upconversion nanoprobes for multiplexed simultaneous in situ molecular mapping of cancer biomarkers. *Nature Communications*, 6, Article ID 6938, 2015.

[129] HJMAA Zijlmans, Jan Bonnet, Jarrett Burton, *et al.* Detection of cell and tissue surface antigens using up-converting phosphors: A new reporter technology. *Analytical Biochemistry*, 267(1):30–36, 1999.

[130] Paul Corstjens, Michel Zuiderwijk, Antoinette Brink, *et al.* Use of up-converting phosphor reporters in lateral-flow assays to detect specific nucleic acid sequences: A rapid, sensitive dna test to identify human papillomavirus type 16 infection. *Clinical Chemistry*, 47(10):1885–1893, 2001.

[131] Johannes Hampl, Michael Hall, Naheed A Mufti, *et al.* Upconverting phosphor reporters in immunochromatographic assays. *Analytical Biochemistry*, 288(2):176–187, 2001.

[132] R Sam Niedbala, Hans Feindt, Keith Kardos, *et al.* Detection of analytes by immunoassay using up-converting phosphor technology. *Analytical Biochemistry*, 293(1):22–30, 2001.

[133] Leyu Wang and Yadong Li. Green upconversion nanocrystals for DNA detection. *Chemical Communications*, 2006(24):2557–2559, 2006.

[134] Henna Pakkila, Minna Yliharsila, Satu Lahtinen, *et al.* Quantitative multianalyte microarray immunoassay utilizing upconverting phosphor technology. *Analytical Chemistry*, 84(20):8628–8634, 2012.

[135] Vishal Kale, Henna Pakkilaa, Jiri Vainio, *et al.* Spectrally and spatially multiplexed serological array-in-well assay utilizing two-color upconversion luminescence imaging. *Analytical Chemistry*, 88(8):4470–4477, 2016.

[136] Yiqing Lu, Jiangbo Zhao, Run Zhang, *et al.* Tunable lifetime multiplexing using luminescent nanocrystals. *Nature Photonics*, 8(1):32–36, 2014.

[137] Katri Kuningas, Terhi Rantanen, Telle Ukonaho, Timo Lövgren, and Tero Soukka. Homogeneous assay technology based on upconverting phosphors. *Analytical Chemistry*, 77(22):7348–7355, 2005. PMID: 16285685.

[138] Terhi Rantanen, Marja-Leena Jarvenpaa, Johanna Vuojola, Riikka Arppe, Katri Kuningas, and Tero Soukka. Upconverting phosphors in a dual-parameter lret-based hybridization assay. *Analyst*, 134(8):1713–1716, 2009.

[139] Peng Zhang, Snezna Rogelj, Khoi Nguyen, and Damon Wheeler. Design of a highly sensitive and specific nucleotide sensor based on photon upconverting particles. *Journal of the American Chemical Society*, 128(38):12410–12411, 2006. PMID: 16984179.

[140] Meng Wang, Wei Hou, Cong-Cong Mi, *et al.* Immunoassay of goat anti-human immunoglobulin g antibody based on luminescence resonance energy transfer between near-infrared responsive $NaYF_4$:Yb, Er upconversion fluorescent nanoparticles and gold nanoparticles. *Analytical Chemistry*, 81(21):8783–8789, 2009. PMID: 19807113.

[141] Yan-Mei Wu, Yao Cen, Li-Jiao Huang, Ru-Qin Yu, and Xia Chu. Upconversion fluorescence resonance energy transfer biosensor for sensitive detection of human immunodeficiency virus antibodies in human serum. *Chemical Communications*, 50(36):4759–4762, 2014.

[142] Yujie Ding, Hao Zhu, Xiaoxia Zhang, *et al.* An upconversion nanocomposite for fluorescence resonance energy transfer based cholesterol-sensing in human serum. *Nanoscale*, 6(24):14792–14798, 2014.

[143] Zhiguo Zhou, He Hu, Hong Yang, *et al.* Up-conversion luminescent switch based on photochromic diarylethene and rare-earth nanophosphors. *Chemical Communications*, 2008(39):4786–4788, 2008.

[144] Carl-Johan Carling, Farahnaz Nourmohammadian, John-Christopher Boyer, and Neil R Branda. Remote-control photorelease of caged compounds using near-infrared light and upconverting nanoparticles. *Angewandte Chemie International Edition*, 49(22):3782–3785, 2010.

[145] Yi-Hsin Chien, Yu-Lin Chou, Shu-Wen Wang, *et al.* Near-infrared light photocontrolled targeting, bioimaging, and chemotherapy with caged upconversion nanoparticles in vitro and in vivo. *ACS Nano*, 7(10):8516–8528, 2013.

[146] John V Garcia, Jianping Yang, Dengke Shen, *et al.* Nir-triggered release of caged nitric oxide using upconverting nanostructured materials. *Small*, 8 (24):3800–3805, 2012.

[147] Lingzhi Zhao, Juanjuan Peng, Qi Huang, *et al.* Near-infrared photoregulated drug release in living tumor tissue via yolk-shell upconversion nanocages. *Advanced Functional Materials*, 24(3):363–371, 2014.

[148] Yanmei Yang, Qing Shao, Renren Deng, *et al.* In vitro and in vivo uncaging and bioluminescence imaging by using photocaged upconversion nanoparticles. *Angewandte Chemie*, 124(13):3179–3183, 2012.

[149] Muthu Kumara Gnanasammandhan Jayakumar, Niagara Muhammad Idris, and Yong Zhang. Remote activation of biomolecules in deep tissues using near-infrared-to-UV upconversion nanotransducers. *Proceedings of the National Academy of Sciences*, 109(22):8483–8488, 2012.

[150] Muthu Kumara Gnanasammandhan Jayakumar, Akshaya Bansal, Kai Huang, Risheng Yao, Bing Nan Li, and Yong Zhang. Near-infrared-light-based nano-platform boosts endosomal escape and controls gene knockdown in vivo. *ACS Nano*, 8(5):4848–4858, 2014.

[151] Akshaya Bansal, Haichun Liu, Muthu Kumara Gnanasammandhan Jayakumar, Stefan Andersson-Engels, and Yong Zhang. Quasi-continuous wave near-infrared excitation of upconversion nanoparticles for optogenetic manipulation of*C. elegans*. *Small*, 12(13):1732–1743, 2016.

[152] Kristian Berg, Pal Kristian Selbo, Lina Prasmickaite, *et al*. Photochemical internalization. *Cancer Research*, 59(6):1180–1183, 1999.

[153] Kristian Berg, Marco Folini, Lina Prasmickaite, *et al*. Photochemical internalization: A new tool for drug delivery. *Current Pharmaceutical Biotechnology*, 8(6):362–372, 2007.

[154] Hai Sheng Qian, Hui Chen Guo, Paul Chi-Lui Ho, Ratha Mahendran, and Yong Zhang. Mesoporous-silica-coated up-conversion fluorescent nanoparticles for photodynamic therapy. *Small*, 5(20):2285–2290, 2009.

[155] Niagara Muhammad Idris, Muthu Kumara Gnanasammandhan, Jing Zhang, Paul C Ho, Ratha Mahendran, and Yong Zhang. In vivo photodynamic therapy using upconversion nanoparticles as remote-controlled nanotransducers. *Nature Medicine*, 18(10):1580, 2012.

[156] Sasidharan Swarnalatha Lucky, Niagara Muhammad Idris, Zhengquan Li, Kai Huang, Khee Chee Soo, and Yong Zhang. Titania coated upconversion nanoparticles for near-infrared light triggered photodynamic therapy. *ACS Nano*, 9(1):191–205, 2015.

[157] Sasidharan Swarnalatha Lucky, Khee Chee Soo, and Yong Zhang. Nanoparticles in photodynamic therapy. *Chemical Reviews*, 115(4):1990–2042, 2015.

[158] Gan Tian, Xiao Zhang, Zhanjun Gu, and Yuliang Zhao. Recent advances in upconversion nanoparticles-based multifunctional nanocomposites for combined cancer therapy. *Advanced Materials*, 27(47):7692–7712, 2015.

[159] Haichun Liu, Can T Xu, Gökhan Dumlupinar, Ole B Jensen, Peter E Andersen, and Stefan Andersson-Engels. Deep tissue optical imaging of upconverting nanoparticles enabled by exploiting higher intrinsic quantum yield through use of millisecond single pulse excitation with high peak power. *Nanoscale*, 5(20):10034–10040, 2013.

[160] Dongmei Yang, Zhiyou Hou, Ziyong Cheng *et al.* Current advances in lanthanide ion (Ln^{3+})-based upconversion nanomaterials for drug delivery. *Chemical Society Reviews*, 44(6):1416–1448, 2015.

Chapter 10

Near-infrared spectroscopy systems for biomedical applications

Shudong Jiang[1] *and Brian Pogue*[1]

10.1 Introduction

10.1.1 History of near-infrared spectroscopy and imaging in tissue

Near-infrared spectroscopy (NIRS) has been developed as a unique noninvasive modality because of the discovery that light in the near infrared (NIR) region (600–1,000 nm) passes through centimeters of tissue with relatively low attenuation, allowing interrogation of the hemoglobin components of blood within the tissue. Most practical uses are spectroscopic in nature, but also measurements at one or more locations can allow the transmitted/reflected portion of the light to be collected and analyzed for mapping out images of attenuation values. These transmission or reflectance values are interpreted with empirical formulas for blood component or used with image reconstruction algorithms based on Monte Carlo simulation or diffusion theory, the functional information of tissue molecular compositions changes, such as oxyhemoglobin, deoxyhemoglobin, water, and lipids can be obtained.

NIRS has the capability to obtain tissue vascular and cellular information noninvasively, so it has a long history for being researched for medical diagnosis and imaging. Early studies began in the mid-1900s when NIR light was first postulated to characterize cancer tumors in diagnosis and surgery [1,2]. However, it was not until the invention and commercialization of compact lasers and light sources, that the use of NIRS for spectral information became widely successful. In 1997, Jobsis described the ability to measure oxygen saturation changes of the blood through thick tissue volumes [3], which was the landmark discovery that rapidly precipitated system developments for NIRS oximetry for arterial oxygen saturation as measured from the pulsatile flow signal in the finger, toe or earlobe [4,5]. These pulse oximeters have become one of the most successful and robust measurement systems available in critical care monitoring [6]. In addition, there are applications of NIRS in other areas such as neonatal monitoring, sports medicine, elite sports training, ergonomics, rehabilitation, urology (bladder contraction), and pharmaceutical research.

[1]Solar Energy Research Institute of Singapore (SERIS), National University of Singapore (NUS), Singapore

10.1.2 Basic principles

The major physical phenomena that dictate laser light interaction with tissue are absorption and elastic scattering. The mathematical analysis of measurements can be described in terms of these two interaction parameters, as listed below.

Beer-Lambert law for calculating the absorption

As is shown in (10.1), the Beer-Lambert law is a linear equation to represent the relationship between observed absorbance, A, and concentration of an absorbing sample, c.

$$A = \varepsilon(\lambda) * L * C \tag{10.1}$$

where $\varepsilon(\lambda)$ is the molar absorption coefficient dependent upon wavelength, λ, L is the path length of travel of the light through tissue, and C is the absorber's concentration. When absorber concentration is in Molar units, the molar absorption coefficient is then in units of M^{-1} cm^{-1}.

The absorbance of a sample is usually measured through the transmittance (T), which is defined as $T = I/I_o$, where I_o and I are the light intensity before and after passing through the sample. Then, the relation between A and measured light intensity before and after the sample can be displayed as

$$A = -\log T = -\log(I/I_o) \tag{10.2}$$

Note that while this description is based on the chemical absorption of light from a molecule at concentration, C, the light transport in tissue is often described as by the bulk attenuation coefficient, $\mu_a(\lambda)$ in units of cm^{-1}, which is defined in terms of natural logarithms, and so is related by:

$$\mu_a(\lambda) = 2.303\varepsilon(\lambda) * C \tag{10.3}$$

Figure 10.1 illustrates this NIR spectrum of the absorption coefficient for the major absorbers natively present within tissue.

10.1.2.1 Scattering

In the NIR wavelength range, most elastic scattering events in tissue can be considered to come from Mie-like particles which have a physical size similar to the wavelength of the light (100s of microns). In this case, the spectrum of the scattering coefficient is anisotropic, and so is best characterized over longer distance scales by a reduced scattering coefficient, $\mu_s'(\lambda)$, which is the effective scattering coefficient under the assumption that the scattering was isotropic. The difference between isotropic versus nonisotropic scatter can only really be discerned over small distances (2–3 mm or less) and so tissue spectroscopy over longer distances utilizes this reduced scattering coefficient parameter. Within tissue this parameter has been shown to fit well to an empirical approximation to Mie scattering theory [7,8] given by,

$$\mu_s'(\lambda) = a\lambda^{-b} \tag{10.4}$$

where a and b are constants, with scatter amplitude, a, and scatter power, b, often in wavelength in the units of μm. The coefficient μ_s' has unit mm^{-1} and b is

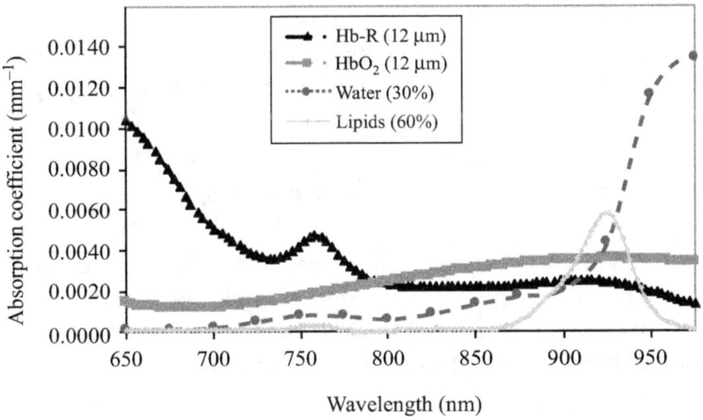

Figure 10.1 NIR Spectrum of the absorption coefficients for the main absorbers present within tissue, including reduced hemoglobin (Hb-R), oxygenated hemoglobin (HbO_2), water, and lipids, displayed at relative physiological concentrations, C

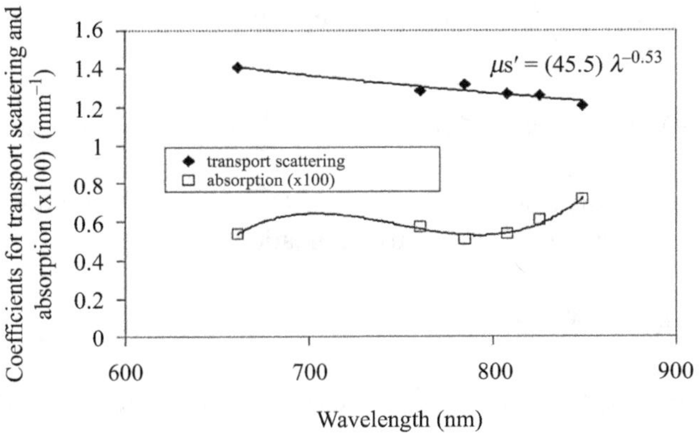

Figure 10.2 Measured absorption and transport (or reduced) scattering coefficient spectrum of a single subject's breast tissue

dimensionless, so that *a* has unusual units, but represents the amplitude. Scattering power is governed by the shape or slope of scattering which is predominantly affected by the size and the amplitude relates to the number density of the scatters.

Figure 10.2 shows a representative spectrum of the reduced scattering coefficient as well as the absorption coefficient of human tissue [9]. The scattering coefficient is more than 200 times higher than absorption coefficient indicating that light transport in tissue is dominated by scattering rather than absorption in NIR wavelength range, and thus the long distance transport ($L > 2$–3 mm) is diffuse in nature. As a result, the well-developed methods of diffusion theory modeling have been widely implemented to understand and characterize how light propagates in tissue.

10.1.2.2 Diffusion equation

In case of a scattering dominated propagation process, the following diffusion equation can be used to describe NIR light traveling through long distances of tissue:

$$-\nabla \bullet D(r)\nabla\Phi(r,\omega) + \left[\mu_a(r) + \frac{i\omega}{c}\right]\Phi(r,\omega) = q_0(r,\omega) \tag{10.5}$$

where $\Phi(r,\omega)$ is the isotropic fluence at intensity modulation frequency ω, and c is the speed of light in tissue, $q_0(r,\omega)$ is an isotropic light source at position r. The two tissue-dependent parameters in the diffusion equation are the absorption coefficient and diffusion coefficient, $D(r)$. Here, $D(r)$ can be written as

$$D(r) = \frac{1}{3\left[\mu_a(r) + \mu_s'(r)\right]} \tag{10.6}$$

where $\mu_s'(r)$ is the reduced scattering coefficient. Since in diffusion theory, $\mu_s' >> \mu_a$, $D(r)$ is often simplified to be $D = 1/(3\mu_s')$. The measurement data at the tissue surface, Φ, includes two parts: the amplitude and phase shift of transmitted light modulated at frequency ω. In CW measurement with charge-coupled device (CCD)-based spectrometers, only the intensity information is recorded with $\omega = 0Hz$, yielding this simplified equation:

$$-\frac{1}{3\mu_s'(r)}\nabla^2\Phi(r) + \mu_a(r)\Phi(r) = q_0(r) \tag{10.7}$$

Solving for the distribution of $\mu_a(r)$ and $D(r)$ is a nonlinear inverse problem, which has been described in numerous publications [10,11].

10.2 Imaging systems

10.2.1 Geometries: reflectance, transmission, and circular/lunate

According to the different light source-detector geometry, NIRS tomography systems can be divided into three basic types: reflectance, transmittance, and circular/arc (combined reflectance and transmittance). Figure 10.3 shows the scheme and pictures of these three types.

10.2.1.1 Subsurface reflectance

As shown in Figure 10.3(a), in the reflectance system, the light source and detectors are held on the same side of the tissue. The light source can be either near-infrared laser or white light source. Avalanche photodiode (APD) or photomultiplier tube (PMT) or CCD detectors are used to detect the reflected diffuse light from tissue. The distance between the light source and the detector can be adjusted to commentate the detection depth of the tissue. The advantage of a reflective optical imaging system is that the remitted light intensity is relatively high, and there is no specific requirement for the size of targeted tissue and the density of the light.

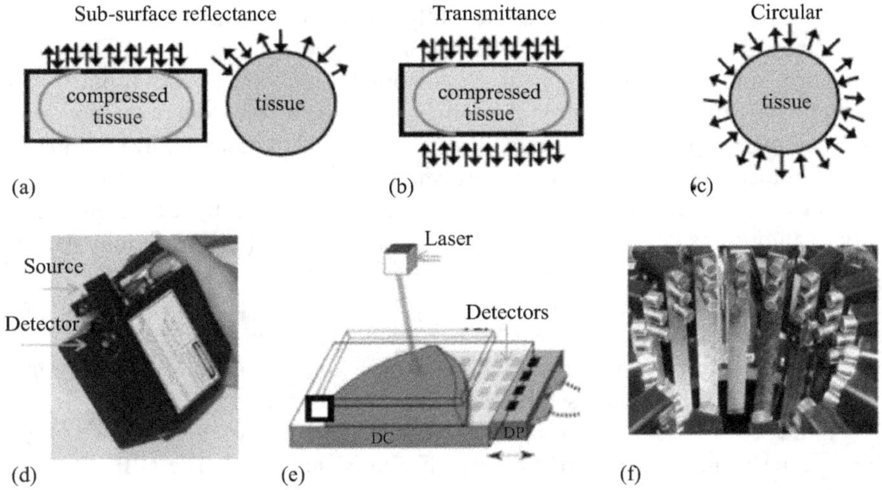

Figure 10.3 Basic types of image system geometries (a) subsurface reflectance; (b) transmittance; and (c) circular/lunate. Below are realizations of these with (d) picture of a hand-held probe for subsurface reflectance measurement, (e) noncontact light scanning approach used in digital tomosynthesis combined with transmittance breast cancer imaging, and (f) a circular type breast imaging patient interface with three circular rings of source-detector fiber bundles

However, the substantial disadvantage is that the sensitivity and specificity decays exponentially with depth into the tissue. This geometry was pioneered by Dr. Britton Chance and colleagues [12,13], for both brain and breast imaging. It remains the main approach to study brain function, changes in blood volume and oxygen saturation in response to trauma induced changes such as stroke or infarct, because of the limitations for light access into the brain [14]. In addition, a sub-surface reflectance system has been used for wound healing monitoring and surgical guidance [15,16]. Figure 10.3(d) is a reflective optical imaging probe used in the Tromberg group [17]. While near-infrared laser and white light are used as the light source. In this realization, an APD is used to detect diffuse light, an optical fiber bundle is used to collect the diffuse light induced by the white light, and deliver it into a spectrometer for spectral analysis. The distance between the light source and APD/detector fiber bundles can be adjusted to commentate the targeted tissue depth.

10.2.1.2 Transmittance

The parallel plate imaging geometry mimics toe geometry of x-ray radiography of tissue. Since the light source and detectors are in the opposite side of the tissue, the detected diffuse light actually contains tissue information of the entire volume that it passed through. The signal is exponentially lost with depth, and so in this parallel plate transmission system there is often some compression of the tissue to reduce attenuation, and the detected optical signal covering the optical path (optical path) are relatively simple. The quality of the image may be affected by projection of the tissue heterogeneities along the axial direction, as with all radiography systems, but in NIRS transmission there is also significant blurring from the light diffusion. The first and most widely discussed was the system developed by Carl Zeiss, using analysis algorithms developed by Fantini and Franceschini [18,19]. Similar systems have been developed at the University of Pennsylvania and Massachusetts General Hospital [20,21]. The advantage of this type of system is that it is easy to integrate the light source and the detector to other imaging systems, and can provide the relatively high signal-to-noise ratio (SNR), only needs moderate dynamic range, and works well if the tissue is compressed to some degree.

Figure 10.3(e) is a picture of the patient interface of a combined digital Tomosynthesis and optical NIRS breast imaging system [22]. The NIRS laser can be raster scanned over the tissue on top, using a pair of the galvanometer scanner mirrors. Underneath the tissue, is a detector board that consists of 75 large area silicon photodiodes (PDs) (S9270, Hamamatsu Corporation, Japan).

10.2.1.3 Circular/Arc

Compared with the above two kinds of geometric structures, circular or annular geometry system combines both of the reflected light and transmitted light detection for imaging (Figure 10.3(c)). The tissue being imaged maintains the natural shape and the detected optical signal covers the more abundant optical path to improve the image quality. Dartmouth College, has pursued this circular geometry since 1995, largely due to the fact that this geometry provides the most symmetric data collection geometry and provides the best sampling of the image space from multiple angles for tomographic reconstruction [23]. The disadvantage of this system is that the detected optical signal intensity is limited by the size and density of the tissue; the tissue has to be reformed to the circular; and to change the diameter of the circle to adapt to the different size of the tissue also need to be more complex than the first two geometries. Additionally this geometry requires an extremely large dynamic range capability from the detectors since the near-source detection might just be an order of magnitude down from the source, whereas the furthest transmission detectors could be 8–10 orders of magnitude down in intensity. This must be achieved with dynamic gain setting of the detectors.

Figure 10.3(f) is an example of a circular geometry patient interface [24]. 48 optical bundles with the detection effective diameter of 6 mm optical are evenly distributed in the three circular planes. The vertical distance of each plane is 1.5 cm. The height of the planes can be adjusted by the vertical motor, and the circular plane diameter is controlled synchronously by 16 radial motors. The distal end of each bundle is uniformly connected to a circular rotating table. One of the 16 bundles in the same plane is couple to the laser source and the other 15 are coupled each to a PMT for detecting diffuse light at 15 different locations around the tissue. For each plane, data is collected at 16×15 source-detector positions for image reconstruction.

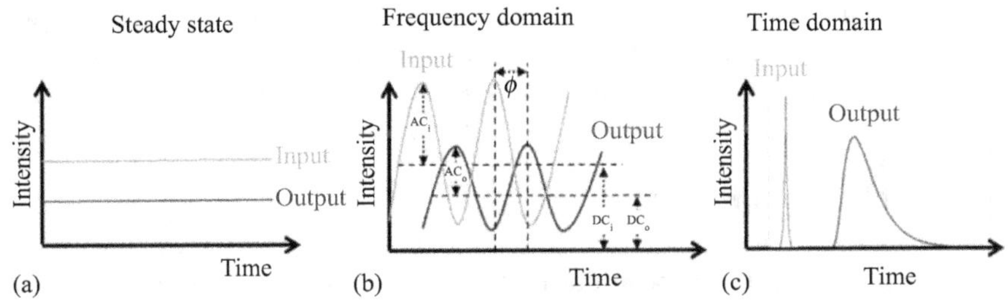

Figure 10.4 Light input/output of optical signal for the three instrument types used in DOT: (a) steady state (CW); (b) FD; and (c) TD

Table 10.1 The comparison of the cost, complicity and signal strength

	Steady state	Frequency domain	Time domain
Cost	Inexpensive	Moderate	Expensive
Complexity	Low	Medium	High
Measurement	High SNR	Medium SNR	Low SNR

10.2.2 Signal detection types

Three different measurement types are used to obtain NIRS data, including: (1) steady state (SS) [also known as continuous wave (CW)]; (2) frequency domain (FD); and (3) time domain (TD). These types differ in their complexity as well as the amount of tissue information that may be retrieved from the measurements. The light input and measured output signal for these three different types are shown in Figure 10.4, and the comparison of the cost, complicity, and signal strength is shown in Table 10.1.

10.2.2.1 CW measurement

The simplest and most economical approach to NIRS is CW. In this approach, the light from a light source (white light, LED, or laser) is incident on tissue while the attenuated reflected and/or transmitted light is measured. The light source may be modulated at low frequency (from DC up to several kHz) for lock-in detection to increase the SNR. In this case, since the output through the lock-in amplifier is only the attenuated light at the same frequency as the source, the noise from the ambience room/background light can be eliminated. In addition, by modulating sources and/or wavelengths at difference modulation frequencies allow simultaneous acquiring the diffused light at multiple sources and/or multiple wavelengths. The advantages of the CW measurement are the system is simple and low cost, and only moderately fast response optical detectors are required for the detection, while the drawback is the absorption and scattering of tissue cannot be separated [25]. CW imaging systems can be used in the applications that only required the relative values of the excellular and vascular properties. In this type of the applications, a constant scattering property throughout tissue is assumed [26]. There is work showing that by multispectral fitting to CW data, that the spectrally constrained absorption and scattering coefficients can be estimated to a reasonable accuracy, and this realization has gained wider use in recent years [27].

10.2.2.2 TD measurement

TD systems can provide the most information about the nature of the diffuse photons and thus the tissue composition. In TD systems, a narrow laser pulse illuminates the tissue and the light flux on the boundary is measured. As the laser pulse travels through tissue, it is attenuated, delayed, and gets broadened with respect to the original pulse. Absorption and scattering coefficients may be determined by measuring the time it takes for the pulse to reach the detector and the shape of the pulse. For example, photons in a high scattering medium will take longer to exit the tissue as they keep scattering, which will lead to a broadening of the input laser pulse as shown in Figure 10.4(c).

Streak cameras, time-correlated photon counting systems (TCSPC), and time-gated intensified CCD imagers are used with TD measurements and are able to generate images with improved image quality [28]. Furthermore, TD systems can be used with FDOT where images of fluorescent yields and lifetimes of multi-components can be obtained to further enhance DOT [29]. TCSPC systems are discussed further in Chapter 4.

The advantage of time-resolved techniques (TD and FD) over SS is that these methods are able to measure the distance a photon travels in addition to the number of photons that exit the tissue (i.e. light intensity). FD provides an average path-length of the photons while TD systems provide a precise measure of the path-length. This additional data adds to the uniqueness of the solution to recovering the absorption and scattering coefficients and enables a better image reconstruction.

10.2.2.3 FD measurement

FD spectroscopy has the advantage of being a relatively simple, stable system with the ability to obtain absolute scattering and absorption properties of tissue. In this type of the measurement, the light source is modulated with frequencies of 50–500 MHz. Through the use of the measured phase and amplitude of the light signal through the tissue, the scattering properties can be separated and the extracellular and vascular properties can be obtained from scattering and absorption, respectively. Demodulation of the signal from the MHz range down into the kHz and lower ranges is commonly done to make the signal acquisition less expensive. This is achieved by homodyne detection, heterodyne detection and single-side band detection can be used to measure the amplitude and phase of the optical signal.

10.2.2.4 Combined FD and CW measurement

Many studies have shown that the water, lipid, and collagen concentrations are the important biomarkers for differentiating cancer and normal tissues [17,30,31], and these markers have spectral signatures which are better differentiated in the highest wavelength ranges. As shown in above, FD measurements using intensity-modulated sources, are extremely stable and cost effective, but have the inherent limitation that highest dynamic range FD systems are based upon PMT detection. Unfortunately PMT detectors in general have low photon detection efficiency with the highest gain PMTs have wavelength response curves that fall off dramatically above 825 nm, and so while these systems provide excellent separation of μ_a and μ_s', they do not provide for the best wavelength coverage for NIRS. The use of SS CW (or low-frequency modulated) light to measure the attenuation of light at longer wavelengths (above 825 nm) can supplement FD measurements. Combined FD and CW system use both PMT and solid-state silicon PD detectors simultaneously to detect the diffuse light. The tissue scattering properties can be obtained by the FD measurement at the wavelength range of 660–850 nm, while the spectral absorption properties can be obtained over the wider wavelength range of 660–1024 nm, which covers all absorption peaks of oxy-hemoglobin, deoxyhemoglobin, water, lipid, and collagen. The technology to scan or acquire signals in parallel over this spectral band are also far better in CW detection, versus FD or TD solutions.

10.3 Light sources and detectors

10.3.1 Light source: laser diodes, LEDs, and other lasers

The laser diode (LD) is a common light source using in NIRS system due to its large variety of the wavelength, the easy high power modulation by current, and compact size. The typical power of LD used for NIRS for tissue diagnosis and imaging is several mW to 100 mW. Since the tissue absorption bandwidth is usually in the range of 1–5 nm, either single or multi-mode LD works for NIRS. These are common choices for fiber-delivered light, where coupling into a fiber is the key need, and wavelength switching or temporal modulation are required.

LEDs are used for lower cost, yet effective, CW NIRS systems that the light directly illuminate to and/or are used at the tissue interface without using fiber delivery. However because of the nature of these systems being lower cost, LEDs are often used in the systems that acquire the signal with limited depth in tissue because the detectors are also designed to be limited cost and therefore limited dynamic range. However, LEDs can obviously supply large power, although this is only for CW delivery. Prototype systems for FD imaging with LEDs have been demonstrated, but are typically limited to well near 50 MHz intensity modulation.

Besides LDs and LEDs, Ti-sapphire (Ti:S) lasers and Supercontinuum Lasers have been used as the light source to provide signals for continuously tunable wavelengths. However, the advantage of obtaining the spectroscopic data at many wavelengths were traded off by the relatively low power stability when it has been tuned over the wide wavelength range [32,33].

10.3.2 Photon detectors

For light in the NIR wavelength range, tissue absorption may reduce optical signals by approximately a factor of 10 per centimeter of tissue [34]. Thus, the detected optical signal in these systems depends on the tissue type, source strength, and

Table 10.2 Summary of photon detectors properties

	Ideal	**PD/PIN**	**APD**	**PMT**
Quantum efficiency	1	… 80%	… 80%	10% … 50%
Collection efficiency	1	1	1	0.9
Wavelength (nm)	300–1700	300–1700	300–1700	115–850
Bandwidth	High	~100 kHz	~GHz	~300 MHz
ENF	1	1	2	1.3
NEP ($fW/\sqrt{Hz}$)	–	100	10	0.1
Gain	10^6	1	10–1,000	10^4–10^9
Min. detectable PE	1	100–300	10–20	1
Operation volt (V)	<10 V	10–100	100–500	1k–3k
Miniaturization	Yes	Yes	Yes	No
Large area	Yes	Yes	No	Yes
Noise	None	Low	Medium	Low
Bandwidth	High	~100 kHz	~GHz	~100 MHz
Dynamic range	Single photon	Good	Good (4– 5OD)	Superior – single photon (4–6OD)
Production cost	–	Low	Low	Medium

source-detector separation and may range from milliwatts down to femto-watts. This constitutes a total dynamic range of 10^{12}. The summary of the typical detectors is shown in Table 10.1, as well as the ideal specs for NIRS systems.

10.3.2.1 PDs and avalanche photodiodes

A PD is a fast, highly linear photon detector with high quantum efficiency. The output photocurrent on the light power can be very linear over six or more orders of magnitude of the light power, e.g., in a range from a few nanowatts to tens of milliwatts for a silicon PD with an active area of a few mm^2. As shown in Table 10.2, the detection wavelength range is covered entire Visible-NIR range. However, the disadvantage of PD is that they provide no internal gain mechanism and the temporal response is inversely proportional to the area size of the photoactive area. Therefore, PD must use an amplifier circuit in order to amplify the output signal and detect low light levels. The bandwidth of the amplifier for limiting the noise level further reduce the temporal response of PDs. Considering the large numerical aperture (NA) and high sensitivity that is needed for picking up this very weak diffused light, the applications that use PDs as the detector are limited to CW detection NIRS system and/or the surface imaging of the tissue. However for systems where the wavelength is scanned on the sources side a PD based detection system can be inexpensive and effective, with high stability.

An APD can be considered for a detector with a built-in first stage of gain through avalanche multiplication. In addition to the merits of perhaps the fastest speed, it also has equally high linearity as a PD does, and the APD has much higher sensitivity (×100) than a PD. However, due to the avalanche multiplication nature and requirement of small physical size, the biomedical applications of APDs are limited by the smaller detection area, and high bias voltage (~100 V), and associated cooling devices. There is also a slightly higher dark current as compared to a PD. So as a result, the number of situations where the APD is the right choice is relatively few.

10.3.2.2 Photomultiplier tube

For situations which require the most sensitive signal detection at short acquisition times, there is a need for high gain and high light capture. The ability to detect single photons give the photomultiplier distinct advantages over other light detectors when the signal is low to extremely low, and with picosecond to nanosecond response times. These detectors can be large in physical size with the highest gain levels possible and can be coupled efficiently to fiber pick up. A single photon striking a PMT detector generates anywhere between 10^6 and 10^9 electrons at the highest gains. With the proper resistor terminated at the output of the PMT, this current pulse is converted to voltage in the mV range which can easily be measured.

Typical noise-equivalent power (NEP) values for a bandwidth of 1 Hz for a PMT are between 10^{-15} and 10^{-16} W [35]. This translates to detection of 400 to 4000 photons per second for an 800 nm photons. Cooling the PMT can improve the NEP value by reducing dark current. Cooled PMT detectors are able to reach NEP values of 10^{-21} W/Hz$^{-0.5}$ [36].

The fast temporal response of PMTs means that they can detect FD light modulated up to 300 MHz while maintaining the sensitivity as low as 10^{-12} W. The main limitations of these devices are the moderate stability over time and the limited wavelength range of capabilities for higher gain PMTs.

10.4 Applications

10.4.1 Pulse oximeter

The pulse oximeter is perhaps the most successful application of NIRS in the biomedical field. The initial concept of the pulse oximetry was established in 1972, by Takuo Aoyagi *et al.* in Japan [37]. Since then, it has been commercialized beginning in 1981, and now the pulse oximeter has been used in nearly every operating room, recovery room, intensive care unit, neonatal unit, and wherever the patients may have suspect heart rate or oxygen supply. Figure 10.5 shows a photo and schemes of a standard pulse oximeter. As shown in Figure 10.5(a), it is a compact monitor that attaches to a fingertip to monitor the amount of oxygen carried in the human body. The pulse oximeter is based on measuring the absorbance difference at two wavelengths due to the pulsing arterial blood (Figure 10.5(b)). The most common type of pulse oximeter is using two light-emitting diodes (LED) with the wavelengths at red (660 nm) and Infrared (940 nm) with a PD as a single detector (Figure 10.5(c)). The two LEDs are alternated on and the intensity of the transmitted diffused light is picked up by a PD. By using the empirical formula, the tissue oxygen can be estimated through arterial blood hemoglobin oxygen saturation

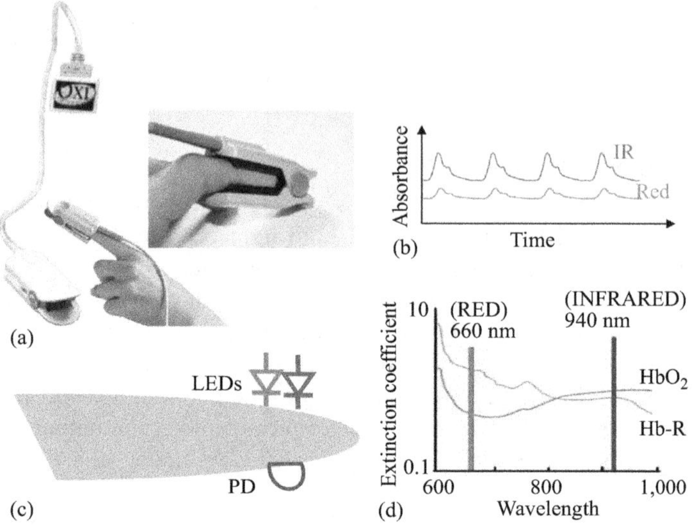

Figure 10.5 The photos and schemes of a pulse oximeter: (a) photograph; (b) temporal pulse curves of the tissue absorbance at IR and red wavelengths; (c) schematic; and (d) absorption spectra of the oxy-hemoglobin (HbO_2) and deoxyhemoglobin (HbR) in red wavelength (660 nm) and infrared (940 nm) wavelength ranges

(S_aO_2), $SaO_2 = [HbO_2]/([HbR] + [HbO_2])$, as shown in Figure 10.5(d)). The limitation of the pulse oximeter is that it may not completely reflect the respiratory and/or circulatory sufficiency, due to insufficient blood flow or anemia.

10.4.2 Breast cancer detection and treatment monitoring

Breast cancer is the most common newly diagnosed cancer among women in the United States with an estimated 249,260 new cases in 2016 and over 1 million cases diagnosed worldwide [38]. Adding to the conventional imaging modalities such as mammogram, ultrasound and magnetic resonance imaging (MRI), NIRS tomography has been widely studied for its ability of breast cancer detection and treatment monitoring, due to NIRST's non-invasive and relatively low cost nature.

10.4.2.1 MRI-guided near-infrared breast cancer detection

Based on clinical research, long-term survival odds are greatly increased by early detection and treatment prior to tumor metastasis [39]. In the group of the high-risk women, a yearly breast contrast-MR scan is recommended because of MR's good sensitivity to breast tumors. However, breast contrast-MR yields a high number of false positives that lead to expensive and sometimes unnecessary biopsy procedures. On the other hand, diffuse optical spectroscopy (DOS) can provide the functional information, which is correlated with tumor malignancy but limited its usefulness as a single diagnostic system due to its low spatial resolution.

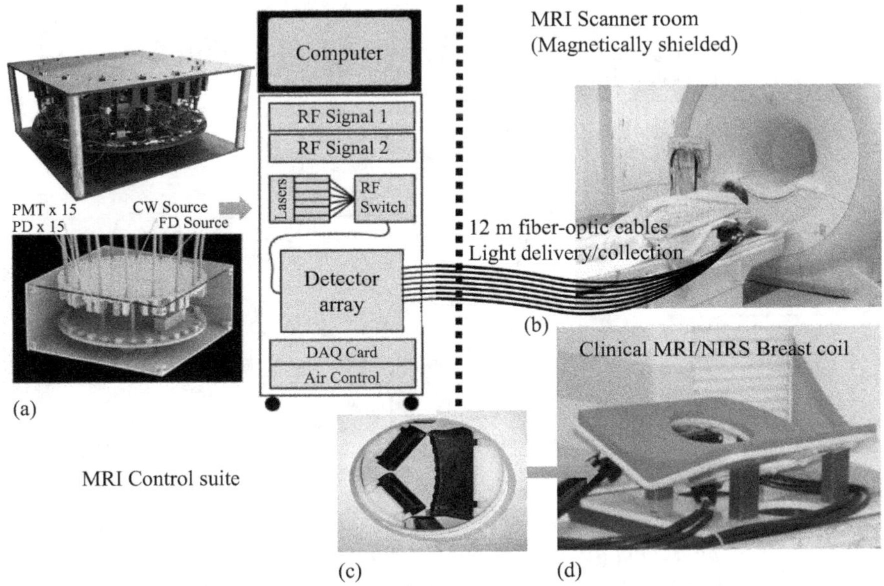

Figure 10.6 Photos of a hybrid MRI-DOST imaging system: (a) system components; (b) MRI-DOST imaging setting; (c) and (d) DOST fiber interface attached to a clinical MRI coil

Therefore, a combined MRI-DOST (diffuse optical spectroscopic tomography) tomographic system has been developed to ultimate the breast cancer detection with matched high sensitivity and specificity.

As it mentioned in Section earlier, due to the limited sensitivity of PMT detectors, most DOST systems use light between 650 nm and 850 nm [40–42]. For covering the larger lipid and water absorption peaks lie at 930 nm and 975 nm, respectively, [43], a hybrid imaging system, shown in Figure 10.6, was designed to accommodate FD and CW measurements [44]. All the FD/CW imaging components along with rotary switch array and the data acquisition and control software are housed inside a single portable rack (Figure 10.6(a)) in MRI Control Suite. The rotary switch contains each set of 15 PMT and PD modules; and each of CW and FD laser source coupler. Sixteen fiber optic bundles held into thick cables connect the light sources and detectors to the patient breast interface, in MRI scanner room (magnetically shielded) (Figure 10.6(b)). A DOST patient interface (Figure 10.6(c)) is used to hold 16 fiber bundles to the conventional MRI coil (Figure 10.6(d)). The DOST patient interface can be adjusted in 3D dimensions to accommodate the breast sizes and the tumor locations.

In the rotary switch, sixteen fiber bundles are held on the top fixed plate. One pair of FD and CW laser source couplers and fifteen pairs of PMT and PD are uniformly hang in the rotary stage. In each source position, one fiber bundle is connected to a source coupler (FD or CW) and other 15 bundles are connected each of PMTs or PDs. Six FD lasers and three CW lasers are sequentially connected to

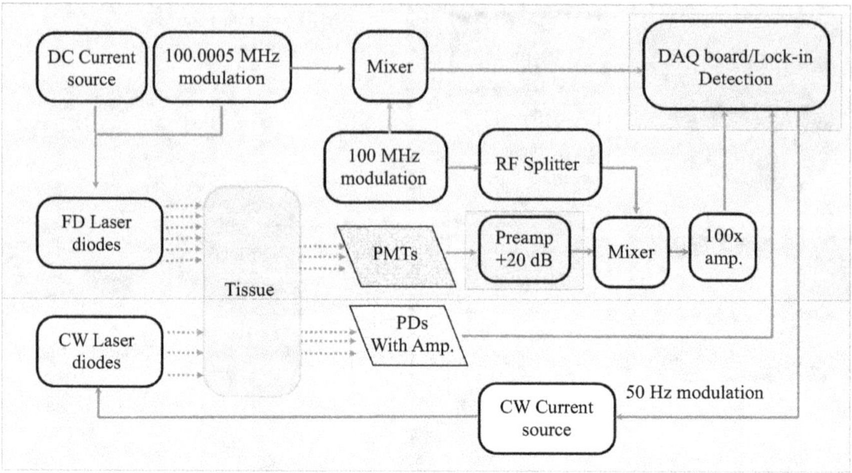

Figure 10.7 Diagram of an FD+CW breast imaging system, with the FD components above and the CW components below, both used for interrogating the same tissue, and complementing each other's information stream. The combined system allows separation of absorption and scatter (with FD) and wide spectral coverage (from CW)

the source couplers and for each laser, the data at $16 \times 15 = 240$ source-detector pairs are collected.

A diagram of this MRI-DOST system is shown Figure 10.7 [44]. For FD measurement portion (top half portion in Figure 10.7), the light source consists of six intensity-modulated (100 MHz) laser diodes that cover the main spectral range of 660, 735, 785, 808, 826, and 849 nm. A bias-T combines DC bias current from an ILX Lightwave™ laser driver with a 100 MHz AC signal from a function generator in order to drive the laser diodes. The modulation depth of the lasers is 25%–40%. The lasers are activated one at a time by digitally controlling a 6-way RF switch, which passes the biased current to the selected laser diode. Each laser diode is mounted on a fiber launch module where the output of each laser is directed into the input of a 6×1 fiber optic combiner. The output fiber is then directed to the rotating stage, where it acts as the source fiber in the system.

PMT (H9305-3, Hamamatsu, Japan) detectors with a 1.4 ns rise time are used to acquire the attenuated laser light that is transmitted through tissue. The spectral response range of the PMTs is 185 to 900 nm with peak sensitivity at 450 nm. The sensitivity of the PMT decreases rapidly at wavelengths longer than 850. Power to all PMTs is supplied from a single power supply. Thin film filters (Kodal Wratten Red 25, Edmund Optics) are installed on the detectors window in order to block visible light below 600 nm from entering the PMTs. The RF output of the PMT detectors is amplified using a 20 dB RF low-noise preamplifier (Minicircuits)

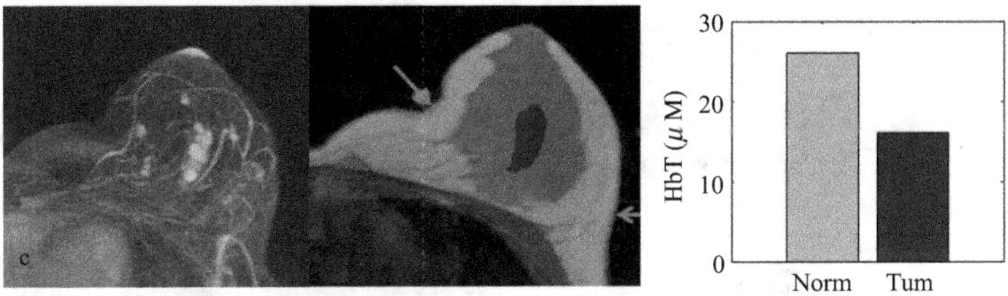

(a) (b) (c)

Figure 10.8 MRI/optical results for a patient: (a) contrast-enhanced MRI image shows abnormality location and enhancement pattern; (b) MRI used to guide optical reconstruction and green arrows show locations of fiber bundles; and (c) graph of DOST HbT contrast correctly characterize the lesions as benign based on a lesion to normal surrounding tissue contrast less than 1× [45]

which also filters out the DC component of the signal. The output of the pre-amplifier is then heterodyned down to a signal with the frequency of 500 Hz and read by a computer through a DAQ card. Finally, the phase and amplitude of the attenuated signal detected by the PMT can be obtained through the digital lock-in detection via Labview program.

The CW portion of the system (bottom portion of Figure 10.7) consists of three laser diodes (903, 912, and 948 nm) and 15 silicon PD modules. The output of the laser diodes is intensity-modulated at 50 Hz in order to obtain better SNR through software-based lock-in detection.

The detectors used are high-precision silicon PD modules which incorporate a current-to-voltage amplifier within the module (C10439-03, Hamamatsu, Japan). These modules have an active area of 1 cm^2 and a large spectral response range (190–1,100 nm) with peak sensitivity at 960 nm. The sensitivity of the PD modules is switchable between a low and a high range. The two PDs nearest to the laser source are set to low sensitivity in order to avoid light saturation of the modules while the remaining modules are set to high sensitivity. Thin film filters (Kodak-87 Wratten, Edmund Optics) are installed on the detectors' window in order to block light below 800 nm. All PDs are driven by a single dual-output power supply. The output of the PD's is connected directly to the data acquisition cards.

Figure 10.8 is a representative case example of MRIDOST image data from a patient with a histological 10 × 20 × 19mm breast lesion [45]. Contrast-enhanced MR image shows the lesion is malignant (BIRADS 4, Figure 10.8(a)). Guided by the structure information and fiber location that MRI given (Figure 10.8(b)), DOST image is reconstructed. Opposed MRI diagnosis, the contrast of the tumor to the normal surrounding breast tissue of 0.62 suggested the lesion is benign. The final surgical pathologic results indicate the lesion is benign (Fibroadenoma) and confirmed MRI-DOST combined imaging modality have the potential to improve the specificity of MRI breast cancer diagnosis.

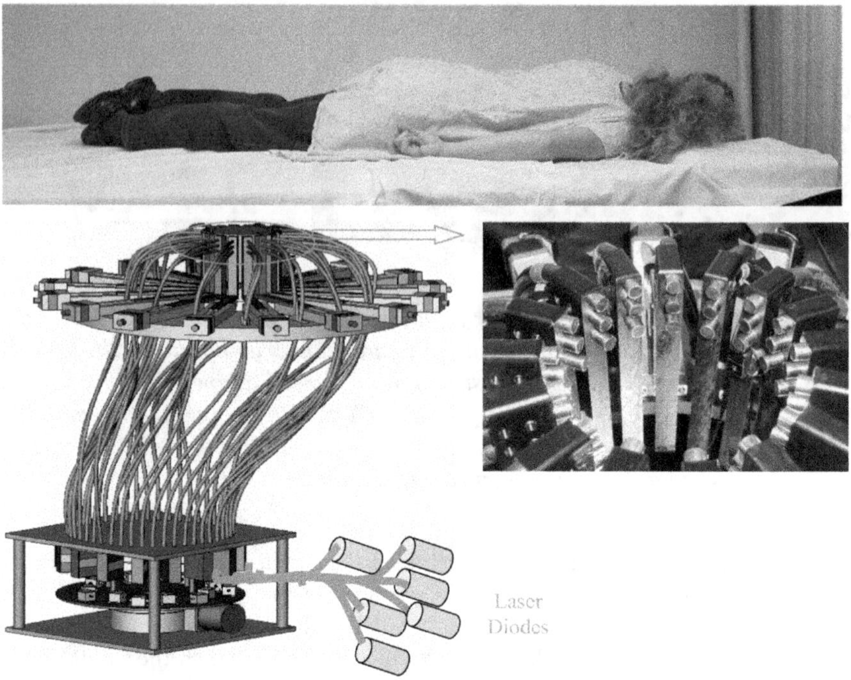

Figure 10.9 DOST system for predicating and monitoring breast tumor response to neoadjuvant chemotherapy

10.4.2.2 Monitoring and predicating the tumor response to neoadjuvant chemotherapy

The traditional treatment strategy of surgery followed by adjuvant chemotherapy is largely being replaced by neoadjuvant chemotherapy (NAC) prior to surgery when tumor size is larger or under certain hormone receptor status. Clinical studies have shown that patients with a complete pathological response (pCR) to NAC experience longer disease-free survival [46,47]. However, the pCR rate is only about 35%–23% [48], and the delay in definitive local therapy that may occur from a complete course of NAC can be long (up to 8 months) [49]. Thus, prediction of pCR from early treatment response to stratify disease management is likely to improve the outcomes of patients and their long-term survival. In recent years, DOST has been caught the attention in monitoring and prediction of breast tumor response to NAC [50–52]. Compared with other existing medical imaging systems, DOS optical imaging is low cost, no radiation, and can image the patient for unlimited times. In addition, DOS systems are portable and have the potential to allow patients with severe illness to monitor themselves at home and transmit data to the doctor via the Internet. The system used for imaging is shown in Figure 10.9. Since the tumor size is large in this study, 48 short fiber-bundles are used to image 3 cm breast tissue from chest-wall towards nipple and only FD measurements are taken with each of six FD lasers. Figure 10.10 is an example of the patient images

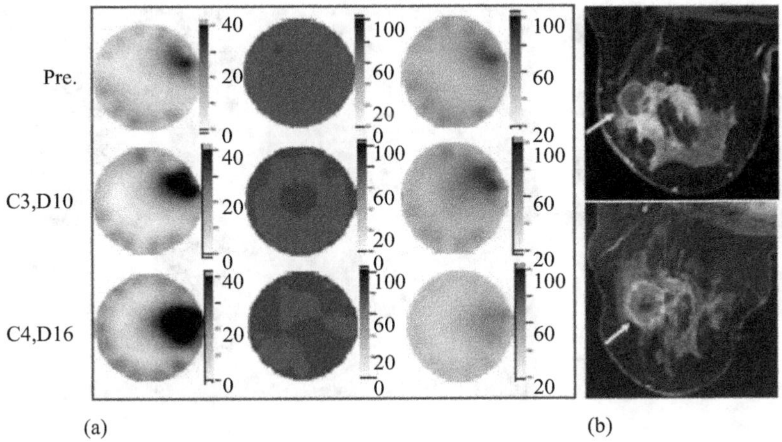

Figure 10.10 Representative DOST, MR images from a subject with non-pCR: (a) DOST images of abnormal breast are shown at three time points during the course of treatment. The tumor remains visible in the upper right of the breast views presented. The plane diameter of the images is 80 mm. In (b), the contrast-enhanced MR images before (top) and after (bottom) treatment indicate substantial disease remained during therapy

before, during and after NAC. The results of a clinical trial at Dartmouth college shown that the ratio of total hemoglobin inside of the tumor to the normal surrounding breast tissue before chemotherapy can distinguish pCR from non-pCR groups [51]. This result also shown the possibility of DOST may expedite the optimization of the clinical validation process for new chemotherapy regimens by reducing the number of patients required and the duration of treatment follow-up.

10.4.3 Monitoring brain activities

Near infrared spectral measurements in the brain are largely focused around maximizing spatial and temporal data, and most sacrifice spectral coverage for the former information streams. As such, one or two wavelengths of coverage are common with either LDs or LEDs, and maximal sampling in time is critical to be able to discern vascular pulses in the cortical surface. The physiological changes in response to external stimuli are largely small vaso-dilations and changes in flow at the capillary level, which are triggered nearly instantaneously from neuronal signaling. This functional NIRS data (fNIRS) is used to assess brain function or pathophysiology from well characterized functional tests. Numerous studies have been developed in rodent and mammal models, and today there are dozens of commercial systems available for human use. Illustrations of the geometry and the type of data are shown in the Figure 10.11.

In terms of technology used in these systems, the fiber coupled systems tend to be focused around pairs of LDs driven at two NIR wavelengths, around the isosbestic point

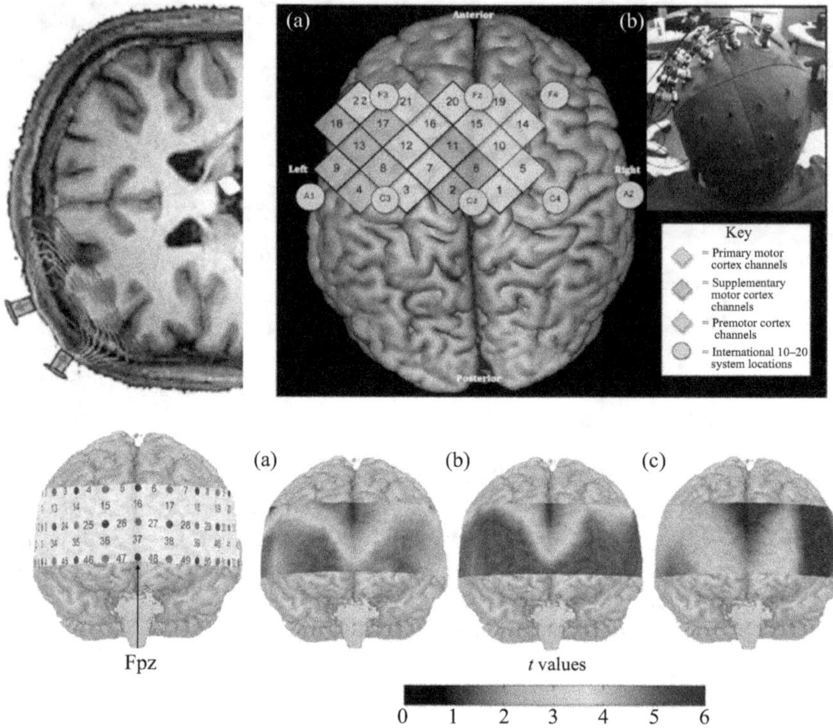

Figure 10.11 An illustration of NIRS light diffusion into the brain through the skin, scalp and skull is shown, and a mapping of an array of optical sources and detectors is shown overlaid on the brain functional network [53]. Then in the bottom row, and band style network of sources and detectors is shown with functional stimuli from mathematical calculations revealed StO_2-activated regions with the lower t values, showing the areas of simulation with statistical validity [54]

of hemoglobin and oxyhemoglobin, and the contact puck type systems with lower cost tend to focus on LEDs which can be directly applied to the surface and instrumented into the headgear. Both are commonly used in CW mode, for topography measurements as shown in the figure above. Because of the high attenuation coefficient of the brain, due to high hemoglobin content, it is not really possible to image through the entire human brain, although research systems exist to image through the entire neonate brain as shown in Figure 10.12 [55]. As such, adult brain topography systems tend to utilize LEDs if they are looking to measure superficial signals, but LDs if they are attempting to measure deeper more attenuating regions of the brain. One of the most problematic areas of this field of study though has been the interference of scalp blood flow changes into the desired brain signal measurements, and it is now well established that superficially measurements only can be completely corrupted by this skin/scalp signal [55].

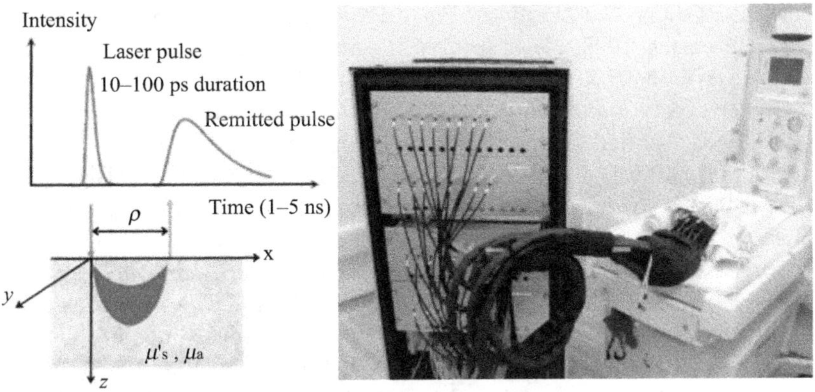

(a) (b)

Figure 10.12 The optical imaging scanner for neonates developed by the Biomedical Optics Research Laboratory at the University College London: (a) illumination of the time domain measurement and (b) photo of the system (https://neolabresearch.com/research/fast-optical-tomography-onneonates-foton-study/)

On the detection side, APDs have been employed as surface detectors, with direct contact onto the scalp, or via fiber delivery to a detector box with banks of APDs installed. These have the advantage of moderate cost and robust operation when the gain settings have been carefully calibrated. Alternative realizations have used PMTs and PDs but each have their limitations and benefits. The most advanced systems have used single photon counting PMTs for maximal signal penetration and maximal dynamic range, however because of cause and complexity of operation, few of these have been commercialized [56].

References

[1] Cutler, M., *Transillumination as an aid in the diagnosis of breast lesions.* Surgery, Gynaecology & Obstetrics, 1929. **48**: p. 721–729.

[2] Cartwright, C.H., *Infra-red transmission of the flesh.* Journal of the Optical Society of America, 1930. **20**: p. 81–84.

[3] Jobsis, F.F., *Non-invasive, infra-red monitoring of cerebral and myochardial oxygen sufficiency and circulatory parameters.* Science, 1977. **198**: p. 1264–1267.

[4] Middleton, P.M. and J.A. Henry, *Pulse oximetry: evolution and directions.* International Journal of Clinical Practice., 2000. **54**(7): p. 438–444.

[5] Severinghaus, J.W., *History and recent developments in pulse oximetry.* Scandinavian Journal of Clinical & Laboratory Investigation - Supplement., 1993. **214**: p. 105–111.

[6] Webster, J.G., *The Design of Pulse Oximeters*. Medical Science Series, ed. J.G. Webster. 1997, New York, NY: Taylor & Francis Group, LLC.

[7] van Staveren, H.J., C.J.M. Moes, J. van Marle, S.A. Prahl, and J.C. van Gemert, *Light scattering in Intralipid - 10% in the wavelength range of 400–1100 nm.* Applied Optics, 1991. **30**(31): p. 4507–4514.

[8] Mourant, J.R., T. Fuselier, J. Boyer, T.M. Johnson, and I.J. Bigio, *Predictions and measurements of scattering and absorption over broad wavelength ranges in tissue phantoms.* Applied Optics, 1997. **36**(4): p. 949–957.

[9] Srinivasan, S., B.W. Pogue, S. Jiang, *et al.*, *Interpreting hemoglobin and water concentration, oxygen saturation, and scattering measured by near-infrared tomography of normal breast in vivo*. Proceedings of the National Academy of Sciences of the USA, 2003. **100**(21): p. 12349–12354.

[10] Arridge, S.R., J.C. Hebden, M. Schwinger, *et al.*, *A method for three-dimensional time-resolved optical tomography.* International Journal of Imaging Systems and Technology, 2000. **11**(1): p. 2–11.

[11] Paulsen, K.D. and H.B. Jiang, *Spatially varying optical property reconstruction using a finite-element diffusion equation approximation.* Medical Physics, 1995. **22**(6): p. 691–701.

[12] Gratton, E., W.W. Mantulin, M.J. van de Ven, J.B. Fishkin, and M.B. Maris, *A novel approach to laser tomography.* Bioimaging, 1993. **1**: p. 40–46.

[13] Nioka, S., Miwa, M., Orel, S., Shnall, M., Haida, M., Zhao, S., Chance, B., *Optical imaging of human breast cancer.* Advances in Experimental Medicine and Biology, 1994. **361**: p. 171–179.

[14] Gratton, G., A. Chiarelli, and M. Fabiani, *From brain to blood vessels and back: a noninvasive optical imaging approach.* Neurophotonics, 2017. **4**(3): p. 031208.

[15] Jayachandran, M., S. Rodriguez, E. Solis, J. Lei, and A. Godavarty, *Critical review of noninvasive optical technologies for wound imaging.* Advances in Wound Care, 2016. **5**(8): p. 349–359.

[16] Evers, D.J., R. Nachabé, H.M. Klomp, *et al.*, *Diffuse reflectance spectroscopy: a new guidance tool for improvement of biopsy procedures in lung malignancies.* Clinical Lung Cancer, 2012. **13**(6): p. 424–431.

[17] Cerussi, A., N. Shah, D. Hsiang, A. Durkin, J. Butler, and B.J. Tromberg, *In vivo absorption, scattering, and physiologic properties of 58 malignant breast tumors determined by broadband diffuse optical spectroscopy.* Journal of Biomedical Optics, 2006. **11**(4): p. 044005.

[18] Fantini, S., Francecshini, M. A., Gaida, *et al.*, *Frequency-domain optical mammography: edge effect corrections.* Medical Physics, 1996. **23**: p. 149–157.

[19] Jess, H., Erdl, H., Moesta, T., *et al. Intensity modulated breast imaging: Technology and clinical pilot study results. in Proceedings of the Advances in Optical Imaging and Photon Migration*. 1996. Orlando, FL: Opt. Soc. Am.

[20] Ban, H., M. Schweiger, V. Kavuri, *et al.*, *Heterodyne frequency-domain multispectral diffuse optical tomography of breast cancer in the parallel-plane transmission geometry.* Medical Physics, 2016. **43**(7): p. 4383–4395.

[21] Zimmermann, B.B., B. Deng, B. Singh, *et al.*, *Multimodal breast cancer imaging using coregistered dynamic diffuse optical tomography and digital breast tomosynthesis.* Journal of Biomedical Optics, 2017. **22**(4): p. 046008.

[22] Krishnaswamy, V., K.E. Michaelsen, B.W. Pogue, *et al.*, *A digital x-ray tomosynthesis coupled near infrared spectral tomography system for dual-modality breast imaging.* Optics Express, 2012. **20**(17): p. 19125.

[23] Pogue, B.W., T. McBride, U. Osterberg, and K. Paulsen, *Comparison of imaging geometries for diffuse optical tomography of tissue.* Optics Express, 1999. **4**(8): p. 270–286.

[24] Jiang, S., B.W. Pogue, T.O. McBride, and K.D. Paulsen, *Quantitative analysis of near-infrared tomography: sensitivity to the tissue-simulating precalibration phantom.* Journal of Biomedical Optics, 2003. **8**(2): p. 308–315.

[25] Arridge, S.R. and W.R. Lionheart, *Nonuniqueness in diffusion-based optical tomography.* Optics Letters, 1998. **23**(11): p. 882–884.

[26] Enfield, L.C., A.P. Gibson, J.C. Hebden, and M. Douek, *Optical tomography of breast cancer-monitoring response to primary medical therapy.* Targeted Oncology, 2009. **4**(3): p. 219–233.

[27] Michaelsen, K.E., V. Krishnaswamy, L. Shi, *et al.*, *Calibration and optimization of 3D digital breast tomosynthesis guided near infrared spectral tomography.* Biomedical Optics Express, 2015. **6**(12): p. 4981–4991.

[28] Gao, F., H. Zhao, and Y. Yamada, *Improvement of image quality in diffuse optical tomography by use of full time-resolved data.* Applied Optics, 2002. **41**(4): p. 778–791.

[29] Gao, F., H. Zhao, Y. Tanikawa, and Y. Yamada, *A linear, featured-data scheme for image reconstruction in time-domain fluorescence molecular tomography.* Optics Express, 2006. **14**(16): p. 7109–7124.

[30] Pifferi, A., J. Swartling, E. Chikoidze, *et al.*, *Spectroscopic time-resolved diffuse reflectance and transmittance measurements of the female breast at different interfiber distances.* Journal of Biomedical Optics, 2004. **9**(6): p. 1143–1151.

[31] Wang, J., B.W. Pogue, S. Jiang, and K.D. Paulsen, *Near-infrared tomography of breast cancer hemoglobin, water, lipid and scattering properties using a combined frequency domain and continuous wave measurement system.* Optics Letters, 2010. **35**(1): p. 82–84.

[32] Wang, J., S. Jiang, Z. Li, *et al.*, *In vivo quantitative imaging of normal and cancerous breast tissue using broadband diffuse optical tomography.* Medical Physics 2010. **37**(7): p. 3715–3724.

[33] Krishnaswamy, V., A.M. Laughney, K.D. Paulsen, and B.W. Pogue, *A dark-field scanning in situ spectroscopy platform for broadband imaging of resected tissue.* Optics Letters, 2011. **36**(10): p. 1911–1913.

[34] Luker, G.D. and K.E. Luker, *Optical imaging: current applications and future directions.* Journal of Nuclear Medicine, 2008. **49**(1): p. 1–4.

[35] Hamamatsu Photonics, K. and E. Committee, *Photomultiplier tubes-basics and applications.* Edition 3a, 2006.

[36] Kleinfeld, D., *Noise and signal-to-noise in photodetector systems* 1979, UC San Diego.

[37] Severinghaus, J.W. and Y. Honda, *History of blood gas analysis. VII. Pulse oximetry.* Journal of Clinical Monitoring, 1987. **3**(2): p. 135–138.

[38] *American-Cancer-Society. Cancer Facts and Figures 2016.* 2016.

[39] Tabar, L., M.-F. Yen, B. Vitak, H.-H.T. Chen, R.A. Smith, and S.W. Duffy, *Mammography service screening and mortality in breast cancer patients: 20-year follow-up before and after introduction of screening.* Lancet, 2003. **361**(9367): p. 1405–1410.

[40] Nissila, I., T. Noponen, K. Kotilahti, *et al.*, *Instrumentation and calibration methods for the multichannel measurement of phase and amplitude in optical tomography.* Review of Scientific Instruments, 2005. **76**(4): p. 044302–044302–10.

[41] McBride, T.O., B.W. Pogue, S. Jiang, U.L. Osterberg, and K.D. Paulsen, *A parallel-detection frequency-domain near-infrared tomography system for hemoglobin imaging of the breast in vivo.* Review of Scientific Instruments, 2001. **72**(3): p. 1817–1824.

[42] Culver, J., R. Choe, M. Holboke, *et al.*, *Three-dimensional diffuse optical tomography in the parallel plane transmission geometry: evaluation of a hybrid frequency domain/continuous wave clinical system for breast imaging.* Medical Physics, 2003. **30**: p. 235.

[43] Wang, J., S. Jiang, Z. Li, *et al.*, *In vivo quantitative imaging of normal and cancerous breast tissue using broadband diffuse optical tomography.* Medical Physics, 2010. **37**: p. 3715.

[44] El-Ghussein, F., M.A. Mastanduno, S. Jiang, B.W. Pogue, and K.D. Paulsen, *Hybrid photomultiplier tube and photodiode parallel detection array for wideband optical spectroscopy of the breast guided by magnetic resonance imaging.* Journal of Biomedical Optics, 2014. **19**(1): p. 011010.

[45] Mastanduno, M.A., J. Xu, F. El-Ghussein, *et al.*, *MR-guided near infrared spectral tomography increases diagnostic performance of breast MRI.* Clinical Cancer Research, 2015. **21**(17): p. 3906–3912.

[46] Le-Petross, C.H. and N. Hylton, *Role of breast MR imaging in neoadjuvant chemotherapy.* Magnetic Resonance Imaging Clinics of North America, 2010. **18**: p. 249–258.

[47] Baek, H.-M., J.-H. Chen, K. Nie, *et al.*, *Predicting pathologic response to neoadjuvant chemotherapy in breast cancer by using MR imaging and quantitative 1H MR spectroscopy.* Radiology, 2009. **251**(3): p. 653–662.

[48] Dawood, S., K. Broglio, S.-W. Kau, *et al.*, *Triple receptor–negative breast cancer: the effect of race on response to primary systemic treatment and survival outcomes.* Journal of Clinical Oncology, 2009. **27**(2): p. 220–226.

[49] Neal, A.J., *Neo-adjuvant chemotherapy for early breast cancer.* CME BREAST, 2001. **1**(1): p. 10–13.

[50] Zhu, Q., P.A. DeFusco, A.J. Ricci, *et al.*, *Assessing response to neoadjuvant chemotherapy by using US-guided Near-Infrared tomography.* Radiology, 2013. **226**(2): p. 433–442.

[51] Jiang, S., B.W. Pogue, P.A. Kaufman, *et al.*, *Predicting breast tumor response to neoadjuvant chemotherapy with diffuse optical spectroscopic tomography prior to treatment.* Clinical Cancer Research, 2014. **20**(23): p. 6006–6015.

[52] Tromberg, B.J., Z. Zhang, A. Leproux, *et al.*, *Predicting pre-surgical neoadjuvant chemotherapy response in breast cancer using diffuse optical spectroscopic imaging (DOSI): Results from the ACRIN 6691 study* Cancer Research, 2016. **76**(20): p. 5933–5944.

[53] Bajaj, S., D. Drake, A.J. Butler, and M. Dhamala, *Oscillatory motor network activity during rest and movement: an fNIRS study.* Frontiers in systems Neuroscience, 2014. **8**(February 14): p. Article 13.

[54] Verner, M., M.J. Herrmann, S.J. Troche, C.M. Roebers, and T.H. Rammsayer, *Cortical oxygen consumption in mental arithmetic as a function of task difficulty: a near-infrared spectroscopy approach.* Frontiers in human Neuroscience, 2013. **7**(May): p. Article 217.

[55] Lee, C.W., R.J. Cooper, and T. Austin, *Diffuse optical tomography to investigate the newborn brain.* Pediatric Research, 2017. **82**: p. 376–386.

[56] Torricellia, A., D. Continia, A. Pifferia, *et al.*, *Time domain functional NIRS imaging for human brain mapping.* NeuroImage, 2014. **85**(1): p. 28–50.

Chapter 11

Semiconductor lasers for bio-medical Raman spectroscopy

Bavishna B. Praveen[1], *Praveen C. Ashok*[1], *and Kishan Dholakia*[1]

Raman spectroscopy (RS) is a powerful spectroscopic tool that has secured an important role in the field of biomedical diagnostics and fundamental biological studies. RS, which is based on inelastic scattering of light, is capable of yielding detailed and specific chemical fingerprint from biological materials. The advent of narrow linewidth lasers and efficient spectrographic detectors have played a crucial role in extending the application of this 90-year-old technique to advanced applications in biophotonics in the last two decades. It has been shown that this technique may be an indicator for early cancer detection, particularly in cells and tissue both in *in vitro* and *in vivo* settings. Currently a major international effort is underway in translating this technique into both the surgical and histopathological environments that clearly will benefit from developing compact fiber-based Raman analysis systems. A key prerequisite to achieving this is the development of compact laser sources to realize this vision. Biomedical RS typically requires a stable narrow linewidth (preferably below 50 MHz) laser. In the early embodiments of Raman systems for biomedical applications relied on gas or dye lasers that satisfied the spectral requirements. However, the physical footprint of these systems was a key barrier to using these for Raman studies in a clinical environment. The advent of stable, narrow linewidth semiconductor lasers, including those based on external cavity geometries or Bragg gratings, have played a crucial role in making Raman systems more compact, portable and enabling particular embodiments in the microfluidic domain. This chapter discusses various semiconductor laser systems that can be used for RS and explores how such lasers have enabled a paradigm shift for compact but powerful forms of RS.

11.1 Introduction

Bio-photonics is a field which has emerged as an amalgam of photonics, biological science and engineering. It utilizes the breadth of the electromagnetic spectrum for imaging, identifying, analysis, and diagnosing diseases and to solve both

[1]School of Physics & Astronomy, University of St Andrews, Scotland

fundamental and applied problems to increase our understanding of biological processes at all scale levels. The role of light in biology and medicine is not new to mankind and may be tracked back to the pre-historic times. The invention of microscope in the seventeenth century contributed towards proactive research in biology and medicine where light could be used passively to observe micro-scaled cells which eventually led to the development of cell theory by Theodor Schwann, Matthias Jakob Schleiden, and Rudolf Virchow [1]. The advent of lasers opened up a myriad of opportunities in biology where, in addition to powerful forms of imaging, light could be used as an active tool for diagnosis [2], micro-manipulation [3,4], nano-surgery [5], and even healing [6].

Spectroscopy has emerged as a powerful tool in bio-photonics where the spectral signature of the light absorbed-emitted or scattered reveals the chemical composition of biological materials. This information can be used for fundamental biological studies as well as in disease diagnosis. This is mainly due to the fact that any change in the biochemistry of the sample as a result of the onset of a disease would subtly change the spectral signature. In contrast to other spectroscopic techniques such as fluorescence spectroscopy and absorption spectroscopy, Raman spectroscopy (RS) relies on the inelastic scattering of light where the spectral bands are narrower or in other words more specific to the constituent biomolecules making it a highly sensitive technique and hence highly desirable for disease diagnosis. The Raman spectrum obtained from a biological sample can provide information, both quantitative and qualitative regarding the structure, composition and even interaction of the macromolecular components such as the proteins, lipids, DNA and RNA. This allows researchers or clinicians to pinpoint the specific spectral bands which differ when comparing normal to neoplastic or diseased samples. However, the cross section of Raman process is very weak when compared to that of the luminescence processes [7] since only 1 in every 10^6 photons that is inelastically scattered to yield the Raman signal. This makes it necessary to have specialized sources and detectors to perform bio-medical RS. This chapter focuses on the role of semiconductor lasers in revolutionizing the technique of RS especially for biomedical applications. The chapter will discuss the basic principles of RS and the state of the art applications in bio-photonics. The laser source requirements for RS will be discussed in detail with a special focus on semiconductor lasers. While it is beyond the scope of this chapter to have a detailed discussion on the applications of RS in bio-medicine, a few specific applications of recent note will be discussed to demonstrate how the advent and progress of semiconductor lasers technology is creating a paradigm shift in biomedical RS.

11.2 Raman spectroscopy

When light-interacts with matters, light can either be absorbed or scattered. Raman scattering occurs when energy exchange takes place between the incident photon and the interacting molecule (inelastic scattering) leading to the vibrational energy transitions of the molecules. When the incident photon losses energy to the interacting

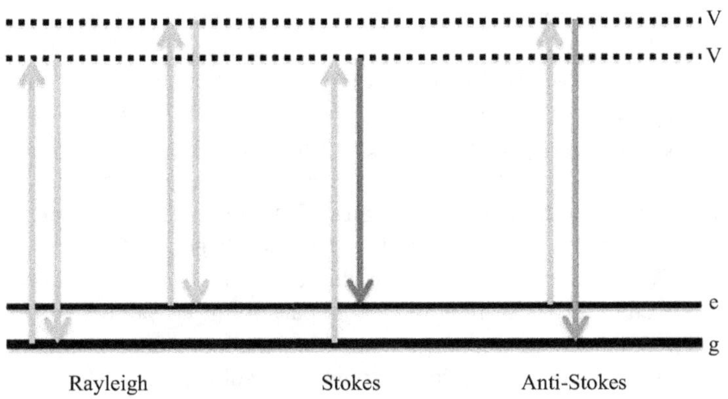

Figure 11.1 Diagram showing the schematic of Rayleigh scattering and Raman scattering processes during one vibration. g—ground energy state, e—excited energy state, v—virtual energy state

molecule, which is usually the case since the molecules would be in their ground state at normal conditions, the scattering is called stokes scattering. If the incident photon gains energy then the scattering is termed the anti-Stokes scattering. However, majority of the photons scatter elastically, a process which is called Rayleigh scattering. In other words, the light scattered off a sample consist of two parts, one is the strong Rayleigh scattering at a frequency same as that of the incident light (ν_0) and the other is a very weak Raman scattering at frequencies $\nu_0 \pm \nu_m$, where ν_m is the vibrational frequency of the molecule. The scattering corresponding to the frequencies $\nu_0 - \nu_m$ and $\nu_0 + \nu_m$ are the Stokes and the anti-Stokes scattering, respectively, shifted from the incident beam frequency ν_0. The schematic of the scattering process is depicted in Figure 11.1 [8].

The Raman spectrum comprises of lines from both the Stokes and Anti-Stokes processes described above and is plotted as a function of energy difference between the incident and the scattered photons which corresponds to the vibrational energy level of the molecules. Typically in RS, the Stokes lines are recorded with most molecules owing to the larger ground state occupancy of molecules at room temperature as dictated by Maxwell-Boltzmann distribution as shown in (11.1) [9].

$$\frac{N_\mathrm{n}}{N_\mathrm{m}} = \frac{g_\mathrm{n}}{g_\mathrm{m}} \exp\left[\frac{-(E_\mathrm{n} - E_\mathrm{m})}{kT}\right] \tag{11.1}$$

where k is the Boltzmann's constant (1.3807×10^{-23} JK^{-1}), T is the given temperature, N_n is the number of molecules in the excited vibrational energy level n, N_m is the number of molecules in the ground vibrational energy level m, g is the degeneracy of the levels in n and m, $E_n - E_m$ is the energy difference between the vibrational energy levels. When the incident excitation wavelength or when the scattered light is close to or coincides with the electronic transition levels of the molecule—that is when the virtual level is close to an actual "real" energy state—this causes an enhancement in the intensity of the Raman peak. This phenomenon is termed "resonance Raman scattering" [10]. The Raman bands are typically a few wavenumbers (10–20 cm^{-1} for biological samples) in width. Their characteristics such as the spectral position, intensity of the band and the width yield specific information regarding the chemical composition, amount of the constituent chemical and the quality/purity of the sample, respectively. Overall this leads to a very specific molecular fingerprint of the sample. Even in complex samples such as the tissues it is still possible to identify the unique Raman bands of specific molecules, which is the key for using this technique in disease diagnosis [11].

Typically RS requires a narrow linewidth laser for Raman excitation. A discussion on lasers used in RS is detailed in the later part of this chapter. A monochromator and a sensitive CCD are required for collecting and recording Raman signatures. Biomedical application of RS typically requires a resolution of $<10\ cm^{-1}$ [12]. The effective resolution of a Raman system depends on factors such as the line width of the laser, slit width and line density of the monochromator and size of the detector. Another factor which needs to be tackled in a Raman system is the luminescence background generated at the system. In a free-space Raman system, a dichroic mirror is used to split the excitation photons and collected Stokes photons. A laser line filter is used in the excitation path, before the dichroic mirror to filter out luminescence generated from the optical components in the excitation path and the amplified spontaneous emission (ASE) from the laser. A Rayleigh rejection filter is used in the collection path, after the dichroic mirror to block Rayleigh-scattered photons, avoiding further luminescence excitation in the collection path. A photograph of a typical Raman system with the different filters along the optical path is shown in Figure 11.2. Since the Raman effect is a weak process, it is very important to optimize the throughput to achieve an efficient RS system.

11.2.1 Raman spectroscopy for bio-medical applications

RS is now considered a potential tool in diagnostic applications and there are several reports validating its use for noninvasive or minimally invasive *in vivo* studies even in humans [13]. The suitability of RS in biology is due to several factors. Being a scattering process, RS allows flexibility in choosing the excitation wavelength which allows the use of Near-IR or Deep-IR excitations. It should be noted that the scattered intensity varies inversely with the fourth power of the observed wavelength ($1/\lambda^4$) for Raman scattering and hence as we approach towards NIR, the collected Raman intensity gets reduced significantly in comparison to the Raman signal collected in the visible regime. However, the choice of NIR and deep- IR are within the therapeutic or diagnostic window of 700–1,100 nm where minimal photodamage occurs and hence is suitable for biological samples. This choice also reduces the fluorescence emission contribution as a consequence of the lower fluorescence cross-section outside the visible regime of the electromagnetic spectra. Apart from this, standard RS does not require any optical labeling of the sample, ruling out any damage to the sample induced by external agents such as the fluorescent dyes. Thus, RS allows noninvasive, nondestructive and rapid assessment of the cellular biochemistry. Water, which is an abundant component in biological

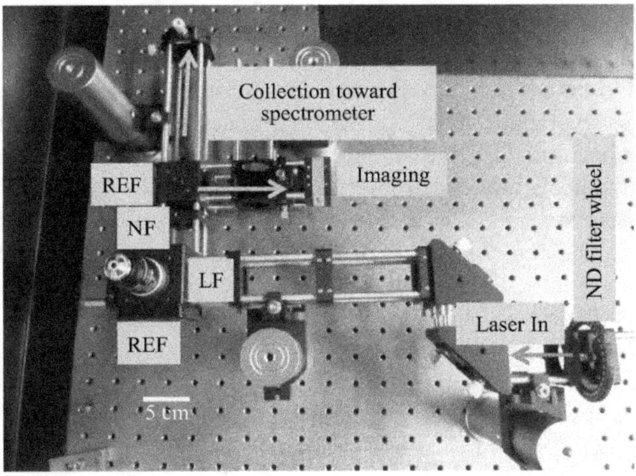

Figure 11.2 Photograph of a typical micro-Raman system. A collimated laser beam is passed through a laser line filter that discards ASE from the laser and fluorescence generated in the optical path. The beam then directed toward an objective using a dichroic mirror. The collected Raman signal passes through the dichroic mirror to the collection arm of the system. The collected beam is passed through another notch filter to filter out Rayleigh scattered photons and coupled into an optical fiber which acts as a confocal aperture. REF-Razor edge filter, LF-line filter, NF-notch filter, ND-neutral density. Arrows show the direction of beam propagation. On the bottom right: arrow denoting the laser in, top left: horizontal arrow denotes the imaging path and top left: vertical arrow denotes the Raman photons

samples, is a weak Raman scatterer and thus gives minimal interference to the Raman signals collected from biological samples, even from those suspended in aqueous media. However, we do remark that choice of the material of the substrate upon which cells and or tissue reside can affect the Raman signature. Either suitable choice of the substrate material or optical trapping with Raman can readily assist in this regard. Typically RS requires practically minimal alteration to the environment of the biological sample thus allowing studies ranging from *in vitro* to *in vivo.*

The other vital factor which makes RS an appealing factor for biological applications is its compatibility with imaging modalities such as confocal microscopy, micro-manipulation modalities such as optical tweezers, and high throughput platforms such as microfluidics. This makes RS an ideal tool for multimodal approaches.

However, the main obstacle with this technique is its weak nature in comparison to the strong fluorescence process occurring in the same spectral window. For biomedical applications, a key major goal remains to suppress this unnecessary background and to enhance the signal to noise ratio of the weak Raman signal. The following section would focus on the laser sources for bio-medical RS and the role of the semiconductor laser to assist in the translation of RS from research laboratories to the clinic.

11.3 Laser sources for bio-medical Raman spectroscopy

As mentioned in the section above, being a scattering process RS allows the use of any excitation wavelength. Thus we have a choice of source dictated not only by budget but application. The Raman detectors depend on the excitation source used and based on the necessary spectrometers can be broadly classified into dispersive and nondispersive systems. Lasers operating in the UV, and in the visible region up to 900 nm, use dispersive systems in their detectors and lasers operating at and above 1,064 nm usually use a nondispersive Fourier transform (FT) Raman systems [14]. A discussion on nondispersive FT Raman systems is beyond the scope of this chapter as this chapter mainly focuses upon semiconductor lasers as the Raman excitation source.

Raman studies in the past utilized gas lasers (Ar^+, Kr^+, He-Ne) and solid-state lasers (ND:YAG, Ti:Sapphire) as excitation sources. These traditional laser sources were well suited for Raman studies owing to their notable frequency stability, relatively high output power (up to 500 mW), near-diffraction limited Gaussian beam profile [15,16]. However, the main drawbacks with these systems are their bulky size and the extensive electronic and cooling systems that are required to maintain their frequency and temperature stability. These features make them less practical particularly for clinical applications. Optical techniques that successfully reach the end-user, in this case clinics demand miniaturization and portability as key features for which advancements in the side of excitation sources and detectors is essential. This brings in the relevance of diode lasers to bio-medical RS [14].

Semiconductor diode lasers were first demonstrated in the early 1960s for applications in optical communication. The next two decades witnessed rapid advancements in laser design making it more practical in terms of the low injection currents and operability at room temperature conditions. The advent of new diode laser materials whose emission wavelengths covers from UV to far-infrared regime has made a direct impact in the field of RS. Typical examples of these laser materials include indium gallium nitride for the UV region, aluminum gallium indium phosphide for the visible region and aluminum gallium arsenide/indium gallium arsenide for NIR to far-IR regimes [17].

Diode lasers are traditionally compact and its output beam is usually elliptical owing to the rectangular shape of the output facet with inherent astigmatism. The output wavelength of a diode laser is dictated by band gap of the semiconductor material, the junction temperature and current density in the device. Temperature of the system and running the operating current can therefore play a strong role in the output characteristics, crucial for Raman studies [19]. Standard diode lasers can have linewidths of 10−100s of MHz.

Figure 11.3 shows a typical example for the small footprint of diode lasers. The laser (PD-LD, Raman Boxx) shown here is a 785 nm volume Bragg grating (VBG)-stabilized diode laser with a maximum output of 500 mW. For a given diode laser, the output power, beam profile, output frequency and the line width depends on factors such as the injection current and the temperature of the active region.

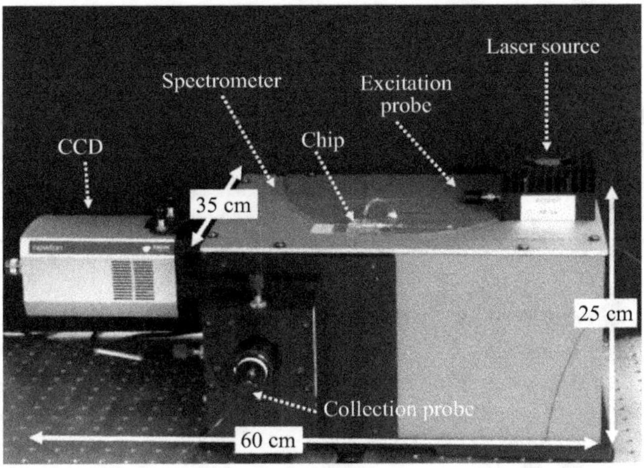

Figure 11.3 Photograph of the probe-based microfluidic Raman detection system ([18] Reprinted with permission from the Optical Society of America). An example showing the small footprint of diode laser. The laser shown here is a 785 nm VBG-stabilized diode laser with a maximum output of 500 mW

Hence diode lasers require accurate control electronics and stabilized current sources to ensure a stable and reliable output [14].

Typically diode lasers are quite susceptible to optical feedback due to a flat gain curve variation with wavelength, and a short lasing cavity with a low finesse. The resultant laser gain thus has a dependency upon wavelength. We see relatively few photons in the cavity, so the introduction of "external" photons can rapidly influence the operating frequency. This behavior, detailed elsewhere [19] is both a blessing and a curse. It does mean we have to be careful with diode lasers as feedback from spurious reflections in the optical train can affect the lasing frequency and thus the Raman signature. As such optical isolators often feature in diode laser-based optical systems.

However with judicious use of external dispersive components, particularly diffraction gratings, these characteristics can be exploited to line narrow and stabilize a diode laser system. The stability and performance of laser diodes was enhanced with the advent of the external cavity diode laser (ECDL). In contrast to a simple laser diode, an ECDL consists of an external cavity. This is achieved typically by implementing an anti- reflection coating on one end of the laser diode chip and completing the cavity on that side with an external diffraction grating in the Littrow geometry. This controlled feedback can "force" the laser to run with a narrow linewidth within its gain curve. This results in better wavelength tuning and narrower line width (typically <1 MHz) which may be desirable for Raman applications. In fact we find there are two main variations based on the Tunable ECDL, namely the Littrow configuration, and the Littman-Metcalf configuration.

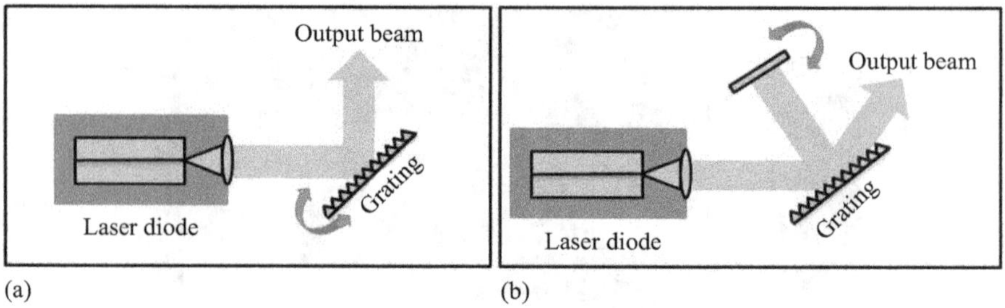

Figure 11.4 Schematic of external cavity diode laser configurations: (a) Littrow configuration. A rotatable grating is used to select the wavelength that allows further narrowing of the laser linewidth. (b) Littman-Metcalf configuration, where an additional external mirror is used for wavelength tuning

The Littrow configuration that has already briefly been mentioned, the diffraction grating is rotated, typically by a PZT, for tuning the laser wavelength, and typically tuning of 10s GHz can be achieved. In the Littman-Metcalf configuration an extra mirror (tuning mirror) shown in Figure 11.4 is rotated for tuning the laser wavelength rendering a relatively small linewidth and better wavelength selectivity as shown in Table 11.1 [20]. The Littman geometry also results in a much more extended tuning range (typically 100s of GHz upwards). Such a diode laser system can assist in advanced RS where we combine this with fluorescence removal. This will be discussed in detail in the later part of this chapter.

The other equally preferred laser diode configurations in the field of RS are the distributed Bragg reflector (DBR) and VBG laser [21]. DBR lasers have at least one distributed Bragg reflector as their laser resonator, providing single frequency at the output. VBR lasers contain Bragg gratings coupled to a laser diode, providing laser wavelength stability independent of temperature variation and narrower line width. Both of these configurations greatly reduce the size of the total laser system making it compact and portable. Most of the laser diode configurations mentioned in this discussion now come with a pigtail option for fiber coupling, thereby further reducing losses due to the elliptical shape and astigmatism associated with the emerging output beam. The specifications of some of the semiconductor laser systems currently available in market, which are suitable for RS are listed in Table 11.1.

11.4 Role of diode lasers in biomedical RS

Over the past two decades, RS has been widely used as a tool for solving biomedical problems. Its applications range from understanding basic biological processes to disease diagnosis. There are a variety of hardware configurations in which RS can be performed depending on the type of problem that needs to be addressed with this technique. This ranges from free-space confocal systems that may be used

Table 11.1 Various semiconductor lasers suitable for Raman spectroscopy that are currently available in market (DBR – Distributed Bragg Reflector, VBG – volume Bragg grating) (Note: this may not be an exhaustive list and the sole purpose of this table is to give some examples.)

Typical wavelengths (nm)	Wavelength stabilizing mode	Linewidth	Beam quality	Power	Maximum mode-hop free tuning range	Source
532	VBG	270 GHz	Multimode	50 mW	NA	www.pd-ld.com
785	VBG	140 GHz	Multimode	300–500 mW	NA	www.pd-ld.com
642, 760, 785, 830, 960	External cavity—Littrow geometry	2 MHz	M2<1.5	100–2000 mW	5 GHz	www.sacher-laser.com
785, 830	External cavity—Littman /Met calf geometry	2 MHz	M2<1.3	100–200 mW	10 GHz	www.sacher-laser.com
780, 850, 940, 1040	External cavity—Littrow geometry	2 MHz	M2<1.5	50–80 mW	30–50 GHz	www.toptica.com
767, 770, 780,785, 795, 852, 895, 920, 976, 1064, 1083	DBR	3 MHz	Multimode	100 mW	NA	www.elliotscientific.com www.vescent.com

for single cellular studies to fiber probe-based systems that allows clinical translation of RS for disease diagnosis. Combining these configurations with fields such as microfluidics allows developing miniaturized, high throughput systems that may be used for fundamental biological studies or for disease diagnosis. This section discusses the role of diode lasers in each of these configurations and how these lasers would help to further the technology and its applications.

11.4.1 Micro-Raman tweezers

Micro-RS or confocal RS is one of the widely accepted approaches that can be used in the field of biology. This technique allows us to record Raman signals from single cells or sub-cellular materials, to an extend that chemical dynamics of the living cells may be observed. When studying single cells or a single sub-cellular component, it is important to confine the sample of interest within the Raman collection region. This is of special importance when Raman spectra need to be obtained from nonadherent cells. Combining micro-RS with optical trapping enables micromanipulation at single cellular or sub-cellular level, resulting in a technique called micro-Raman tweezers, whose feasibility was first demonstrated in the late 1990s on organic micro-droplet samples [22,23]. This combination also allows isolating the cell sample away from the substrate, which will eliminate the background contribution from the substrate in the Raman signature of the cells [22,23]. The ability to manipulate samples at single cellular level while acquiring Raman signal paves the way to developing Raman-activated cell-sorting devices [24,25].

There are two main approaches in realizing confocal micro-Raman tweezers. The first approach uses the same laser beam for trapping and Raman excitation. The second approach is to use a dual-beam system, where one laser beam (typically $\lambda = 1{,}064$ nm) is used for trapping and another laser beam for Raman excitation, as shown in Figure 11.5. This offers more flexibility, since both the power requirement and the time of exposure requirement are different for both trapping and RS.

Another version of the dual-beam approach is to use an orthogonal configuration of the Raman beam and the trapping beam, where the trapping may be realized by two counter-propagating diverging beams emanating from single mode optical fibers and the Raman beam may be delivered orthogonally through a high numerical aperture objective as shown in Figure 11.6 [24,27]. This is more compatible with microfluidic systems for potential Raman flow-based analysis or sorting.

It only requires few tens of mW of power in the IR region to achieve three-dimensional trapping of micro-particles and in the case where cellular dynamics are observed over a longer period of time, trapping cells with such a power level will not create any photo-damage to the cells. Typically, RS requires power levels close to 100 mW to obtain a signal with reasonably good signal to noise from single cells and the acquisition time need to be only few seconds, which again would not induce any photo-damage to the sample of interest. This makes the latter approach more desirable [22].

The first approach, where a single beam would be used for both trapping and RS are usually realized using lasers with excellent beam characteristics (an M^2 value close to 1), since they are required to achieve efficient three-dimensional

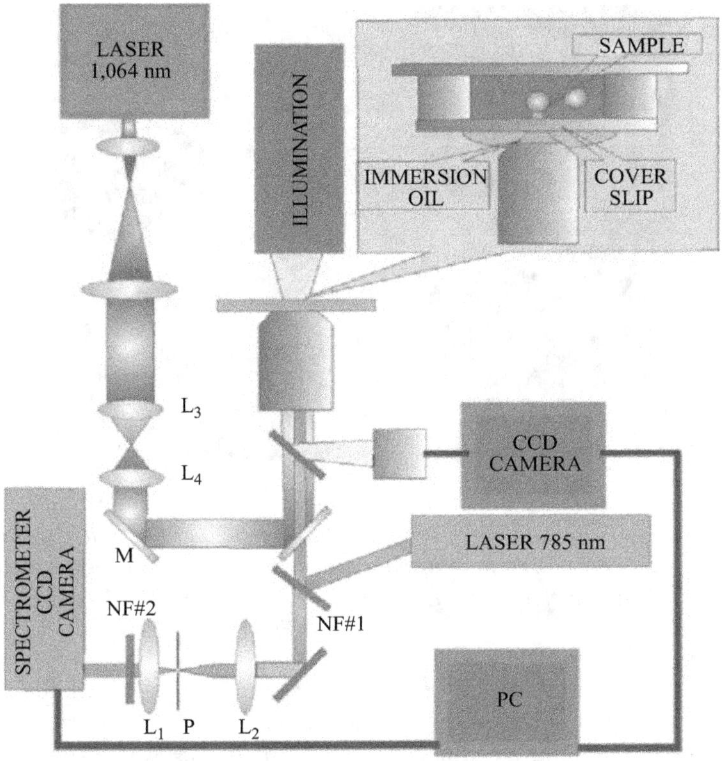

Figure 11.5 The confocal RS setup with dual tweezers ([26] Reprinted with permission from Elsevier). A 1064 nm laser was used for trapping the microparticles/cells and a 785 nm diode laser-provided Raman excitation

optical trapping at relatively low power levels. Hence, diode lasers are not an ideal choice for such embodiments. However one of the first approaches where a single beam was utilized for a combined laser tweezers and Raman spectroscopy utilized a 785 nm wavelength-stabilized, beam shape-circulated semiconductor diode laser. A 100× high numerical aperture objective (N.A = 1.3) was used to form an optical trap and to provide simultaneous Raman excitation. A power-switching scheme was utilized to reduce the effective laser power illuminated while trapping. Spatial filtering was used to enhance the beam quality of the laser beam to achieve efficient trapping [28]. This system was used for observing the dynamics of living cells such as yeast and bacteria to thermal variations in its environment [29,30].

The dual-wavelength approach relaxes the constraints over the beam quality where the inherent advantages of diode lasers may be utilized in realizing compact RSOT systems. Such a dual wavelength RSOT system was realized using a 1,064 nm trapping laser for optical manipulation and a 785 nm diode laser for RS, as shown in Figure 11.5.

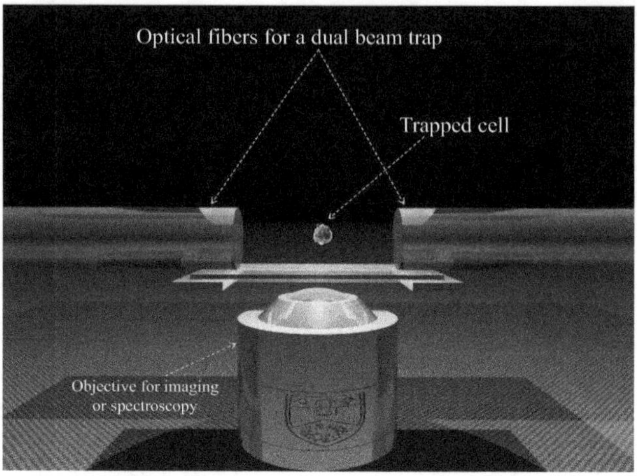

Figure 11.6 Schematic of the dual beam micro-Raman system with dual-beam trap. The counter-propagating diverging outputs from the two optical fibers are used for trapping the sample while RS is performed using a high numerical aperture objective, orthogonal to the trapping beams

Such a system allowed continuous trapping of a single cell for over two hours without incurring photo-damage to the cell [26]. This system was in turn used for a variety of single cellular studies. Some of the examples include real-time detection of hyperosmotic stress response in optically trapped single yeast cells [31] and monitoring of the lag phase and G_1 phase of a single yeast cell [32]. Figure 11.7 shows an example for the typical Raman spectrum acquired from yeast cells.

The technique of micro-Raman tweezers has become a widely used technique for studying the chemical dynamics of a single cell. Research attempts are currently undergoing to develop a Raman-activated cell sorter (RACS) by combining RSOT with microfluidics. Few early prototypes include a system to sort algae using RSOT with a line trap deflector [33], a system to identify and differentiate tumor cells where a dual beam trap in a microfluidic and micro-Raman system was combined for obtaining chemical fingerprint from cells [24] and a system to profile carbon dioxide fixing micro-organisms using resonance RS combined with microfluidics [25]. The prospect of achieving a cell sorter which does not require tagging the samples with external dyes makes RACS a desirable technique. This opens up avenues for sorting cells that cannot be sorted with existing techniques such as fluorescence-activated flow cytometry. However the main challenge in realizing an efficient RACS system is the throughput, which is mainly limited due to relatively high acquisition time (in order of few seconds) required for RS. The current attempts aim to translate the RACS technique from the laboratory to the bench-top. Miniaturization of the device would be one of the main criteria in such a transition. Diode Raman lasers will play a crucial role in such attempts for

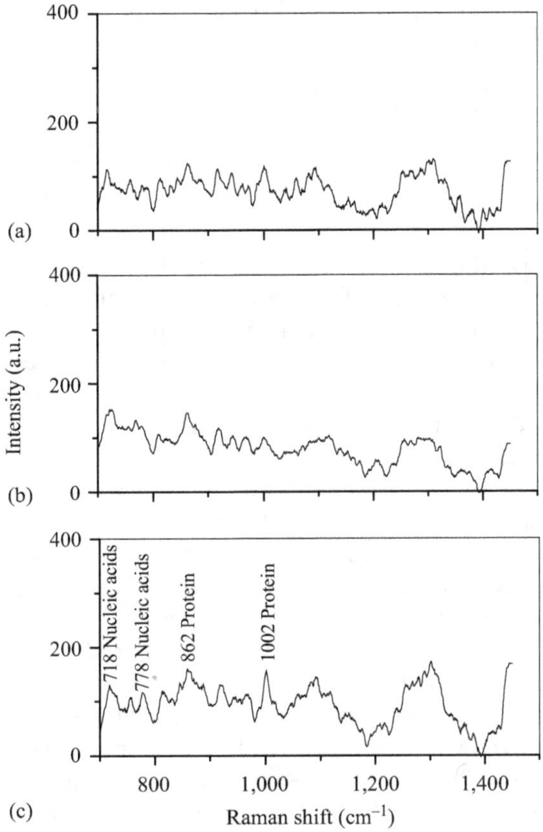

Figure 11.7 *Raman spectra of yeast cells trapped for a total of three: (a) 3; (b) 33 and 126 (c) min, 180 s accumulation time, laser power >10 mW. Spectra are background subtracted using the tweezers method and subsequently baseline subtracted [26] (Reprinted with permission from Elsevier). Typical Raman spectrum acquired from yeast cells*

achieving a miniaturized, bench-top RSOT system capable of cell counting and cell sorting.

11.4.2 Fiber Raman probes

The practical use of any optical technique developed for bio-medical applications are realized when it is translated for real-time clinical studies. The compatibility of RS to a fiber-based system makes it a potential candidate for *in vivo* disease diagnosis. This configuration allows flexibility to access the hard to reach regions of human body. A typical fiber Raman probe would consist of three main parts: one is the fiber end through which the excitation beam is coupled from the laser, the other fiber end delivers the Raman photons scattered from the sample to the spectrometer or monochromator and finally the probe head which essentially consist of the other two terminals of both the excitation and collection fibers which comes in contact with the sample. It should be noted here that while Raman signal acquisition there can be a significant fluorescence background generated within the optical fibers. To avoid this, typically filters are placed at the probe head. A laser line filter at the tip of the excitation fiber will filter out the fluorescence background excited within the excitation fiber. A Rayleigh rejection filter at the collection part of the probe head would avoid excitation wavelength getting coupled into the collection fibers, avoiding any further excitation of fluorescence signal within the collection fiber. Since practically filters are not 100% efficient, the usage of low-OH fibers would further reduce the amount of fluorescence generated within the fiber [34]. Since multimode fibers are used in fiber Raman probes, the beam quality need not be diffraction limited to couple light into them. This makes semiconductor diode lasers, whose output is not typically TEM_{00} mode, readily compatible with fiber Raman probes.

The other major features required for a clinical Raman device are their compact size and portability. The advent of the laser diode and the compact spectrometer has now opened up opportunities to develop such portable diagnostic devices using fiber Raman probes. As mentioned previously, laser diodes come with fiber pigtail options or various option for fiber connectors, allowing plug and play operations for fiber Raman probes, which avoid the need for sensitive fiber coupling. This makes the system compact, rugged and portable. Such diode laser-based compact endoscopy like systems has already been demonstrated for both *ex vivo* and *in vivo* studies on a variety of human tissue samples such as breast [35,36], cervix [37,38], lung [39,40], bladder [41,42], and esophagus [43,44]. One of the examples for such an in vivo study on esophagus demonstrated a sensitivity and specificity > 85 %, with 5 s Raman acquisition time at an excitation power of 70 mW from a 785 nm diode laser [45]. A schematic of a typical compact clinical fiber-based Raman system is shown in Figure 11.8.

11.4.3 Microfluidic Raman spectroscopy

Another embodiment of RS for clinical diagnostics is its combination with Lab-on-a-chip technologies. Lab-on-a-chip technology aims at translating the function of a scientific lab into micro-meter scaled devices that are portable, compact, potentially disposable and can be implemented directly for point of care diagnosis. Microfluidic RS has potential to be used for rapid analysis of body fluids such as saliva, urine and blood and single cells derived from patient's body [46]. Recent advancements in this field have included an all fiber integrated system based on a technique called waveguide-confined Raman spectroscopy (WCRS) which allowed alignment-free, compact system for analyzing few microliters of bio-fluids using RS [47,48]. The photograph of the WCRS-based microfluidic chip is shown in Figure 11.9. As a proof of principle, rapid detection of urea concentration has been demonstrated with sensitivity down to few tens of micro-molars [24,48] concentration. Semiconductor lasers based on VBG have played a crucial role in miniaturizing this to a bench-top portable system.

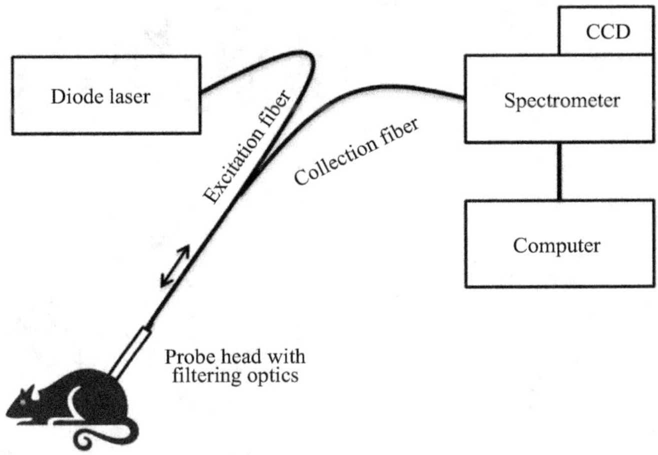

Figure 11.8 Schematic diagram of a typical compact clinical Raman system. The filtering unit is usually housed at the probe head. The specifications of the excitation and collection fiber vary depending on the application. Typically, a single excitation fiber is surrounded by several collection fibers to enhance the collection efficiency

11.4.4 Laser-based fluorescence suppression in Raman spectroscopy

One of the factors that limit the sensitivity of RS is the broad luminescence background that overshadows the weak Raman signal from the sample. Traditionally, the routine to bring down the background in the acquired Raman spectra consist of applying *mathematical methods* such as background-subtraction procedures, polynomial fit of the background. Such an approach is computationally straight forward. However, multiple fluorescence peaks and possibilities for errors in the fit can lead to artifacts in the final processed spectra. Another approach is to suppress fluorescence using a physical principle rather than relying upon a computational approach. There are a variety of techniques which was proposed based on the temporal, polarization, frequency and phase response of Raman process [50–52].

In terms of the topic of this chapter, we focus upon wavelength-modulated Raman spectroscopy (WMRS). This is also known as frequency-modulated Raman spectroscopy, which is a fluorescence rejection technique introduced in the 1,970s, has great potential owing to the ease with which it can be implemented in a conventional Raman spectroscopic system. This technique is based on the fact that when the wavelength of the exciting light is slightly modulated (≤ 5 Å), the Raman bands are also modulated while the fluorescence background which is insensitive to small changes in the excitation remains a constant. This varying Raman signal can then be extracted from the essentially constant fluorescence background.

The advent of this idea in RS began in 1976 when Funfschilling and Williams first demonstrated WMRS using a tunable CW dye laser [53]. An Argon ion laser

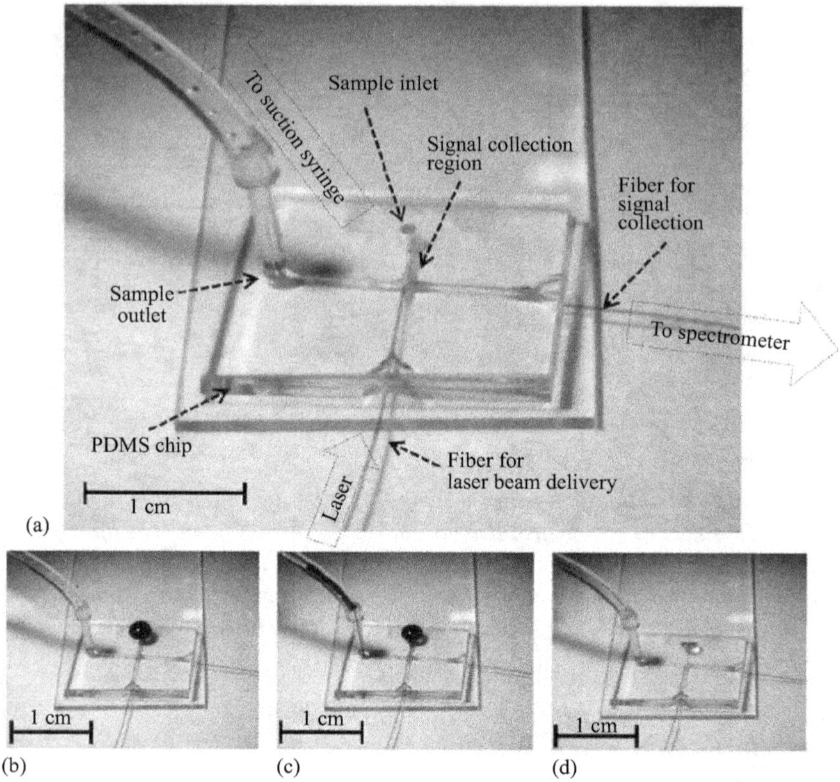

Figure 11.9 (a) Photograph of the WCRS-based microfluidic chip for fluid analysis. (b–d) Sample loading and Raman spectrum acquisition procedure; a food-coloring agent has been used as the sample for better photographic contrast. (b) 20 μL of the sample is placed at the sample inlet of the microfluidic channel (c) The sample is introduced into the microfluidic channel by creating a negative pressure within the channel using a 1 mL syringe attached to the sample outlet port of the chip, followed by Raman acquisition of the analyte for 2s (d) After removing the remaining sample at the inlet, 40 μL of deionized water is sucked into the microfluidic channel to rinse and to avoid contamination of the subsequent sample [49] (Reprinted with permission from the Optical Society of America)

was used to pump the CW dye laser and the wavelength modulation was achieved using an intra-cavity wedge and etalon. An additional diffraction grating was used to eliminate unwanted orders of the etalon. In 1977, Galeener *et al.* demonstrated a similar experiment utilizing a dye laser which constituted of birefringent tuning plates mounted on the shaft of a galvanometer [54]. A function generator was used to activate the galvanometer that produced two slightly different laser frequencies to probe the sample. Both these systems utilized monochromators and phase-lock amplifiers for detection.

The term "wavelength-modulated Raman spectroscopy" was coined in the year 1978 by Levin and Tang, where an electro-optically tuned dye laser was used for fluorescence suppression. This system was much more stable in terms of wavelength reproducibility when compared to the previous mechanically tuned laser sources and monochromators. Even though the proof of principle of WMRS was demonstrated in pure chemicals, owing to the bulk nature of the lasers and stability requirements this technique remained unexploited for any practical studies [55].

The technique has been revisited and significantly refined with its applicability to biology demonstrated in recent years [56,57]. The key to this resurrection was advent of stable external cavity diode lasers with reasonably high mode hop-free tunable range (>150 GHz). These studies utilized discrete diode lasers with closely spaced wavelengths (temperature tuned) or external cavity diode lasers in Littman-Metcalf geometry, where the wavelength was modulated in discrete steps or in a continuous fashion. Further, the marriage of WMRS with multivariate techniques such as the expectation-maximization algorithm and principal component analysis (PCA), to extract the varying Raman information from the constant background fluorescence has eliminated lock-in detection requirements [56,58].

While mathematical methods for fluorescence suppression can introduce artifacts, techniques such as time-resolved RS and polarization modulation RS depend upon the temporal and polarization properties of the Raman signal making it susceptible to loss in Raman information. The laser-based fluorescence suppression technique WMRS, thus stands out as a strong potential candidate in rejecting fluorescence. In contrast to other fluorescence suppression techniques WMRS does not require any complex modification to the Raman apparatus, instead only the Raman laser in the traditional Raman system is replaced with a tunable diode laser.

Another aspect that decides the practicality of a fluorescence suppression technique in RS is the total acquisition time. WMRS allows signal acquisition within only a few seconds potentially enabling high throughput in comparison to other fluorescence suppression techniques. The suitability of WMRS for tissue studies in a fiber-based system was demonstrated recently where power fluctuation and photo-bleaching were observed as a possible derogatory mechanism that could affect the Raman signal extraction in WMRS. Mathematical procedures to simultaneously correct for power fluctuation and photo-bleaching allowed efficient rejection of the fluorescence background thereby allowing optimal extraction of the Raman information, as shown in Figure 11.10 [59]. Importantly, Raman scattering being a weak process makes it necessary to perform Raman studies in a dark environment to prevent interference from any possible ambient stray light sources. This is another major obstacle in taking RS to a clinical environment. However, diode laser-based WMRS is capable of eliminating any signal that is independent of the change in excitation wavelength thereby allowing the acquisition of Raman signals in a clinical setting. The applicability of this technique in the field of biomedical diagnosis is still at its infancy and has wider scope for translating RS to the clinic.

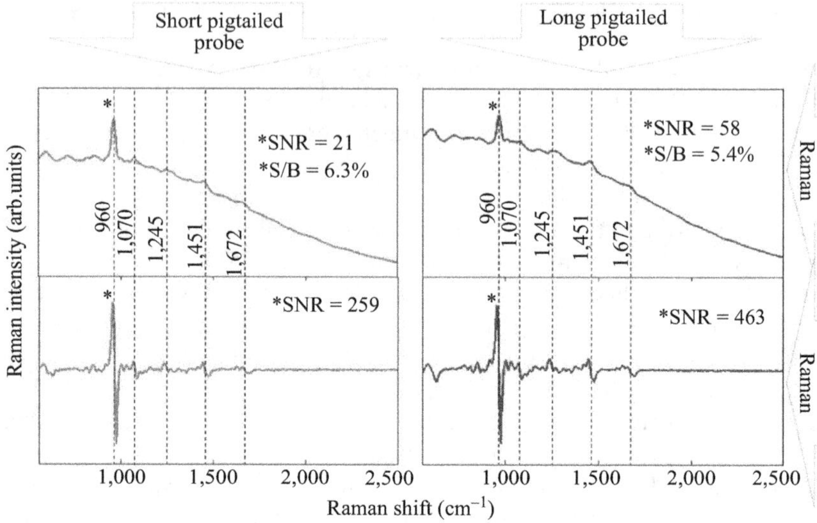

Figure 11.10 *Comparison between the standard and modulated Raman signal of bovine bone tissue recorded using two fiber probes with different fiber pigtail length. The SNR has been estimated from 20 Raman spectra acquired with a total acquisition time of 30 s each. The modulated Raman spectrum has been obtained by applying PCA to six Raman spectra acquired with single acquisition time of 5 s each. Standard Raman spectrum has been obtained by accumulating six Raman spectra acquired with single acquisition time of 5 s each. (Peak assignment: phosphate vl: 960 cm^{-1}, carbonate vl: 1070 cm^{-1}, amide III: 1245 cm^{-1}, CH_2 scissoring: 1451 cm^{-1}, amide I: 1672 cm^{-1}) [59] (Reprinted with permission from SPIE)*

11.5 Conclusion

Over the last two decades, RS has proved its potential to be used for chemical fingerprinting of biological materials. This has enabled this technique to be translated as a novel diagnostic tool for disease diagnosis with high sensitivity and specificity. Raman scattering, being a weak process, relies greatly on the power and spectral quality of the excitation source for its effectiveness. Advent of relatively high power, frequency stabilized semiconductor lasers which has relatively small form factor when compared to its gas laser counterparts enabled developing portable Raman systems for clinical diagnostics. The inherent compatibility of the output from diode lasers with the fiber Raman probe opens up promising avenues in developing a clinical diagnostic platform using RS. Advanced techniques such as microfluidic RS and WMRS also shows potential in extending the applicability and enhancing the sensitivity of this technique for a variety of point of care diagnosis. Semiconductor diode lasers remain at the heart of both innovation and the applicability of RS in the field of biomedicine.

Acknowledgment

We thank the CR-UK/EPSRC/MRC/DoH (England) imaging program for the funding. KD is a Royal Society-Wolfson Merit Award holder. The authors are grateful to Dr. Klaus Metzger for the permission to use Figure 11.6 in this book chapter. We would also like to thank Prof. Andre Riches, Prof. Simon Herrington, and Dr. Michael Mazilu for their useful discussions.

References

[1] McClellan JE, III; Dorn, Harold. *Science and Technology in World History: An Introduction*. The John Hopkins University Press, Baltimore, MD; 1999.

[2] Mahadevan-Jansen A, Richards-Kortum RR. Raman spectroscopy for the detection of cancers and precancers. *Journal of Biomedical Optics* 1996, 1:31–70.3.

[3] Ashkin A, Dziedzic J. Optical trapping and manipulation of viruses and bacteria. *Science* 1987, 235:1517–1520.

[4] Dholakia K, Lee WM, Paterson L, *et al.* Optical separation of cells on potential energy landscapes: Enhancement with dielectric tagging. *IEEE Journal of Selected Topics in Quantum Electronics* 2007, 13:1646–1654.

[5] Praveen BB, Stevenson DJ, Antkowiak M, Dholakia K, Gunn-Moore FJ. Enhancement and optimization of plasmid expression in femtosecond optical transfection. *Journal of Biophotonics* 2011, 4:229–235.

[6] Zeitouni NC, Oseroff AR, Shieh S. Photodynamic therapy for nonmelanoma skin cancers: Current review and update. *Molecular Immunology* 2003, 39:1133–1136.

[7] Krafft C, Sergo V. Biomedical applications of Raman and infrared spectroscopy to diagnose tissues. *Spectroscopy-an International Journal* 2006, 20:195–218.

[8] Ferraro JR, Nakamoto K, Brown CW. Introductory Raman spectroscopy. 2003.

[9] Smith E, Dent G. *Modern Raman Spectroscopy - A Practical Approach*. Wiley, Chichester; 2005.

[10] Efremov EV, Ariese F, Gooijer C. Achievements in resonance Raman spectroscopy: Review of a technique with a distinct analytical chemistry potential. *Analytica Chimica Acta* 2008, 606:119–134.

[11] Chan J, Fore S, Wachsmann-Hogiu S, Huser T. Raman spectroscopy and microscopy of individual cells and cellular components. *Laser & Photonics Reviews* 2008, 2:325–349.

[12] Stone N, Kendall C, Smith J, Crow P, Barr H. Raman spectroscopy for identification of epithelial cancers. *Faraday Discussions* 2004, 126:141–157.

[13] Nijssen A, Koljenovic S, Bakker Schut TC, Caspers PJ, Puppels GJ. Towards oncological application of Raman spectroscopy. *Journal of Biophotonics* 2009, 2:29–36.

[14] McCreery RL. Raman Spectroscopy for Chemical Analysis. In: Winefordner JD, ed. *A Series of Monographs of Analytical Chemistry and Its Applications*. Vol. 157: Wiley-Interscience, New York, NY; 2000.

[15] Peticolas WL. Raman spectroscopy of DNA and proteins. In: Kenneth S, ed. *Methods in Enzymology*. Vol. Volume 246: Academic Press; 1995, 389–416.

[16] Scalfi-Happ C, Udart M, Hauser C, Ruck A. Investigation of lipid bodies in a colon carcinoma cell line by confocal Raman microscopy. *Medical Laser Application* 2011, 26:152–157.

[17] Moloney JV, Hader J, Koch SW. Quantum design of semiconductor active materials: laser and amplifier applications. *Laser & Photonics Reviews* 2007, 1:24–43.

[18] Ashok PC, Singh GP, Tan KM,Dholakia K. Fiber probe based microfluidic Raman spectroscopy. *Optics Express* 2010, 18:7642–7649.

[19] Wieman CE, Hollberg L. Using diode lasers for atomic physics. *Review of Scientific Instruments* 1991, 62:1–20.

[20] Ye C. *Elements for Tunable External Cavity Diode Lasers*, World Scientific Press, https://doi.org/10.1142/5694; 2004.

[21] Hawthorn CJ, Weber KP, Scholten RE. Littrow configuration tunable external cavity diode laser with fixed direction output beam. *Review of Scientific Instruments* 2001, 72:4477–4479.

[22] Petrov DV. Raman spectroscopy of optically trapped particles. *Journal of Optics a-Pure and Applied Optics* 2007, 9:S139–S156.

[23] Ashok PC, Dholakia K. Optical trapping for analytical biotechnology. *Current Opinion in Biotechnology* 2012, 23:16–21.

[24] Dochow S, Krafft C, Neugebauer U, *et al.* Tumour cell identification by means of Raman spectroscopy in combination with optical traps and microfluidic environments. *Lab on a Chip* 2011, 11:1484–1490.

[25] Li M, Ashok PC, Dholakia K, Huang WE. Raman-activated cell counting for profiling carbon dioxide fixing microorganisms. *The Journal of Physical Chemistry A* 2012, 116:6560–6563.

[26] Creely CM, Singh GP, Petrov D. Dual wavelength optical tweezers for confocal Raman spectroscopy. *Optics Communications* 2005, 245:465–470.

[27] Jess P, Garces-Chavez V, Smith D, *et al.* A dual beam fibre trap for Raman micro-spectroscopy of single cells. *Journal of Pathology* 2006, 210: 28–28.

[28] Xie C, Dinno MA, Li Y-q. Near-infrared Raman spectroscopy of single optically trapped biological cells. *Optics Letters* 2002, 27:249–251.

[29] Changan X, Yong-qing L, Tang W, Newton RJ. Study of dynamical process of heat denaturation in optically trapped single microorganisms by near-infrared Raman spectroscopy. *Journal of Applied Physics* 2003, 94:6138–6142.

[30] Xie C, Goodman C, Dinno M, Li Y-Q. Real-time Raman spectroscopy of optically trapped living cells and organelles. *Optics Express* 2004, 12:6208–6214.

[31] Singh GP, Creely CM, Volpe G, Grotsch H, Petrov D. Real-time detection of hyperosmotic stress response in optically trapped single yeast cells using Raman microspectroscopy. *Analytical Chemistry* 2005, 77:2564–2568.

[32] Singh GP, Volpe G, Creely CM, Grötsch H, Geli IM, Petrov D. The lag phase and G1 phase of a single yeast cell monitored by Raman microspectroscopy. *Journal of Raman Spectroscopy* 2006, 37:858–864.

[33] Samek O, Zemanek P, Jonáš A, Telle HH. Characterization of oil-producing microalgae using Raman spectroscopy. *Laser Physics Letters* 2011, 8:701–709.

[34] Utzinger U, Richards-Kortum RR. Fiber optic probes for biomedical optical spectroscopy. *Journal of Biomedical Optics* 2003, 8:121–147.

[35] Keller MD, Majumder SK, Mahadevan-Jansen A. Spatially offset Raman spectroscopy of layered soft tissues. *Optics Letters* 2009, 34:926–928.

[36] Haka AS, Volynskaya Z, Gardecki JA, *et al.* Diagnosing breast cancer using Raman spectroscopy: prospective analysis. *Journal of Biomedical Optics* 2009, 14.

[37] Utzinger U, Heintzelman DL, Mahadevan-Jansen A, Malpica A, Follen M, Richards-Kortum R. Near-Infrared Raman spectroscopy for in vivo detection of cervical precancers. *Applied Spectroscopy* 2001, 55:955–959.

[38] Robichaux-Viehoever A, Kanter E, Shappell H, Billheimer D, Jones Iii H, Mahadevan-Jansen A. Characterization of Raman spectra measured in vivo for the detection of cervical dysplasia. *Applied Spectroscopy* 2007, 61:986–993.

[39] Huang Z, McWilliams A, Lui H, McLean DI, Lam S, Zeng H. Near-infrared Raman spectroscopy for optical diagnosis of lung cancer. *International Journal of Cancer* 2003, 107:1047–1052.

[40] Magee ND, Villaumie JS, Marple ET, Ennis M, Elborn JS, McGarvey JJ. Ex vivo diagnosis of lung cancer using a Raman miniprobe. *Journal of Physical Chemistry B* 2009, 113:8137–8141.

[41] Grimbergen MCM, van Swol CFP, van Moorselaar RJA, Uff J, Mahadevan-Jansen A, Stone N. Raman spectroscopy of bladder tissue in the presence of 5-aminolevulinic acid. *Journal of Photochemistry and Photobiology B: Biology* 2009, 95:170–176.

[42] Draga ROP, Grimbergen MCM, Vijverberg PLM, *et al.* In vivo bladder cancer diagnosis by high-volume Raman spectroscopy. *Analytical Chemistry* 2010, 82:5993–5999.

[43] Puppels GJ, van Aken M, Wolthuis R, *et al.* In-vivo tissue characterization by Raman spectroscopy. *Proceedings of SPIE – The International Society for Optical Engineering*. 3257. 10.1117/12.306086.

[44] Hattori Y, Komachi Y, Asakura T, *et al.* In vivo Raman study of the living rat esophagus and stomach using a micro-Raman probe under an endoscope. *Applied Spectroscopy* 2007, 61:579–584.

[45] Wong Kee Song L-M, Molckovsky A, Wang KK, *et al.* Diagnostic potential of Raman spectroscopy in Barrett's esophagus. In; 2005.

[46] Ashok P, Dholakia K. Microfluidic Raman Spectroscopy for Bio-chemical Sensing and Analysis. In: Fritzsche W, Popp J, eds. *Optical Nano- and*

Microsystems for Bioanalytics. Vol. 10: Springer, Berlin Heidelberg; 2012, 247268.

[47] Ashok PC, De Luca AC, Mazilu M, Dholakia K. Enhanced bioanalyte detection in waveguide confined Raman spectroscopy using wavelength modulation. *Journal of Biophotonics* 2011, 4:514–518.

[48] Ashok PC, Singh GP, Rendall HA, Krauss TF, Dholakia K. Waveguide confined Raman spectroscopy for microfluidic interrogation. *Lab Chip* 2011, 11:1262–1270.

[49] Ashok PC, Praveen BB, Dholakia K. Near infrared spectroscopic analysis of single malt Scotch whisky on an optofluidic chip. *Optics Express* 2011, 19:229822–2992.

[50] Van Duyne RP, Jeanmaire DL, Shriver DF. Mode-locked laser Raman spectroscopy. New technique for the rejection of interfering background luminescence signals. *Analytical Chemistry* 1974, 46:213–222.

[51] Arguello CA, Mendes GF, Leite RCC. Simple technique to suppress spurious luminescence in Raman spectroscopy. *Applied Optics* 1974, 13:1731–1732.

[52] Rusciano G, De Luca AC, Sasso A, Pesce G. Phase-sensitive detection in Raman tweezers. *Applied Physics Letters* 2006, 89:261–116.

[53] Funfschilling J, Williams DF. CW Laser wavelength modulation in Raman and site selection fluorescence spectroscopy. *Applied Spectroscopy* 1976, 30:443–446.

[54] Galeener FL. FM spectroscopy: Raman scattering and luminescence. *Chemical Physics Letters* 1977, 48:7–11.

[55] Levin KH, Tang CL. Wavelength-modulation Raman spectroscopy. *Applied Physics Letters* 1978, 33:817–819.

[56] McCain ST, Willett RM, Brady DJ. Multi-excitation Raman spectroscopy technique for fluorescence rejection. *Optics Express* 2008, 16:10975–10991.

[57] De Luca AC, Mazilu M, Riches A, Herrington CS, Dholakia K. Online fluorescence suppression in modulated Raman spectroscopy. *Analytical Chemistry* 2010, 82:738–745.

[58] Mazilu M, De Luca AC, Riches A, Herrington CS, Dholakia K. Optimal algorithm for fluorescence suppression of modulated Raman spectroscopy. *Optics Express* 2010, 18:11382–11395.

[59] Praveen BB, Ashok PC, Mazilu M, Riches A, Herrington S, Dholakia K. Fluorescence suppression using wavelength modulated Raman spectroscopy in fiber-probe-based tissue analysis. *Journal of Biomedical Optics* 2012, 17:077006–077006.

Chapter 12

Nonlinear imaging applications of high-power lasers: figures of merit

Pablo Loza-Alvarez[1], David Artigas[1,2], and Rodrigo Aviles-Espinosa[1]

Ultrashort pulsed (femtosecond/picosecond) lasers are important tools for imaging due to the advantages they possess. The most significant one is the confinement of induced nonlinear effects to the focal volume (<1 femtoliter). When used for microscopy, it is possible to efficiently produce two-photon excited fluorescence (TPEF), where virtually all the detected fluorescence arrives from the focal point, resulting in the possibility of performing optical sectioning. In addition, most of the generated signal efficiently contributes to the detection, allowing the use of moderate to low excitation average powers. On the contrary, in confocal microscopy, due to the use of a pinhole, lower detected signals are usually compensated by higher excitation powers, often compromising the sample viability. Other additional advantages are related to the fact that the ultrashort pulses normally fall in the near infrared (NIR) spectral wavelengths. This, besides being low energetic photons, allows overcoming some drawbacks inherent in linear imaging techniques such as low penetration depths and sample photodamage/photobleaching occurring out of the focal plane. Furthermore, at NIR wavelengths, Rayleigh scattering, the main limitation to achieve large penetration depths, scales with the inverse of the wavelength, λ^{-n}, with a power n that varies between 2.2 and 4, depending on the material. Therefore, scattering is in general an order of magnitude lower with TPEF than the equivalent one-photon excitation in the visible-UV spectrum. Thus, nonlinear microscopy is routinely used together with conventional linear fluorescence confocal microscopy. Together with TPEF, the same excitation laser can take advantage of other nonlinear-based mechanisms such as second- and third-harmonic generation (SHG and THG, respectively) and can easily be used to perform optical-manipulation techniques such as nanosurgery [1]. SHG and THG go a step forward by providing intrinsic contrast without the need of sample labeling [2–8]. Moreover, these are able to provide structural information at the molecular level [2,3]. Furthermore, they can also be used to obtain high-resolution multiphoton images of the targeted area before and after performing nanosurgery. All these

[1]ICFO-Institut de Ciencies Fotoniques, The Barcelona Institute of Science and Technology, Spain
[2]Department of Signal Theory and Communications, Universitat Politècnica de Catalunya, Spain

applications are sensible to the laser parameters, such as, pulse duration, repetition rate, peak power and average power, requiring different combination of these parameters for an efficient performance. In this chapter, we will make use of an available workstation capable to combine all these techniques and perform real time multi-channel confocal microscopy or epi-fluorescence microscopy while performing femtosecond laser induced manipulation, all of these with different laser sources [9,10]. The different laser systems are compared making use of a figure of merit (FOM) for two-photon interaction and introduce a new FOM for three photon processes. The origin of the FOMs is justified theoretically, becoming a unique tool to select the proper laser system.

12.1 Multiphoton nonlinear microscopy

In this section, we will review the contrast mechanism used in multiphoton nonlinear microscopy. As a reference, we will start with the golden standard in microscopy, linear fluorescence. The nonlinear contrast technique of TPEF will be reviewed afterwards. This will be followed by an introduction to SHG and THG, respectively, contrast mechanisms. The basic set-up of a multiphoton microscope will be then described and the laser FOM for application to microscopy will be established.

12.1.1 One-photon excited fluorescence

Linear fluorescence is the combination of two consecutive effects (see Figure 12.1). First, the absorption of one photon by a molecule or fluorophore, which occurs when the photon energy coincides with the energy differences between the fluorophore electronic energy levels. Second the relaxation to the ground state emitting a photon. In this case the density of excited fluorophores (or absorbed photons) per second, N_{1p}, depends on the probability of being absorbed and the density of excitation photons. The probability of being absorbed can in turn be related to

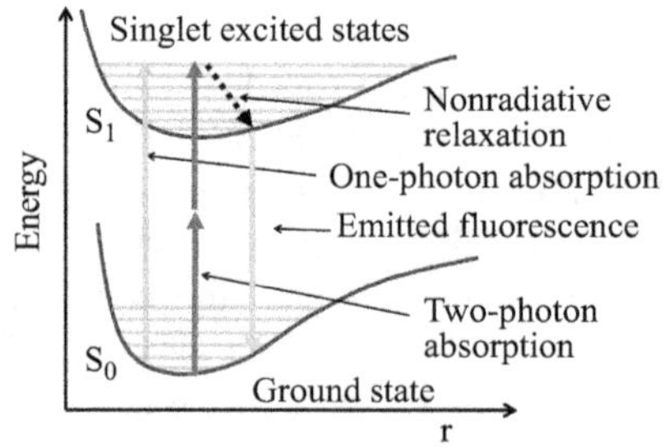

Figure 12.1 Electronic energy level schematic showing the different excitation processes: linear (one-photon) absorption and nonlinear (two-photon) absorption. The emitted fluorescence is the same for both cases. Taken from [11]

the wavelength (absorption bandwidth) and the material capability to perform one-photon absorption. These two features are characterized by the linear (one photon) absorption cross section σ_{1p}. In addition, it must depend on the fluorophore density N_0. Therefore, the density of excited fluorophores per second can be written as

$$N_{1p} = \sigma_{1p} N_0 \frac{I_\nu}{h\nu} \tag{12.1}$$

where the photon density has been written in terms of the excitation laser beam intensity I_ν at the photon frequency ν as $I_\nu/h\nu$, with h being the Planck's constant.

In the second effect, related to the relaxation to the ground state emitting one photon, in most of the cases, different energy levels (usually vibrational levels) are involved in the relaxation process, and only one of the transitions is responsible of the emission. As a result, the emitted photons results in a longer wavelength. The difference between the exciting wavelength and the emitted wavelength is known as the Stokes shift. The Stokes shift is very convenient in microscopy as it allows discriminating between the excitation and emitted wavelengths by using, in the detection system, low-pass filters. The larger the Stokes shift, the more efficient the filtering and the higher the contrast mechanism can be achieved. In addition, we must take into account the efficiency of the emission, determined by the quantum yield $\boldsymbol{\Phi}$, which is defined as the fraction of excited fluorophores that returns to the ground state emitting one fluorescent photon.

From (12.1) and taking into account a laser beam with a section S and the quantum yield $\boldsymbol{\Phi}$, we can obtain the number of emitted photons per second Θ_{em} for an interaction length l as

$$\Theta_{em} = \boldsymbol{\Phi} \cdot N_{1p} \cdot S \cdot l = \boldsymbol{\Phi} \sigma_{1p} N_0 \frac{P}{h\nu} \cdot l \tag{12.2}$$

where the laser power at the sample is defined as $P = I \cdot S$. Note that (12.2) does not depends on S and is valid for any section along the laser beam, as long as the saturation state is not reached ($\Theta_{em} << N_0$). Practically, this means that at the focus of an objective, the generated fluorescence is more intense, but since the area is smaller, the total amount of generated photons is the same than at any other section of the beam (see Figure 12.2(a)). This has two consequences in laser scanning fluorescent microscopy. First, the resulting image is blurred by photons generated out of the focal plane. Second, photodamage is produced all along the path of the laser beam, compromising the sample survival in *in vivo* applications. The first, problem was resolved by adding a pinhole in a conjugated plane, which blocks out off-focus light, resulting in the wide used confocal microscope. However, the photodamage, which is directly related to sample viability, still remains.

12.1.2 Two-photon excited fluorescence

TPEF is the combination of two consecutive effects: two-photon absorption, which is followed by an emitted fluorescence (see Figure 12.1). Two-photon absorption is a nonlinear effect associated to the presence of the third-order susceptibility, $\chi^{(3)}$,

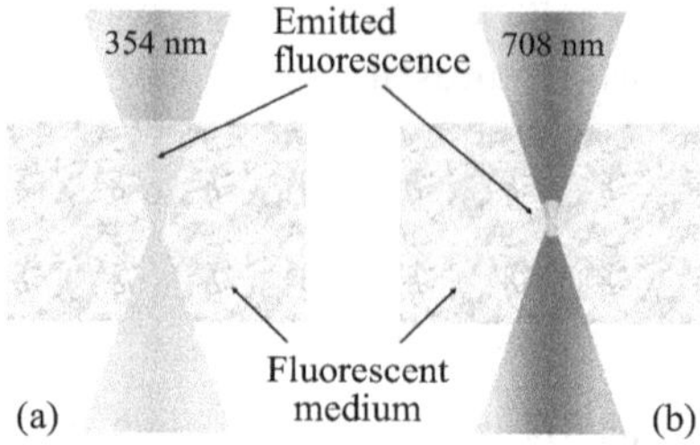

Figure 12.2 Volume at which fluorescence (white) is generated: (a) in one-photon fluorescence, it is generated mainly at the surface since the power of the excitation beam (at λ = 354 nm in this example) is depleted during propagation due to the linear absorption; (b) in two-photon fluorescence, it is generated only at the focal volume where the highest intensity is reached. Here, there is no absorption outside the focal volume and therefore there is not power depletion (in this case at λ = 708 nm, the medium is transparent)

which is present in any material [12]. This effect can occur when the energy difference between electronic levels is similar to the added energy of two excitation photons. In order to observe this effect, the two photons must coincide at the same position within the material. In addition, since the process lasts only for a very short period, in the femtosecond–attosecond scale, both photons must effectively coincide also at the same time. As a consequence, the absorption probability, and thus the efficiency, depends on the photon density, i.e., on the light intensity I_ν. The efficiency also depends on the wavelength (absorption bandwidth) and the material capability to perform two-photon absorption, i.e., on the two-photon absorption cross section σ_{2p}, and the fluorophore density N_0. Then, the density of excited fluorophores per second can be written as:

$$N_{2p} = \sigma_{2p} N_0 \frac{I_\nu^2}{h^2 \nu^2} \tag{12.3}$$

Once a fluorophore has been excited by two photons, the relaxation to the ground state is identical to the situation with one-photon (linear) excitation and its efficiency is determined by the quantum yield Φ. As in the case of the linear fluorescence, it is worth to find the number of photons generated per second for an interaction length l as:

$$\Theta_{em} = N_{2p} S \cdot \Phi = \sigma_{2p} N_0 \Phi \frac{P^2}{h^2 \nu^2 S} l. \tag{12.4}$$

Note that in this case the number of emitted photons depends on the surface S. In a collimated beam this would result in a constant value of Θ_{em} all along the laser beam path. However, in microscopy, the laser beam is tightly focused, and the surface S is the smallest at the objective focus, resulting in the highest generation of photons. The result is that with standard values of objective numerical apertures (NA), around 80% of the photons are absorbed at the focus. Figure 12.2(b) schematically shows what is known as intrinsic optical sectioning for 3D imaging. TPEF has other interesting advantages. Here the radiated wavelength is much shorter than the long wavelength used to produce two-photon absorption. This large stokes shift makes filtering of the near-IR excitation wavelengths from the fluorescent signal wavelengths extremely easy. In addition, the effect of Rayleigh scattering is one order of magnitude lower due to the longer excitation wavelength. Finally, since absorption is only produced at the focal plane, photodamage and photobleaching are absent along the excitation laser beam path.

Finally, nonlinear absorption can also be used to generate ablation by cascading absorption of an excited molecule. This effect occurs above an intensity threshold, associated to the necessary energy to break a chemical bound. When the molecular bounds break there is an increase of the temperature associated to the electron thermalization and gas emission that produces a localized bubble that expands to a certain dimension and then collapses. By controlling the deposited energy, the bubble expansion-contraction effect can be controlled, producing a damage localized to the focus. This damage can be used to perform precise surgery at the submicron scale, known as nanosurgery.

12.1.3 Other nonlinear effects used in microscopy

Two-photon absorption microscopy belongs to a broader family of microscopy techniques making use of nonlinear effects. This family of techniques is known as nonlinear or multiphoton microscopy. Some of the most widely used are SHG [13] and THG microscopy [14]. Since these techniques are based on nonlinear processes they share one of the main characteristics of the TPEF microscopy such as intrinsic optical sectioning. Another interesting nonlinear technique is coherent anti-stokes Raman scattering (CARS) microscopy [15]. In this chapter, and therefore in this subsection, we briefly comment on SHG and THG, leaving CARS to be described elsewhere.

SHG and THG signals are coherent and the generated photons therefore have well-defined polarizations and they are generally generated in the forward direction. The nonlinear interaction at the molecular level is described by a nonlinear electric dipole moment μ. The component ν of the molecule for a single subunit can then be described as:

$$\mu_\nu = \sum_{\kappa\zeta} \beta_{\nu\kappa\zeta} E_\kappa E_\zeta + \sum_{\kappa\zeta\eta} \gamma_{\nu\kappa\zeta} E_\kappa E_\zeta E_\eta \tag{12.5}$$

where E_i is the light excitation electric field component in the ν, κ, and η molecule coordinate system. $\beta_{\nu\kappa\zeta}$ and $\gamma_{\nu\kappa\zeta\eta}$ are the third- and fourth-rank hyperpolarizability tensors. The tensors β and γ are then the ultimate responsible at the molecular level of the SHG and THG signals, respectively. Then, the macroscopic SHG and THG responses within a macromolecular assembly can be described by the coherent addition of β and γ, respectively, in a similar way to an array of microscopic antennas. The result of the coherent addition is a second- $\chi^{(2)}$ and third-order $\chi^{(3)}$ susceptibility tensor, which describes the effect at the macroscopic level and whose efficiency will depend on the macromolecular organization.

In order to obtain SHG, a requirement is that a subunit of the active molecule possesses a noncentrosymmetric symmetry. This usually occurs in peptide bound C-N pair, which provides the main charge electron transfer site. Then the efficiency can be increased by molecular arrangements that in some situation can produce an effect similar to quasi-phasematching. Precisely, quasi-phasematching is likely to be the responsible of SHG generation in the backward direction occurring in collagen. Apart from collagen, other tissues that produce SHG in living samples are muscle (myosin) and microtubules. In addition, other vegetal tissues such as starch and cellulose also produce SHG.

When a laser beam is focused on these tissues, the generated intensity depends on the $\chi^{(2)}$ tensor, which determines the efficiency and the dependence with the excitation polarization, the interaction length l, and the excitation intensity as

$$I_{\mathrm{SHG}} \propto \left(\chi^{(2)}\right)^2 l^2 \cdot I^2 \tag{12.6}$$

As in the previous cases, it is worth to find the number of emitted photons per second as

$$\Theta_{\mathrm{SHG}} \propto \left(\chi^{(2)}\right)^2 l^2 \frac{P^2}{h^2 \nu^2 S} \tag{12.7}$$

Note, that this equation is similar to (12.4), and therefore, TPEF and SHG share similar properties as can be optical sectioning. The main difference however, is that the SHG is a coherent process, and the signal is therefore directional with a square dependence with the interaction length l, while TPEF is omnidirectional and the dependence with l is linear.

On the contrary to SHG, THG is present in any tissue, as noncentrosymmetry is not required. It is associated to the real part of the $\chi^{(3)}$ tensor, while the imaginary part is associated to two-photon absorption. Although THG is present in any material, the phase change occurring along the focus of a Gaussian beam (Gouy shift) produces a back-conversion of the signal to the fundamental wave just after the focus. Therefore, effective THG signal is only produced when the focus coincides with an interface between two media with different refractive indices. Then, a fraction of the generated signal is not reconverted after the focus. As a consequence, by detecting this signal, the corresponding interface can be imaged.

When a laser beam is focused on a tissue, the generated intensity depends on the $\chi^{(3)}$ tensor, which determines the efficiency and the dependence with the excitation polarization, the propagation length l, and the excitation intensity as

$$I_{\mathrm{THG}} \propto \left(\chi^{(3)}\right)^2 l^2 \cdot I^3 \tag{12.8}$$

As in the previous cases, it is worth to find the number of emitted photons per second as

$$\Theta_{THG} \propto \left(\chi^{(3)}\right)^2 l^2 \cdot \frac{P^3}{h^3 \nu^3 S^2} \tag{12.9}$$

Similarly to SHG, THG is a coherent process and therefore the generated signal is directional. However, the dependence with the power and the surface is different to the one featured by TPEF and SHG ((12.4) and (12.7)). This will have important consequences in the definition of the FOM in Subsection 1.5.

12.1.4 The multimodal laser scanning microscope workstation

Nonlinear effects requiring high intensity lasers are currently used in microscopy. In order to preserve the sample against photodamage, mode-locked laser sources able to provide the required high intensity while at the same time maintaining a low averaged power are required. The laser is coupled to the microscope through a scanning system (usually two galvanometric mirrors) and a dichroic mirror, leading the beam to the microscope objective. The excitation beam is then scanned at the sample plane using both galvanometric mirrors. The TPEF is then collected by the same objective and directed to the detection system through the dichroic mirror, coupled in one of the microscope output ports. Additional filters can be used to stop the remaining IR light. For imaging purposes, the detection system is a standard photomultiplier tube (PMT), whose electric signal is recorded and sent a computer for reconstructing the 3D image. SHG and THG are detected using differentiated optics in the forward direction. For this purpose usually a high NA condenser is used, with the corresponding filters to block unwanted signal. As in TPEF, the detection system consists on a PMT and the image is computer generated. A schematic of a scanning microscope is depicted in Figure 12.3. This setup can be built using a commercial confocal inverted microscope

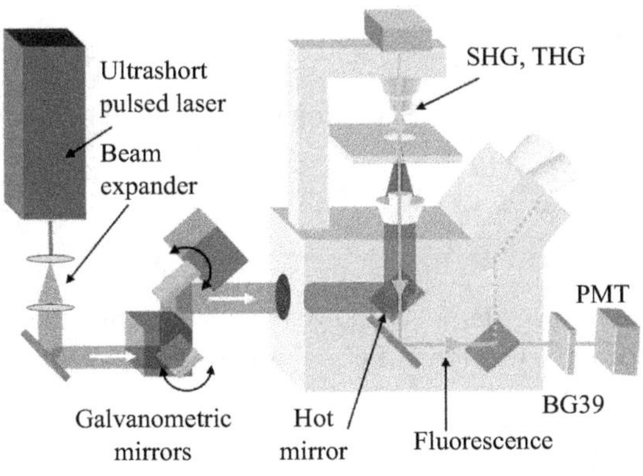

Figure 12.3 Scheme of an inverted nonlinear microscope simultaneously operating with TPEF (epi-detection), SHG and THG (forward detection)

in a way that both the confocal and multiphoton sections can work independently and simultaneously. For that an additional filter turret is used. This allows the use of an additional scanning system and detectors, in both the forward and backward directions, for exciting and collecting the multiphoton signals. Finally a separate control system and software is added for the multiphoton part of the system so that all the flexibility offered by the commercial confocal system is kept operational.

The nonlinear part of the microscope can readily be converted into a nanosurgery station by directing the tightly focused femtosecond laser beam (at intensities above the damage threshold) to a selected target position. This is done with the same laser scanning system used for imaging but in this case, the mirrors remain static. Since the laser beam used for TPEF imaging is the same to be used subsequently in surgery, this way of target selection enables precise determination of the focal plane and a more efficient incision. In addition, the two independent laser scanning systems allow performing simultaneous imaging with the confocal section while, at the same time, the multiphoton part performs the nanosurgery. Furthermore, once that the surgery has been performed, TPEF and/or SHG imaging can be activated if required to analyze the performed damage.

12.1.5 Laser FOM in multiphoton microscopy

For the purpose of this chapter, it is interesting to identify the effect that the laser parameters have on the generation of an image. Mode-locked laser sources are defined by the pulse duration, Δt_p, repetition rate, f_R, and averaged power P_{av}. In order to define a laser FOM, we need to rewrite (12.4), (12.7) and (12.9) taking into account the laser parameters.

In the case of TFEF the power P in (12.4) referred to the peak power, that can be written in terms of the averaged powers as $P = P_{av}/(\Delta t_p f_r)$, while the photon generation Θ_{em} in turn must be multiplied by the pulse duration, Δt_p, and repetition rate, f_R, to take into account the laser duty-cycle, resulting in

$$\Theta_{em} = \sigma_{2p} N_0 \Phi \frac{l}{h^2 \nu^2 S} \frac{P_{av}^2}{\Delta t_p f_R}. \tag{12.10}$$

The situation is similar in SHG microscopy, and (12.7) adopt a similar expression as

$$\Theta_{SHG} \propto \left(\chi^{(2)}\right)^2 \frac{l^2}{h^2 \nu^2 S} \frac{P_{av}^2}{\Delta t_p f_R} \tag{12.11}$$

In both cases, we can define a FOM that determines the efficiency in generating fluorescence or SHG signal, i.e., by a two-photon process, as

$$FOM_{2p} = \frac{P_{av}^2}{\Delta t_p f_R}. \tag{12.12}$$

This FOM clearly shows that short pulses and low repetition rate increases the number of generated photons, and it is an interesting parameter to take into account in the selection/design of the laser source for TPEF and SHG microscopy applications. This FOM is totally equivalent to the one defined before in terms of the pulse peak power P_p as $FOM_{2p} = P_{av} \times P_p$ [16,28]. This FOM was introduced empirically, while here we justify its validity from basic theoretical assumption. A special remark is worth. In (12.12) the pulse duration Δt_p should be the one measured at the microscope sample plane. Typical dispersion introduced by the microscope can be neglected for pulses longer than 100 fs. Below this value, dispersion effects start to be important, and therefore it is necessary to pre-compensate the dispersion so that the transform-limited pulse is obtained at the sample plane.

The FOM changes in THG, since the number of generated photons depends on I^3. Here, the inclusion of the duty-cycle and the laser parameters results in

$$\Theta_{THG} \propto \left(\chi^{(3)}\right)^2 \cdot \frac{l^2}{h^3 \nu^3 S^2} \frac{P_{av}^3}{\Delta t_p^2 f_R^2}. \tag{12.13}$$

We can readily see that in this case, the dependence of the laser parameters changes, and the FOM should be defined accordingly as

$$FOM_{THG} = \frac{P_{av}^3}{\Delta t_p^2 f_R^2}. \tag{12.14}$$

I what follows, the two FOM defined in (12.12) and (12.14) will be used to compare different laser systems, starting by the most widely used Ti:sapphire mode-locked laser. It will also be used to define the threshold conditions to form an image at the sample plane.

12.2 Ti:sapphire laser in microscopy

Titanium:sapphire lasers are the most commonly mode-locked lasers used in microscopy. What has made this laser interesting for biological applications is its tuning wavelength range $\lambda = 700$–900 nm. This spectral range allows exciting most of the visible dyes used in fluorescence microscopy [17], which were specifically developed with a linear absorption in the green-blue-UV spectral region, such as rodamin-B, fluorescein, coumarin, and the green-fluorescent protein present in genetically modified organisms. The tuning wavelength range ($\lambda = 700$–900 nm) it is also well suited for SHG microscopy, since the generated signal falls in the blue region, and both standard optics and detectors can be used. However, Ti:sapphire lasers are not suitable for THG, as the generated signal would fall in the UV region (233–300 nm). At this spectral region, the requirement of special detection optics makes the Ti:sapphire laser unpractical, in addition tissue absorption at the UV can compromise the sample viability.

A standard commercial Ti:sapphire system can deliver ultrashort pulses with typical duration in the range $\Delta t_p = 40$–200 fs, repetition rate around $f_R = 80$ MHz, and average powers of 1–2 W at the central wavelengths. Typical FOM for multiphoton microscopy falls in the range $FOM_{2p} = 62.5$–$1{,}250$ (10^3 W^2). The laser used in this section has a pulse duration of 200 fs and 1 W of averaged power, which results in $FOM_{2p} = 62.5 \times 10^3$ W^2.

We start by showing the imaging capabilities of the Ti:sapphire, in the context of an application for neuron surgery on *Caenorhabditis elegans* worms. The nonlinear part of the microscope can readily be converted into a nanosurgery station by directing the tightly focused femtosecond laser beam (at intensities above the damage threshold) to the neuron. Axotomies were performed on the commissures of the D-type motoneurons on the most ventral/dorsal part of the axon just after its association with the cords (~1–5 μm away). To be able to properly target the neuron axon with the laser, the worms were also expressing CFP in the motoneuron axons. The target position is selected by obtaining first an image with the laser scanning system used for TPEF imaging. Then, using the same scanning system the beam is sent to the target point which will then remain static. Since the laser beam used for TPEF imaging is the same to be used subsequently in the laser surgery, the focal point and a thus the incision is perfectly determined. Once that the neurosurgery has been performed, imaging is again activated to assess any unwanted collateral damage. If this is minimum or nonexisting, the axotomized neurons are then followed for studying any possible spontaneous regeneration process.

So far, evaluation of collateral damage resulting from axotomy has been done by employing fluorescence microscopy to measure the separation of the axon ending tips and the size of the damage caused in the neighboring muscle cell. To do that, both axons and muscles need to be labeled [18,19] so that they can be observed. While this can provide an indication of the size of the damaged areas [20], important endogenous sources of contrast present in the region surrounding the operated axon are not taken into account. In this context, SHG microscopy shows as a promising technique. The energy conservation and label-free characteristics enable minimally invasive imaging, is highly desirable, especially for *in vivo* studies. Biological structures in tissues that produce endogenous SHG signal are based on collagen [21], microtubules [22,23], and muscle (myosin) [24]. Furthermore, these elementary SHG active molecules (hyperpolarizable molecules) or other more complex noncentrosymmetric structures, when excited with different incoming linear polarizations, provide an SHG response that is characteristic of the local active molecular orientation [25]. This polarization-dependent second-harmonic generation (or PSHG) technique can be used to interpret images taken from the *Caenorhabditis elegans* (*C. elegans*) muscles [26]. Furthermore, using PSHG imaging it is possible to determine the orientation, θ_{SHG}, and degree of organization, $\Delta\theta$, of the SHG elementary source inside the muscular tissue. Therefore, this PSHG technique can be used as a novel high-resolution contrast mechanism able to reveal laser axotomy inflicted molecular damage in the muscle of this model organism. Importantly, by taking into account the polarization dependence on the SHG signal, it is possible to determine the orientation and degree of organization of the SHG elementary source (i.e., the hyperpolarizable molecule) inside the muscular tissue. This might be an important source of information about molecular muscle structure before and after the axotomy process.

We employed an oil immersion 60×1.4 NA microscope objective that focuses the beam to a point spread function (PSF) of full-width-at-half-maximum (FWHM) dimensions of 0.45 μm transversal and 1.2 μm axial. While imaging requires a 2–3 mW, to find the appropriate laser power for cutting, the exposure time was set to 200 ms (time controlled by means of an electronic shutter) and the laser power was increased up to the value that led to an effective disruption of axons. This methodology yielded an optimal average power of 90 mW (energy per pulse of 1.2 nJ) which corresponds to a peak intensity of approximately 3×10^{12} W/cm^2 at the sample plane (with a microscope transmission of 30%). The same power was used throughout the experiments. Each worm used in the experiments, although carefully selected and synchronized, is different and can be positioned in different ways, causing, therefore, different types of errors and different degrees of efficiency of the cut (and induced damage). Finally, for comparison purposes, from all the above data, we can find the threshold FOM at the sample plane, $FOM_{2p} \sim 0.25$ W^2 and $FOM_{2p} = 506$ W^2 necessary to perform an image and nano-neurosurgery, respectively.

Figure 12.4 shows a combined simultaneous TPEF and SHG imaging performed at the region before and after the axotomy. In this case, a large collateral damage, evidenced by a big fracture of a thick filament fiber by the TPEF and the SHG image, can be evidenced.

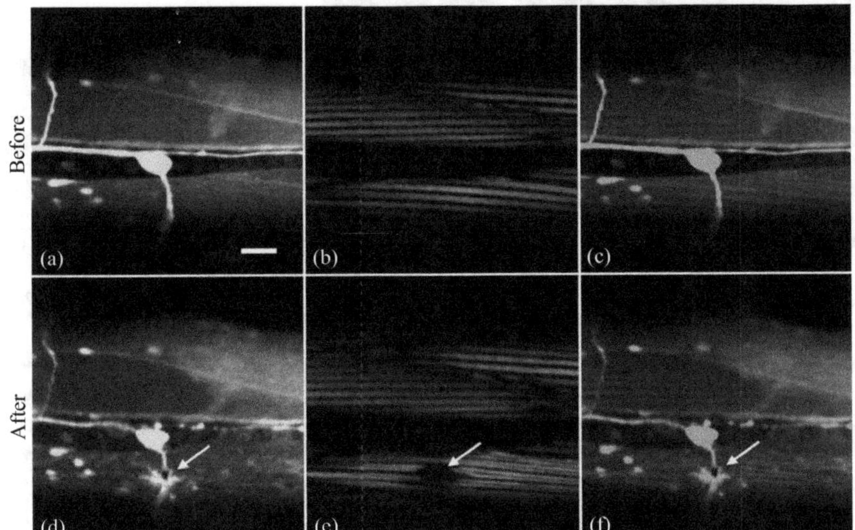

Figure 12.4 Comparison between SHG and TPEF for the muscle damage assessment after the axotomy under large collateral damage conditions. Before the axotomy: (a) TPEF images of YFP marked muscles and axons; (b) SHG signal of the same muscles and (c) merge of TPEF and SHG images. Post-surgical images: (d) TPEF and (e) SHG images showing the laser damage hat is evident over a larger region. (c) Shows a merged TPEF and SHG images. Arrows point to the place of the laser axotomy. Scale bar 10 μm (image taken from https://doi.org/10.1371/journal.pone.0058600)

Then, a reduced collateral damage in which no evident muscle fiber is broken was inflicted (see Figure 12.5). In this case, TPEF from the axons reveals a successfull cut showing a clear gap that interrupts the continuity of the axon. In this case, fluorescence from the muscle (YFP labeled) structure does not reveal any collateral damage. In contrast, SHG imaging shows the appearance of double-band structure in each of the thick filaments around the axotomized region. Upon analysis with the PSHG technique (unpaired two-tailed *t*-test ($n = 200$ pixels for both sets)),

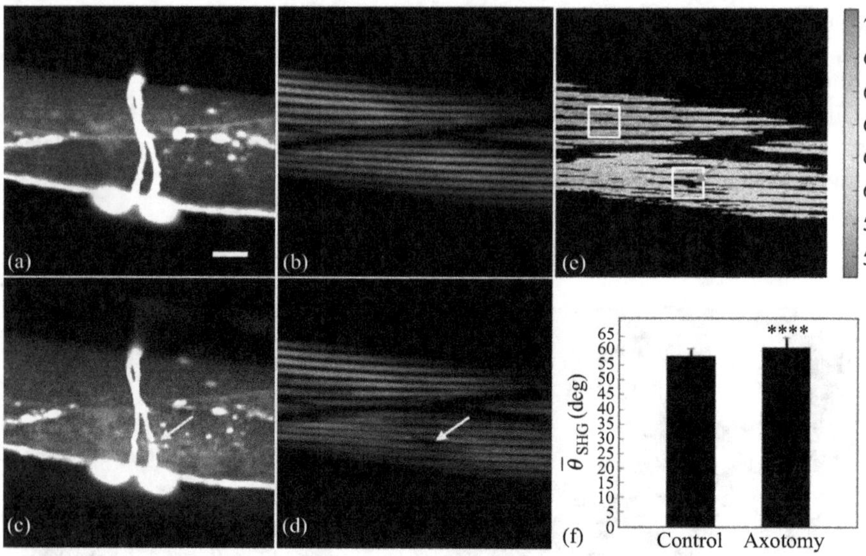

Figure 12.5 Comparison between PSHG and fluorescence for the muscle damage assessment after the axotomy under medium collateral damage conditions. Before the axotomy: (a) TPEF image of YFP marked muscles and axons; (b) SHG image of the muscles in the same region. After the axotomy: (c) TPEF reveals a successfully cut axon. No other damage is apparent. (d) SHG image of the body wall muscle shows the change from SB to DB structure of the sarcomeres. PSHG analysis: (c) post-surgical pixel-resolution mapping of myosin θ_{SHG} (angular orientation of the hyperpolarizable molecule in myosin) at the muscle (gray bar in degrees); (d) θ_{SHG} (mean$\pm$1 standard deviation) for the muscles in the region surrounding the cut and in the control region (white squares). The control region was selected in the adjacent muscular cell far away from the axotomy. Unpaired two-tailed t-test *(*n *= 200 pixels for both sets) yields* p < 0.001 *(****) meaning that θ_{SHG} mean is significantly different between the two regions. Arrows point to the place of the laser axotomy. Scale bar 10 μm. (image taken from https://doi.org/10.1371/journal.pone.0058600)*

we found that the region around the axotomy revealed significant differences ($p < 0.001$ (***)) when compared to an intact sarcomere region.

Finally, we analyzed a case in which TPEF from the axons (GFP labeled) reveals a successfully cut axon (shown as a gap that interrupts the continuity of the axon) but no other damage was apparent. Here a highly confined (with a few pixels) decrease of SHG signal near the axotomized neuron was observed (see Figure 12.6). In this case, there was no double-band structure being produced and our PSHG analysis (unpaired two-tailed *t*-test ($n = 200$ pixels for both sets)) did not find any significant difference ($p > 0.05$) with intact tissue. This suggests that no damage has

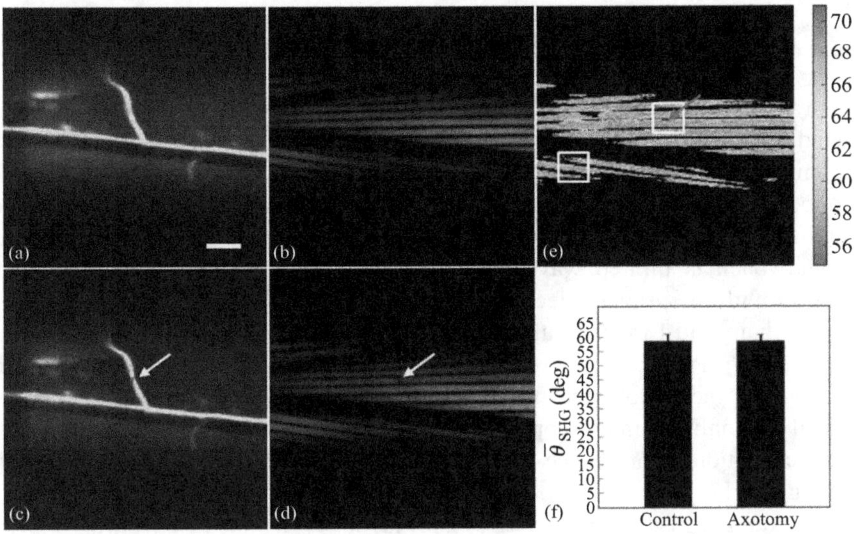

Figure 12.6 Comparison between PSHG and fluorescence for the muscle damage assessment of after the axotomy under minimum collateral damage. Before the axotomy: (a) TPEF image of YFP marked muscles and axons; (b) SHG image of the muscles in the same region. After the axotomy: (c) TPEF reveals a successfully cut axon, in the shape of a gap that interrupts the continuity of the axon, but no other damage is apparent; (d) SHG image of the body wall muscle shows a small signal decrease at the targeted point on the axon. No other change or structural transformation is evident. PSHG analysis: (c) Post-surgical pixel-resolution mapping of myosin θ_{SHG} at the muscle (gray bar in degrees); (d) θ_{SHG} (mean$\pm$1 standard deviation) for the muscles in the region surrounding the cut and in the control region (white squares). The control region was selected in the adjacent muscular cell far away from the axotomy. Unpaired two-tailed t-test ($n = 200$ pixels for both sets) yields $p > 0.05$ meaning that θ_{SHG} mean is not significantly different between the two regions. Arrows point to the place of the laser axotomy. Scale bar 10 μm (image taken from https://doi.org/10.1371/journal.pone.0058600)

been produced or that this has been propagated in a negligible way after the axotomy procedure.

This study shows how a MHz femtosecond Kerr-lens mode-locked Ti:sapphire laser oscillator can be used to precisely sever axons of *C. elegans* D-type motoneurons and image collateral damage. Pre- and post-surgical high-resolution images were used to assess the collateral damage caused to the surrounding tissues. TPEF and SHG provided the anatomical references to precisely locate the laser spot on the target axon. SHG and PSHG microscopy were used *a posteriori* to determine the collateral damage inflicted to the muscle.

12.3 Yb-doped lasers in nonlinear microscopy

In many imaging applications such as the ones previously outlined, there is a high demand for low-cost, compact, tunable, and high power ultrafast lasers. Compared to commercial Ti:sapphire systems typically used in multiphoton imaging, Yb-doped lasers [26] offer ultrashort pulses with high power and direct diode pumping, which drastically lowers the complexity and cost of the laser. However, their drawback is limited wavelength tunability. This lack in tunability can be addressed with the proper selection of a nonlinear fiber [27]. In addition, a center wavelength around 1 micron can be beneficial for biomedical cell surgery and laser axotomy, and TPEF- and SHG-based imaging owing to less tissue absorption and scattering for most biological tissues. Here we show the Yb-based laser system potential for nonlinear microscopy by obtaining high signal-to-noise ratio (SNR) two-photon excitation fluorescence (TPEF) images of different samples labeled with different dyes [28].

An example of a tunable laser system layout is depicted in Figure 12.7. The high power, low-noise, SESAM-mode-locked Yb-doped solid-state laser generates 2 W of average power, 230 fs pulse duration, at 1,040 nm central wavelength, with 40 MHz pulse repetition rate. As a result the Yb oscillator has a FOM of $FOM_{2p} = 435 \times 10^3 W^2$, one order of magnitude higher that the Ti:sapphire laser used in the previous experiments. Therefore, this system is in principle a good candidate to perform both imaging and nanosurgery.

The Yb laser oscillator output is coupled into a micro-structured photonic crystal fiber (PCF) with coupling efficiency >60%. The incoming pump power was controlled by using a half-wavelength plate and Brewster window. As a result, an output spectrum covering the 800–1,250 nm range was obtained. Subsequently, the broadened spectrum is sliced using a "V"-shaped slit that is placed just before of the mirror for selecting the desired spectral region. This was finally compressed by

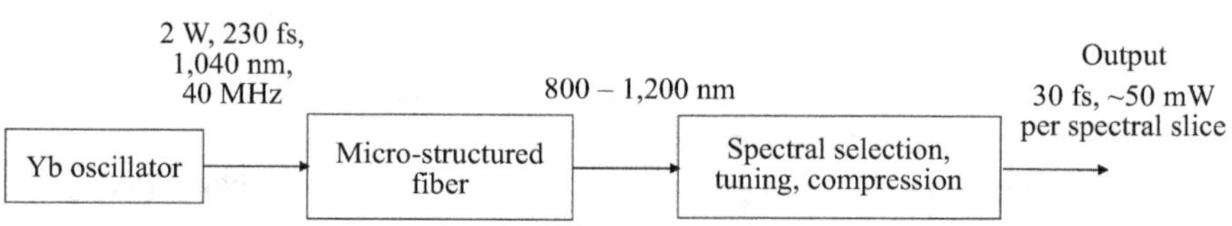

Figure 12.7 Yb-doped tunable laser system layout (taken from Ref. [28])

a prism compressor to 30 fs pulse duration. The average power of each selected spectral slice is up to 80 mW, for a slice bandwidth of 100 nm. The resulting FOM after the fiber and compression setup is then reduced by two orders of magnitude to $FOM_{2p} = 5 \times 10^3 W^2$, a value in principle high enough to perform both imaging and nanosurgery if the microscope shows high transmission efficiency.

Such output beam was coupled to the multimodal workstation using an infrared multifunction port. The beam was focused using a 40 × oil immersion microscope objective (Leica, HCX PL APO) with NA = 1.25. The microscope is equipped with a Scan Head, a software interface which is used to control both scanning and data acquisition units. The output beam alignment to the microscope and compression (at the sample plane) was optimized through a systematic approach, visualizing directly the image obtained from a test sample.

We start comparing directly the Yb 1,040 nm oscillator (without the nonlinear fiber) and the Ti:sapphire laser at 850 nm. The full power of the oscillator after transmission through the different optics was available at the sample plane (1.5 W) but it was reduced to reach 70 mW to avoid photodamage. The average power of the Ti: sapphire was also set to 80 mW at the sample plane, for comparison purposes. The resulting FOM_{2p} at the sample plane was similar in both cases, i.e., 533 W^2 and 309 W^2 for the Yb oscillator and the Ti:sapphire laser. The large wavelength difference between lasers avoided to use a single fluorophore to perform the comparison. Therefore, a sample with different markers was used. In this way GFP and red dyes (mcherry/td:tomato) could be effectively excited with the Ti:sapphire and the Yb oscillator. 500–550 nm and 596–661 nm filters were used for the GFP and red dyes, respectively. Very good-quality images were easily obtained as can be seen from the images of the Yolk region and otic (ear) cavity in Zebra fish Embryo (see Figure 12.8, left). In this case, each dye was exited with a different laser, but simultaneously imaged, showing a similar image quality due to the similar FOM_{2p}. This experiment

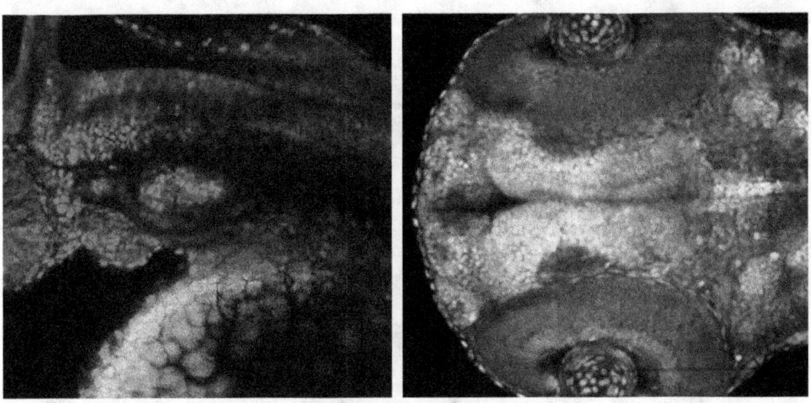

Figure 12.8 Left: Zebra fish Embryo. Yolk region (mcherry), exited with the Ytterbium oscillator and otic cavity (green fluorescent protein (GFP)), excited with a Ti:sapphire laser. Right: Zebra fishEmbryo head labeled using Td: tomato obtained using the Yb oscillator (taken from Ref. [28])

demonstrated the good performance of the Yb oscillator to excite red markers. As an example, Figure 12.8 (right) shows the results with Td:tomato markers.

The next step was to optimize the coupling of the pump beam into the PCF. We obtained ~ 60% coupling efficiency (200 mW output from the fiber). Then, we proceeded to find the positions of the prisms, waveplates and slit in order to get the desired spectral range. The positions were adjusted and at the same time the spectrum was monitored with a spectrum analyzer placed out of the laser system. The average power of selected output pulses was measured using a broadband power meter.

In order to gain a better knowledge about the appropriate configuration and considering the results obtained, we use this time an SHG signal, being this not susceptible for photobleaching. For that, a sample containing barium titanate ($BaTiO_3$) nano-particles was used. SHG images were acquired varying the wave-plates positions. The total image intensity of each image was then determined. All the images were acquired from the same zone. In Figure 12.9 set of SHG images

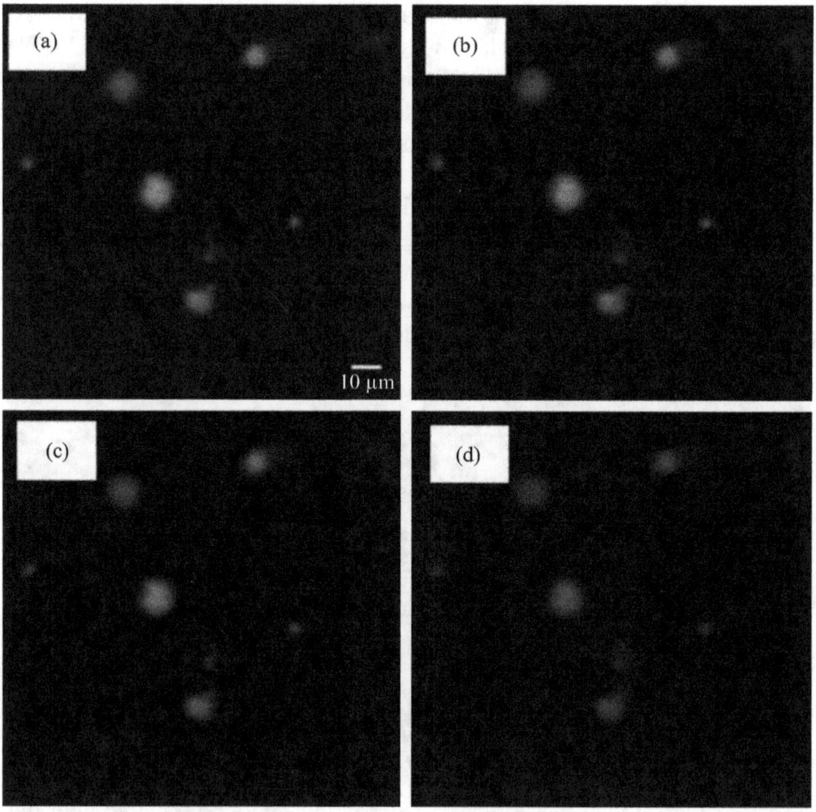

Figure 12.9 SHG images using barium titanate particles for different waveplates positions where it is possible to see a drop in image intensity from image (a) to (d)

are shown for four different half-quarter waveplates positions using the barium titanate sample. The average power at the sample plane was 1.7 mW, resulting in a $FOM_{2p} = 2.41\ W^2$. Although imaging $BiTiO_3$ nanoparticles was possible with this FOM_{2p} (thanks to the high $\chi^{(2)}$value), it is not enough to produce SHG images in biological samples.

Once this was performed, we selected the settings for 970 nm as it enables to efficiently excite GFP. Next, Figure 12.10 shows a TPEF image of fluorescent beads obtained with the optimum configuration of prisms and the half- and quarter-wave wave plates. We used a Leica 20 ×/1.25 oil immersion microscope objective and the average power in the sample plane was 2.6 mW. This results in a $FOM_{2p} = 5.63\ W^2$, higher than the one used with the Ti:sapphire for imaging purposes in the nanosurgery experiment, and therefore it should result in good images.

Maintaining the configuration at 970 nm, with 50 mW of average laser output power, we obtained TPEF images of a mouse intestine section stained with different markers. The average power at the microscope sample plane was 5 mW, resulting in a $FOM_{2p} = 20.8\ W^2$. The acquired images without (Figure 12.11 left) and with 10 frames averaging (Figure 12.11 right) feature a high signal-to-noise ratio, due to the system compressor ability to adjust for the highest peak power at the sample plane of the microscope. The ability to obtain sufficiently high SNR images without averaging demonstrates the laser system potential for nonlinear imaging.

This enabled to demonstrate high SNR TPEF imaging of a mouse intestine section using such ultrafast laser system with 30 fs pulse duration, 50 mW average power, at 970 nm. The relatively simple and potentially low-cost laser system consists of a high power, low-noise, SESAM-mode-locked Yb-doped solid-state laser, coupled into a micro-structured fiber, followed by a simple prism compressor

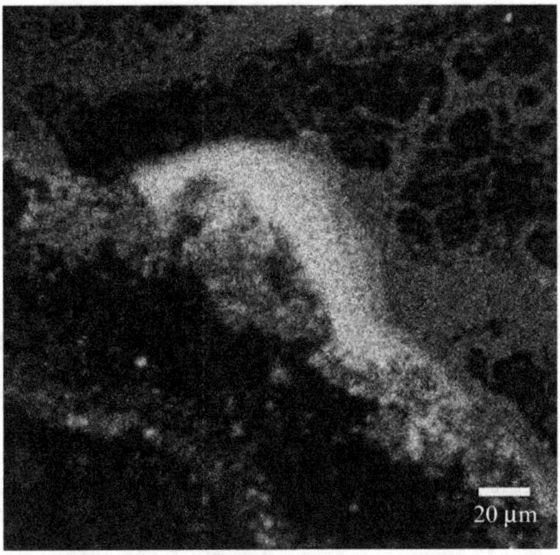

Figure 12.10 TPEF image using a sample with fluorescent beads of 1 μm. We used a Leica 20 × /1.25 oil immersion microscope objective. The average power in the sample plane was 2.6 mW

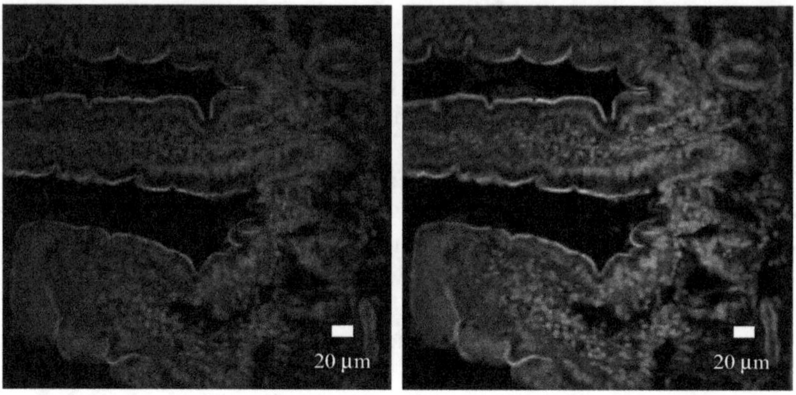

Figure 12.11 *Spectrally resolved mouse intestine section without 10 frames averaging (left) and with 10 frames averaging (right). Mouse intestine was labeled with Alexa Fluor 350 WGA (mucus of goblet cells), Alexa Fluor 568 phalloidin (filamentous actin prevalent in the brush border), and SYTOX Green nucleic acid stain (nuclei of goblet cells) (taken from Ref. [28])*

performing also the spectral selection. Compressibility of the output laser pulses to 30 fs within the tunable range of 800 to 1,200 nm demonstrates the high degree of coherence in the generated continuum.

12.4 Compact novel lasers for TPEF and SHG

On the previous sections, we have evaluated the use of two different lasers for nonlinear (TPEF and SHG) imaging. This section outlines the applicability of a more compact, semiconductor disk laser (SDL) or vertical extended cavity surface-emitting laser (VECSEL) for such task.

Ultrashort pulsed SDLs or VECSEL can be mode-locked by using semiconductor saturable absorber mirrors (SESAMs) [29]. In this way, these compelling sources can be used for multiphoton microscopy, as they combine key features such as excellent beam quality, output power, short pulse durations, amplitude stability, and can be made to operate at a large set of wavelengths, well matched to key TPEF dyes, while maintaining simplicity and ease of operation. In addition, the repetition rate of these lasers can be adjusted into the hundreds of MHz to 1 GHz, resulting in a compact laser cavity, but still working in a range where the peak power can be reasonable without requiring excessively large average powers. SDLs also have the potential for low-cost manufacturing, as their gain element is based on wafer technology and because they have a fairly simple laser cavity design requiring only relatively low-cost and low-brightness pump diodes.

The tested portable SDL had a footprint of only 140 × 240 × 70 mm. It is based on a V-shaped cavity centered on the VECSEL chip mounted on top of a heat sink.

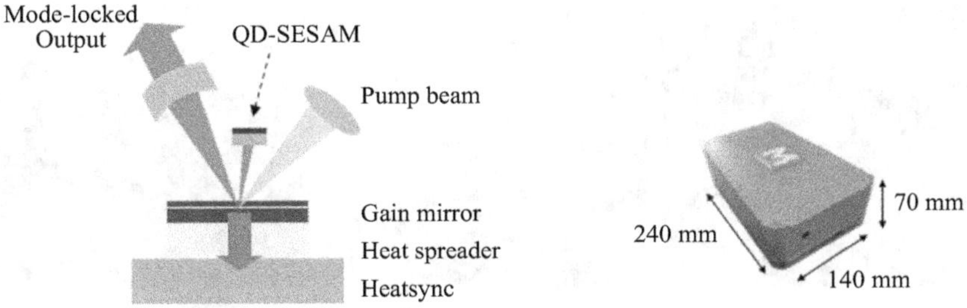

Figure 12.12 Left panel is the schematic of the V-shaped cavity inside the SDL system. The right panel is the SDL prototype (Taken from Ref. [30])

Table 12.1 Normalized two-photon action cross sections of different fluorescent markers (GFPσ_{2P} ~ 178 GM) [data taken from Ref. [16]]

Fluorescent dye	**Expected signal compared to GFP TPEF cross section**
Di-8-ANEPPS in EtOH	0.03
Bodipy (in water)	0.05
Alexa 488 (in water)	0.20
dsRed	0.47
eGFP	1.00
mVenus	1.00
YFP	1.15
mEGFP	1.50
mCitrine	1.54

The pump beam of this laser is focused at an angle of 45° towards the VECSEL. A curved output coupling mirror and a quantum-dot SESAM form the end-mirrors of the laser cavity (see Figure 12.12). The VECSEL gain chip is mode-locked using a quantum-dot SESAM [30–32] to produce 1.5 ps pulses at 500 MHz with an output average power of 287 mW corresponding to 0.4 KW peak power. This results in a FOM at the output of the laser of $FOM_{2p} = 110\ W^2$. Although this value is more than two orders of magnitude lower than the Ti:sapphire laser previously used, the SDL can be designed with output wavelengths that match the two-photon absorption cross section of the green fluorescent protein (GFP), one of the most widely used fluorescent markers for bio-imaging research [33–36]. Excitation of GFP at 965 nm substantially relaxes the requirements for NL imaging. This allows for the reduction of the average and peak power values needed for such task, and therefore for a lower imaging FOM_{2p} threshold.

To show the utility of this laser for TPEF imaging, the capability to efficiently excite different dyes was assessed theoretically and experimentally. The estimations were based on the two-photon action cross section of different dyes relative to GFP at 965 nm (~180 GM) (data taken from Ref. [16]). This is shown in Table 12.1.

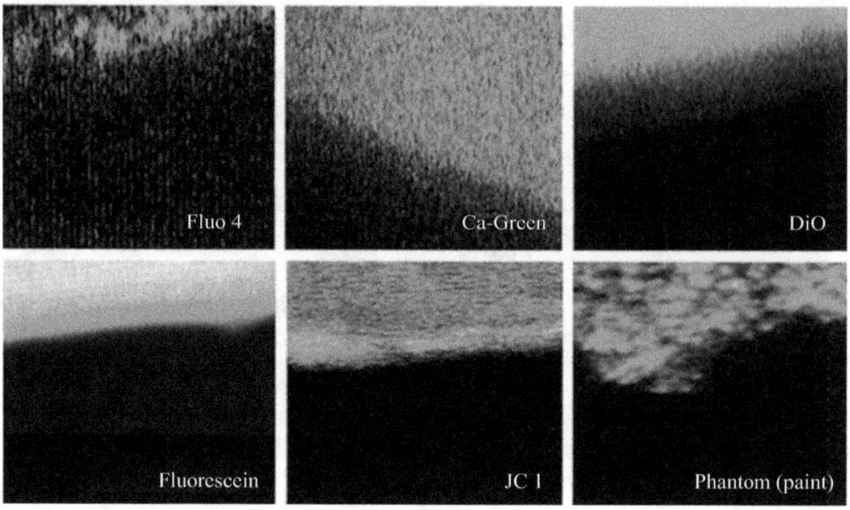

Figure 12.13 TPEF images from different dyes in solution. All the images are 500 × 500 pixels. PMT voltages are Fluo 4: 819 V, Ca-Green: 757 V, DiO: 711 V, Fluorescein: 510 V, JC1: 478 V and Phantom (paint): 572 V. The field of view of all the images is of 40 × 40 μm 15–36 (taken from Ref. [16])

This theoretical calculation was experimentally verified using different dyes in solutions. In each case the excitation average power was kept to a constant value of ~10 mW (measured at the sample plane). Then the relative two-photon signal intensities were recorded using the PMT voltage value. For each image, the background signal was first set to zero and then the emitted signal was adjusted so that only a few pixels were saturated. In each case, low TPEF signal resulted in high PMT voltages and vice versa [16,37,38].

Figure 12.13 shows images at the edge of drops of different dyes that were placed on top of a cover glass. In each image, the voltages resulting from the excitation with the SDL laser system were recorded. It is important to mention that, the precise detector setting may vary not only between instruments but also between samples. However, the excitation/emission general trend should not change, meaning that the PMT values shown here are not absolute but they are relative guides to help in understanding the imaging capabilities of our laser.

From the above, we can observe that the laser can be indeed used for exciting different biological markers once they would be incorporated into the samples. We started by using 1 μm fluorescent beads (Figure 12.14(a)) and a mouse intestine section labeled with different dyes (Figure 12.14(b)). To obtain these images the average excitation power, measured at the sample plane was maintained to ~10 mW, resulting in a FOM_{2p}= 0.13 W^2.

The beads shown in Figure 12.14(a) contain green fluorescing firefly dye, which can be excited at 468 nm using linear excitation. In the case of the SDL,

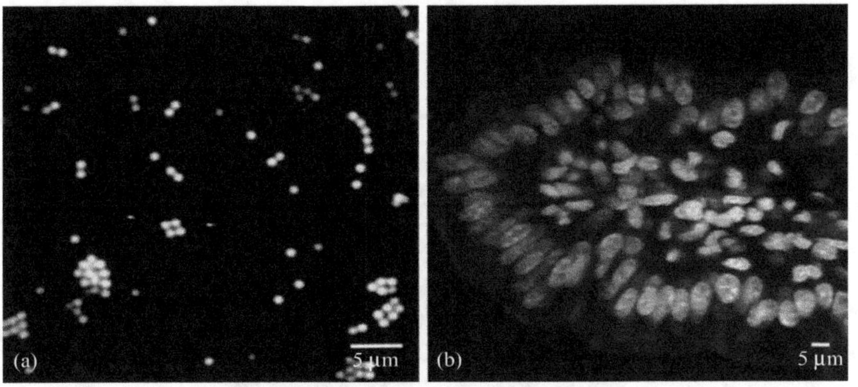

Figure 12.14 TPEF images from (a) green fluorescent beads and (b) mouse intestine section labeled with Alexa Fluor 350 WGA (mucus of goblet cells), Alexa Fluor 568 phalloidin (filamentous actin prevalent in the brush border), and SYTOX Green nucleic acid stain (nuclei of goblet cells). All the images are 500×500 pixels [25,59] (Image taken from Ref. [16])

nonlinear excitation can be efficiently produced due to the two-photon absorption characteristics of the sample. From Figure 12.14(b), it is possible to see the nuclei of goblet cells labeled with SYTOX green, and the border surrounding these cells labeled with Alexa Fluor 568 phalloidin. In this sample, SYTOX green possess a two photon absorption peak centered at 940 nm. This is very close to the operating wavelength of our laser and therefore, cells are visualized very efficiently. For the cell structure labeled with Alexa Fluor 568 (filamentous actin in the brush border) the image region is very faintly observed as the dye is more efficiently excited in the 720–840 nm wavelength range. Finally, for the cells labeled with Alexa Fluor 350 (mucus of goblet cells) there is no TPEF signal observed as the excitation wavelength range is restricted to the 720–800 nm wavelength range.

We can compare the performance of the compact system with the Ti:sapphire laser. To do that, images from different samples (including living *C. elegans* nematodes) were obtained from each system. This was carried out for TPEF and SHG imaging modalities employing different samples. All the experiments were performed using the Ti:sapphire laser at 860 nm and the SDL prototype working at 965 nm. The same amount of averages was applied for both acquired signals. This was performed to allow for a fair comparison between both sources. It is important to mention that to interpret such images; the wavelength difference directly related with the two-photon action-cross section needs to be considered for the TPEF case according to the two-photon action cross section of each employed wavelength.

The first trial was performed employing a fluorescent beads sample. A single frame was taken using both sources (see Figure 12.15 left). As it can be observed the obtained signal from both lasers has a similar quality. However, we needed to almost double the peak power of the Ti:sapphire laser as its two-photon action

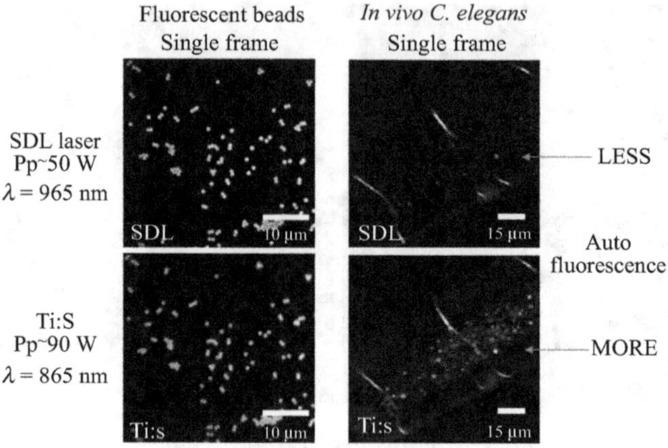

Figure 12.15 Left: Single fames taken from a fluorescent beads sample (TPEF). Top panel depicts the image taken with the SDL prototype. Bottom panel depicts the image obtained using the Ti:sapphire laser. Right: TPEF signal taken from an in vivo C. elegans sample showing a single frame take using both the SDL (top) and the Ti: sapphire (bottom) lasers. The arrows point out the intestine of the C. elegans nematode, where the auto fluorescence signal is generated. (Taken from Ref. [16])

cross section is three times less due to its operating wavelength. However, the higher efficiency in the fluorescent conversion is not capable of compensating for the longer pulse duration and repetition rate of the SDL laser, which results in FOM_{2p} of 0.13 W^2 and 1.93 W^2 for the Ti:sapphire and SDL laser, respectively.

In a similar way, we imaged a living *C. elegans* nematode expressing GFP in motor neurons. In this trial we performed the same comparison of both signals, also accounting for the two-photon action cross section of the Ti:sapphire laser. Figure 12.15 (right) depicts the obtained results acquired using both lasers. In this case it is also evident that both lasers are able to deliver images having a similar quality. However, we can observe an additional signal in the bottom row (Ti:S laser) which is due to the auto-fluorescence (see arrows) from the intestine of the *C. elegans* sample. This type of signal is often unwanted as sometimes interferes with the signal obtained from the structure under study. In this sense the signal provided by SDL laser given its operating wavelength is cleaner. As before, the peak power of the Ti:S laser had to be doubled to compensate for the lower two-photon action cross section.

12.5 Lasers for THG

Similarly to SHG, THG is a nonlinear scattering process that deposits no energy to the tissue, and therefore, enables minimally invasive imaging, which is highly desirable especially for *in vivo* studies. In this case, the generated signal wavelength is one third of the laser excitation wavelength. Therefore Ti:sapphire lasers are not

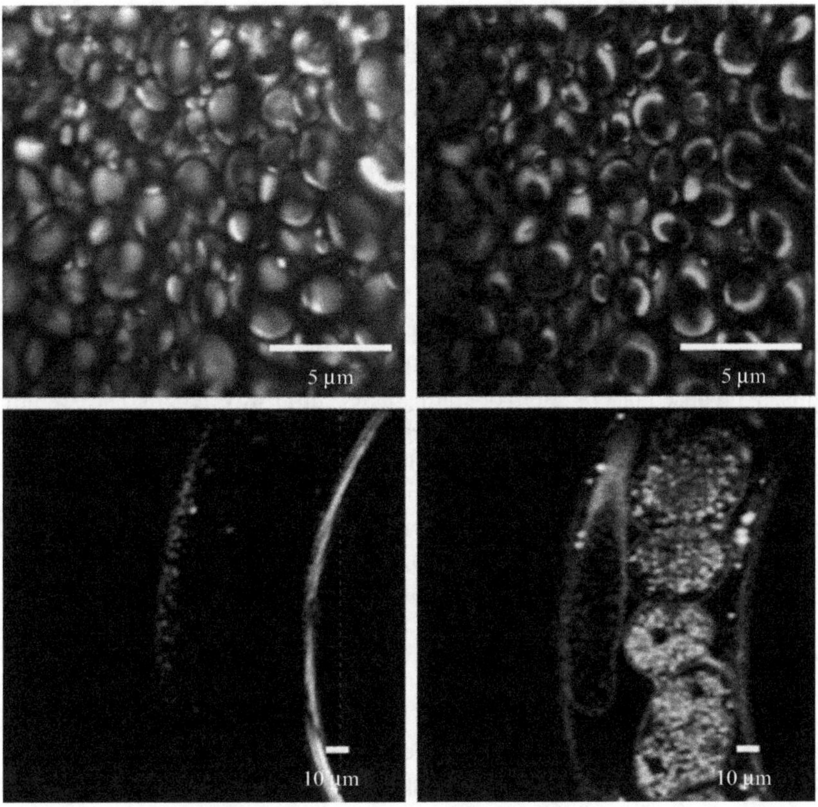

Figure 12.16 SHG image of starch granules taken using a Ti:sapphire laser at 810 nm (top left panel) and with the Er fiber system operating at 1,550 nm (top right panel). SHG images taken at the uterus region containing embryos at cell division stage inside of a C. elegans nematode taken with the Ti:S laser at 810 nm (bottom left panel) and THG images taken using Er. fiber laser system at 1,550 nm

recommended since the generated wavelength would fall in the UV part of the spectrum (shorter than 300 nm). This shorter wavelength requires the use of specialized UV optics and in addition, the high energy of the generated photons can induce collateral damage. Therefore, laser systems with longer wavelengths are desired. In this chapter we will review some of the available alternatives.

A first straightforward approach is to use Er fiber system operating at 1,550 nm. This wavelength is very convenient since the generated signal (516 nm) is at the center of the transmission window of the microscope optical elements. A first test that can be performed to assess the performance of this laser is to compare in our microscopy setup SHG images obtained with the Ti:sapphire laser with THG images obtained with the Er fiber system, using for that starch granules and *C. elegans* embryos. These images are shown in Figure 12.16. The idea is to enable the observation of two different but complementary label free nonlinear imaging modalities (based on scattering processes). In the case of SHG noncentrosymmetric structures like starch granules and muscles enable the production of an efficient signal. In a complementary way THG is generated at the interfaces of the samples, therefore the border of starch granules can be observed. In the case of the *C. elegans*, SHG images only show the body walls while THG is capable to show the cell dividing embryos inside the uterus region. Importantly, the cell nuclei are clearly identified due to the generated contrast. In this case, nucleus does not produce THG signal (it appears as a black spot) as it has a homogeneous composition in contrast with the surrounding structures. The bright spots located within the volume of the cell are lipid depositions [39,40].

We can proceed to compare complementary THG images performed using other lasers operating at a different wavelength to that of 1,550 nm already tested. For this, a Yb based laser system operating at a central wavelength of 1,028 nm was used. In this set of experiments *C. elegans* embryos were imaged outside the nematode. The images obtained from THG at 1,028 and 1,550 can be observed in Figure 12.17. In this case the homogeneous structures such as the cell nuclei does not produce THG signal. This result helped to conclude that THG can be used for label free cell lineage studies, simplifying the imaging procedure and eliminating the requirement of complex algorithms used to identify the dividing nuclei [41–43].

Several experiments were performed to find out the similarities and differences between both setups and the capabilities of each wavelength paying special attention to the FOM_{THG}. Table 12.2 summarizes the laser parameters required to have an efficient THG signal. For the case of THG at 1,028 nm, the use of UV grade optics and a lock-in amplification scheme were required, limiting the image acquisition in terms of speed and thus requiring seven times more average power at

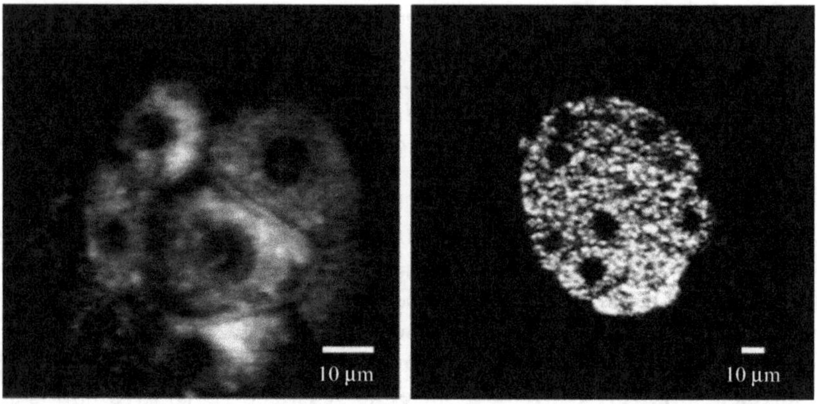

Figure 12.17 THG images from C. elegans embryos at the cell division stage. This image shows each cell and a clearly defined nucleus (dark round areas). The left panel is the image taken with the 1,028 nm system and the right panel is the image taken with the 1,550 nm system. (Taken from Ref. [44])

Table 12.2 THG signal and power requirements comparison at 1,550 nm and 1,028 nm

Laser-operating wavelength	Excitation/ Collection NA	Repetition rate (MHz)	Δτ (fs)	Avg. power (mW)	Peak power (kW)	FOM_{THG} (10^3 W^3)
1,028	0.8/1.4	50	00	35	3.5	428
1,550	1.3/1.4	100	00	4.9	0.45	1.18

the sample plane (35 mW). On the other hand, THG at 1,550 requires the use of only 4.9 mW at the sample plane. These very low average power requirements can be attributed to the fact that THG is emitted within the visible range (516 nm), therefore there is no need of specialized collection optics or amplification schemes as absorption is minimized in comparison to the emitted UV wavelengths. In addition most of the detectors have its peak sensitivity within visible range, thus allowing efficient signal detection. As a result, the threshold FOM necessary to have an image was more than two orders of magnitude larger in the Yb system than in the Er fiber system.

The above experiments show that THG is capable to image embryos in an efficient way using ultrashort pulses operating at 1,550 nm. In addition, since it does not deposit energy on the sample, it results in the ideal technique to perform long-term exposure studies on living samples, specifically embryogenesis. To show this possibility and the effects of imaging at long IR wavelengths, different long-term time-lapse experiments were performed. We imaged *C. elegans* nematodes and recorded its evolution starting from the cell division stage until the L1 larvae stage for periods up to six hours. Figure 12.18 depicts all the developmental stages of a *C. elegans* embryo.

On the top panel of Figure 12.18, we can clearly identify the dividing cell embryos where cell rearrangement and migration takes place. In all cases, it is possible to identify structures such as cell membranes, lipid droplets (see Figure 12.18 bright spots) and the cell nuclei [39,44,45]. As mentioned before this structure has a homogeneous composition in contrast to the surrounding structures, therefore it does not produce any THG signal. On the bottom panel of Figure 12.18, the transition from the early bean-shaped embryo onto the long thin pretzel-shaped is depicted. During this period different morphological process such as elongation and tissue differentiation in the embryo take place. Given that cells have already differentiated into different tissues, it is possible to observe that the THG signal originates mainly from the outer layers of the embryo.

It is known that IR wavelengths around 1,500 nm are associated with increased water absorption in tissue and may lead to unwanted heating of biological specimens. In the previous experiments, the power was kept bellow the loss of cell viability threshold [46] and effects related to sample heating were not observed in agreement to other works [44,47–49]. It is worth to stress that due to the employed wavelength, the lateral and axial resolutions are reduced approximately by a factor

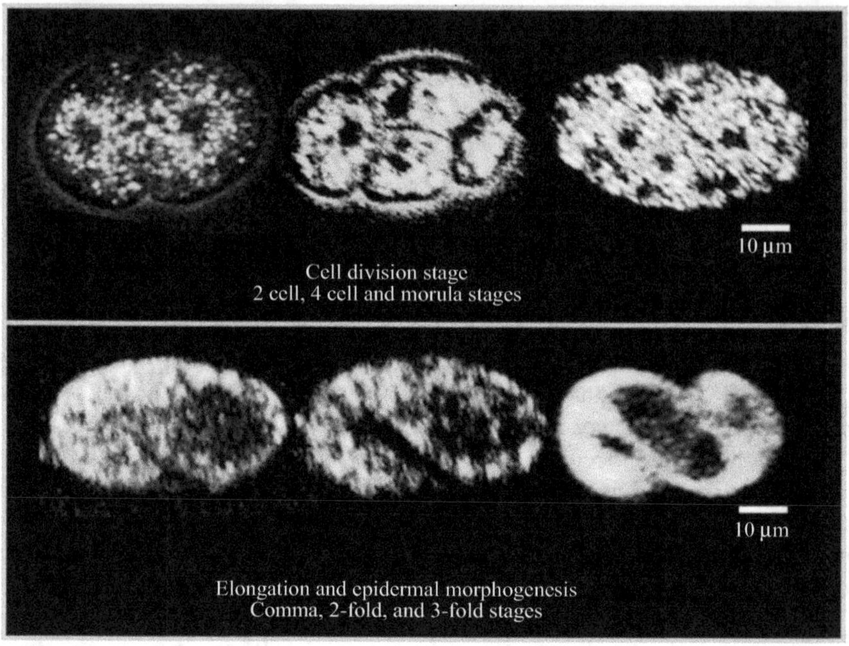

Figure 12.18 THG images of C. elegans embryos at different development stages. Top panel depicts the early developmental stages corresponding to cell division. Bottom panel depicts the later developmental stages were tissue differentiation occurs. (Taken from Ref. [45])

of 2 when compared to that obtained with a Ti:sapphire laser [50] (at 800 nm and objective lens, NA = 1.3). In any case, by employing this laser these work demonstrated that this wavelength in combination with THG is suitable to perform high resolution, long term time-lapse imaging in living samples, e.g., *C. elegans* even in one of the most sensitive stages of its life, such as morphogenesis. This indicates that the use of shorter wavelengths in the IR could also be successfully used for *in vivo* imaging of other biological samples.

12.6 Conclusions

In this chapter, we have reviewed the use of ultrashort pulsed lasers systems used in microscopy. We started with the standard Ti:sapphire laser that is used through all the chapters as a golden standard to be compared with. To do that, we have defined a FOM for two- and three-photon nonlinear processes. This allows a more objective comparison of the usability of the different laser systems for imaging. We found that Yb systems are a good and affordable alternative to conventional Ti:sapphire lasers when red markers are utilized in biological samples. To overcome the lack of tunability of the laser, a super-continuum generation based system, in combination

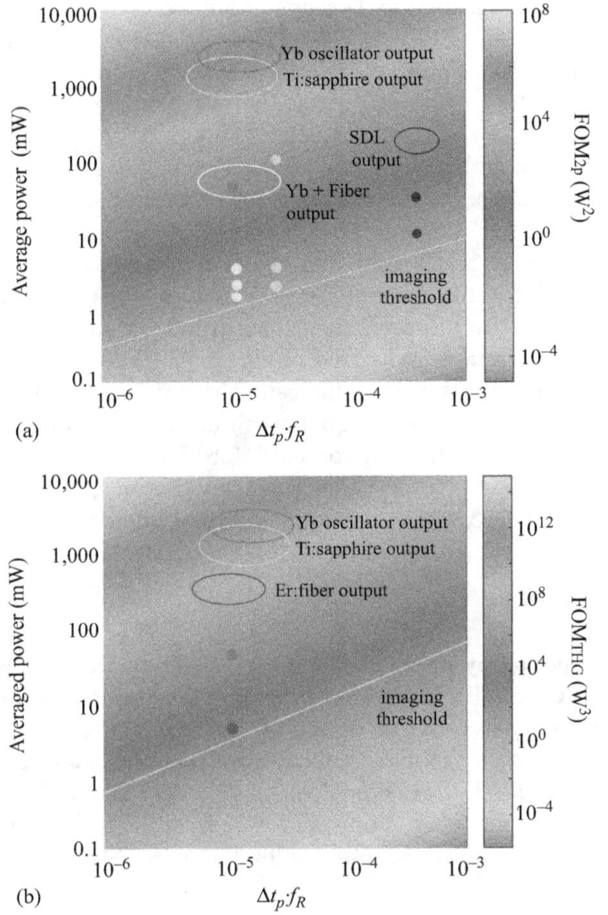

Figure 12.19 Figures of merit for the TPEF and SHG excitation, FOM_{2p} (a) and THG, FOM_{THG} (b). The areas with surrounded by an elliptical line shows in (a) the FOM_{2p} at the laser output for typical Ti:sapphire laser, an ytterbium oscillator, the combination of the Ytterbium oscillator with the micro-structured fiber and the SDL, and in (b) the typical FOM_{THG} for the Ti:sapphire laser, the ytterbium oscillator and the Er:fiber laser. Dots shows the actual FOM for the experiments reviewed in the chapter

with a pulse compression unit, can be used. With this whole system, a large variety of fluorescent markers could be excited using TPEF. However, this also resulted in lower average output powers, and therefore in a strong decrease of the FOM. Since the SHG efficiency only depends on the FOM and not on the wavelength, its applicability to SHG imaging was limited.

Another studied alternative to Ti:sapphire are the compact SDL. This system was designed to target the TPEF of GFP. Here, the performance of the laser was similar to that of the Ti:sapphire despite the low FOM. This is due to the higher two-photon cross section at the SDL excitation wavelengths. However, as before, its performance on SHG was limited.

We then compare an Er fiber laser with a Ti:sapphire laser for SHG applications and to Yb based lasers for THG imaging. In the first case, both systems are comparable as the FOM_{2p} are similar. In the case of THG, even the fact that the FOM_{THG} is smaller than that for the Yb, this is compensated by the fact that the THG wavelengths fall in the spectral range. This means that the collection optics and detection systems are more efficient than in the UV. The FOM of the laser analyzed in this chapter are summarized in Figure 12.19.

Finally, the work of this chapter indicates that the FOMs here defined are an important parameter to take into account when designing and choosing the right laser for any specific application. Importantly, for SHG and THG applications, the higher the FOM, the better. However, for TPEF applications, besides the FOM, the two-photon absorption cross section must also be considered.

Acknowledgments

This work is supported by the Generalitat de Catalunya (2009-SGR-159), FAST DOT, and NEXPRESSO EU project for providing the laser prototypes used in this work. Authors also acknowledge the support from Laserlab-Europe (BIOPTICHAL EU-FP7 284464) and the Photonics4Life network. This research has been partially supported by Fundació Cellex Barcelona and has been conducted at ICFO's "Super-Resolution Light Microscopy and Nanoscopy Facility" (SLN@ICFO). Authors also acknowledge Time Bandwidth Products, M2 Limited, TOPTICA Photonics AG.

References

[1] M. F. Yanik, H. Cinar, H. N. Cinar, A. D. Chisholm, Y. Jin, and A. Ben-Yakar, "Neurosurgery: functional regeneration after laser axotomy," *Nature* **432**: 822 (2004).

[2] P. J. Campagnola, and L. M. Loew, "Second-harmonic imaging microscopy for visualizing biomolecular arrays in cells, tissues and organisms," *Nat. Biotechnol.* **21**, :1356–1360 (2003).

[3] S. Psilodimitrakopoulos, S. I. C. O. Santos, I. Amat-Roldan, A. K. N. Thayil, D. Artigas, and P. Loza-Alvarez, "In vivo, pixel-resolution mapping of thick filaments' orientation in nonfibrilar muscle using polarization-sensitive second harmonic generation microscopy," *J. Biomed. Opt.* **14**: 014001 (2009).

[4] W. Denk, J. H. Strickler, and W. W. Webb, "Two-photon laser scanning fluorescence microscopy," *Science* **248**: 73–76 (1990).

[5] J. Squier, M. Muller, G. Brakenhoff, and K. R. Wilson, "Third harmonic generation microscopy," *Opt. Express* **3**: 315–324 (1998).

[6] D. Yelin, and Y. Silberberg, "Laser scanning third-harmonic-generation microscopy in biology," *Opt. Express* **5**: 169–175 (1999).

[7] E. J. Gualda, G. Filippidis, G. Voglis, M. Mari, C. Fotakis, and N. Tavernarakis, "In vivo imaging of cellular structures in *Caenorhabditis elegans* by combined TPEF, SHG and THG microscopy," *J. Microsc.* **229**: 141–150 (2008).

[8] E. J. Gualda, G. Filippidis, M. Mari, *et al.*, "In vivo imaging of neurodegeneration in *Caenorhabditis elegans* by third harmonic generation microscopy," *J. Microsc.* **232**: 270–275 (2008).

[9] M. Mathew, S. I. C. O. Santos, D. Zalvidea, and P. Loza-Alvarez, "Multimodal optical workstation for simultaneous linear, nonlinear microscopy and nanomanipulation: upgrading a commercial confocal inverted microscope," *Rev. Sci. Instrum.* **80**: 073701 (2009).

[10] S. I. C. O. Santos, M. Mathew, O. E. Olarte, S. Psilodimitrakopoulos, P. Loza-Alvarez, "Femtosecond laser axotomy in *Caenorhabditis elegans* and collateral damage assessment using a combination of linear and nonlinear imaging techniques," PLoS ONE 8(3): e58600. doi:10.1371/journal.pone.0058600 (2013).

[11] O. E. Olarte, J. Andilla, E. J. Gualda, and P. Loza-Alvarez, "Light-sheet microscopy: a tutorial," *Adv. Opt. Photon.* **10**: 111–179 (2018).

[12] R. W. Boyd, Nonlinear Optics. Boston: Academic Press (1992).

[13] J. N. Gannaway and C. Sheppard, "2nd-harmonic imaging in scanning optical microscope," *Opt. Quantum Electron.* **10**: 435–439 (1978).

[14] Y. Barad, H. Eisenberg, M. Horowitz, and Y. Silberberg, "Nonlinear scanning laser microscopy by third harmonic generation," *Appl. Phys. Lett.* **70**: 922–924 (1997).

[15] A. Zumbusch, G. R. Holtom, and X. S. Xie, "Three-dimensional vibrational imaging by coherent anti-Stokes Raman scattering," *Phys. Rev. Lett.* **82**: 4142–4145 (1999).

[16] R. Aviles-Espinosa, G. Filippidis, C. Hamilton, *et al.*, "Compact ultrafast semiconductor disk laser: targeting GFP based nonlinear applications in living organisms," *Biomed. Opt. Express*, **2**: 739–747 (2011).

[17] C. Xu and W. W. Webb, "Measurement of two-photon excitation cross sections of molecular fluorophores with data from 690 to 1050 nm," *J. Opt. Soc. Am.* B: 481–491 (1996).

[18] M. F. Yanik, H. Cinar, H. N. Cinar, *et al.*, "Nerve regeneration in *Caenorhabditis elegans* after femtosecond laser axotomy," *Sel. Topics in Quantum Electronics, IEEE* **12**: 1283–1291 (2006). doi:10.1109/JSTQE.2006.879579.

[19] F. Bourgeois, and A. Ben-Yakar, "Femtosecond laser nanoaxotomy properties and their effect on axonal recovery in *C. elegans*," *Opt Express* **15**: 8521–8531 (2007).

[20] S. I. C. O. Santos, M. Mathew, and P. Loza-Alvarez, "Real time imaging of femtosecond laser induced nano-neurosurgery dynamics in *C. elegans*,". *Opt Express* **18**: 364–377 (2010).

[21] S. Psilodimitrakopoulos, D. Artigas, G. Soria, I. Amat-Roldán, A. M. Planas, and P. Loza-Álvarez, "Quantitative discrimination between endogenous SHG sources in mammalian tissue, based on their polarization response," *Opt. Express* **17**: 10168–10176 (2009).

[22] S. Psilodimitrakopoulos, V. Petegnief, G. Soria, *et al.*, "Estimation of the effective orientation of the SHG source in cortical neurons," *Opt. Express* **17**: 14418–14425 (2009).

[23] S. Psilodimitrakopoulos, V. Petegnief, N. de Vera, *et al.*, "Quantitative imaging of microtubule alteration as an early marker of axonal degeneration after ischemia in neurons," *Biophys. J.* **104**: 968–975 (2013).

[24] S. Psilodimitrakopoulos, S. I. C. O. Santos, I. Amat-Roldan, A. K. Thayil, D. Artigas, P. Loza-Alvarez, "In vivo, pixel-resolution mapping of thick filaments' orientation in nonfibrilar muscle using polarization-sensitive second harmonic generation microscopy," *J. Biomed. Opt.* **14**: 014001 (2009).

[25] S. V. Plotnikov, A. C. Millard, P. J. Campagnola, and W. A. Mohler, "Characterization of the myosin-based source for second-harmonic generation from muscle sarcomeres," *Biophys. J.* **90**: 693–703 (2006).

[26] F. Brunner, G. J. Spuehler, J. Aus der Au, *et al.*, "Diode-pumped femtosecond Yb:KGd(WO4)2 laser with 1.1-W average power, " *Opt. Lett.* **25**: 1119–1121 (2000).

[27] P. S. J. Russell, "Photonics crystal fibers," *Science* **299**: 358–362 (2003).

[28] B. Resan, R. Aviles-Espinosa, S. Kurmulis, *et al.*, "Two-photon fluorescence imaging with 30 fs laser system tunable around 1 micron," *Opt. Express* **22**: 16456–16461 (2014).

[29] U. Keller and A. C. Tropper, "Passively modelocked surface-emitting semiconductor lasers," *Phys. Rep.* **429**: 67–120 (2006).

[30] A. McWilliam, A. A. Lagatsky, C. T. A. Brown, *et al.*, "Quantum-dot-based saturable absorber for femtosecond mode-locked operation of a solid-state laser," *Opt. Lett.* **31**: 1444–1446 (2006).

[31] U. Keller, K. J. Weingarten, F. X. Kartner, *et al.*, "Semiconductor saturable absorber mirrors (SESAM's) for femtosecond to nanosecond pulse generation in solid-state lasers," *IEEE J. Selec. Topics Quantum Electron.* **2**: 435–453 (2016).

[32] D. Maas *et al.*, "Growth parameter optimization for fast quantum dot SESAMs," *Opt. Express*, **16**: 18646–18656 (2008).

[33] Imaging Developmental Recourse for Biophysical Optoelectronics. Two photon action cross sections 2010; Available from: http://www.drbio.cornell.edu/cross_sections.htm.

[34] E. Spiess, F. Bestvater, A. Heckel-Pompey, *et al.*, "Two-photon excitation and emission spectra of the green fluorescent protein variants ECFP, EGFP and EYFP," *J. Microscop.* **217**: 200–204 (2005).

[35] G. A. Blab, P. H. M. Lommers, L. Cognet, G. S. Harms, and T. Schmidt, "Two-photon excitation action cross-sections of the autofluorescent proteins," *Chem. Phys. Lett.* **350**(1–2): 71–77 (2001).

[36] R. Heim, A. B. Cubitt, and R. Y. Tsien, "Improved green fluorescence," *Nature* **373**: 663–664 (1995).

[37] T. R. Neu, U. Kuhlicke, and J. R. Lawrence, "Assessment of fluorochromes for two-photon laser scanning microscopy of biofilms," *Appl. Environmen. Microbiol.* **68**: 901–909 (2002).

[38] R. Aviles-Espinosa, G. Filippidis, C. Hamilton, *et al.* Portable semiconductor disk laser for in vivo tissue monitoring: a platform for the development of clinical applications. in *Conference on Medical Laser Applications and Laser-Tissue Interactions V.* 2011. Munich, Germany.

[39] D. Debarre, W. Supatto, A. M. Pena, *et al.*, "Imaging lipid bodies in cells and tissues using third-harmonic generation microscopy," *Nat. Methods* **3**: 47–53 (2006).

[40] R. Aviles-Espinosa, S. I. C. O. Santos, A. Brodschelm, *et al.*, "Third harmonic generation for the study of C. elegans embryogenesis," *J. Biomed. Opt.* **15**: 046020 046020–7 (2010).

[41] S. Hamahashi, S. Onami, and H. Kitano, "Detection of nuclei in 4-D Nomarski DIC microscope images of early *Caenorhabditis elegans* embryos using local image entropy and object tracking," *BMC Bioinf.* **1**: 86–93 (2005).

[42] S. Hamahashi and S. Onami, "Objective measurement of spindle orientation in early *Caenorhabditis elegans* embryo," *Genome Informatics* **6**: 1–15 (2005).

[43] Z. Bao, J. I. Murray, T. Boyle, S. L. Ooi, M. J. Sandel, and R. H. Waterson, "Automated cell linage tracing in *Caenorhabditis elegans*," *PNAS* **103**: 2707–2712 (2006).

[44] R. Aviles-Espinosa, G. J. Tserevelakis, S. I. C. O. Santos, *et al.*, "Cell division stage in C. elegans imaged using third-harmonic generation microscopy," in *Proc. Biomedical Optics, Optical Society of America*, Washington, D.C., paper. BTuD78 (2010).

[45] R. Aviles-Espinosa, S. I. C. O. Santos, Andreas Brodschelm, *et al.*, "In-vivo third-harmonic generation microscopy at 1550 nm: three-dimensional long-term time-lapse studies in living *C. elegans* embryos," *Proc. Photonics West*, San Francisco C.A. (USA), Paper 7904–50 (2011).

[46] K. Konig, T. W. Becker, P. Fischer, I. Riemann, and K.-J. Halbhuber, "Pulse-length dependence of cellular response to intense near-infrared laser pulses in multiphoton microscopes," *Opt. Lett.* **24**: 113–115 (1999).

[47] W. Supatto, D. Débarre, B. Moulia, *et al.*, " *In vivo* modulation of morphogenetic movements in Drosophila embryos with femtosecond laser pulses," *PNAS* **102**: 1047–1052 (2005).

[48] D. Debarre, W. Supatto, E. Farge, B. Moulia, M. C. Schanne-Klein, and E. Beaurepaire, "Velocimetric third-harmonic generation microscopy: micrometer-scale quantification of morphogenetic movements in unstained embryos," *Opt. Lett.* **29**: 2881–2883 (2004).

[49] A. C. Millard, P. W. Wiseman, D. N. Fittinghoff, K. R. Wilson, J. A. Squier, and M. Muller, "Third-harmonic generation microscopy by use of a compact, femtosecond fiber laser source," *Appl. Opt.* **38**: 7393–7397 (1999).
[50] W. R. Zipfel, R. M. Williams, and W. W. Webb, "Nonlinear magic: multiphoton microscopy in the biosciences," *Nat. Biotechnol.* **21**: 1369–1377 (2003).

Chapter 13
Optical coherence tomography

Sebastian Marschall[1] *and Peter E. Andersen*[1]

Optical coherence tomography (OCT) is a non-invasive imaging technique providing real-time two- and three-dimensional images of light-scattering samples with micrometer resolution [1]. Mapping the local reflectivity, OCT visualizes the internal structure of the sample (Figure 13.1) and it can provide information on different sample properties, such as birefringence [2,3], motion [4–9], or material composition [10,11]. Due to the unique ability to image the morphology of biologic tissues non-invasively, OCT has become a well-established tool for biomedical research and clinical diagnostics [12]. The main field of application is ophthalmology, where OCT is accepted as a clinical standard for diagnosing a number of retinal pathologies and for monitoring the treatment [13–15]. It is also being applied for examining the anterior eye segment [16,17], e.g., for preparing and following up on corneal surgeries. OCT is becoming an important instrument for clinical cardiology, where endoscopic OCT systems enable assessment of atherosclerotic plaques in cardiac blood vessels and provide valuable information for implanting coronary stents [18–21], OCT is used in many other fields of medical and biologic research [12], but also for technical purposes, such as non-destructive testing or contact-free metrology [22].

During the past quarter century, since the invention of OCT, there have been dramatic improvements in performance [23]. These impressive developments have greatly benefited from the possibilities given by semiconductor-based light sources, as the performance of an OCT system to a large extent depends on its light source. Semiconductor technology enabled the development of specialized broadband emitters and broadband tunable lasers covering a wide range of the wavelength regions of interest for biomedical OCT imaging. Many mature manufacturing techniques could be adopted from the telecommunications industry, which accelerated the process. Semiconductor devices can directly convert electrical energy into light and thus allow for implementing efficient, compact and easy-to-handle light sources. Furthermore, they can be mass-produced at low cost, which was an important factor for the rapid adoption of OCT in the clinical environment. The continuing developments in semiconductor light source technology will be a driving force for OCT spreading into a growing number of applications.

In this chapter, we will provide a brief explanation of the principle of OCT imaging and highlight the influence of the light source properties (Section 13.1).

[1]Department of Photonics Engineering, Technical University of Denmark, Denmark

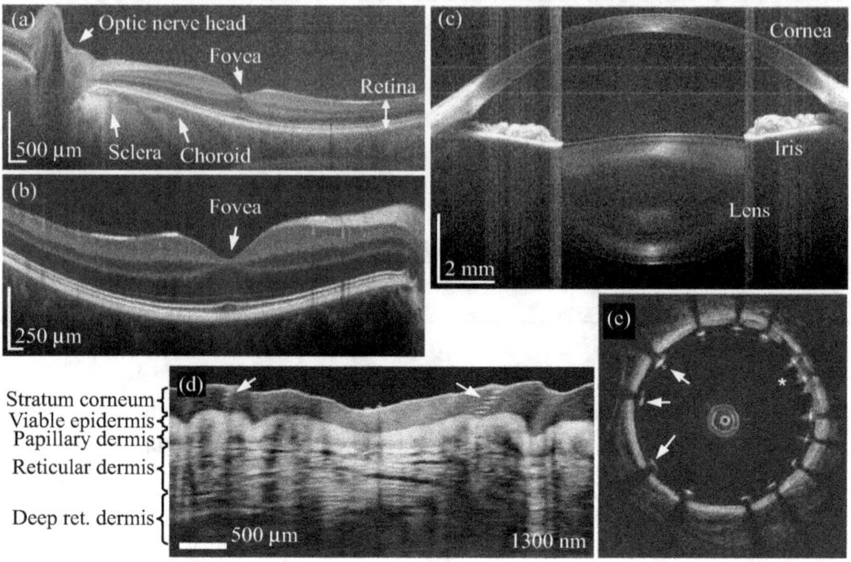

*Figure 13.1 OCT can provide in vivo cross-sectional images of many parts of the human body: (a) wide-field image of the human retina and the layers beneath; (b) ultrahigh-resolution image of the central retina (Courtesy of Dr. A. Unterhuber, Medical University of Vienna); (c) cross section through the anterior segment of the human eye; (d) upper skin layers on the inner side of a finger with visible sweat ducts (arrows); and (e) endoscopic image of a stented porcine coronary artery showing the stent struts (arrows) along the blood vessel's circumference and a tissue prolapse (*) between the struts. ((a) and (c) reprinted with permission from [39], (d) reprinted with permission from [59], and (e) reprinted with permission from [18])*

For a comprehensive discussion of all aspects related to OCT, we may refer the reader to the review by Fercher [24] or to the books by Brezinski [25] or Drexler and Fujimoto [26]. In the following sections, we will discuss the application of semiconductors in specialized light sources for OCT, namely, in broadband light sources (Section 13.2) and in broadband wavelength-swept light sources (Section 13.3). We will conclude with an outlook on the continuing development of light source technologies for OCT (Section 13.4).

13.1 Fundamentals

13.1.1 The principle of OCT

An OCT system locates reflecting or back-scattering structures within a sample by probing it with a beam of light and determining the roundtrip time of the fraction of light returning from the sample. Direct measurement of the light's roundtrip time is not possible, but interferometry enables measuring the relative time delay between the probing light and light traveling along a reference path. One such measurement yielding the depth-resolved reflectivity profile along the probing beam axis is called an A-scan in analogy to ultrasound imaging. By scanning the beam laterally over the sample, a two- or three-dimensional image can be assembled from a number of adjacent A-scans.

There are two major approaches to acquire the signal for an A-scan, namely acquisition in the *time domain* or in the *frequency domain*. Here we will only discuss frequency-domain OCT (FD-OCT) in detail, which is nowadays the preferred method for most biomedical applications, as it offers a superior sensitivity [23,27–29] and up to three orders of magnitude faster image acquisition [23,30]. However, the important relationships between the light source properties and the system performance are also valid for time-domain OCT (TD-OCT).

To understand principle of signal acquisition we consider a simple Michelson interferometer (Figure 13.2(a)). The incident light with the power spectral density $S_{in}(\omega)$ is split into two fractions, one of which is directed to a reference mirror with reflectivity R_R and one to a sample consisting of multiple weakly reflective layers. Each layer is characterized by its reflectivity, R_i, and the time delay, τ_i, relative to the roundtrip time in the reference path. The power spectrum at the interferometer exit port is then

$$\begin{aligned} S_{\text{out}}(\omega) &= \frac{1}{4} S_{\text{in}}(\omega) \left(R_R + \sum_i R_i \right) \text{ “DC baseline”} \\ &+ \frac{1}{4} S_{\text{in}}(\omega) \sum_{i \neq j} \sqrt{R_i R_j} \cos\left(\omega\left[\tau_i - \tau_j\right]\right) \text{ “sample auto-correlation”} \\ &\pm \frac{1}{2} S_{\text{in}}(\omega) \sum_i \sqrt{R_R R_i} \cos(\omega \tau_i), \text{ “sample/reference cross-correlation”} \end{aligned} \tag{13.1}$$

where ω denotes the angular optical frequency.

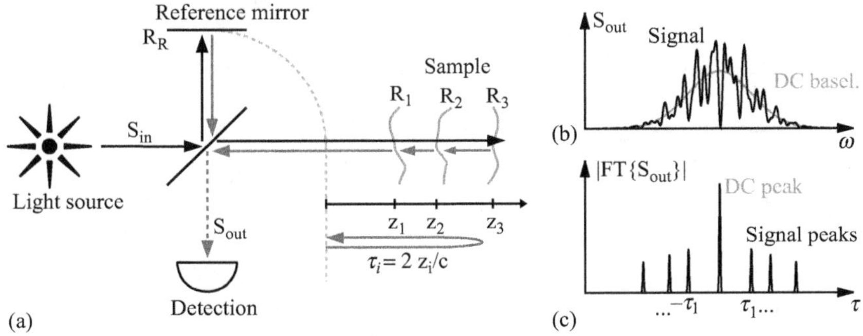

Figure 13.2 (a) An OCT system determines the time delay, τ_i, between light reflected from different depths, z_i, in the sample and light that has traveled along a reference path; (b) interference between the sample beam and the reference beam leads to a modulation of the output spectrum, S_{out}; and (c) the contributions with different delays, τ_i, can be separated by applying a Fourier transform (FT) to the spectrum

The first term, the summed spectrum of all light reflected from the reference mirror and from all layers in the sample, can be viewed as a "DC baseline" carrying the following interference signal terms. The second term, commonly called "auto-correlation", originates from the interference of the reflections from all sample layers with each other. Since typically $R_i \ll R_R$, this contribution is very small and most of the time negligible. The signal of interest is contained in the third term, the "cross-correlation", which arises from interference between the reference beam and the sample reflections. If we look at the signal as a function of ω, each reflection contributes an oscillating term $\cos(\omega\tau_i)$, where τ_i corresponds to the oscillation frequency. Hence, the longer the time delay is, the higher is the fringe frequency of the corresponding spectral interference pattern (Figure 13.2(b)). By Fourier transforming this signal, we can separate the contributions with different time delays (Figure 13.2(c)). Since the amplitude of each term scales with $\sqrt{R_i}$, we can obtain the reflectivity profile of the sample along the probing beam axis by taking the square of the Fourier-transformed signal.

As the detected signal is also proportional to $\sqrt{R_R}$, the typically weak reflections from the sample get "amplified" by the more intense reference light. Hence, the coherent detection scheme makes OCT a highly sensitive imaging technique.

The $\pm$ sign of the interference term represents the complementary intensity distribution between the signals obtained from both output ports of the interferometer[1]. This can be exploited in an OCT system with dual balanced detection [31,32]. By measuring the differential intensity between both output ports, the baseline and auto-correlation terms cancel out, yielding ideally the pure OCT signal.

FD-OCT systems measure the spectrum of the light from the interferometer exit port, using either a broadband light source combined with a spectrometer or a tunable narrowband light source in conjunction with a fast photo-detector [33,34] (Figure 13.3) In the latter scheme, the light source performs rapid sweeps over a

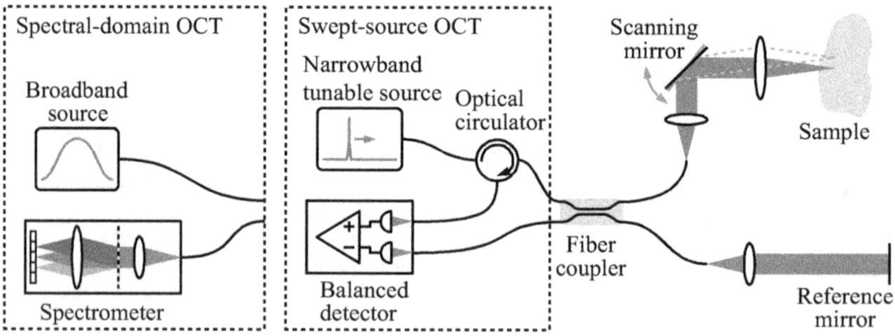

Figure 13.3 A common OCT system architecture is based on a fiber-optical Michelson interferometer. In SS-OCT systems, dual-balanced detection can yield a significant improvement of the signal-to-noise ratio.

[1]In a Michelson interferometer, the input port is at the same time an output port.

broad range of optical frequencies [35,36], hence this method is commonly termed swept-source OCT (SS-OCT). We will refer to the spectrometer-based method as spectral-domain OCT (SD-OCT)[2].

The two primary parameters that govern the performance of a frequency-domain OCT system are the spectral bandwidth covered by the light source and the spectral resolution. These determine the depth resolution and the effective depth range of the system.

To analyze the influence of the light properties, let us consider an SS-OCT system[3]. The swept source has an instantaneous line spectrum, S_l, of narrow, but non-zero width with a varying center frequency, ω_l. The photo-detector at the interferometer exit port measures the integrated intensity of the transmitted power spectrum:

$$I_{\text{det}}(\omega_l) = \int_0^\infty S_{\text{out}}(\omega_l, \omega) d\omega, \tag{13.2}$$

where we obtain $S_{\text{out}}(\omega_l,\omega)$ by replacing $S_{\text{in}}(\omega)$ in (13.1) with $S_l(\omega_l,\omega)$. The line center frequency, ω_l, varies as the light source is tuned over its frequency range. Let us assume that the amplitude, but not the shape of the line spectrum may depend on ω_l[4]. S_l can then be separated into the integrated instantaneous intensity of the light source, I_{src}, and the normalized line spectrum $\overline{S}_l$:

$$S_l(\omega_l, \omega) = I_{\text{src}}(\omega_l)\overline{S}_l(\omega - \omega_l) \tag{13.3}$$

$$I_{\text{src}}(\omega_l) = \int_0^\infty S_l(\omega_l, \omega) d\omega. \tag{13.4}$$

By combining (13.1)–(13.4), while omitting all terms that do not scale with R_R (since $R_i \ll R_R$), we can express the detected intensity as

$$I_{\text{det}}(\omega_l) = \frac{1}{4} I_{\text{src}}(\omega_l) R_R \pm \frac{1}{2} I_{\text{src}}(\omega_l) \cdot \overline{S}_l(\omega_l) * \sum_i \sqrt{R_R R_i} \cos(\omega_l \tau_i), \tag{13.5}$$

where * denotes a convolution. The resulting detected intensity is practically a weighted average over that interval of the interference signal, $\cos(\omega_l\tau_i)$, which is covered by the line spectrum, $\overline{S}_l$. As a result, the interference terms with a short fringe period relative to the width of the line spectrum, i.e., those with large τ_i, exhibit a decreased fringe contrast. The same effect occurs also in SD-OCT systems

[2]Throughout the literature, there is no uniform terminology for the OCT methods performing the depth scan in the frequency domain. The terms "frequency-domain OCT" and "spectral-domain OCT" may be interchanged, or "Fourier-domain OCT" may be used for either one. "Swept-source OCT" is also called "optical frequency domain imaging" (OFDI), while spectrometer-based OCT was formerly referred to as "spectral radar".

[3]After redefining some quantities, the same derivation is also applicable to SD-OCT systems.

[4]In practice, the instantaneous line shape may change during tuning, but this simplification is useful to illustrate the concept.

due to the finite resolving power of the spectrometer and the spatial averaging by the detector array. The intensity spectrum of the light source, I_{src}, acts as an envelope function and determines thus the spectral range (or bandwidth) within which the interference signal can be measured.

The Fourier transform of the detected signal yields then

$$\mathrm{FT}\{I_{det}(\omega_l)\}(\tau) = \frac{1}{4}R_R P(\tau) \pm \frac{1}{2}P(\tau) * \gamma(\tau) \cdot \sum_i \sqrt{R_R R_i}[\delta(\tau-\tau_i)+\delta(\tau+\tau_i)]$$

$$= \frac{1}{4}R_R P(\tau) \pm \frac{1}{2}\sum_i \sqrt{R_R R_i}\gamma(\tau_i)[P(\tau-\tau_i)+P(\tau+\tau_i)], \tag{13.6}$$

where $P = \mathrm{FT}\{I_{src}\}$, $\gamma = \mathrm{FT}\{\overline{S}_l\}$, and δ represents the Dirac delta function. Each reflection in the sample thus results in a peak at τ_i on the time delay scale with a mirror peak at $-\tau_i$ (Figure 13.2(c)). These peaks have a certain width, because the spectral interferogram can only be measured with finite bandwidth. This is expressed by the convolution of the δ-functions with $P(\tau)$, which is termed the *point spread function.* The factor $\gamma(\tau)$ leads to a decreasing amplitude of the signal components with large τ_i. This effect is called the *signal roll-off* or *sensitivity roll-off.*

To generate OCT images, the Fourier-transformed signal is typically converted into logarithmic gray-scale values representing the depth-resolved relative reflectivity, R, in decibels:

$$\frac{R(z)}{\mathrm{dB}} = 10 \cdot \log_{10}\left(\left|\mathrm{FT}\{I_{det}\}(z)\right|^2\right). \tag{13.7}$$

$z = c\tau/2$ is the *probing depth,* i.e., the optical distance along the probing beam axis relative to the zero-delay point.

Using these theoretical considerations, we will in the following section analyze the influence of the light source properties on the OCT performance. Particularly important figures are the bandwidth of the spectrum, I_{src}, and in case of a swept source the width of the dynamic line spectrum, S_l, which determine the depth resolution (Section 13.1.2.1) and the signal roll-off (Section 13.1.2.2), respectively.

13.1.2 Requirements for light sources

In this section, we will discuss how the performance of an OCT system is linked to the properties of the light source. As described in the previous section, the different OCT methods require different types of light sources. Spectral-domain OCT (SD-OCT) and time-domain OCT (TD-OCT) systems apply broadband light sources that emit a wide spectrum of wavelengths simultaneously. In swept-source OCT (SS-OCT) the light source emits only a narrow spectral line at a time which is then tuned over a broad wavelength range. Swept sources are considerably more complex than typically used broadband sources. However, SS-OCT offers a number of advantages, such as lower susceptibility to artifacts caused by sample motion [37], a long imaging depth range [38–44], and the possibility of ultra-high speed image acquisition [30,45,46].

13.1.2.1 Bandwidth and spectral shape

The primary parameter of an OCT light source is its bandwidth, i.e., the spectral range it covers, which determines the axial resolution. As we see from (13.6), each isolated reflection appears as a peak in the obtained reflectivity profile, whose shape is given by the point spread function (PSF), $P(z)$. The axial resolution is commonly defined as the full width at half-maximum (FWHM) of the PSF. Since the PSF is linked to the light source spectrum, I_{src}, through the Fourier transform, the axial resolution is inversely proportional to the bandwidth.

The axial resolution of OCT does not depend on the focusing of the probing beam. Instead, it stays virtually constant over the entire imaging depth range, even in strongly scattering samples, such as biological tissues. OCT thus has a depth range advantage over other high-resolution optical imaging techniques that rely on tight focusing, for instance confocal microscopy.

Preferably, the PSF should have low side-lobes, as these could obscure adjacent weaker signals and thus limit the dynamic range of the system. Hence, also the exact shape of I_{src} is important, since it acts similar to a window function in spectral analysis [47]. A Gaussian spectrum is considered ideal, as it would also yield a Gaussian PSF free of side-lobes. However, since real light sources can only have a finite bandwidth, the spectrum has to provide an adequate compromise between axial resolution and dynamic range. Generally, a smooth bell- or dome-shaped spectrum without gaps or notches is beneficial [48–54].

For a given spectral shape, we can calculate the axial resolution, δz, as a function of the bandwidth. For example, for a Gaussian spectrum with FWHM $\Delta\nu$, the resolution is

$$\delta z = \frac{2\ln 2}{\pi}\frac{c}{\Delta\nu} \tag{13.8}$$

$$\approx \frac{2\ln 2}{\pi}\frac{\lambda_c^2}{\Delta\lambda} \approx 0.44\frac{\lambda_c^2}{\Delta\lambda}, \tag{13.9}$$

where $\nu = \omega/(2\pi)$ is the ordinary optical frequency and c the speed of light. Equation (13.9) is often used to express this relationship in terms of the wavelength bandwidth, $\Delta\lambda$, and the center wavelength, λ_c. However, real-world broadband light sources often feature complex spectral shapes, yielding strongly side-lobed PSFs. In these cases, (13.9) can at best provide a coarse estimate of the axial resolution[5]. An accurate prediction of the axial resolution from the light source spectrum requires numerical computation of the PSF [54,55].

Since the optical distance, z, within a material equals the physical distance times its index of refraction, n, the effective resolution within that material is $\delta z/n$. Hence, OCT literature usually states both the measured resolution in air ($n \approx 1$) and the calculated resolution in the material under inspection (e.g., $n \approx 1.4$ in biological tissue).

[5]In the literature, (13.9) is still frequently applied to non-Gaussian spectra. Therefore, care must be taken with statements about the "theoretical" resolution. Note also that the bandwidth of broadband light sources is commonly defined as the FWHM of the spectrum, whereas in case of swept sources this term often refers to the total tuning range, making a direct comparison difficult.

An axial resolution of ~15 μm in air, which can be considered low to moderate for OCT imaging in biological samples, requires a bandwidth of ~10 THz. Typical light sources based on a single semiconductor emitter provide a bandwidth of 15–30 THz, thus enabling depth resolutions as high as ~5 μm in air. Even higher resolutions can be achieved by combining several emitters with shifted spectra.

13.1.2.2 Instantaneous linewidth

For swept sources, a second parameter is highly important: the instantaneous linewidth or dynamic linewidth of the emitted narrowband spectrum, S_l, during tuning operation. This parameter determines the signal roll-off in a similar relationship as the bandwidth determines the axial resolution. Because the shape and the width of the line spectrum can change during the sweep cycle and because direct measurement is challenging [56,57], typically just the signal roll-off as a function of the probing depth is measured for characterizing a swept source. A common figure of merit is the probing depth, z_r, at which the squared PSF amplitude has decreased by 6 dB, corresponding to a 50% reduction of the interference fringe amplitude. For instance, in case of a Lorentzian line spectrum with FWHM $\delta\nu$ the signal rolls off exponentially with

$$z_r = \frac{\ln 2}{\pi}\frac{c}{\delta\nu}. \tag{13.10}$$

The dynamic linewidth of a swept source is significantly broader than the linewidth of classical continuous-wave lasers. Up to several 10 GHz may be acceptable, which permits an imaging depth range of a few millimeters. This is often sufficient, since attenuation of the probing light limits the penetration depth into most tissues to less than 2 mm [58,59]. Some applications may, however, require a longer depth range, for example, imaging of semi-transparent structures such as the anterior eye segment [60–62], measurement of the axial eye length [39] or endoscopic imaging of hollow organs where the distance between the probe and the tissue surface can vary [38].

Sometimes, the coherence length, $l_c = 2z_r$, is stated as the figure of merit. Obviously, the instantaneous line spectrum of a swept source should have a long coherence length. Confusion may arise, because in case of a broadband source for SD- or TD-OCT the term coherence length refers to the broadband power spectrum and is thus equivalent to the axial resolution which should be as small as possible.

As stated above (Section 13.1.1), in SD-OCT the roll-off is determined by the spectrometer's resolving power and thus independent of the light source properties. Since state-of-the-art swept sources can achieve a considerably finer spectral resolution than typical spectrometers, SS-OCT enables correspondingly longer imaging depth ranges.

13.1.2.3 Sweep repetition rate

Another figure of merit specifically for swept sources is the sweep repetition rate, which determines the image acquisition speed. Unless an SS-OCT system scans the sample with multiple beams simultaneously [30,45,61,63–65], it performs one A-scan per frequency sweep.

High speed is necessary for many in-vivo applications, where the time for image acquisition is often limited to a couple of seconds. In ophthalmology, for instance, prolonged eye measurements would be affected by gaps or distortions in the datasets, caused by the blinking reflex and by involuntary eye motion. As another example, during intravascular imaging of the coronary arteries the blood is typically displaced by a saline flush, as it would otherwise scatter the probing light beam. Obviously, the blood flow must not be interrupted for longer than a few seconds in order to avoid ischemia.

The first generation of clinical high-speed OCT systems (SD-OCT) features A-scan rates of $\sim$50–70 kHz. These allow the acquisition of moderately large 3D-datasets. Typical scan protocols yield either a low number of B-scans (a few dozens) composed of many densely sampled A-scans ($\sim$1,000) or a similar sampling density of A-scans and B-scans (a few hundred each) covering relatively small volumes [66].

For gathering comprehensive diagnostic information, researchers seek to acquire large sets of 3D volume data with high sampling density of both A-scans and B-scans. These can easily comprise several hundred thousands or millions of A-scans [46,64,67,68]. Correspondingly, this requires A-scan rates of hundreds of kilohertz or even several megahertz, which have become possible with the development of modern swept sources [30,41,61,64,69–71]. These ultrahigh acquisition speeds also enable the study of temporal dynamics [72–75], complex averaging schemes for generating large high-quality volume datasets [45], or methods for increasing the performance with redundant data collection [76].

13.1.2.4 Output power

The light source of an OCT system must provide sufficient power to achieve a good signal-to-noise ratio (SNR). Furthermore, as typical point-scanning OCT systems require near-diffraction-limited light beams for generating a high-contrast signal, the light source must deliver its output power with very high beam quality or coupled into a single-mode fiber, respectively.

Biological samples are normally scanned with a sample beam power close to the maximum permissible exposure (MPE) level. The MPE is defined by the standards IEC 60825 and ANSI Z136.1 and varies considerably depending on the target tissue and exposure time, wavelength range and other parameters. For imaging of the eye, where damage to the photo-receptors must be avoided, the MPE is typically in the range between 0.5 mW and 2 mW, while skin and other tissues can tolerate higher exposure.

The power required from the light source is often considerably higher than the actual probing beam power. A large fraction of the light ($\geq$ 50%) is usually not directed to the sample, but into the reference path of the interferometer, and intrinsic losses in all optical components cause additional attenuation. Certain techniques may also increase the power demand, for instance simultaneous scanning with multiple beams [30,63,65] or improving the light retrieval from the sample at the expense of directing a lower fraction of light into the sample beam [64,77]. Preferably, the light source should emit an average power of at least 10 mW for typical OCT applications.

13.1.2.5 Noise

As a coherent detection technique, OCT is highly sensitive and able to detect very weak signals. The sensitivity of an OCT system denominates the smallest detectable reflectivity of a sample structure, commonly defined as the reflectivity that results in an SNR of 1. High-quality images of biologic samples typically require a sensitivity > 90 dB, whereas sensitivity in excess of 100 dB has been reported in various cases.

A number of noise sources affect the image quality in an OCT system. Shot noise arises inherently from the quantum nature of light. Other sources are the detection system, which adds noise arising from dark current or from the readout process, and the light source, whose intensity fluctuations manifest themselves as excess noise in the system. These types of noise depend differently on the reference power. If significantly higher power is received from the reference path than from the sample $\left(R_R \gg \sum_i R_i\right)$, the SNR behaves as follows in each regime where one of the noise sources is dominating [27,28]:

$$\mathrm{SNR}_{\mathrm{shot}} \propto I_{\mathrm{src}} R_i, \tag{13.11}$$

$$\mathrm{SNR}_{\mathrm{detector}} \propto I_{\mathrm{src}}^2 R_i R_R, \tag{13.12}$$

$$\mathrm{SNR}_{\mathrm{excess}} \propto R_i R_R^{-1}. \tag{13.13}$$

Since shot noise sets the fundamental limit to the achievable SNR, one strives to operate an OCT system in the shot-noise-limited regime. Hence, the reference power typically gets adjusted sufficiently high, such that shot noise dominates over detector noise. However, if the light source exhibits too strong intensity noise, the SNR falls off due to excess noise before the shot-noise limit can be reached.

Relative intensity noise (RIN), i.e., spectral and temporal fluctuations of I_{src}, is present in all components of the detected signal (13.5). However, the actual interference signal (third term) is typically only a small modulation on top of the baseline (first term). Hence, in case of strong RIN the fluctuations on the baseline may have similar or even higher magnitude as the interference fringes and thus lead to a high noise floor in the Fourier-transformed signal. Especially high-speed swept sources may exhibit strong RIN [78]. However, SS-OCT systems can make use of a balanced detection scheme which cancels out the first and second term in (13.5), and thus eliminates a major fraction of the RIN [28]. Balanced detection is possible in SD-OCT, too [79,80], but it is complex and costly, since it requires an additional spectrometer. Instead, SD-OCT normally relies on low-noise light sources and careful adjustment of the reference power [27].

13.1.2.6 Spectral range

The optical properties of biological tissues are highly wavelength-dependent. Hence, an appropriate choice of the operating wavelength range is crucial for good imaging performance. The critical properties are absorption and scattering, which attenuate the probing light, thereby reducing the penetration depth [81].

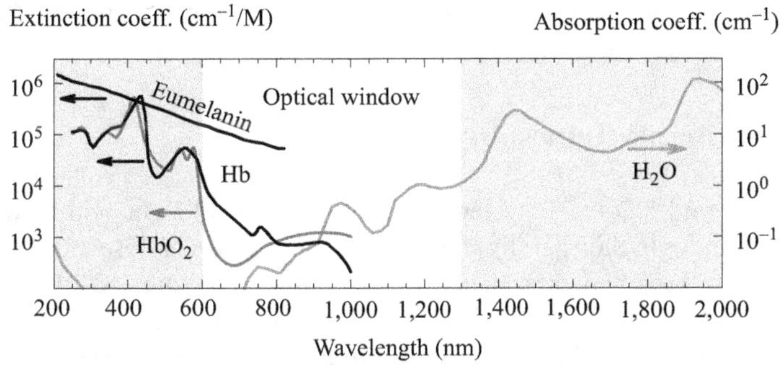

Figure 13.4 Most OCT systems operate within the optical window of biological tissue. This spectral range permits deep penetration, due to relatively low attenuation by typical absorbers, such as water (H_2O), oxy- and deoxy-hemoglobin (HbO_2, Hb), and melanin. Data: [199–201]

Between 600 nm and 1,300 nm, the absorption by water and other typical tissue constituents like melanin or hemoglobin is low (Figure 13.4) which gives rise to the so-called *optical window* [82]. Strong scattering, however, limits the maximum imaging depth [83]. Scattering arises from variations in the index of refraction between different tissue constituents and the surrounding medium [84]. It is dependent on the size and shape of present scatterers, hence the macroscopic properties of different tissue types can vary considerably. Generally, scattering coefficients decrease with increasing wavelength, due to contributions in the Rayleigh regime.

The selection of an operating wavelength within the optical window depends on the requirements of the application. Light at 1,300 nm permits deep penetration into most tissues, such as skin, arterial wall, or mucosa [59,83], because absorption by melanin and scattering decrease with increasing wavelength [85–88]. Longer wavelengths, for example, around 1,550 nm or 1,700 nm, can enable even deeper penetration for applications where the influence of water absorption is less significant [58,89–91].

For imaging the human retina in vivo, however, wavelengths above 1,130 nm are not suited due to strong signal attenuation by water [55], which is the main constituent of the eye [92]. Commonly, the 800 nm band, where absorption by water is very low, is used for retinal imaging. But also the intermediate range around 1,060 nm with moderate water absorption has found great interest. Weaker absorption in the melanin-rich retinal pigment epithelium and decreased scattering compared to the shorter wavelengths permit deep penetration into the structures behind the retina (choroid, sclera) [93–95]. The 1,060 nm band proved also beneficial for imaging in eyes affected by a cataract [96], and it exhibits minimal chromatic dispersion in water [97], which would otherwise degrade the depth resolution if not properly compensated [98–101]. However, the neighboring water absorption bands around 970 nm and 1,150–1,200 nm limit this spectral window for retinal imaging [102]. Since the parts of a light source spectrum overlapping with these absorption bands would get severely attenuated, a total bandwidth exceeding 200 nm would barely be useful [55].

13.2 Broadband light sources

Both SD-OCT and TD-OCT systems require broadband light with low temporal coherence. Although thermal light sources, such as incandescent lamps or arc lamps, emit very broad spectra, they are not suitable for typical OCT systems. The emitted light is spatially incoherent, and only a small fraction of it can be focused into a diffraction-limited beam or into a single-mode fiber. Instead, a variety of amplified spontaneous emission (ASE) sources can efficiently generate light with near-diffraction-limited beam quality, adequate power, and sufficient bandwidth to achieve high axial resolution. Due to their beneficial properties, semiconductor *superluminescent diodes* (SLDs) are the most commonly used ASE sources, especially in commercial OCT systems. On the other hand, research-grade ultrahigh-resolution OCT systems typically exploit the ultrabroad bandwidths provided by femtosecond pulse lasers or supercontinuum sources [103–106].

In this section, we will discuss how OCT applications benefit from SLDs, and how the performance has improved with continuing light source development (Section 13.2.1). Furthermore, we will present an emerging application of semiconductor diode-lasers for direct pumping of titanium:sapphire femtosecond lasers for ultrahigh-resolution OCT (Section 13.2.2).

13.2.1 Superluminescent diodes

Semiconductor-based ASE light sources, i.e., SLDs, are very attractive for OCT applications due to a number of unique properties. SLDs can generate light with broad spectral bandwidth in a wide range of wavelength regions. Different material systems and active-layer structures offer many degrees of freedom for tailoring the emission spectrum and for adjusting the center wavelength. The light being generated in a single-mode waveguide is emitted with high beam quality and can be coupled into optical fibers with high efficiency. SLDs are very compact and—due to direct electrical pumping—also highly efficient and easy to handle.

SLDs became interesting for OCT applications, when their bandwidth was broadened by narrowing the active layer down into a quantum-well (QW) structure [107,108]. Although SLDs based on bulk heterostructures typically feature nice near-Gaussian ASE spectra, their limited bandwidth < 10 THz does not permit high axial resolution [109]. In comparison to bulk active layers, the injected charge carriers in QWs occupy a wider range of energy levels due to the step-function-like density of states [107], allowing for a bandwidth of 20–30 THz. However, the excitation of states in the next higher-energy subband at high injection currents gives rise to a higher and narrower peak at the short-wavelength side of the ASE spectrum [110]. The resulting typical two-lobed spectrum of QW-SLDs leads in OCT to a point spread function with higher side-lobes [54]. Hence, the injection current needs to be adjusted to a value that provides the best compromise between bandwidth and spectral shape. In some cases, this effect can be exploited, for example, for compensating higher attenuation at the short wavelengths [102].

Since the emission wavelength of a QW can be tuned by varying the thickness of the active layer, further spectral broadening is possible by incorporating multiple QWs with different widths [111,112]. With this method an SLD with a bandwidth of 42 THz (92 nm at 810 nm) has been demonstrated [113].

State-of-the-art commercially available QW-SLDs provide bandwidths of 20-40 THz (up to 85 nm at 830 nm, 115 nm at 930 nm, 110 nm at 1,060 nm, 120 nm at 1,300 nm) with output powers ranging from a few milliwatts up to 35 mW coupled into a single-mode fiber.

Further confinement of the charge carriers by growing an active layer consisting of quantum dots (QDs) enables even greater control over the emission spectrum. Each single QD emits light only at discrete wavelengths, but an ensemble of differently sized QDs can cover a broad spectral width [114,115]. A variety of QD-SLDs with wide bandwidths ranging from 20 THz up to 45 THz have been developed [111,116–125], and a bandwidth of even 60 THz was achieved with a hybrid QD/QW structure [126]. However, many of these broadband devices have not provided more than 1–2 mW of optical power, with few exceptions [118,119,122]. The emission spectra of most presented QD-SLDs fall into the spectral range from 1,100 nm to 1,300 nm, often overlapping with the water absorption band between 1,150 nm and 1,200 nm which limits their feasibility for biomedical OCT applications. To date only few OCT applications using QD-SLDs have been demonstrated, for example, skin imaging with a QD-SLD providing adequate power (18 mW) with a moderate bandwidth of ~15 THz (85 nm at 1,280 nm) [127,128].

The emission spectrum of QW- and QD-SLDs depends strongly on the injection current. Generally an increasing current leads to a spectral broadening towards shorter wavelengths. Due to the short excited-state lifetime in semiconductors—on the order of nanoseconds—they have a fast response to modulation of the current. This may be exploited in SD-OCT systems which detect the light with time-integrating CMOS or CCD sensor arrays. If the SLD's spectral shape is changed during the integration time of the detector array, then the acquired signal will have an envelope given by the time-averaged spectral shape. Hence, by modulating the injection current synchronized with the integration cycle the spectrum can be optimized for a good compromise between depth resolution and dynamic range [129].

In order to improve the axial resolution further, light sources combining the emission of several SLDs with different spectra have found their application in OCT [110,130]. Multiplexing of up to four SLDs has been demonstrated, yielding bandwidths as high as 100 THz (300 nm at 940 nm) [111]. However, these multiplexed spectra often exhibit deep notches between the intensity peaks of the single SLDs which lead to a point spread function with increased side lobes [48,53,54], Nevertheless, OCT imaging with 3–4 μm axial resolution in air has been successfully demonstrated [131–133], and a variety of multi-SLD light sources are now commercially available.

Due to their various beneficial properties, SLDs have become the light source of choice for TD- and SD-OCT systems. Continuous development of broadband devices has led to greatly improved axial resolution. Moreover, multiplexed SLD modules permit OCT imaging in the ultrahigh-resolution regime with a comparably simple and affordable light source.

13.2.2 Diode-pumped femtosecond lasers

Originally developed for studying ultra-fast physical phenomena, Kerr-lens mode-locked lasers can generate intense light pulses with durations down to a few femtoseconds. The corresponding coherence lengths of a few micrometers makes these lasers attractive for ultrahigh-resolution OCT (UHR-OCT). Especially titanium: sapphire lasers became popular [103,134], as they provide an axial resolution down to 1 μm in the 800 nm band, ideal for retinal imaging [135]. As these lasers are significantly more complex and expensive than SLD-based light sources, they have up to now only been used in OCT research rather than in wide-spread clinical applications. A major cost item is the required pump light source, typically a frequency-doubled solid-state laser emitting green light in the range around 530 nm.

Direct pumping with a semiconductor laser can simplify the system, increase the wall-plug efficiency and lower the cost considerably. However, direct generation of green light with sufficient power and high beam quality, as required for optimal pumping, still remains a challenge [136], since state-of-the-art green single-mode laser diodes emit only up to a few hundred milliwatts. As an alternative approach, the high-power emission of *tapered laser diodes* in the 1,060 nm range can be efficiently up-converted by second-harmonic generation or sum-frequency generation [137–140], yielding several watts of green light in a diffraction-limited beam. Pumping of a femtosecond laser with a frequency-doubled tapered diode laser was successfully demonstrated [141], generating an output spectrum with 112 nm bandwidth (FWHM). This prototype light source applied in an UHR-OCT system provided similar axial resolution (~3 μm) and image quality as a commercial femtosecond laser specially designed for OCT [142] (Figure 13.1 (b)).

By reducing the cost and complexity of titanium:sapphire lasers, semiconductor laser technology will enable a broader use not only of UHR-OCT, but also of a variety of non-linear optical imaging techniques that benefit from ultrashort laser pulses [143,144], such as multi-photon tomography, fluorescence life-time imaging microscopy (FLIM), second-harmonic generation (SHG) microscopy, or coherent anti-Stokes Raman scattering (CARS) spectroscopy. Combining the deep-tissue morphological imaging capability of OCT with the chemical specificity of non-linear imaging modalities has found growing interest. Intense research is therefore aiming for multi-modal imaging systems sharing a single titanium:sapphire laser for economically acquiring a wealth of complementary diagnostic information [145–151].

13.3 Wavelength-swept light sources

With only few exceptions, light sources for SS-OCT are tunable lasers in specialized configurations that meet the requirements for OCT. Virtually all of them utilize semiconductors as gain material. The key advantage of SS-OCT—high-speed image acquisition with A-scan rates up to several megahertz—could only be realized with semiconductor gain media enabling high-speed broadband tunable lasers.

In the following, we will explain the basic setup of a tunable laser (Section 13.3.1) and discuss why semiconductor gain media play a crucial role for swept sources (Section 13.3.2). We will then present different configurations of commercial and research-grade swept sources, covering current state-of-the-art light sources as well as novel devices that have been developed in recent years (Section 13.3.3). Furthermore, we recommend chapters 18 and 20–24 in reference [26] as well as the review on high-speed swept sources by Klein and Huber [152] which extend the discussion beyond the scope of this chapter.

13.3.1 Tunable lasers

Common to all lasers is a gain medium, which amplifies light by stimulated emission, placed in a resonator. Spontaneously emitted light from the gain medium circulates inside the resonator and gets amplified multiple times. A fraction of it is extracted from the resonator, for example, through a semi-transparent mirror. Light at certain frequencies, corresponding to the longitudinal resonator modes, can add up over many roundtrips and builds up intense coherent laser radiation. These mode frequencies, ν_{m}, are determined by the optical roundtrip length of the resonator, L_{rt}.

$$\nu_{\mathrm{m}} = k\,\delta\nu_{\mathrm{m}} \quad k \in 1, 2 \ldots \tag{13.14}$$

$$\delta\nu_{\mathrm{m}} = \frac{c}{L_{\mathrm{rt}}}. \tag{13.15}$$

$\delta\nu_{\mathrm{m}}$ is called the mode spacing, and c denotes the speed of light.

Laser light can be generated at the mode frequencies that exhibit positive net gain, i.e., those for which amplification in the gain medium is higher than the losses in the resonator. Hence, a tunable laser must provide positive net gain for the entire tuning range. Furthermore, it needs a means of selecting at which frequency laser light is generated. If the laser resonator is so short that only a single mode frequency falls into the gain bandwidth, tuning is possible by changing the resonator length (Figure 13.5(a)). In longer resonators, where many modes fall into the bandwidth of the gain medium, a tunable bandpass filter incorporated into the feedback loop can be used to select the laser frequency (Figure 13.5(b)).

13.3.2 Semiconductor laser gain media

Due to a number of beneficial properties, semiconductor diodes have become the preferred optical gain media for swept sources. All advantages that make semiconductors attractive as ASE sources for SD-OCT (Section 13.2.1) make them also favorable for SS-OCT applications. They allow broadband tuning with 20–30 GHz bandwidth in many different wavelength ranges and provide large small-signal gain (> 25 dB) in a very small volume. Direct electrical pumping permits implementing compact, efficient and robust swept sources.

Semiconductor gain media are especially suitable for high-speed broadband swept sources due to their short excited-state life time in the nanosecond regime.

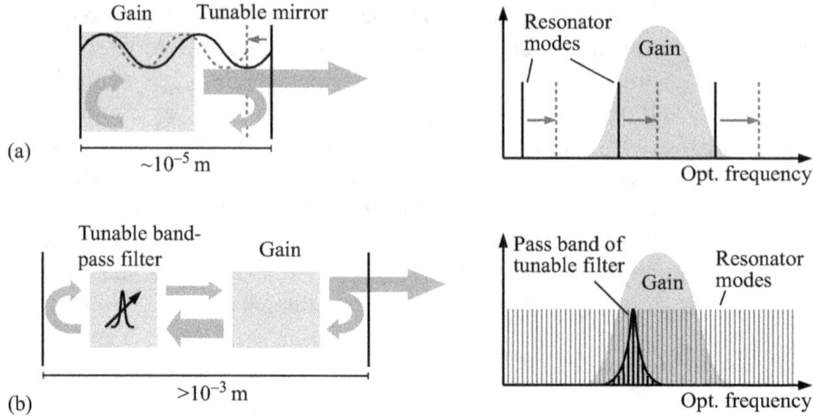

Figure 13.5 Two different laser tuning mechanisms are applied in the common swept source configurations: (a) a short resonator supporting only one mode within the gain bandwidth can be tuned by changing its length; or (b) a long resonator with densely spaced modes can be equipped with a tunable frequency-selective element

Only this made the high tuning speeds of state-of-the-art swept sources (10^7–10^8 nm/s) possible. Other swept source configurations based upon doped crystals or fibers did not show good performance at high sweep rates [36,153,154]. Swept sources using an alternative fast gain mechanism, fiber Raman amplification, have been demonstrated in the 1,550 nm band [155,156]. However, the Raman gain bandwidth in silica fiber is limited to < 15 THz, and the required fiber length of several kilometers renders good performance at high sweep rates difficult to achieve.

The fast modulation response of semiconductor gain materials, which follows from the short excited-state life time, enables an uncomplicated and flexible way of shaping the light source spectrum. Simply by modulating the injection current during the sweep cycle, one can arbitrarily shape the spectrum [52]. For a given sweep range, one can thus optimize the spectrum for high dynamic range or axial resolution [64,157,158], or generate a spectrum that compensates for wavelength-dependent attenuation of the probing light [55]. In contrast, shaping a broadband light source spectrum with similar flexibility requires complex spectral filters [49].

Typically, optical amplification in semiconductor gain elements exhibits strong polarization-dependence due to different confinement factors for the two orthogonal axes in the waveguide structure. This can become a disadvantage if applied in a fiber-optic setup where bend-induced birefringence can cause an unpredictable wavelength-dependent change of the propagating light's polarization state. In a fiber-based swept source, this may impair or even suppress the laser activity at some wavelengths, leading to a notched or discontinuous light source spectrum. A polarization controller can to some extent compensate for this effect, but good adjustment for a broad wavelength range can be difficult in practice. On the other hand, semiconductor-based lasers often emit light with an inherently high degree of polarization which can be beneficial for certain applications, such as polarization-sensitive OCT [159].

13.3.3 Swept source configurations

The first generation of high-speed swept sources are *external-cavity tunable lasers* (ECTLs) using a semiconductor optical amplifier (SOA) as gain medium and a tunable bandpass filter (Figure 13.5(b)) in either a free-space or fiber-based setup. Free-space resonators (Figure 13.6) can be very compact [160–163], especially with tunable filters based on micro-electro-mechanical systems (MEMS). Although the emission spectra of these lasers span many longitudinal modes at a time, a narrow pass band of the tunable filter in conjunction with high gain and a short resonator allows for a relatively narrow dynamic linewidth. Commercially available swept sources of this type provide 5–12 mm coherence length with tuning ranges up to 30 THz (>100 nm at 1,060 nm) and sweep rates up to 200 kHz [61]. Alternatively, there are also models featuring coherence lengths up to 80–100 mm at a reduced tuning range.

Partly or fully fiber-based ECTLs (Figure 13.7(a)) are popular within the research community, since fiber optics permits uncomplicated implementation of stable, alignment-free light sources [164–166]. Furthermore, when using a long fiber resonator (up to several kilometers) one can synchronize the sweep rate of the tunable filter with the light's roundtrip frequency in the resonator [69]. This technique, called *Fourier domain mode-locking* (FDML), allows overcoming the tuning speed limitation of standard ECTLs [166]. In a conventional ECTL coherent laser

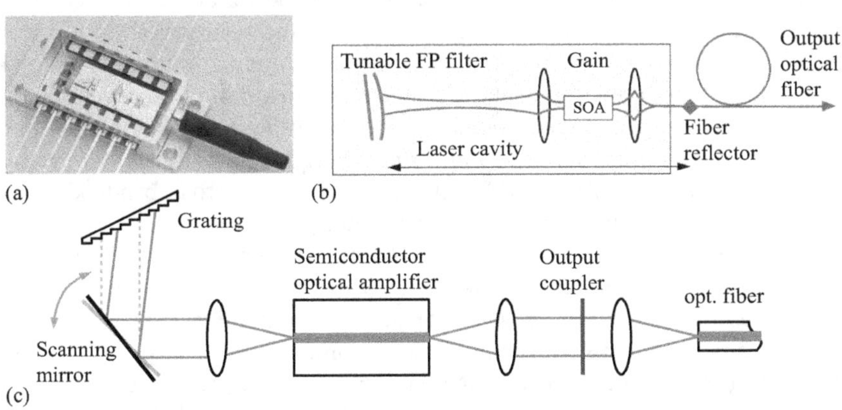

Figure 13.6 (a) Free-space swept source configurations can be implemented on a small footprint, especially with a tunable filter based on MEMS technology. Commonly used tuning elements are (b) a reflective Fabry-Perot (FP) filter, and (c) a diffraction grating in conjunction with a scanning mirror. ((a) and (b) reprinted with permission from [161])

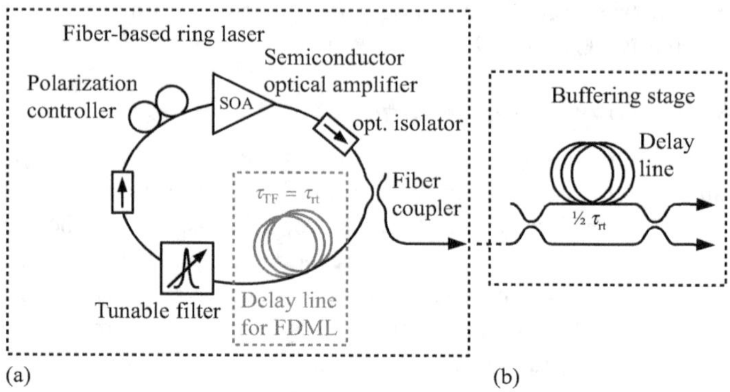

Figure 13.7 (a) Swept sources based on fiber-optic components are simple to assemble and to maintain. Long fiber resonators (~10^2-10^4 m) enable FDML, i.e., synchronizing the filter tuning period, τ_{TF}, with the resonator roundtrip time, τ_{rt}, for ultra-fast sweeping. (b) Fiber delay lines can also be used for buffering, i.e., for multiplying sweeps with a duty cycle < 50% for optimized speed and performance

radiation has to build up from spontaneous emission during each single sweep cycle. Optimal performance in terms of output power and coherence is achieved, if the light can circulate sufficiently many times through the gain medium until the gain is saturated at all wavelengths [166]. If the pass-band of the tunable filter sweeps too rapidly through the wavelengths, the light gets interrupted before the gain is fully saturated, which leads to reduced output power and coherence length. This effect occurs most prominently at the edges of the laser spectrum where the gain is lowest. Hence, increasing the tuning speed beyond the saturation limit also leads to a narrowing of the light source's spectral bandwidth. If, however, the filter tuning period matches the light's roundtrip time, then the light can circulate in the resonator without interruption, just as in a single-wavelength continuous-wave laser in steady-state operation. Thus FDML allows to maintain a broad bandwidth and a long coherence length at ultrahigh sweep rates [38,62]. Whereas conventional ECTLs with short resonators have been demonstrated with sweep rates up to 400 kHz [70], more than 5 MHz could be achieved with an FDML laser [30]. Commercial FDML swept sources with ~1.5 MHz sweep rate have entered the market in 2014, and have since then been adopted by the research community [73–76]. FDML lasers have shown great potential for ocular imaging, where the effects of eye motion need to be minimized [45,66,67], for intravascular imaging, where large datasets need to be acquired in a short time period [46], for studying temporal dynamics [72–75], and generally for applications that benefit from densely sampled volume data [76].

Optimal FDML operation with long coherence length requires a laser resonator with low chromatic dispersion, i.e., the net roundtrip time should be equal for all wavelengths within the laser spectrum. However, optical fiber, which is necessary for implementing the long laser resonator, always introduces chromatic dispersion. In the 1,300 nm range, standard optical fiber exhibits relatively low dispersion, permitting fairly good FDML performance with several millimeters coherence length without special efforts [69]. To achieve longer coherence lengths or good FDML performance at other wavelengths proper dispersion compensation is necessary [38,45]. Record coherence lengths of several hundred millimeters at A-scan rates beyond 3 MHz have been demonstrated in an FDML laser built from different fiber types in conjunction with a chirped fiber Bragg grating and a sophisticated temperature control system [44].

There are two main approaches to implement rapidly tunable filters with broad tuning range and narrow linewidth. One is to use a diffraction grating in conjunction with a beam scanner, such as a galvanometer- or MEMS-actuated mirror [35,162,163], or a rotating polygon mirror [164,167]. Also swept sources using an electro-optic beam deflector have been demonstrated [168–170]. As an alternative approach, piezo- or MEMS-actuated Fabry-Perot filters provide good performance and can be designed to operate either in transmission or in reflection [161,165].

The widely used Fabry-Perot filters and oscillating scanning mirrors can support broadband sweeping with several 100 kHz repetition rate, but only with a bidirectional, typically sinusoidal, sweep. In conjunction with semiconductor gain media this is a disadvantage, as non-linear effects in the SOA lead to strongly asymmetric performance in terms of output power and coherence length between the two sweep directions [165,166,171]. Because an SOA introduces a small negative frequency shift while amplifying the light, a conventional ECTL operates best in the sweep direction of decreasing frequency—or increasing wavelength, respectively—which is commonly termed the "forward" sweep direction [165,166]. At higher tuning speeds, the backward sweep often is no longer useful, effectively reducing the light source duty cycle and thus the possible imaging speed by 50%. In an FDML laser this asymmetry ideally becomes very small [38,69,78], unless it operates in a wavelength range where the long fiber resonator exhibits strong chromatic dispersion [172].

The fast modulation response of semiconductors can be exploited to overcome this performance asymmetry. If the SOA is switched off during the low-performance sweep, half of the laser light of the high-performance sweep can be sent through a delay line corresponding to half a tuning cycle, and both fractions can be recombined thereafter [173] (Figure 13.7(b)). This technique, called "buffering", even allows to increase the sweep rate beyond the maximum tuning frequency of the tunable filter. The filter sweep amplitude is increased, such that the filter pass band sweeps through the laser gain spectrum within a short fraction of the tuning cycle, and the SOA is only active during that short period. The resulting short and fast sweep can then be multiplied in cascaded buffering stages [30,64,70]. Up to sixteen-fold multiplication of the filter tuning frequency has been successfully demonstrated, enabling OCT imaging at multi-megahertz A-scan rates [30,45,62,64].

The tuning range of a swept source can be extended beyond the bandwidth of a single SOA by multiplexing several SOAs with shifted gain spectra. However, this is considerably more complicated than multiplexing broadband emitters, such as SLDs, and has so far only been demonstrated for two SOAs. A simple serial arrangement within the resonator is not feasible, because semiconductors absorb light at frequencies above their own gain band. A parallel configuration is possible if both SOAs provide sufficient gain to compensate for the additional losses in the resonator. This has been demonstrated with fiber-based ring lasers in the 1,300 nm band [174–176], where pairs of SOAs yielded a combined sweep range up to 43 THz (> 220 nm). In this configuration, interference noise may occur in the overlapping region of both SOAs' gain spectra. However, this can be avoided by interlaced operation, as one can quickly switch the SOAs on and off in turn [176]. In a different approach, the emission of two separate laser resonators sharing a common polygon scanner to synchronize their sweeps were incoherently combined [177], in this case covering a total sweep range of 25 THz (145 nm).

State-of-the-art SOA-based ECTLs can provide several tens of milliwatts of output power in a single-transverse-mode beam, which is sufficient for many OCT applications. If however required, higher power can be generated by using an additional optical amplifier either as an external booster or as an extra gain element inside the laser resonator. Different configurations have been demonstrated employing additional standard SOAs [63,166], tapered SOAs [172], or rare-earth-doped fiber amplifiers [64,65].

The ECTL has proven itself as a very successful approach to implement swept sources for OCT. However, two promising alternative technologies adopted from the telecommunication market have emerged in recent years. Both have brought forth compact semiconductor lasers with integrated tuning mechanism, aiming for cost-effective mass production.

One approach relies on a *vertical-cavity surface-emitting laser* (VCSEL) with a MEMS-actuated suspended top mirror for fast tuning of the resonator length (Figure 13.8(a)). These devices, which can be produced with standard semiconductor fabrication methods, operate inherently on a single, continuously tunable optical frequency (Figure 13.5(a)), and feature hence a narrow dynamic linewidth. In 2011, a VCSEL-based swept source for the 1,300 nm band was demonstrated [178] and a device operating at 1,060 nm was presented in the following year [179,180]. After further development sweep ranges up to ~27 THz (100 nm at 1,060 nm, 150 nm at 1,300 nm) have been demonstrated, as well as tuning frequencies up to 600 kHz [68,71]. These swept sources exhibit near-symmetric performance for both sweep directions, enabling A-scan rates up to 1.2 MHz. In 2016, also VCSELs operating at 850 nm were demonstrated, achieving a bandwidth of 10 THz (36 nm) at 429 kHz tuning frequency [181].

Tunable VCSELs can reach long coherence lengths up to several tens of centimeters [39,40,42] or even many meters [71,182]. Full-length imaging of the human eye from the corneal apex to the retina at 50 kHz A-scan rate was demonstrated with a 1,060 nm VCSEL [39], and meter-scale objects were imaged using a 1,300 nm VCSEL at 100 kHz [42].

Due to their high speed and long coherence length, VCSEL-based swept sources have become a popular new tool in the research community [183,184]. However, they will in the foreseeable future not make ECTLs obsolete in cost-sensitive medical or industrial applications. VCSEL-based swept sources are more

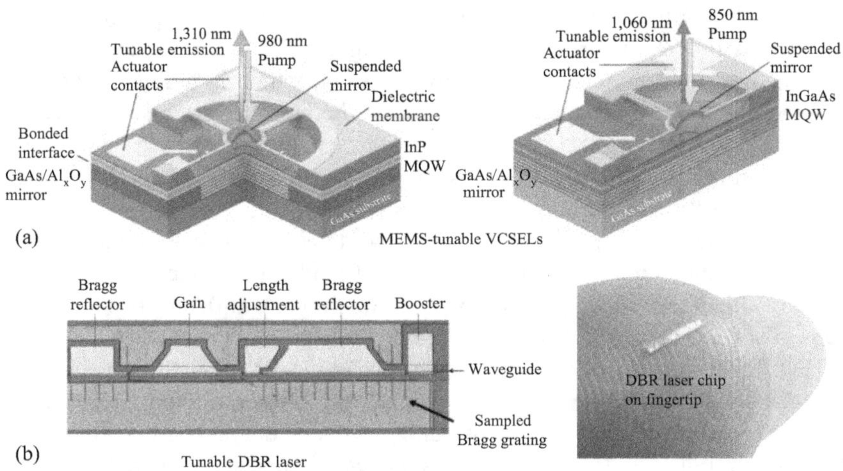

Figure 13.8 Two novel swept source technology platforms promise very high sweep rates and long coherence lengths in a compact setup: (a) VCSELs with a MEMS-mirror enable pure single-mode laser operation with mode-hop free tuning. (b) DBR lasers are free of moving parts and integrate the entire laser including tuning elements and a booster in a single chip. ((a) reprinted with permission from [71], (b) reprinted with permission from [187])

complex than state-of-the-art ECTLs, and thus more costly, for two reasons. VCSELs in sweep operation often cannot emit sufficient output power for OCT imaging applications, which makes at least one additional optical amplifier necessary to boost the light [39,68,71]. Furthermore, most devices currently available are optically pumped with a separate laser diode, since direct electrical pumping of VCSELs is difficult. The charge carriers have to move in the same direction as the light propagates, but doping the material for electrical conductivity increases at the same time the optical losses, which impairs the laser performance. However, electrically pumped tunable VCSELs operating at 850 nm and 1,060 nm with sweep ranges up to ~17 THz (38 nm at 850 nm; 64 nm at 1,060 nm) have been demonstrated, and current research aims for improving their performance [181,182,185,186]. Further progress may eventually lead to less complex and more affordable VCSEL-based swept sources.

The other approach for novel swept sources is based on a monolithic *distributed Bragg reflector* (DBR) *laser,* that incorporates a gain section, tuning elements and a booster section in a single semiconductor chip [187] (Figure 13.8(b)). The electrically tunable DBRs act as wavelength-selective resonator mirrors enclosing the gain section. Changing the refractive index in the DBR sections by current injection allows arbitrary tuning of the laser wavelength in the single-mode regime without any moving parts. This purely electrical tuning mechanism opens interesting possibilities for OCT. With appropriately chosen tuning steps the DBR laser can perform a near-perfect linear frequency sweep, eliminating the need for clocked digitization or computationally intensive re-sampling of the signal. The exact frequency steps can be repeated very accurately from sweep to sweep, which is highly beneficial for techniques comparing the interferogram phase between consecutive A-scans, for instance Doppler-OCT [188,189]. Achieving the same level of phase stability with a mechanical tuning system requires closed-loop control of the tunable filter or computational correction of the acquired signal [183,184].

Different DBR lasers developed for the telecommunication bands around 1,550 nm and 1,600 nm have been shown in OCT applications since 2005 [190–192], The bandwidth of the individual devices was < 6 THz (~50 nm), but concatenating the sweeps of several devices was demonstrated, yielding combined sweep ranges up to 17.5 THz (143 nm) [191–193]. Initially, stepwise tuning of the laser frequency allowed A-scan rates only up to a few kilohertz. Since 2008, however, intense research went after high-speed piecewise-continuous tuning without degradation of the dynamic linewidth [187,194,195]. In 2013, OCT imaging was demonstrated with different DBR lasers operating at 1,550 nm and 1,310 nm featuring A-scans rates up to 200 kHz, bandwidths up to 10 THz (80 nm at 1,550 nm), and coherence lengths > 100 mm [41,196]. After further development of the technology, new 1,310 nm DBR lasers with sweep ranges up to 17.5 THz (100 nm) have become available in 2016 [188,189,197], and in 2018 a DBR laser operating at 1,060 nm with a bandwidth of 16 THz (60 nm) was demonstrated [198]. Hence, swept sources based on DBR lasers enable high-speed, long-range OCT imaging with a depth resolution approaching that of other swept source technologies.

Although DBR lasers have not yet reached the maximum bandwidth achievable with state-of-the-art ECTLs, this technology offers many advantages. The purely electrical tuning mechanism and the full integration of the tunable laser plus amplifier in one monolithic semiconductor chip will open new possibilities for high-speed long-range SS-OCT systems at lower cost.

13.4 Outlook

Due to their various beneficial properties, semiconductor-based light sources are the preferred choice for all variants of OCT. The ability to generate broadband light with a compact, efficient and robust device makes them ideally suited for a wide variety of applications from clinical diagnostics over biomedical research to industrial sensing.

The development of SLDs based on quantum-well and quantum-dot structures during the past three decades has led to devices with formidable bandwidth. By combining several diodes with shifted spectra, the regime of ultrahigh axial resolution down to a few micrometers has become accessible with robust and affordable light sources. Since multiplexed SLD light sources are now commercially available, we expect them to be widely used in the next generation of commercial SD-OCT systems. Although the emission spectra are not ideally shaped, some improvement may be made in the near future with advanced spectral shaping techniques, for example, by current modulation.

The successful demonstration of pumping a titanium:sapphire femtosecond laser directly with a diode laser shows the potential to reduce the cost and complexity of high-end ultrahigh-resolution OCT systems considerably. Although femtosecond lasers will unlikely be applied in stand-alone clinical OCT systems, they are attractive for combining OCT with imaging techniques relying on non-linear optical effects, such as multi-photon fluorescence, second-harmonic generation, or coherent anti-Stokes Raman scattering.

Compact external-cavity tunable lasers (ECTLs), now available from a number of manufacturers, have laid the foundation for the wide-spread application of SS-OCT. These light sources are being used in OCT systems for intravascular imaging in the 1,300 nm band, where they outperform current infrared spectrometer technology in terms of A-scan rate and depth range. New ophthalmic OCT systems profit from ECTLs operating in the 1,060 nm band. These provide the necessary penetration depth for comprehensive imaging of the retina and sub-retinal blood vessels, enable high-resolution imaging of the entire anterior segment, and they have proven valuable for performing measurements through dense cataracts.

Swept sources employing vertical-cavity surface emitting lasers (VCSELs) will enable the development of OCT systems with even higher image acquisition speed and depth range. The first commercial VCSEL-based swept sources at 1,300 nm have been available for a couple of years, while devices for the 1,060 nm band have recently entered the market. Even though VCSEL light sources will likely remain more costly than ECTLs, they will be adopted in new applications that benefit from the superior speed and imaging depth range.

Tunable distributed-Bragg-reflector (DBR) lasers promise similar benefits while being less complex to manufacture. This swept source technology has become available for the 1,300 nm and 1,060 nm bands, and has shown its potential in various research applications. Although these lasers have not yet achieved the same broad bandwidths as ECTL- or VCSEL-based swept sources, this technology will likely become widely adopted in SS-OCT applications, as it promises to reduce the hardware cost considerably.

FDML lasers will in the foreseeable future be unrivaled in offering the highest imaging speeds (multi-megahertz A-scan rates). These light sources are significantly more complex than the other types of swept sources on the market, but fully automated turn-key systems are now commercially available and have made this technology accessible to a wide field of users. FDML lasers have gained great popularity within the research community, and they will keep their pioneering role in exploring new possibilities opened by ultrahigh-speed OCT imaging. However, for the wide spread application in the medical sector or in industry the aforementioned less costly swept source technologies will remain in favor. As we expect, ECTLs, VCSELs, and DBR lasers will within the next couple of years sufficiently cover the currently most interesting range of A-scan rates between 100 kHz and 1 MHz. Yet, the FDML laser will remain the light source of choice for more demanding OCT applications requiring the fastest imaging speeds with high resolution and long depth range. Until now, FDML is the only technology that can fully exploit the beneficial properties of semiconductor laser gain materials enabling ultrahigh-speed broadband sweeping with long coherence length.

Semiconductor technology has influenced the development of OCT in various ways, by enabling the implementation of robust, affordable, high-performance light sources, and has thereby laid the foundation for the impressive success of OCT. We may expect that ongoing development in industry and academia will continue to improve the performance, lower the cost, and hence broaden the range of applications.

References

[1] D. Huang, E. A. Swanson, C. P. Lin, *et al.*, "Optical coherence tomography," *Science* **254**(5035), pp. 1178–1181, 1991.
[2] M. R. Hee, D. Huang, E. A. Swanson, and J. G. Fujimoto, "Polarization-sensitive low-coherence reflectometer for birefringence characterization and ranging," *J. Opt. Soc. Am. B* **9**(6), pp. 903–908, 1992.
[3] J. F. de Boer and T. E. Milner, "Review of polarization sensitive optical coherence tomography and stokes vector determination," *J. Biomed. Opt.* **7**(3), pp. 359–371, 2002.
[4] J. A. Izatt, M. D. Kulkarni, S. Yazdanfar, J. K. Barton, and A. J. Welch, "In vivo bidirectional color Doppler flow imaging of picoliter blood volumes using optical coherence tomography," *Opt. Lett.* **22**, pp. 1439–1441, 1997.
[5] Z. Chen, T. E. Milner, D. Davé, and J. S. Nelson, "Optical Doppler tomographic imaging of fluid flow velocity in highly scattering media," *Opt. Lett.* **22**, pp. 64–66, 1997.
[6] Y. Zhao, Z. Chen, C. Saxer, S. Xiang, J. F. de Boer, and J. S. Nelson, "Phase-resolved optical coherence tomography and optical Doppler tomography for imaging blood flow in human skin with fast scanning speed and high velocity sensitivity," *Opt. Lett.* **25**, pp. 114–116, 2000.
[7] R. K. Wang, S. L. Jacques, Z. Ma, S. Hurst, S. R. Hanson, and A. Gruber, "Three dimensional optical angiography," *Opt. Express* **15**, pp. 4083–4097, 2007.
[8] M. S. Mahmud, D. W. Cadotte, B. Vuong, *et al.*, "Review of speckle and phase variance optical coherence tomography to visualize microvascular networks," *J. Biomed. Opt.* **18**(5), p. 050901, 2013.
[9] R. A. Leitgeb, R. M. Werkmeister, C. Blatter, and L. Schmetterer, "Doppler optical coherence tomography," *Prog. Retin. Eye Res.* **41**(Supplement C), pp. 26–43, 2014.
[10] S. A. Boppart, A. L. Oldenburg, C. Xu, and D. L. Marks, "Optical probes and techniques for molecular contrast enhancement in coherence imaging," *J. Biomed. Opt.* **10**(4), p. 041208, 2005.
[11] C. Yang, "Molecular contrast optical coherence tomography: A review," *Photochem. Photobiol.* **81**(2), pp. 215–237, 2005.
[12] S. Marschall, B. Sander, M. Mogensen, T. M. Jorgensen, and P. E. Andersen, "Optical coherence tomography—current technology and applications in clinical and biomedical research," *Anal. Bioanal. Chem.* **400**(9), pp. 2699–2720, 2011.

[13] N. Tanno and S. Kishi, "Optical coherence tomographic imaging and clinical diagnosis," *Medical Imaging Technology* **17**(1), pp. 3–10, 1999.

[14] M. E. van Velthoven, D. J. Faber, F. D. Verbraak, T. G. van Leeuwen, and M. D. de Smet, "Recent developments in optical coherence tomography for imaging the retina," *Prog. Retin. Eye Res.* **26**(1), pp. 57–77, 2007.

[15] W. Drexler and J. G. Fujimoto, "State-of-the-art retinal optical coherence tomography," *Prog. Retin. Eye Res.* **27**(1), pp. 45–88, 2008.

[16] M. Doors, T. T. J. M. Berendschot, J. de Brabander, C. A. B. Webers, and R. M. M. A. Nuijts, "Value of optical coherence tomography for anterior segment surgery," *J. Cataract Refr. Surg.* **36**(7), pp. 1213–1229, 2010.

[17] E. B. Rodrigues, M. Johanson, and F. M. Penha, "Anterior segment tomography with the cirrus optical coherence tomography," *J. Ophthalmol.* **2012**, p. 806989, 2012.

[18] S. H. Yun, G. J. Tearney, B. J. Vakoc, *et al.*, "Comprehensive volumetric optical microscopy in vivo.," *Nat. Med.* **12**(12), pp. 1429–1433, 2006.

[19] G. J. Tearney, S. Waxman, M. Shishkov, *et al.*, "Three-dimensional coronary artery microscopy by intracoronary optical frequency domain imaging," *J. Am. Coll. Cardiol. Img.* **1** (6), pp. 752–761, 2008.

[20] F. Prati, E. Regar, G. S. Mintz, *et al.*, "Expert review document on methodology, terminology, and clinical applications of optical coherence tomography: physical principles, methodology of image acquisition, and clinical application for assessment of coronary arteries and atherosclerosis," *Eur. Heart J.* **31**(4), pp. 401–415, 2010.

[21] E. Regar, J. Ligthart, N. Bruining, and G. van Soest, "The diagnostic value of intracoronary optical coherence tomography," *Herz* **36**(5), pp. 417–429, 2011.

[22] D. Stifter, "Beyond biomedicine: a review of alternative applications and developments for optical coherence tomography," *Appl. Phys. B-Lasers O.* **88**(3), pp. 337–357, 2007.

[23] J. F. de Boer, R. Leitgeb, and M. Wojtkowski, "Twenty-five years of optical coherence tomography: the paradigm shift in sensitivity and speed provided by Fourier domain OCT," *Biomed. Opt. Express* **8**, pp. 3248–3280, 2017.

[24] A. F. Fercher, "Optical coherence tomography - development, principles, applications," *Z. Med. Phys.* **20**(4), pp. 251–276, 2010.

[25] M. E. Brezinski, *Optical Coherence Tomography - Principles and Applications*, Amsterdam, Elsevier, 2006.

[26] W. Drexler, and J. G. Fujimoto, eds., *Optical Coherence Tomography - Technology and Applications*, Basel, Springer International Publishing, 2nd ed., 2015.

[27] R. Leitgeb, C. K. Hitzenberger, and A. F. Fercher, "Performance of Fourier domain vs. time domain optical coherence tomography," *Opt. Express* **11**(8), pp. 889–894, 2003.

[28] M. Choma, M. Sarunic, C. Yang, and J. Izatt, "Sensitivity advantage of swept source and Fourier domain optical coherence tomography," *Opt. Express* **11**(18), pp. 2183–2189, 2003.

[29] J. F. de Boer, B. Cense, B. H. Park, M. C. Pierce, G. J. Tearney, and B. E. Bouma, "Improved signal-to-noise ratio in spectral-domain compared with time-domain optical coherence tomography," *Opt. Lett.* **28**(21), pp. 2067–2069, 2003.

[30] W. Wieser, B. R. Biedermann, T. Klein, C. M. Eigenwillig, and R. Huber, "Multi-megahertz OCT: High quality 3D imaging at 20 million A-scans and 4.5 Gvoxels per second," *Opt. Express* **18**(14), pp. 14685–14704, 2010.

[31] K. Takada, A. Himeno, and K. Yukimatsu, "Phase-noise and shot-noise limited operations of low coherence optical time domain reflectometry," *Appl. Phys. Lett.* **59**(20), pp. 2483–2485, 1991.

[32] A. G. Podoleanu, "Unbalanced versus balanced operation in an optical coherence tomography system," *Appl. Opt.* **39**, pp. 173–182, 2000.

[33] A. F. Fercher, C. K. Hitzenberger, G. Kamp, and S. Y. El-Zaiat, "Measurement of intraocular distances by backscattering spectral interferometry," *Opt. Commun.* **117**(1–2), pp. 43–48, 1995.

[34] U. Haberland, W. Riitten, V. Blazek, and H. J. Schmitt, "Investigation of highly scattering media using near-infrared continuous wave tunable semiconductor laser," *Proc. SPIE* **2389**, pp. 503–512, 1995.

[35] S. R. Chinn, E. A. Swanson, and J. G. Fujimoto, "Optical coherence tomography using a frequency-tunable optical source," *Opt. Lett.* **22**(5), pp. 340–342, 1997.

[36] B. Golubovic, B. E. Bouma, G. J. Tearney, and J. G. Fujimoto, "Optical frequency-domain reflectometry using rapid wavelength tuning of a Cr^{4+}:forsterite laser," *Opt. Lett.* **22**(22), pp. 1704–1706, 1997.

[37] S.-H. Yun, G. J. Tearney, J. F. de Boer, and B. E. Bouma, "Motion artifacts in optical coherence tomography with frequency-domain ranging," *Opt. Express* **12**(13), pp. 2977–2998, 2004.

[38] D. C. Adler, W. Wieser, F. Trepanier, J. Schmitt, and R. A. Huber, "Extended coherence length Fourier domain mode locked lasers at 1310 nm," *Opt. Express* **19**, pp. 20930–20939, 2011.

[39] I. Grulkowski, J. J. Liu, B. Potsaid, *et al.*, "Retinal, anterior segment and full eye imaging using ultrahigh speed swept source OCT with vertical-cavity surface emitting lasers," *Biomed. Opt. Express* **3**, pp. 2733–2751, 2012.

[40] I. Grulkowski, J. J. Liu, B. Potsaid, *et al.*, "High-precision, high-accuracy ultralong-range swept-source optical coherence tomography using vertical cavity surface emitting laser light source," *Opt. Lett.* **38**, pp. 673–675, 2013.

[41] M. Bonesi, M. P. Minneman, J. Ensher, *et al.*, "Akinetic all-semiconductor programmable swept-source at 1550 nm and 1310 nm with centimeters coherence length," *Opt. Express* **22**, pp. 2632–2655, 2014.

[42] Z. Wang, B. Potsaid, L. Chen, *et al.*, "Cubic meter volume optical coherence tomography," *Optica* **3**, pp. 1496–1503, 2016.

[43] J. Xu, S. Song, and R. K. Wang, "Ultralong-range optical coherence tomography-based angiography by akinetic swept source," *Proc. SPIE* **10053**, p. 100531W, 2017.

[44] T. Pfeiffer, W. Draxinger, W. Wieser, T. Klein, M. Petermann, and R. Huber, "Analysis of FDML lasers with meter range coherence," *Proc. SPIE* **10053**, p. 100531T, 2017.

[45] T. Klein, W. Wieserm, L. Reznicek, A. Neubauer, A. Kampik, and R. Huber, "Multi-MHz retinal OCT," *Biomed. Opt. Express* **4**(10), pp. 1890–1908, 2013.

[46] T. Wang, W. Wieser, G. Springeling, *et al*, "Intravascular optical coherence tomography imaging at 3200 frames per second," *Opt. Lett.* **38**, pp. 1715–1717, 2013.

[47] F. Harris, "On the use of windows for harmonic analysis with the discrete Fourier transform," *Proc. IEEE* **66**, pp. 51–83, 1978.

[48] Y. Zhang, M. Sato, and N. Tanno, "Numerical investigations of optimal synthesis of several low coherence sources for resolution improvement," *Opt. Commun.* **192**(3–6), pp. 183–192, 2001.

[49] A. C. Akcay, J. P. Rolland, and J. M. Eichenholz, "Spectral shaping to improve the point spread function in optical coherence tomography," *Opt. Lett.* **28**, pp. 1921–1923, 2003.

[50] R. Tripathi, N. Nassif, J. S. Nelson, B. H. Park, and J. F. de Boer, "Spectral shaping for non-Gaussian source spectra in optical coherence tomography," *Opt. Lett.* **27**, pp. 406–408, 2002.

[51] M. Szkulmowski, M. Wojtkowski, T. Bajraszewski, *et al.*, "Quality improvement for high resolution in vivo images by spectral domain optical coherence tomography with supercontinuum source," *Opt. Commun.* **246**(4–6), pp. 569–578, 2005.

[52] B. R. Biedermann, W. Wieser, C. M. Eigenwillig, *et al.*, "Real time en face Fourier-domain optical coherence tomography with direct hardware frequency demodulation," *Opt. Lett.* **33**, pp. 2556–2558, 2008.

[53] P. Jansz, S. Richardson, G. Wild, and S. Hinckley, "Modeling of low coherence interferometry using broadband multi-Gaussian light sources," *Photonic Sensors* **2**(3), pp. 247–258, 2012.

[54] P. V. Jansz, S. Richardson, G. Wild, and S. Hinckley, "Characterizing the resolvability of real superluminescent diode sources for application to optical coherence tomography using a low coherence interferometry model," *J. Biomed. Opt.* **19**(8), p. 085003, 2014.

[55] S. Marschall, C. Pedersen, and P. E. Andersen, "Investigation of the impact of water absorption on retinal OCT imaging in the 1060 nm range," *Biomed. Opt. Express* **3**, pp. 1620–1631, 2012.

[56] B. R. Biedermann, W. Wieser, C. M. Eigenwillig, T. Klein, and R. Huber, "Direct measurement of the instantaneous linewidth of rapidly wavelength-swept lasers," *Opt. Lett.* **35**, pp. 3733–3735, 2010.

[57] T. von Niederhäusern C. Meier, M. Duelk, and P. Vorreau, "Instantaneous coherence length measurement of a swept laser source using a Mach-Zehnder interferometer," *Proc. SPIE* **7889**, p. 78892R, 2011.

[58] B. R. Biedermann, W. Wieser, C. M. Eigenwillig, and R. Huber, "Recent developments in Fourier domain mode locked lasers for optical coherence

tomography: Imaging at 1310 nm vs. 1550 nm wavelength," *J. Biophotonics* **2**(6–l), pp. 357–363, 2009.

[59] A. Alex, B. Považay, B. Hofer, *et al.*, "Multispectral in vivo three-dimensional optical coherence tomography of human skin," *J. Biomed. Opt.* **15**(2), p. 026025, 2010.

[60] M. Góra, K. Karnowski, M. Szkulmowski, *et al.*, "Ultra high-speed swept source OCT imaging of the anterior segment of human eye at 200 kHz with adjustable imaging range," *Opt. Express* **17**, pp. 14880–14894, 2009.

[61] B. Potsaid, B. Baumann, D. Huang, *et al.*, "Ultrahigh speed 1050nm swept source / Fourier domain OCT retinal and anterior segment imaging at 100,000 to 400,000 axial scans per second," *Opt. Express* **18**, pp. 20029–20048, 2010.

[62] W. Wieser, T. Klein, D. C. Adler, *et al.*, "Extended coherence length megahertz FDML and its application for anterior segment imaging," *Biomed. Opt. Express* **3**, pp. 2647–2657, 2012.

[63] M. K. K. Leung, A. Mariampillai, B. A. Standish, *et al.*, "High-power wavelength-swept laser in Littman telescope-less polygon filter and dual-amplifier configuration for multichannel optical coherence tomography," *Opt. Lett.* **34**(18), pp. 2814–2816, 2009.

[64] T. Klein, W. Wieser, C. M. Eigenwillig, B. R. Biedermann, and R. Huber, "Megahertz OCT for ultrawide-field retinal imaging with a 1050 nm Fourier domain mode-locked laser," *Opt. Express* **19**, pp. 3044–3062, 2011.

[65] M. K. Harduar, A. Mariampillai, B. Vuong, X. Gu, B. A. Standish, and V. X. D. Yang, "Dual-core ytterbium fiber amplifier for high-power 1060 nm swept source multichannel optical coherence tomography imaging," *Opt. Lett.* **36**, pp. 2976–2978, 2011.

[66] R. Poddar, J. V. Migacz, D. M. Schwartz, J. S. Werner, and I. Gorczynska, "Challenges and advantages in wide-field optical coherence tomography angiography imaging of the human retinal and choroidal vasculature at 1,7-MHz A-scan rate," *J. Biomed. Opt.* **22**(10), p. 106018, 2017.

[67] C. Blatter, T. Klein, B. Grajciar, *et al.*, "Ultrahigh-speed non-invasive widefield angiography," *J. Biomed. Opt.* **17**(7), p. 070505, 2012.

[68] T.-H. Tsai, B. Potsaid, Y. K. Tao, *et al.*, "Ultrahigh speed endoscopic optical coherence tomography using micromotor imaging catheter and VCSEL technology," *Biomed. Opt. Express* **4**, pp. 1119–1132, 2013.

[69] R. Huber, M. Wojtkowski, and J. G. Fujimoto, "Fourier domain mode locking (FDML): a new laser operating regime and applications for optical coherence tomography," *Opt. Express* **14**(8), pp. 3225–3237, 2006.

[70] W.-Y. Oh, B. J. Vakoc, M. Shishkov, G. J. Tearney, and B. E. Bouma, ">400 kHz repetition rate wavelength-swept laser and application to high-speed optical frequency domain imaging," *Opt. Lett.* **35**, pp. 2919–2921, 2010.

[71] V. Jayaraman, B. Potsaid, J. Jiang, *et al.*, "High-speed ultra-broad tuning MEMS-VCSELs for imaging and spectroscopy," *Proc. SPIE* **8763**, p. 87630H, 2013.

[72] W. Wieser, W. Draxinger, T. Klein, S. Karpf, T. Pfeiffer, and R. Huber, "High definition live 3D-OCT in vivo: design and evaluation of a 4D OCT engine with 1 GVoxel/s," *Biomed. Opt. Express* **5**, pp. 2963–2977, 2014.

[73] Z. Zhi, W. Qin, J. Wang, W. Wei, and R. K. Wang, "4D optical coherence tomography-based micro-angiography achieved by 1.6-MHz FDML swept source," *Opt. Lett.* **40**, pp. 1779–1782, 2015.

[74] S. Wang, M. Singh, A. L. Lopez, *et al.*, "Direct four-dimensional structural and functional imaging of cardiovascular dynamics in mouse embryos with 1.5 MHz optical coherence tomography," *Opt. Lett.* **40**, pp. 4791–4794, 2015.

[75] M. Singh, C. Wu, C.-H. Liu, *et al.*, "Phase-sensitive optical coherence elastography at 1.5 million A-lines per second," *Opt. Lett.* **40**, pp. 2588–2591, 2015.

[76] S. Elahi, L. Thrane, A. M. Rollins, and M. W. Jenkins, "Extended dynamic range of Doppler OCT by application of a new method to high density B-scans using a MHz FDML swept laser source," *Proc. SPIE* **10053**, p. 100531U, 2017.

[77] B. Potsaid, I. Gorczynska, V. J. Srinivasan, *et al.*, "Ultrahigh speed spectral / Fourier domain OCT ophthalmic imaging at 70,000 to 312,500 axial scans per second," *Opt. Express* **16**, pp. 15149–15169, 2008.

[78] B. R. Biedermann, W. Wieser, C. M. Eigenwillig, T. Klein, and R. Huber, "Dispersion, coherence and noise of Fourier domain mode locked lasers," *Opt. Express* **17**(12), pp. 9947–9961, 2009.

[79] A. Bradu and A. Podoleanu, "Evaluating the benefits of using balance detection in Fourier domain optical coherence tomography," *Proc. SPIE* **8802**, p. 880206, 2013.

[80] W.-C. Kuo, C.-M. Lai, Y.-S. Huang, C.-Y. Chang, and Y.-M. Kuo, "Balanced detection for spectral domain optical coherence tomography," *Opt. Express* **21**, pp. 19280–19291, 2013.

[81] J. M. Schmitt, A. Knüttel, M. Yadlowsky, and M. A. Eckhaus, "Optical-coherence tomography of a dense tissue: statistics of attenuation and back-scattering," *Phys. Med. Biol.* **39**(10), pp. 1705–1720, 1994.

[82] J. A. Parrish, "New concepts in therapeutic photomedicine; photochemistry, optical targeting and the therapeutic window," *J. Invest. Dermatol.* **77**, pp. 45–50, 1981.

[83] M. E. Brezinski, G. J. Tearney, B. E. Bouma, *et al.*, "Optical coherence tomography for optical biopsy: Properties and demonstration of vascular pathology," *Circulation* **93**(6), pp. 1206–1213, 1996.

[84] B. Wilson and S. Jacques, "Optical reflectance and transmittance of tissues: principles and applications," *Quant. Elect. IEEE J.* **26**, pp. 2186–2199, 1990.

[85] M. L. Wolbarsht, A. W. Walsh, and G. George, "Melanin, a unique biological absorber,', *Appl. Opt.* **20**, pp. 2184–2186, 1981.

[86] P. Parsa, S. L. Jacques, and N. S. Nishioka, "Optical properties of rat liver between 350 and 2200 nm," *Appl. Opt.* **28**, pp. 2325–2330, 1989.

[87] S. J. Matcher, M. Cope, and D. T. Delpy, "In vivo measurements of the wavelength dependence of tissue-scattering coefficients between 760 and 900 nm measured with time-resolved spectroscopy," *Appl. Opt.* **36**, pp. 386–396, 1997.

[88] S. Srinivasan, B. W. Pogue, S. Jiang, *et al.*, "Interpreting hemoglobin and water concentration, oxygen saturation, and scattering measured in vivo by near-infrared breast tomography," *P. Natl. Acad. Sci. USA* **100**(21), pp. 12349–12354, 2003.

[89] U. Sharma, E. W. Chang, and S. H. Yun, "Long-wavelength optical coherence tomography at 1.7 μm for enhanced imaging depth," *Opt. Express* **16**, pp. 19712–19723, 2008.

[90] M. Tanaka, T. Okuno, H. Obi, *et al.*, "Performance improvement by a broadband super-luminescent diode light source in 1.7-μm spectroscopic spectral-domain optical coherence tomography for lipid distribution imaging in a coronary artery," *Proc. SPIE* **8926**, p. 89262T, 2014.

[91] Y. Li, J. Jing, E. Heidari, J. Zhu, Y. Qu, and Z. Chen, "Intravascular optical coherence tomography for characterization of atherosclerosis with a 1.7 micron swept-source laser," *Sci. Rep.* **7**, p. 14525, 2017.

[92] T. J. van den Berg and H. Spekreijse, "Near infrared light absorption in the human eye media," *Vis. Res.* **37**(2), pp. 249–253, 1997.

[93] B. Považay, K. Bizheva, B. Hermann, *et al.*, "Enhanced visualization of choroidal vessels using ultrahigh resolution ophthalmic OCT at 1050 nm," *Opt. Express* **11** (17), pp. 1980–1986, 2003.

[94] A. Unterhuber, B. Považay, B. Hermann, H. Sattmann, A. Chavez-Pirson, and W. Drexler, "In vivo retinal optical coherence tomography at 1040 nm - enhanced penetration into the choroid," *Opt. Express* **13**(9), pp. 3252–3258, 2005.

[95] S.-P. Chee, N. C. S. Wen, and A. Jap, "Comparison of enhanced depth imaging and swept source optical coherence tomography in assessment of Vogt-Koyanagi-Harada disease," *J. Clin. Exp. Ophthalmol.* **6**(3), p. 1000432, 2015.

[96] B. Považay, B. Hermann, A. Unterhuber, *et al.*, "Three-dimensional optical coherence tomography at 1050 nm versus 800 nm in retinal pathologies: enhanced performance and choroidal penetration in cataract patients," *J. Biomed. Opt.* **12**(4), p. 041211, 2007.

[97] Y. Wang, J. S. Nelson, Z. Chen, B. J. Reiser, R. S. Chuck, and R. S. Windeler, "Optimal wavelength for ultrahigh-resolution optical coherence tomography," *Opt. Express* **11** (12), pp. 1411–1417, 2003.

[98] C. K. Hitzenberger, A. Baumgartner, W. Drexler, and A. F. Fercher, "Dispersion effects in partial coherence interferometry: Implications for intraocular ranging," *J. Biomed. Opt.* **4**(1), pp. 144–151, 1999.

[99] J. F. de Boer, C. E. Saxer, and J. S. Nelson, "Stable carrier generation and phase-resolved digital data processing in optical coherence tomography," *Appl. Opt.* **40**, pp. 5787–5790, 2001.

[100] M. Wojtkowski, V. Srinivasan, T. Ko, J. Fujimoto, A. Kowalczyk, and J. Duker, "Ultrahigh-resolution, high-speed, Fourier domain optical coherence tomography and methods for dispersion compensation," *Opt. Express* **12**, pp. 2404–2422, 2004.

[101] T. Hillman and D. Sampson, "The effect of water dispersion and absorption on axial resolution in ultrahigh-resolution optical coherence tomography," *Opt. Express* **13**, pp. 1860–1874, 2005.

[102] S. Hariri, A. A. Moayed, A. Dracopoulos, C. Hyun, S. Boyd, and K. Bizheva, "Limiting factors to the OCT axial resolution for in-vivo imaging of human and rodent retina in the 1060 nm wavelength range," *Opt. Express* **17**, pp. 24304–24316, 2009.

[103] W. Drexler, U. Morgner, F. X. Kärtner, *et al.*, "In vivo ultrahigh-resolution optical coherence tomography," *Opt. Lett.* **24**, pp. 1221–1223, 1999.

[104] I. Hartl, X. D. Li, C. Chudoba, *et al.*, "Ultrahigh-resolution optical coherence tomography using continuum generation in an air-silica microstructure optical fiber," *Opt. Lett.* **26**(9), pp. 608–610, 2001.

[105] B. Povazay, K. Bizheva, A. Unterhuber, *et al.*, "Submicrometer axial resolution optical coherence tomography," *Opt. Lett.* **27**(20), pp. 1800–1802, 2002.

[106] W. Drexler, "Ultrahigh-resolution optical coherence tomography," *J. Biomed. Opt.* **9**(1), pp. 4774, 2004.

[107] T. R. Chen, L. Eng, Y. H. Zhuang, A. Yariv, N. S. Kwong, and P. C. Chen, "Quantum well superluminescent diode with very wide emission spectrum," *Appl. Phys. Lett.* **56**(14), pp. 1345–1346, 1990.

[108] S. Kondo, H. Yasaka, Y. Noguchi, K. Magari, M. Sugo, and O. Mikami, "Very wide spectrum multiquantum well superluminescent diode at 1.5 μm," *Electron. Lett.* **28**(2), pp. 132–133, 1992.

[109] V. R. Shidlovski, "Superluminescent diode light sources for OCT," in *Optical Coherence Tomography - Technology and Applications*, W. Drexler and J. G. Fujimoto, eds., ch. 9, pp. 280–299, Heidelberg, Springer-Verlag, 2008.

[110] D. S. Adler, T. H. Ko, A. K. Konorev, *et al.*, "Broadband light source based on quantum-well superluminescent diodes for high-resolution optical coherence tomography," *Quantum Electron.* **34**(10), p. 915, 2004.

[111] E. V. Andreeva, P. I. Lapin, V. V. Prokhorov, V. R. Shidlovski, M. V. Shramenko, and S. D. Yakubovich, "Novel superluminescent diodes and SLD-based light sources for optical coherence tomography," *Proc. SPIE* **6627**, p. 662703, 2007.

[112] I. G. Tooley, D. T. D. Childs, B. J. Stevens, K. M. Groom, and R. A. Hogg, "Simulation of broad spectral bandwidth emitters at 1060 nm for optical coherence tomography," *Proc. SPIE* **9767**, p. 97671W, 2016.

[113] C.-F. Lin and B.-L. Lee, "Extremely broadband AlGaAs/GaAs superluminescent diodes," *Applied Physics Letters* **71**(12), pp. 1598–1600, 1997.

[114] Z.-Z. Sun, D. Ding, Q. Gong, W. Zhou, B. Xu, and Z.-G. Wang, "Quantum-dot superluminescent diode: A proposal for an ultra-wide output spectrum," *Opt. Quant. Electron.* **31**(12), pp. 1235–1246, 1999.

[115] Z. Zhang, Z. G. Wang, B. Xu, P. Jin, Z. Z. Sun, and F. Q. Liu, "High-performance quantum-dot superluminescent diodes," *IEEE Photonic Tech. L.* **16**(1), pp. 27–29, 2004.

[116] L. Li, M. Rossetti, A. Fiore, L. Occhi, and C. Velez, "Wide emission spectrum from superluminescent diodes with chirped quantum dot multilayers," *Electronics Letters* **41**(1), pp. 41–43, 2005.

[117] Y. C. Xin, A. Martinez, T. Saiz, *et al.*, "1.3-μm quantum-dot multisection superluminescent diodes with extremely broad bandwidth," *IEEE Photonic Tech. L.* **19**(7), pp. 501–503, 2007.

[118] S. K. Ray, T. L. Choi, K. M. Groom, *et al.*, "High-power and broadband quantum dot superluminescent diodes centered at 1250 nm for optical coherence tomography," *IEEE J Sel. Top. Quant. Elect.* **13**, pp. 1267–1272, 2007.

[119] Z. Y. Zhang, R. A. Hogg, B. Xu, P. Jin, and Z. G. Wang, "Realization of extremely broadband quantum-dot superluminescent light-emitting diodes by rapid thermal-annealing process," *Opt. Lett.* **33**, pp. 1210–1212, 2008.

[120] Z. Y. Zhang, Q. Jiang, I. J. Luxmoore, and R. A. Hogg, "A p-type-doped quantum dot superluminescent LED with broadband and flat-topped emission spectra obtained by post-growth intermixing under a GaAs proximity cap," *Nanotechnology* **20**(5), p. 055204, 2009.

[121] P. D. L. Greenwood, D. T. D. Childs, K. Groom, B. Stevens, M. Hopkinson, and R. Hogg, "Tuning superluminescent diode characteristics for optical coherence tomography systems by utilizing a multicontact device incorporating wavelength-modulated quantum dots," *IEEEJ. Sel. Top. Quant.* **15**(3), pp. 757–763, 2009.

[122] X. Li, P. Jin, Q. An, *et al.*, "A high-performance quantum dot superluminescent diode with a two-section structure," *Nanoscale Res. Lett.* **6**, p. 625, 2011.

[123] I. Tsubaki, Y. Harada, and T. Kita, "High-resolution optical coherence tomography using broadband light source with strain-controlled InAs/GaAs quantum dots," *Phys. Status Solidi C* **9**(12), pp. 2473–2476, 2012.

[124] H. Shibata, N. Ozaki, T. Yasuda, *et al.*, "Imaging of spectral-domain optical coherence tomography using a superluminescent diode based on InAs quantum dots emitting broadband spectrum with Gaussian-like shape," *Jpn. J. Appl. Phys.* **54**(4S), p. 04DG07, 2015.

[125] N. Ozaki, D. T. D. Childs, J. Sarma, *et a1.*, "Superluminescent diode with a broadband gain based on self-assembled InAs quantum dots and segmented contacts for an optical coherence tomography light source," *J. Appl. Phys.* **119**(8), p. 083107, 2016.

[126] S. Chen, W. Li, Z. Zhang, *et al.*, "GaAs-based superluminescent light-emitting diodes with 290-nm emission bandwidth by using hybrid quantum well/quantum dot structures," *Nanoscale Res. Lett.* **10**(1), p. 340, 2015.

[127] P. D. L. Greenwood, D. T. D. Childs, K. Kennedy, *et al.*, "Quantum dot superluminescent diodes for optical coherence tomography: Device engineering," *IEEE J. Sel. Top. Quant.* **16**(4), pp. 1015–1022, 2010.

[128] N. Krstajic, L. Smith, S. Matcher, *et al.*, "Quantum dot superluminescent diodes for optical coherence tomography: Skin imaging," *IEEE J. Sel. Top. Quant.* **16**(4), pp. 748–754, 2010.

[129] A. T. Semenov, V. K. Batovrin, I. Garmash, V. R. Shidlovsky, M. V. Shramenko, and S. D. Yakubovich, "(GaAl)As SQW superluminescent diodes with extremely low coherence length," *Electron. Lett.* **31**(4), pp. 314–315, 1995.

[130] J. M. Schmitt, S. L. Lee, and K. M. Yung, "An optical coherence microscope with enhanced resolving power in thick tissue," *Opt. Commun.* **142** (4–6), pp. 203–207, 1997.

[131] T. Ko, D. Adler, J. Fujimoto, *et al.*, "Ultrahigh resolution optical coherence tomography imaging with a broadband superluminescent diode light source," *Opt. Express* **12**, pp. 2112–2119, 2004.

[132] D. Zhu, M. Shen, H. Jiang, *et al.*, "Broadband superluminescent diode-based ultrahigh resolution optical coherence tomography for ophthalmic imaging," *J. Biomed. Opt.* **16**(12), p. 126006, 2011.

[133] M. D. Bayleyegn, H. Makhlouf, C. Crotti, K. Plamann, and A. Dubois, "Ultrahigh resolution spectral-domain optical coherence tomography at 1.3 μm using a broadband superluminescent diode light source," *Opt. Commun.* **285**(24), pp. 5564–5569, 2012.

[134] B. Bouma, G. J. Tearney, S. A. Boppart, M. R. Hee, M. E. Brezinski, and J. G. Fujimoto, "High-resolution optical coherence tomographic imaging using a mode-locked Ti:Al_2O_3 laser source," *Opt. Lett.* **20**, pp. 1486–1488, 1995.

[135] W. Drexler, U. Morgner, R. K. Ghanta, F. X. Kartner, J. S. Schuman, and J. G. Fujimoto, "Ultrahigh-resolution ophthalmic optical coherence tomography," *Nat. Med.* **7**(4), pp. 502–507, 2001.

[136] A. Müller, S. Marschall, O. B. Jensen, *et al.*, "Diode laser based light sources for biomedical applications," *Laser Photonics Rev.* **7**, pp. 605–627, 2012.

[137] O. B. Jensen, P. E. Andersen, B. Sumpf, K.-H. Hasler, G. Erbert, and P. M. Petersen, "1.5 W green light generation by single-pass second harmonic generation of a single-frequency tapered diode laser," *Opt. Express* **17**, pp. 6532–6539, 2009.

[138] A. Müller, O. B. Jensen, K.-H. Hasler, *et al.*, "Efficient concept for generation of diffraction-limited green light by sum-frequency generation of spectrally combined tapered diode lasers," *Opt. Lett.* **37**, pp. 3753–3755, 2012.

[139] O. B. Jensen, A. K. Hansen, A. Müller, *et al.*, "Power scaling of nonlinear frequency converted tapered diode lasers for biophotonics," *IEEE J. Sel. Top. Quant.* **20**, pp. 307–321, 2014.

[140] O. B. Jensen, A. K. Hansen, A. Müller, B. Sumpf, P. M. Petersen, and P. E. Andersen, "Efficient generation of 3.5 W laser light at 515 nm by frequency doubling a single-frequency high power DBR tapered diode laser," *Opt. Commun.* **392**, pp. 167–170, 2017.

[141] A. Müller, O. B. Jensen, A. Unterhuber, *et al.*, "Frequency-doubled DBR-tapered diode laser for direct pumping of Ti:sapphire lasers generating sub-20 fs pulses," *Opt. Express* **19**, pp. 12156–12163, 2011.

[142] A. Unterhuber, B. Považay, A. Müller, *et al.*, "Simultaneous dual wavelength eye-tracked ultrahigh resolution retinal and choroidal optical coherence tomography," *Opt. Lett.* **38**(21), pp. 4312–4315, 2013.

[143] K. König, P. Andersen, T. Le, and H. G. Breunig, "Multiphoton imaging with a novel compact diode-pumped Ti:sapphire oscillator," *Microsc. Res. Tech.* **78**(12), pp. 1154–1158, 2015.

[144] P. Piksarv, D. Marti, T. Le, *et al.*, "Integrated single and two-photon light sheet microscopy using accelerating beams," *Scientific Reports* **7**(1), p. 1435, 2017.

[145] S. Tang, Z. Chen, B. J. Tromberg, and T. B. Krasieva, "Combined multiphoton microscopy and optical coherence tomography using a 12-fs broadband source," *J. Biomed. Opt.* **11**(2), p. 020502, 2006.

[146] C. Hoffmann, B. Hofer, A. Unterhuber, B. Poavzay, U. Morgner, and W. Drexler, "Combined OCT and CARS using a single ultrashort pulse Ti:sapphire laser," *Proc. SPIE 1892*, p. 78920B, 2011.

[147] S. P. Chong, T. Lai, Y. Zhou, and S. Tang, "Tri-modal microscopy with multiphoton and optical coherence microscopy/tomography for multi-scale and multi-contrast imaging," *Biomed. Opt. Express* **4**, pp. 1584–1594, 2013.

[148] T. Le, A. Unterhuber, B. Povazay, and H. G. Breunig, "Multi-modal imaging using a novel and portable ultrafast laser source," in *2013 Conference on Lasers and Electro-Optics (CLEO)*, p. JM20.6, Optical Society of America, 2013.

[149] A. J. Verhoef, A. Fernandez, T. Kamali, *et al.*, "A high power directly diode pumped Ti:sapphire laser with synchronized Yb-fiber amplifier for nonlinear optical microscopy and optical coherence tomography," in *2014 Conference on Lasers and Electro-Optics (CLEO)*, p. ATh1O.7, Optical Society of America, 2014.

[150] T. Kamali, B. Považay, S. Kumar, *et al.*, "Hybrid single-source online Fourier transform coherent anti-Stokes Raman scattering/optical coherence tomography," *Opt. Lett.* **39**, pp. 5709–5712, 2014.

[151] S. Kumar, T. Kamali, J. M. Levitte, *et al.*, "Single-pulse CARS based multimodal nonlinear optical microscope for bioimaging," *Opt. Express* **23**, pp. 13082–13098, 2015.

[152] T. Klein and R. Huber, "High-speed OCT light sources and systems," *Biomed. Opt. Express* **8**, pp. 828–859, 2017.

[153] F. D. Nielsen, L. Thrane, J. Black, K. Hsu, A. Bjarklev, and P. E. Andersen, "Swept-wavelength sources for optical coherence tomography in the 1 μm range," *Proc. SPIE* **5861**, p. 58610H, 2005.

[154] M. K. Harduar, A. Mariampillai, B. Vuong, *et al.*, "Dual core ytterbium doped fiber ring laser in Fourier domain mode locked operation for swept-source optical coherence tomography," *Proc. SPIE* **7580**, p. 75802S, 2010.

[155] T. Klein, W. Wieser, B. R. Biedermann, C. M. Eigenwillig, G. Palte, and R. Huber, "Raman-pumped Fourier-domain mode-locked laser: analysis of operation and application for optical coherence tomography," *Opt. Lett.* **33**, pp. 2815–2817, 2008.

[156] J. Xu, H. Ou, X. Xu, V. X. D. Yang, P. C. Chui, and K. K.-Y. Wong, "Wideband Raman-pumped wavelength-swept laser for optical coherence tomography application," *Appl. Phys. Express* **6**(6), p. 062701, 2013.

[157] S. Marschall, T. Klein, W. Wieser, *et al.*, "Broadband Fourier domain mode-locked laser for optical coherence tomography at 1060 nm," *Proc. SPIE* **8213**, p. 82130R, 2012.

[158] T. Torzicky, S. Marschall, M. Pircher, *et al.*, "Retinal polarization-sensitive optical coherence tomography at 1060 nm with 350 kHz A-scan rate using a Fourier domain mode locked laser," *J. Biomed. Opt.* **18**, p. 026008, 2013.

[159] S. Marschall, T. Torzicky, T. Klein, *et al.*, "High-speed polarization-sensitive OCT at 1060 nm using a Fourier domain mode-locked swept source," *Proc. SPIE* **8427**, p. 84271 D, 2012.

[160] B. D. Goldberg, S. M. R. M. Nezam, P. Jillella, B. E. Bouma, and G. J. Tearney, "Miniature swept source for point of care optical frequency domain imaging," *Opt. Express* **17**, pp. 3619–3629, 2009.

[161] M. Kuznetsov, W. Atia, B. Johnson, and D. Flanders, "Compact ultrafast reflective Fabry-Perot tunable lasers for OCT imaging applications," *Proc. SPIE* **1554**, p. 75541 F, 2010.

[162] K. Totsuka, K. Isamoto, T. Sakai, A. Morosawa, and C. Chong, "MEMS scanner based swept source laser for optical coherence tomography," *Proc. SPIE* **7554**, p. 75542Q, 2010.

[163] S. Gloor, A. H. Bachmann, M. Epitaux, *et al.*, "High-speed miniaturized swept sources based on resonant MEMS mirrors and diffraction gratings," *Proc. SPIE* **8571**, p. 85712X, 2013.

[164] S.-H. Yun, C. Boudoux, G. J. Tearney, and B. E. Bouma, "High-speed wavelength-swept semiconductor laser with a polygon-scanner-based wavelength filter," *Opt. Lett.* **28**(20), pp. 1981–1983, 2003.

[165] M. A. Choma, K. Hsu, and J. A. Izatt, "Swept source optical coherence tomography using an all-fiber 1300-nm ring laser source," *J. Biomed. Opt.* **10**(4), p. 044009, 2005.

[166] R. Huber, M. Wojtkowski, K. Taira, J. G. Fujimoto, and K. Hsu, "Amplified, frequency swept lasers for frequency domain reflectometry and OCT imaging: Design and scaling principles," *Opt. Express* **13**(9), pp. 3513–3528, 2005.

[167] W.-Y. Oh, S.-H. Yun, G. J. Tearney, and B. E. Bouma, "115 kHz tuning repetition rate ultrahigh-speed wavelength-swept semiconductor laser," *Opt. Lett.* **30**(23), pp. 3159–3161, 2005.

[168] S. Yagi, K. Naganuma, T. Imai, *et al.*, "A mechanical-free 150-kHz repetition swept light source incorporated a KTN electro-optic deflector," *Proc. SPIE* **7889**, p. 78891J, 2011.

[169] Y. Okabe, Y. Sasaki, M. Ueno, *et al.*, "High-speed optical coherence tomography system using a 200-kHz swept light source with a KTN deflector," *Optics and Photonics Journal* **3**(2), pp. 190–193, 2013.

[170] M. Fujimoto, M. Yamada, K. Yamamoto, *et al.*, "Stable wavelength-swept light source designed for industrial applications using KTN beam-scanning technology," *Proc. SPIE* **10110**, p. 101100Q, 2017.

[171] A. Bilenca, S. H. Yun, G. J. Tearney, and B. E. Bouma, "Numerical study of wavelength-swept semiconductor ring lasers: the role of refractive-index nonlinearities in semiconductor optical amplifiers and implications for biomedical imaging applications," *Opt. Lett.* **31**, pp. 760–762, 2006.

[172] S. Marschall, T. Klein, W. Wieser, *et al.*, "Fourier domain mode-locked swept source at 1050 nm based on a tapered amplifier," *Opt. Express* **18** (15), pp. 15820–15831, 2010.

[173] R. Huber, D. C. Adler, and J. G. Fujimoto, "Buffered Fourier domain mode locking: unidirectional swept laser sources for optical coherence tomography imaging at 370,000 lines/s," *Opt. Lett* **31**(20), pp. 2975–2977, 2006.

[174] M. Y. Jeon, J. Zhang, Q. Wang, and Z. Chen, "High-speed and wide bandwidth Fourier domain mode-locked wavelength swept laser with multiple SOAs," *Opt. Express* **16**(4), pp. 2547–2554, 2008.

[175] J. Zhang, G. Liu, and Z. Chen, "Ultra broad band Fourier domain mode locked swept source based on dual SOAs and WDM couplers," *Proc. SPIE* **1554**, p. 755411, 2010.

[176] S. H. Kassani, M. Villiger, N. Uribe-Patarroyo, *et al.*, "Extended bandwidth wavelength swept laser source for high resolution optical frequency domain imaging," *Opt. Express* **25**, pp. 8255–8266, 2017.

[177] W.-Y. Oh, S.-H. Yun, G. J. Tearney, and B. E. Bouma, "Wide tuning range wavelength-swept laser with two semiconductor optical amplifiers," *IEEE Photonic Tech. L.* **17**(3), pp. 678–680, 2005.

[178] V. Jayaraman, J. Jiang, H. Li, *et al.*, "OCT imaging up to 760 kHz axial scan rate using single-mode 1310 nm MEMS-tunable VCSELs with >100 nm tuning range," in *2011 Conference on Lasers and Electro-Optics (CLEO)*, p. PDPB2, Optical Society of America, 2011.

[179] B. Potsaid, V. Jayaraman, J. G. Fujimoto, J. Jiang, P. J. S. Heim, and A. E. Cable, "MEMS tunable VCSEL light source for ultrahigh speed 60 kHz-1 MHz axial scan rate and long range centimeter class OCT imaging," *Proc. SPIE* **8213**, p. 82130M, 2012.

[180] V. Jayaraman, G. Cole, M. Robertson, *et al.*, "Rapidly swept, ultra-widely-tunable 1060 nm MEMS-VCSELs," *Electron. Lett.* **48**(21), pp. 1331–1333, 2012.

[181] D. D. John, B. Lee, B. Potsaid, *et al.*, "Single-mode and high-speed 850 nm MEMS-VCSEL," *in Lasers Congress 2016 (ASSL, LSC, LAC)*, p. ATh5A.2, Optical Society of America, 2016.

[182] D. D. John, C. B. Burgner, B. Potsaid, *et al.*, "Wideband electrically pumped 1050-nm MEMS-tunable VCSEL for ophthalmic imaging," *J. Lightwave Technol.* **33**, pp. 3461–3468, 2015.

[183] W. Choi, B. Potsaid, V. Jayaraman, *et al.*, "Phase-sensitive swept-source optical coherence tomography imaging of the human retina with a vertical cavity surface-emitting laser light source," *Opt. Lett.* **38**, pp. 338–340, 2013.

[184] W. Chen, J. You, X. Gu, C. Du, and Y. Pan, "High-speed swept source optical coherence Doppler tomography for deep brain microvascular imaging," *Sci. Rep.* **6**, p. 38786, 2016.

[185] T. Ansbaek, I. S. Chung, E. S. Semenova, and K. Yvind, "1060-nm tunable monolithic high index contrast subwavelength grating VCSEL," *IEEE Photonics Tech. L.* **25**, pp. 365–367, 2013.

[186] K. Li, C. Chase, P. Qiao, and C. J. Chang-Hasnain, "Widely tunable 1060-nm VCSEL with high-contrast grating mirror," *Opt. Express* **25**, pp. 11844–11854, 2017.

[187] M. P. Minneman, J. Ensher, M. Crawford, and D. Derickson, "All-semiconductor high-speed akinetic swept-source for OCT," *Proc. SPIE* **8311**, p. 831116, 2011.

[188] Z. Chen, M. Liu, M. Minneman, *et al.*, "Phase-stable swept source OCT angiography in human skin using an akinetic source," *Biomed. Opt. Express* **7**, pp. 3032–3048, 2016.

[189] J. Xu, S. Song, W. Wei, and R. K. Wang, "Wide field and highly sensitive angiography based on optical coherence tomography with akinetic swept source," *Biomed. Opt. Express* **8**, pp. 420–435, 2017.

[190] D. Choi, T. Amano, H. Hiro-Oka, *et al.*, "Tissue imaging by OFDR-OCT using an SSG-DBR laser," *Proc. SPIE* **5690**, pp. 101–113, 2005.

[191] N. Fujiwara, R. Yoshimura, K. Kato, *et al.*, "140-nm quasi-continuous fast sweep using SSG-DBR lasers," *IEEE Photonic Tech. L.* **20**(12), pp. 1015–1017, 2008.

[192] D. Choi, R. Yoshimura, and K. Ohbayashi, "Tuning of successively scanned two monolithic Vernier-tuned lasers and selective data sampling in optical comb swept source optical coherence tomography," *Biomed. Opt. Express* **4**, pp. 2962–2987, 2013.

[193] B. George and D. Derickson, "High-speed concatenation of frequency ramps using sampled grating distributed Bragg reflector laser diode sources for OCT resolution enhancement," *Proc. SPIE* **1554**, p. 75542O, 2010.

[194] D. Derickson, M. Bernacil, A. DeKelaita, *et al.*, "SGDBR single-chip wavelength tunable lasers for swept source OCT," *Proc. SPIE* **6847**, p. 68472P, 2008.

[195] S. O'Connor, M. A. Bernacil, A. DeKelaita, B. Maher, and D. Derickson, "100 kHz axial scan rate swept-wavelength OCT using sampled grating distributed bragg reflector lasers," *Proc. SPIE* **7168**, pp. 716825–716828, 2009.

[196] B. Zabihian, M. Minneman, M. Bonesi, *et al.*, "Akinetic swept-source technology for high speed OCT with unprecedented coherence length," in *Optics in the Life Sciences*, p. BT3A.2, Optical Society of America, 2013.

[197] S. Song, J. Xu, and R. K. Wang, "Long-range and wide field of view optical coherence tomography for in vivo 3D imaging of large volume object based on akinetic programmable swept source," *Biomed. Opt. Express* 7, pp. 4734–4748, 2016.

[198] M. Salas, M. Augustin, F. Felberer, *et al.*, "Compact akinetic swept source optical coherence tomography angiography at 1060 nm supporting a wide field of view and adaptive optics imaging modes of the posterior eye," *Biomed. Opt. Express* **9**, pp. 1871–1892, 2018.

[199] G. M. Hale and M. R. Querry, "Optical constants of water in the 200-nm to 200-μm wavelength region," *Appl. Opt.* **12**, pp. 555–563, 1973.

[200] Hemoglobin extinction data from W. B. Gratzer, Med. Res. Council Labs, Holly Hill, London, and N. Kollias, Wellman Laboratories, Harvard Medical School, Boston, compiled by S. Prahl, Oregon Medical Laser Center, Portland. omlc.org/spectra/hemoglobin/summary.html

[201] T. Sarna and R. C. Sealy, "Photoinduced oxygen consumption in melanin systems, action spectra and quantum yields for eumelanin and synthetic melanin," *Photochem. Photobiol.* **39**(1), pp. 69–74, 1984.

Index

www.ingramcontent.com/pod-product-compliance
Lightning Source LLC
LaVergne TN
LVHW011900200926
848484LV00002B/494

* 9 7 8 1 7 8 5 6 1 2 7 2 5 *